# orthogonal functions, moment theory, and continued fractions

# LECTURE NOTES IN PURE AND APPLIED MATHEMATICS

1. *N. Jacobson,* Exceptional Lie Algebras
2. *L.-Å. Lindahl and F. Poulsen,* Thin Sets in Harmonic Analysis
3. *I. Satake,* Classification Theory of Semi-Simple Algebraic Groups
4. *F. Hirzebruch et al., Differentiable Manifolds and Quadratic Forms*
5. *I. Chavel,* Riemannian Symmetric Spaces of Rank One
6. *R. B. Burckel,* Characterization of C(X) Among Its Subalgebras
7. *B. R. McDonald et al.,* Ring Theory
8. *Y.-T. Siu,* Techniques of Extension on Analytic Objects
9. *S. R. Caradus et al.,* Calkin Algebras and Algebras of Operators on Banach Spaces
10. *E. O. Roxin et al.,* Differential Games and Control Theory
11. *M. Orzech and C. Small,* The Brauer Group of Commutative Rings
12. *S. Thomier,* Topology and Its Applications
13. *J. M. Lopez and K. A. Ross,* Sidon Sets
14. *W. W. Comfort and S. Negrepontis,* Continuous Pseudometrics
15. *K. McKennon and J. M. Robertson,* Locally Convex Spaces
16. *M. Carmeli and S. Malin,* Representations of the Rotation and Lorentz Groups
17. *G. B. Seligman,* Rational Methods in Lie Algebras
18. *D. G. de Figueiredo,* Functional Analysis
19. *L. Cesari et al.,* Nonlinear Functional Analysis and Differential Equations
20. *J. J. Schäffer,* Geometry of Spheres in Normed Spaces
21. *K. Yano and M. Kon,* Anti-Invariant Submanifolds
22. *W. V. Vasconcelos,* The Rings of Dimension Two
23. *R. E. Chandler,* Hausdorff Compactifications
24. *S. P. Franklin and B. V. S. Thomas,* Topology
25. *S. K. Jain,* Ring Theory
26. *B. R. McDonald and R. A. Morris,* Ring Theory II
27. *R. B. Mura and A. Rhemtulla,* Orderable Groups
28. *J. R. Graef,* Stability of Dynamical Systems
29. *H.-C. Wang,* Homogeneous Branch Algebras
30. *E. O. Roxin et al.,* Differential Games and Control Theory II
31. *R. D. Porter,* Introduction to Fibre Bundles
32. *M. Altman,* Contractors and Contractor Directions Theory and Applications
33. *J. S. Golan,* Decomposition and Dimension in Module Categories
34. *G. Fairweather,* Finite Element Galerkin Methods for Differential Equations
35. *J. D. Sally,* Numbers of Generators of Ideals in Local Rings
36. *S. S. Miller,* Complex Analysis
37. *R. Gordon,* Representation Theory of Algebras
38. *M. Goto and F. D. Grosshans,* Semisimple Lie Algebras
39. *A. I. Arruda et al.,* Mathematical Logic
40. *F. Van Oystaeyen,* Ring Theory
41. *F. Van Oystaeyen and A. Verschoren,* Reflectors and Localization
42. *M. Satyanarayana,* Positively Ordered Semigroups
43. *D. L Russell,* Mathematics of Finite-Dimensional Control Systems
44. *P.-T. Liu and E. Roxin,* Differential Games and Control Theory III
45. *A. Geramita and J. Seberry,* Orthogonal Designs
46. *J. Cigler, V. Losert, and P. Michor,* Banach Modules and Functors on Categories of Banach Spaces
47. *P.-T. Liu and J. G. Sutinen,* Control Theory in Mathematical Economics
48. *C. Byrnes,* Partial Differential Equations and Geometry
49. *G. Klambauer,* Problems and Propositions in Analysis
50. *J. Knopfmacher,* Analytic Arithmetic of Algebraic Function Fields
51. *F. Van Oystaeyen,* Ring Theory
52. *B. Kadem,* Binary Time Series
53. *J. Barros-Neto and R. A. Artino,* Hypoelliptic Boundary-Value Problems
54. *R. L. Sternberg et al.,* Nonlinear Partial Differential Equations in Engineering and Applied Science
55. *B. R. McDonald,* Ring Theory and Algebra III
56. *J. S. Golan,* Structure Sheaves Over a Noncommutative Ring
57. *T. V. Narayana et al.,* Combinatorics, Representation Theory and Statistical Methods in Groups

58. *T. A. Burton,* Modeling and Differential Equations in Biology
59. *K. H. Kim and F. W. Roush,* Introduction to Mathematical Consensus Theory
60. *J. Banas and K. Goebel,* Measures of Noncompactness in Banach Spaces
61. *O. A. Nielson,* Direct Integral Theory
62. *J. E. Smith et al.,* Ordered Groups
63. *J. Cronin,* Mathematics of Cell Electrophysiology
64. *J. W. Brewer,* Power Series Over Commutative Rings
65. *P. K. Kamthan and M. Gupta,* Sequence Spaces and Series
66. *T. G. McLaughlin,* Regressive Sets and the Theory of Isols
67. *T. L. Herdman et al.,* Integral and Functional Differential Equations
68. *R. Draper,* Commutative Algebra
69. *W. G. McKay and J. Patera,* Tables of Dimensions, Indices, and Branching Rules for Representations of Simple Lie Algebras
70. *R. L. Devaney and Z. H. Nitecki,* Classical Mechanics and Dynamical Systems
71. *J. Van Geel,* Places and Valuations in Noncommutative Ring Theory
72. *C. Faith,* Injective Modules and Injective Quotient Rings
73. *A. Fiacco,* Mathematical Programming with Data Perturbations I
74. *P. Schultz et al.,* Algebraic Structures and Applications
75. *L Bican et al.,* Rings, Modules, and Preradicals
76. *D. C. Kay and M. Breen,* Convexity and Related Combinatorial Geometry
77. *P. Fletcher and W. F. Lindgren,* Quasi-Uniform Spaces
78. *C.-C. Yang,* Factorization Theory of Meromorphic Functions
79. *O. Taussky,* Ternary Quadratic Forms and Norms
80. *S. P. Singh and J. H. Burry,* Nonlinear Analysis and Applications
81. *K. B. Hannsgen et al.,* Volterra and Functional Differential Equations
82. *N. L. Johnson et al.,* Finite Geometries
83. *G. I. Zapata,* Functional Analysis, Holomorphy, and Approximation Theory
84. *S. Greco and G. Valla,* Commutative Algebra
85. *A. V. Fiacco,* Mathematical Programming with Data Perturbations II
86. *J.-B. Hiriart-Urruty et al.,* Optimization
87. *A. Figa Talamanca and M. A. Picardello,* Harmonic Analysis on Free Groups
88. *M. Harada,* Factor Categories with Applications to Direct Decomposition of Modules
89. *V. I. Istrăţescu,* Strict Convexity and Complex Strict Convexity
90. *V. Lakshmikantham,* Trends in Theory and Practice of Nonlinear Differential Equations
91. *H. L. Manocha and J. B. Srivastava,* Algebra and Its Applications
92. *D. V. Chudnovsky and G. V. Chudnovsky,* Classical and Quantum Models and Arithmetic Problems
93. *J. W. Longley,* Least Squares Computations Using Orthogonalization Methods
94. *L. P. de Alcantara,* Mathematical Logic and Formal Systems
95. *C. E. Aull,* Rings of Continuous Functions
96. *R. Chuaqui,* Analysis, Geometry, and Probability
97. *L. Fuchs and L. Salce,* Modules Over Valuation Domains
98. *P. Fischer and W. R. Smith,* Chaos, Fractals, and Dynamics
99. *W. B. Powell and C. Tsinakis,* Ordered Algebraic Structures
100. *G. M. Rassias and T. M. Rassias,* Differential Geometry, Calculus of Variations, and Their Applications
101. *R.-E. Hoffmann and K. H. Hofmann,* Continuous Lattices and Their Applications
102. *J. H. Lightbourne III and S. M. Rankin III,* Physical Mathematics and Nonlinear Partial Differential Equations
103. *C. A. Baker and L. M. Batten,* Finite Geometrics
104. *J. W. Brewer et al.,* Linear Systems Over Commutative Rings
105. *C. McCrory and T. Shifrin,* Geometry and Topology
106. *D. W. Kueke et al.,* Mathematical Logic and Theoretical Computer Science
107. *B.-L. Lin and S. Simons,* Nonlinear and Convex Analysis
108. *S. J. Lee,* Operator Methods for Optimal Control Problems
109. *V. Lakshmikantham,* Nonlinear Analysis and Applications
110. *S. F. McCormick,* Multigrid Methods
111. *M. C. Tangora,* Computers in Algebra
112. *D. V. Chudnovsky and G. V. Chudnovsky,* Search Theory
113. *D. V. Chudnovsky and R. D. Jenks,* Computer Algebra
114. *M. C. Tangora,* Computers in Geometry and Topology
115. *P. Nelson et al.,* Transport Theory, Invariant Imbedding, and Integral Equations
116. *P. Clément et al.,* Semigroup Theory and Applications

117. *J. Vinuesa,* Orthogonal Polynomials and Their Applications
118. *C. M. Dafermos et al.,* Differential Equations
119. *E. O. Roxin,* Modern Optimal Control
120. *J. C. Díaz,* Mathematics for Large Scale Computing
121. *P. S. Milojevič,* Nonlinear Functional Analysis
122. *C. Sadosky,* Analysis and Partial Differential Equations
123. *R. M. Shortt,* General Topology and Applications
124. *R. Wong,* Asymptotic and Computational Analysis
125. *D. V. Chudnovsky and R. D. Jenks,* Computers in Mathematics
126. *W. D. Wallis et al.,* Combinatorial Designs and Applications
127. *S. Elaydi,* Differential Equations
128. *G. Chen et al.,* Distributed Parameter Control Systems
129. *W. N. Everitt,* Inequalities
130. *H. G. Kaper and M. Garbey,* Asymptotic Analysis and the Numerical Solution of Partial Differential Equations
131. *O. Arino et al.,* Mathematical Population Dynamics
132. *S. Coen,* Geometry and Complex Variables
133. *J. A. Goldstein et al.,* Differential Equations with Applications in Biology, Physics, and Engineering
134. *S. J. Andima et al.,* General Topology and Applications
135. *P Clément et al.,* Semigroup Theory and Evolution Equations
136. *K. Jarosz,* Function Spaces
137. *J. M. Bayod et al.,* *p*-adic Functional Analysis
138. *G. A. Anastassiou,* Approximation Theory
139. *R. S. Rees,* Graphs, Matrices, and Designs
140. *G. Abrams et al.,* Methods in Module Theory
141. *G. L. Mullen and P. J.-S. Shiue,* Finite Fields, Coding Theory, and Advances in Communications and Computing
142. *M. C. Joshi and A. V. Balakrishnan,* Mathematical Theory of Control
143. *G. Komatsu and Y. Sakane,* Complex Geometry
144. *I. J. Bakelman,* Geometric Analysis and Nonlinear Partial Differential Equations
145. *T. Mabuchi and S. Mukai,* Einstein Metrics and Yang–Mills Connections
146. *L. Fuchs and R. Göbel,* Abelian Groups
147. *A. D. Pollington and W. Moran,* Number Theory with an Emphasis on the Markoff Spectrum
148. *G. Dore et al.,* Differential Equations in Banach Spaces
149. *T. West,* Continuum Theory and Dynamical Systems
150. *K. D. Bierstedt et al.,* Functional Analysis
151. *K. G. Fischer et al.,* Computational Algebra
152. *K. D. Elworthy et al.,* Differential Equations, Dynamical Systems, and Control Science
153. *P.-J. Cahen, et al.,* Commutative Ring Theory
154. *S. C. Cooper and W. J. Thron,* Continued Fractions and Orthogonal Functions
155. *P. Clément and G. Lumer,* Evolution Equations, Control Theory, and Biomathematics
156. *M. Gyllenberg and L. Persson,* Analysis, Algebra, and Computers in Mathematical Research
157. *W. O. Bray et al.,* Fourier Analysis
158. *J. Bergen and S. Montgomery,* Advances in Hopf Algebras
159. *A. R. Magid,* Rings, Extensions, and Cohomology
160. *N. H. Pavel,* Optimal Control of Differential Equations
161. *M. Ikawa,* Spectral and Scattering Theory
162. *X. Liu and D. Siegel,* Comparison Methods and Stability Theory
163. *J.-P. Zolésio,* Boundary Control and Variation
164. *M. Křížek et al.,* Finite Element Methods
165. *G. Da Prato and L. Tubaro,* Control of Partial Differential Equations
166. *E. Ballico,* Projective Geometry with Applications
167. *M. Costabel et al.,* Boundary Value Problems and Integral Equations in Nonsmooth Domains
168. *G. Ferreyra, G. R. Goldstein, and F. Neubrander,* Evolution Equations
169. *S. Huggett,* Twistor Theory
170. *H. Cook et al.,* Continua
171. *D. F. Anderson and D. E. Dobbs,* Zero-Dimensional Commutative Rings
172. *K. Jarosz,* Function Spaces
173. *V. Ancona et al.,* Complex Analysis and Geometry
174. *E. Casas,* Control of Partial Differential Equations and Applications
175. *N. Kalton et al.,* Interaction Between Functional Analysis, Harmonic Analysis, and Probability
176. *Z. Deng et al.,* Differential Equations and Control Theory

177. *P. Marcellini et al.* Partial Differential Equations and Applications
178. *A. Kartsatos*, Theory and Applications of Nonlinear Operators of Accretive and Monotone Type
179. *M. Maruyama*, Moduli of Vector Bundles
180. *A. Ursini and P. Aglianò*, Logic and Algebra
181. *X. H. Cao et al.*, Rings, Groups, and Algebras
182. *D. Arnold and R. M. Rangaswamy*, Abelian Groups and Modules
183. *S. R. Chakravarthy and A. S. Alfa*, Matrix-Analytic Methods in Stochastic Models
184. *J. E. Andersen et al.*, Geometry and Physics
185. *P.-J. Cahen et al.*, Commutative Ring Theory
186. *J. A. Goldstein et al..*, Stochastic Processes and Functional Analysis
187. *A. Sorbi*, Complexity, Logic, and Recursion Theory
188. *G. Da Prato and J.-P. Zolésio*, Partial Differential Equation Methods in Control and Shape Analysis
189. *D. D. Anderson*, Factorization in Integral Domains
190. *N. L. Johnson*, Mostly Finite Geometries
191. *D. Hinton and P. W. Schaefer*, Spectral Theory and Computational Methods of Sturm–Liouville Problems
192. *W. H. Schikhof et al.*, *p*-adic Functional Analysis
193. *S. Sertöz*, Algebraic Geometry
194. *G. Caristi and E. Mitidieri*, Reaction Diffusion Systems
195. *A. V. Fiacco*, Mathematical Programming with Data Perturbations
196. *M. Křížek et al.*, Finite Element Methods: Superconvergence, Post-Processing, and A Posteriori Estimates
197. *S. Caenepeel and A. Verschoren*, Rings, Hopf Algebras, and Brauer Groups
198. *V. Drensky et al.*, Methods in Ring Theory
199. *W. B. Jones and A. Sri Ranga*, Orthogonal Functions, Moment Theory, and Continued Fractions
200. *P. Newstead*, Algebraic Geometry

*Additional Volumes in Preparation*

# orthogonal functions, moment theory, and continued fractions
## theory and applications

edited by

William B. Jones
***University of Colorado at Boulder***
***Boulder, Colorado***

A. Sri Ranga
***Universidade Estadual Paulista***
***São Paulo, Brazil***

CRC Press
Taylor & Francis Group
Boca Raton London New York

CRC Press is an imprint of the
Taylor & Francis Group, an **informa** business

CRC Press
Taylor & Francis Group
6000 Broken Sound Parkway NW, Suite 300
Boca Raton, FL 33487-2742

First issued in hardback 2017



ISBN 13: 978-1-138-41326-9 (hbk)
ISBN 13: 978-0-8247-0207-6 (pbk)



**Visit the Taylor & Francis Web site at**
**http://www.taylorandfrancis.com**

**and the CRC Press Web site at**
**http://www.crcpress.com**

**Library of Congress Cataloging-in-Publication Data**

Orthogonal functions, moment theory, and continued fractions: theory and applications / edited by William B. Jones, A. Sri Ranga.
p. cm.— (Lecture notes in pure and applied mathematics; v. 199)
Includes bibliographical references.
ISBN 0-8247-0207-7
1. Functions, Orthogonal. 2. Continued fractions. 3. Moment problems (Mathematics)
I. Jones, William B. (William Branham) II. Ranga, A. Sri.
III. Series.
QA404.5.072 1998
515'.55—dc21

98-24470
CIP

Research conference participants and guests: June 19–28, 1996 Hotel Fazenda Solar das Andorinhas, Campinas, SP, Brazil
First row (l. to r.): 1. Cathleen Craviotto, 2. H. Waadeland, 3. Vigdis Petersen, 4. Peggy Magnus, 5. Miguel Piñar Gonzalez, 6. Teresa Pérez Fernandez, 7. Eliana X.L. de Andrade, 8. A. Sri Ranga, 9. Catherine Bonan-Hamada, 10. Adhamar Bultheel, 11. Gerry Lange. Second row (l. to r.): 12. Walter Van Assche, 13. Sandra C. Cooper, 14. Martha Jones, 15. William B. Jones, 16. Phil Gustafson, 17. Lisa Lorentzen, 18. Frode Rønning, 19. Andrea Cristina Breti, 20. L.J. Lange, 21. Olav Njåstad, 22. Nancy J. Wyshinski, 23. Jorge M.V. Capela, 24. Brian Haglar, 25. Luis Carlos Matioli.

[illegible]

# Preface

This volume contains the proceedings of a research conference/workshop held at Campinas, São Paulo, Brazil. The conference talks were focused on a constellation of interrelated topics:

- Orthogonal functions
- Moment theory
- Continued fractions
- Padé approximants
- Gaussian quadrature
- Stieltjes transforms
- Linear functionals

Most of these topics are currently studied without reference to continued fractions. Yet many of them received their impetus from the analytic theory of continued fractions developed in the nineteenth and early twentieth centuries. Thus in his 1939 volume *Orthogonal Polynomials*, G. Szegö asserted that "... historically the orthogonal polynomials $\{p_n(x)\}$ originated in the theory of continued fractions. This relationship is of great importance and is one of the possible starting points of the treatment of orthogonal polynomials."

Primary emphasis at the conference was on analytic aspects of the subject; nevertheless, substantial attention was given to applications in other areas such as special functions, frequency analysis (digital signal processing) and related computation. These interests are reflected in the conference proceedings. The conference sessions were devoted not only to reports on recent research, but also to the development of new results (made during the conference) and to the formulation of problems to be investigated.

The Conference/Workshop was organized by A. Sri Ranga and Eliana X. L. de Andrade of the São Paulo State University at São José do Rio Preto, SP, Brazil, and by William B. Jones of the University of Colorado, Boulder. The conference was made possible by grants from the:

- Brazilian National Council for Scientific and Technological Development,
- São Paulo State Research Foundation,
- Norwegian Research Council,
- U. S. National Science Foundation,

and by financial support for individual participants from universities in Belgium, Brazil, Norway, Spain and the United States. We gratefully acknowledge these contributions.

We wish to thank the director and staff of the Hotel Fazenda Solar das Andorinhas, Campinas, SP, Brazil for providing excellent working facilities and a cordial atmosphere for the Conference/Workshop. We would like to thank Maria Allegra, Acquisitions Editor and Manager at Marcel Dekker, Inc., for accepting this volume for publication. We are grateful for her able and friendly assistance. Through our joint efforts these 21 manuscripts have been transformed into a book that is accessible to a wide range of scientists, engineers and other interested readers.

William B. Jones
A. Sri Ranga

# Contents

# Contributors

**Eliana X. L. de Andrade** São Paulo State University, São José do Rio Preto, São Paulo, Brazil

**Catherine M. Bonan-Hamada** Mesa State College, Grand Junction, Colorado

**C. F. Bracciali** São Paulo State University, São José do Rio Preto, São Paulo, Brazil

**Adhemar Bultheel** Catholic University Leuven, Heverlee, Belgium

**J. M. V. Capela** São Paulo State University, Araraquara, São Paulo, Brazil

**Lyle Cochran** Whitworth College, Spokane, Washington

**S. Clement Cooper** Washington State University, Pullman, Washington

**Cathleen M. Craviotto** University of Northern Colorado, Greeley, Colorado

**Dimitar K. Dimitrov** São Paulo State University, São José do Rio Preto, São Paulo, Brazil

**Pablo González-Vera** University of La Laguna, Canary Islands, Spain

**Philip E. Gustafson** Emporia State University, Emporia, Kansas

**Brian A. Hagler** University of Colorado, Boulder, Colorado

**Erik Hendriksen** University of Amsterdam, Amsterdam, The Netherlands

**William B. Jones** University of Colorado, Boulder, Colorado

**L. J. Lange** University of Missouri, Columbia, Missouri

**Xin Li** University of Central Florida, Orlando, Florida

**María Álvarez de Morales** University of Granada, Granada, Spain

**Olav Njåstad** Norwegian University of Science and Technology, Trondheim, Norway

**Teresa E. Pérez** University of Granada, Granada, Spain

**Vigdis Petersen** Sör-Trondelag College, Trondheim, Norway

**Miguel A. Piñar** University of Granada, Granada, Spain

**A. Sri Ranga** São Paulo State University, São José do Rio Preto, São Paulo, Brazil

**André Ronveaux** Facultés Universitaires N.D. de la Paix, Namur, Belgium

**Guoxiang Shen** University of Colorado, Boulder, Colorado

**W. J. Thron** University of Colorado, Boulder, Colorado

**Walter Van Assche** Catholic University Leuven, Heverlee, Belgium

**Haakon Waadeland** Norwegian University of Science and Technology, Dragvoll, Trondheim, Norway

**Nancy J. Wyshinski** Trinity College, Hartford, Connecticut

# Participants

ELIANA X.L. DE ANDRADE, Departamento de Ciências de Computação, e Estatistica, IBILCE, UNESP (Universidade Estadual Paulista), 15054-000 São José do Rio Preto, SP, Brasil.

CATHERINE M. BONAN-HAMADA, Department of Computer Science, Mathematics and Statistics, Mesa State College, P.O. Box 2647, Grand Junction, Colorado 81502, USA.

ADHEMAR BULTHEEL, Department of Computer Science, Katholieke Universiteit Leuven, Celestijnenlaan 200A, B-3001 Heverlee, Belgium.

JORGE M.V. CAPELA, Departamento de Físico-Química, Instituto de Química, UNESP (Universidade Estadual Paulista), 14800-900 Araraquara, SP, Brasil.

SANDRA CLEMENT COOPER, Department of Pure and Applied Mathematics, Washington State University, Pullman, WA 99164-3113, USA.

CATHLEEN M. CRAVIOTTO, Department of Mathematical Sciences, University of Northern Colorado, Greeley, CO 80639, USA.

DIMITAR K. DIMITROV, Departamento de Ciências de Computaçõ e Estatistica, IBILCE, UNESP (Universidade Estadual Paulista), 15054-000 São José do Rio Preto, SP, Brasil.

PHILIP GUSTAFSON, Division of Mathematics and Computer Science, Emporia State University, Emporia, KS 66801-5087, USA.

BRIAN A. HAGLER, Department of Mathematics, University of Colorado, Boulder, CO 80309-0395, USA.

WILLIAM B. JONES, Department of Mathemtics, University of Colorado, Boulder, CO 80309-0395, USA.

L. JEROME LANGE, Department of Mathematics, University of Missouri, Columbia, MO 65211, USA.

LISA LORENTZEN, Department of Mathematical Sciences, Norwegian University of Science and Technology, N-7034 Trondheim, Norway.

PEGGY MAGNUS, 708 Dartmouth Trail, Ft. Collins, CO 80525, USA.

LUIS CARLOS MATIOLI, Departamento de Matemática, Universidade Federal do Paraná, 80620-250 Curitiba, PR, Brasil.

OLAV NJÅSTAD, Department of Mathematical Sciences, Norwegian University of Science and Technology, N-7034 Trondheim, Norway.

TERESA E. PÉREZ FERNÁNDEZ, Departamento de Mathemática Aplicada, Facultad de Ciencias, Universidad de Granada, 18071 Granada, Spain.

VIGDIS PETERSEN, School of Teacher Education, Sør-Trøndelag College, Rotvoll Allé, N-7005 Trondheim, Norway.

MIGUEL A. PIÑAR GONZÁLEZ, Departamento de Matemática Aplicada, Facultad de Ciencias, Universidad de Granada, 18071 Granada, Spain.

A. SRI RANGA, Departamento de Ciências de Computação e Estatistica, IBILCE, UNESP (Universidade Estadual Paulista), 15054-000 São José do Rio Preto, SP, Brasil.

FRODE RØNNING, School of Teacher Education, Sør-Trøndelag College, N-7005 Trondheim, Norway.

WALTER VAN ASSCHE, Department of Mathematics, Katholieke Universiteit Leuven, Celestijnenlaan 200B, B-3001 Heverlee, Belgium.

HAAKON WAADELAND, Department of Mathematics and Statistics, Norwegian University of Science and Technology, N-7005 Dragvoll, Norway.

NANCY J. WYSHINSKI, Department of Mathematics, Trinity College, Hartford, CT 06106, USA.

# orthogonal functions, moment theory, and continued fractions

# Chebyshev–Laurent Polynomials and Weighted Approximation

ELIANA X. L. DE ANDRADE and DIMITAR K. DIMITROV [1] Departamento de Ciências de Computação e Estatística, Instituto de Biociências, Letras e Ciências Exatas, Universidade Estadual Paulista, 15054-000 - São José do Rio Preto, SP, Brazil

Let $(a, b) \subset (0, \infty)$ and for any positive integer $n$, let $S_n$ be the Chebyshev space in $[a, b]$ defined by $S_n := span\{x^{-n/2+k}, k = 0, \ldots, n\}$. The unique (up to a constant factor) function $\tau_n \in S_n$, which satisfies the orthogonality relation $\int_a^b \tau_n(x)q(x)$ $(x(b-x)(x-a))^{-1/2}\, dx = 0$ for any $q \in S_{n-1}$, is said to be the orthogonal Chebyshev $S_n$-polynomials. This paper is an attempt to exibit some interesting properties of the orthogonal Chebyshev $S_n$-polynomials and to demonstrate their importance to the problem of approximation by $S_n$-polynomials. A simple proof of a Jackson-type theorem is given and the Lagrange interpolation problem by functions from $S_n$ is discussed. It is shown also that $\tau_n$ obeys an extremal property in $L_q$, $1 \le q \le \infty$. Natural analogues of some inequalities for algebraic polynomials, which we expect to hold for the $S_n$-polynomials, are conjectured.

## 1 INTRODUCTION

The distribution $d\psi(x)$ is said to be a strong distribution in $(a, b) \subset (0, \infty)$ if $\psi(x)$ is a real bounded nondecreasing function on $(a, b)$ with infinitely many points of increase there, and furthermore, all the moments

$$\mu_k = \int_a^b x^k d\psi(x), \quad k = 0, \pm 1, \pm 2, \ldots \qquad \text{exist.}$$

---

Research supported by the Brazilian foundations FAPESP and CNPq and the Bulgarian Science Foundation under Grant MM-414.

[1] On leave from The University of Rousse, Bulgaria.

For any given strong distribution in $(a, b)$ there exists a unique, up to a nonzero constant factor normalization, sequence of polynomials $\{B_n\}_0^\infty$ such that $B_n$ is a polynomial of precise degree $n$ and $B_n$ satisfies the relations

$$\int_a^b x^{-n+k} B_n(x) d\psi(x) = 0, \quad k = 0, \ldots, n-1. \tag{1.1}$$

These polynomials may be called *orthogonal Laurent polynomials* or simply *orthogonal L-polynomials*. They were studied in details by Jones, Thron and Waadeland [9] in connection with the so-called strong Stieltjes moment problem.

Let $b_n(x) := x^{-n/2} B_n(x)$. Obviously $b_n \in S_n$, where

$$S_n := span\{x^{-n/2+k}, k = 0, \ldots, n\}. \tag{1.2}$$

In what follows the elements of $S_n$ will be called $S_n$-polynomials. Now the orthogonality relations (1.1) may be rewritten in the following more succinct form:

$$\int_a^b b_n(x) q(x) x^{-1/2} d\psi(x) = 0 \quad \text{for any} \quad q \in S_{n-1}. \tag{1.3}$$

This latter orthogonality relation provides a natural relation between the spaces $S_n$ and the orthogonal $S_n$-polynomials $b_n$. The function $b_n$, which obeys the orthogonal property (1.3) with $d\psi(x) = ((b-x)(x-a))^{-1/2} dx$ will be called *orthogonal Chebyshev $S_n$-polynomial.*

An explicit representation for the orthogonal Chebyshev $S_n$-polynomials was obtained by Sri Ranga [15]. He proved that the $S_n$-polynomial $\tau_n$ defined by

$$\tau_n(x) := T_n\left(\frac{1}{\sqrt{b} - \sqrt{a}} \frac{x - \sqrt{ab}}{\sqrt{x}}\right), \tag{1.4}$$

where $T_n$ is the n-th Chebyshev polynomial of the first kind, satisfies the orthogonal properties (1.3). The orthogonal Chebyshev $S_n$-polynomial $\tau_n$ will be called simply *Chebyshev $S_n$-polynomial.*

Let $H_n = span\{g_0, g_1, \ldots, g_n\}$ be a Chebyshev space in $(a, b)$. Then any element of $H_n$ is called a $H_n$-polynomial. The problem of approximation by $H_n$-polynomials is stated as follows:

PROBLEM 1 Given a continuous function $f$, find the quantity

$$E_{H_n}(f) := \inf_{g \in H_n} \|f - g\|_\infty,$$

where $\|\cdot\|_\infty$ is the uniform norm in $(a, b)$.

If this problem has a unique solution $g^* \in H_n$ such that $E_{H_n}(f) = \|f - g^*\|_\infty$, the $H_n$-polynomial $g^*$ is said to be the best approximant of $f$. Another related problem is the problem of weighted approximation with varying weight, which can be stated in the following way:

PROBLEM 2 Given a continuous function $f$ and a weight function $W$ in $(a, b)$, find the quantity

$$E_{H_n}(W; f) := \inf_{g \in H_n} \|f - W^n g\|_\infty.$$

Again, if the latter problem has a unique solution $g_W^* \in H_n$ such that $E_{H_n}(W; f) = \|f - W^n g_W^*\|_\infty$, then the function $W^n g_W^*$ is said to be the best weighted approximant of $f$. The problem of weighted approximation with varying weights has been of recent interest. A thorough study of this topic was given by Totik [18].

In this paper we consider Problem 1 for the Chebyshev space $H_n = S_n$, where $S_n$ is defined by (1.2). It is easy to see that in this case Problem 1 is equivalent to Problem 2 with weight $W(x) = 1/\sqrt{x}$ and Chebyshev space $H_n = \mathbb{P}_n$, the space of algebraic polynomials of degree not exceeding $n$. A very simple proof of a Jackson-type theorem is given in Section 3. The Lagrange interpolation problem by $S_n$-polynomials at the zeros of $\tau_n$ is also discussed there. In Section 4 we prove that the polynomials $\tau_n$ are solutions of an extremal problem in $L_p$. There we state also some conjectured inequalities concerning $S_n$-polynomials, which we find natural analogs of the classical Bernstein, Markov and Szegö's inequalities for algebraic polynomials.

## 2 BASIC PROPERTIES OF THE CHEBYSHEV $S_n$-POLYNOMIALS

In this section we list some properties which follow immediately from the corresponding properties of the Chebyshev polynomial $T_n$. The books of Rivlin [14] and Szegö [17] contain comprehensive information about $T_n$.

Since $\tau_n(x) = T_n(u(x))$, where

$$u(x) = \frac{1}{\sqrt{b} - \sqrt{a}} \frac{x - \sqrt{ab}}{\sqrt{x}}, \tag{2.1}$$

we need some information about the transformation $x \to u(x)$. Note that

$$u'(x) = \frac{1}{2(\sqrt{b} - \sqrt{a})} \frac{x + \sqrt{ab}}{x^{3/2}} > 0, \quad \text{for} \quad x \in [a, b] \tag{2.2}$$

and

$$u''(x) = -\frac{1}{4(\sqrt{b} - \sqrt{a})} \frac{x + 3\sqrt{ab}}{x^{5/2}} < 0, \quad \text{for} \quad x \in [a, b]. \tag{2.3}$$

Inequality (2.2) and the equalities $u(a) = -1$ and $u(b) = 1$ show that $x \to u(x)$ is a one-to-one mapping between $[a, b]$ and $[-1, 1]$. The well-known repesentations $T_n(u) = \cos(n \arccos u)$, $u \in [-1, 1]$, and $T_n(u) = \dfrac{(u + \sqrt{u^2 - 1})^n + (u - \sqrt{u^2 - 1})^n}{2}$ yield

$$\tau_n(x) = \cos\left(n \arccos(\frac{1}{\sqrt{b}-\sqrt{a}} \frac{x-\sqrt{ab}}{\sqrt{x}})\right), \quad x \in [a,b]$$

and

$$\begin{aligned}\tau_n(x) &= \frac{x^{-n/2}}{2(\sqrt{b}-\sqrt{a})^n}\left((x-\sqrt{ab}+\sqrt{(x-a)(x-b)})^n\right. \\ &\left. +(x-\sqrt{ab}-\sqrt{(x-a)(x-b)})^n\right).\end{aligned} \tag{2.4}$$

The latter representation shows that the polynomial

$$B_n(x) := \left(x-\sqrt{ab}+\sqrt{(x-a)(x-b)}\right)^n + \left(x-\sqrt{ab}-\sqrt{(x-a)(x-b)}\right)^n$$

obeys the orthogonal property (1.1) for $d\psi(x) = ((b-x)(x-a))^{-1/2}dx$.

The recurrence relation for the Chebyshev polynomials $T_n$ yields the following recurrence relation for the Chebyshev $S_n$-polynomials:

$$\begin{aligned}\tau_0(x) &= 1, \\ \tau_1(x) &= \frac{1}{\sqrt{b}-\sqrt{a}} \frac{x-\sqrt{ab}}{\sqrt{x}}, \\ \tau_{n+1}(x) &= \frac{2}{\sqrt{b}-\sqrt{a}} \frac{x-\sqrt{ab}}{\sqrt{x}} \tau_n(x) - \tau_{n-1}(x), \quad n = 1,2,\ldots.\end{aligned}$$

Recall that $|T_n(u)| \le 1$ for $u \in [-1,1]$, equality is attained only for $\eta_k = \cos(k\pi/n)$, $k = 0,\ldots,n$, and $T_n(\eta_k) = (-1)^k$. Thus

$$|\tau_n(x)| \le 1, \quad \text{for} \quad x \in [a,b],$$

equality is attained only for

$$\tilde{\eta}_k = \frac{1}{4}\left((\sqrt{b}-\sqrt{a})\cos(k\pi/n) + \sqrt{(\sqrt{b}-\sqrt{a})^2\cos^2(k\pi/n) + 4\sqrt{ab}}\right)^2, k = 0,\ldots,n,$$

and $\tau_n(\tilde{\eta}_k) = (-1)^k$.

The zeros $\xi_k$ of $T_n$ are $\xi_k = \cos((2k-1)\pi/(2n)), k = 1,\ldots,n$. Hence the zeros $\tilde{\xi}_k$, $k = 1,\ldots,n$, of $\tau_n$ are

$$\tilde{\xi}_k = \frac{1}{4}\left((\sqrt{b}-\sqrt{a})\cos\frac{(2k-1)\pi}{2n} + \sqrt{(\sqrt{b}-\sqrt{a})^2\cos^2\frac{(2k-1)\pi}{2n} + 4\sqrt{ab}}\right)^2.$$

Clearly $\tilde{\xi}_k$ are simple zeros of $\tau_n$ and $\tilde{\xi}_k \in (a,b)$, $k = 1, \ldots, n$. Similarly $\tilde{\eta}_k \in [a,b]$, $k = 0, \ldots, n$.

Performing the change of variables $u \to x$ in the differential identities $(1-u^2)(T_n'(u))^2 + n^2T_n^2 = n^2$ and $(1-u^2)T_n''(u) - uT_n'(u) + n^2T_n(u) = 0$, we conclude that $\tau_n$ is a solution of the differential equations

$$\frac{4(b-x)(x-a)x^2}{(x+\sqrt{ab})^2}(y')^2 + n^2y^2 = n^2 \tag{2.5}$$

and

$$\begin{aligned} 4x^2(b-x)(x-a)(x &+ \sqrt{ab})y'' - 2x\Big(2x^3 + (4\sqrt{ab} - a - b)x^2 \\ &- 3\sqrt{ab}(a+b)x + 2(ab)^{3/2}\Big)\, y' + n^2(x+\sqrt{ab})^3 y = 0. \end{aligned} \tag{2.6}$$

It is well-known that $|T_n'(u)| \le n^2$ for $u \in [-1,1]$ and equality is attained only for $u = \pm 1$. This fact together with (2.2) and (2.3) imply that

$$|\tau_n'(x)| \le \frac{\sqrt{b}+\sqrt{a}}{2a(\sqrt{b}-\sqrt{a})}n^2, \quad \text{for} \quad x \in [a,b] \tag{2.7}$$

and that equality in (2.7) is attained only for $x = a$.

We conclude this section with an identity for the Chebyshev $S_n$-polynomials for the particular case when $ab = 1$. Observe that the method that Sri Ranga used to obtained the explicit expression (1.4) implies that when $ab = 1$ the corresponding orthogonal L-polynomials $B_n$ are self-inverse. Recall that a polynomial $p$ is called self-inverse if for any of its zeros $z_i$ its inverse $1/\bar{z}_i$ is also a zero of $p$. O'Hara and Rodriguez [13] proved that for every self-inverse polynomial of degree $n$ the equality $|np(z)/(zp'(z)) - 1| = 1$ holds for each z on the unit circle in the complex plane. This implies:

*If $ab = 1$ and the Chebyshev $S_n$-polynomial $\tau_n$ is defined by (1.4), then*

$$|n\tau_n(z) - 2z\tau_n'(z)| = |n\tau_n(z) + 2z\tau_n'(z)| \quad \textit{for each} \quad z,\ |z| = 1.$$

## 3 APPROXIMATION AND INTERPOLATION BY $S_n$-POLYNOMIALS

Consider the space $S_n$, defined by (1.2), and the distinct points $x_1, x_2, \ldots, x_{n+1}$, in the interval $[a,b]$, $0 < a < b < \infty$. Let the determinant $\Delta_n$ be defined by

$$\Delta_n := \begin{vmatrix} x_1^{-n/2} & x_1^{-n/2+1} & \dots & x_1^{n/2} \\ x_2^{-n/2} & x_2^{-n/2+1} & \dots & x_2^{n/2} \\ \vdots & \vdots & & \vdots \\ x_{n+1}^{-n/2} & x_{n+1}^{-n/2+1} & \dots & x_{n+1}^{n/2} \end{vmatrix}.$$

Then

$$\Delta_n = \prod_{i=1}^{n+1} x_i^{-n/2} \prod_{1\le i<j\le n+1} (x_j - x_i) \neq 0. \tag{3.1}$$

Therefore, since $x^{-n/2+k} \in C[a,b]$, $S_n$ is a Chebyshev system in $[a,b]$.

For any positive integer $n$, we define the space $\bar{S}_n$ by $\bar{S}_n := span\{1 \cup S_n\}$. Observe that $\bar{S}_n$ coincides with $S_n$ when $n$ is even. The modulus of continuity of a function $f$, which is continuous in $[a,b]$, is defined by $\omega(f;\delta) := \sup\{|f(x) - f(y)| : x, y \in [a,b], |x-y| \le \delta\}$.

THEOREM 1 Let $f \in C[a,b]$. Then for every $n \in I\!N$ there exists $s_n^* \in \bar{S}_n$ such that

$$\|f - s_n^*\|_\infty \le D\omega(f; 1/\sqrt{n}),$$

where the constant $D$ depends on $a$ and $b$.

Proof: Recall that (see [6]) the general Chebyshev polynomial $h_{H_n}$ associated with the Chebyshev space $H_n := span\{g_0, \dots, g_n\}$ is the linear form

$$h_{H_n} := c\left(g_n - \sum_{i=0}^{n-1} c_i g_i\right),$$

where $c_i$ are chosen so that $\sum_0^{n-1} c_i g_i$ is the best uniform approximation on $[a,b]$ to $g_n$ from $span\{g_0, \dots, g_{n-1}\}$, and where $c$ is chosen such that $\|h_{H_n}\|_\infty = 1$ and $h_{H_n}(b) = 1$. This uniquely determines $h_{H_n}$. It is known that $h_{H_n}$ has exactly $n$ zeros $x_1 < \dots < x_n$ in $[a,b]$. If $x_0 := a$ and $x_n := b$, then we define the mesh of $h_{H_n}$ by

$$M_{H_n} := \max_{1\le i\le n+1} |x_i - x_{i-1}|.$$

Using the properties of $\tau_n$ listed in Section 2 one can easily see that this is the general Chebyshev polynomial associated with $S_n$ and the mesh $M_{S_n}$ of $\tau_n$ is given by

$$M_{S_n} := \max_{1\le i\le n+1} |\tilde{\xi}_i - \tilde{\xi}_{i-1}|,$$

where $\tilde{\xi}_0 = a$ and $\tilde{\xi}_{n+1} = b$. Borwein and Saff [7, Lemma 1 a)] proved that if the above Chebyshev space $H_n$ satisfies: 1) $g_i \in C'[a,b]$ and 2) $H'_n := span\{g'_0, \ldots, g'_n\}$ is also a Chebyshev space in $[a,b]$ then for any $f \in C[a,b]$ there exists $\bar{h}_n \in \bar{H}_n := span\{1, g_0, \ldots, g_n\}$, such that

$$\|f - \bar{h}_n\|_\infty \le D\omega(f; \sqrt{M_{H_n}}),$$

where the constant $D$ depends on $a$ and $b$. Observe that all the requirements of the above statement are satisfied if $H_n = S_n$. Indeed, $x^{-n/2+k} \in C^1[a,b]$ and $S'_n := span\{(x^{-n/2+k})', k = 0, \ldots, n\}$ is a Chebyshev space in $[a,b]$. Futhermore, we have $\bar{H}_n = \bar{S}_n$. On applying Borwein and Saff's lemma we conclude that for any $f \in C[a,b]$ there exists $s_n^* \in \bar{S}_n$ such that

$$\|f - s_n^*\|_\infty \le D\omega(f; \sqrt{M_{S_n}}),$$

where the constant $D$ depends on $a$ and $b$. Observe that

$$\tilde{\xi}_i - \tilde{\xi}_{i-1} = u^{-1}(\xi_i) - u^{-1}(\xi_{i-1}) = \frac{du^{-1}(\theta)}{d\xi}(\xi_i - \xi_{i-1}), \quad \theta \in (\xi_i, \xi_{i-1}),$$

where $u^{-1}$ is the inverse mapping of the mapping $u$ defined by (2.1). Since $u^{-1}$ has a bounded derivative in $[a,b]$ and $|\xi_i - \xi_{i-1}| \le const\ n^{-1}$, then $\sqrt{M_{S_n}} \le const\ 1/\sqrt{n}$. Now the statement of the theorem follows from the well-known property $\omega(f;\lambda\delta) \le (\lambda+1)\omega(f;\delta)$ of the modulus of continuity. □

Let

$$E_{\bar{S}_n}(f;[a,b]) := \inf_{s \in \bar{S}_n} \|f - s\|_\infty \tag{3.2}$$

be the best approximation of $f$ by $\bar{S}_n$-polynomials. Theorem 1 implies that for any $f \in C[a,b]$ we have $E_{\bar{S}_n}(f;[a,b]) \longrightarrow 0$ as $n \to \infty$. In particular, $E_{S_n}(f;[a,b]) \longrightarrow 0$ as $n \to \infty$ when $n$ is even.

We now consider the Lagrange interpolation problem by $S_n$-polynomials: Given a continuous function $f(x)$ and distinct points $x_1, x_2, \ldots, x_{n+1}$ in $[a,b]$, determine a $S_n$-polynomial $\mathcal{L}_n(f;x)$ such that $\mathcal{L}_n(f;x_i) = f_i := f(x_i)$ for $i = 1, \ldots, n+1$.

This problem has a unique solution since $\Delta_n \ne 0$. Moreover, $\mathcal{L}_n(f;x)$ can be explicitly represented in the form

$$\mathcal{L}_n(x) := \mathcal{L}_n(f;x) := \sum_{k=1}^{n+1} f_k\ \ell_k(x). \tag{3.3}$$

Here

$$\ell_k(x) = \frac{\Omega(x)}{(x - x_k)\Omega'(x_k)}, k = 1, \ldots, n+1, \tag{3.4}$$

where $\Omega(x) = x^{-n/2}\omega(x)$ and $\omega(x) = \prod_{i=1}^{n+1}(x - x_i)$, are the basic $S_n$-polynomials, for which $\ell_k(x_j) = \delta_{ij}$.

THEOREM 2 Let $f(x)$ be defined in $[a,b] \subset (0,\infty)$ and $x^{n/2}f(x) \in C^n[a,b]$. Suppose that $\frac{d^{n+1}(x^{n/2}f(x))}{dx^{n+1}}$ exists everywhere in $(a,b)$. Then

$$f(x) - \mathcal{L}_n(x) = \frac{\Omega(x)}{(n+1)!}\left[\frac{d^{n+1}}{dx^{n+1}}(x^{n/2}f(x))\right]_{x=\theta},$$

where $\min(x, x_1, \ldots, x_{n+1}) < \theta < \max(x, x_1, \ldots, x_{n+1})$. The point $\theta$ depends on $x, x_1, \ldots, x_{n+1}$ and $f$. Moreover, if the nodes of interpolation coincide with the zeros $\{\tilde{\xi}_i\}_1^{n+1}$ of $\tau_{n+1}$, then the error of the interpolation

$$R_n(f;x) = \|f(x) - \mathcal{L}_n(f;x)\|_\infty \tag{3.5}$$

can be estimated in the following way:

$$\begin{aligned} R_n(f;x) &\leq \frac{\|\Omega(x)\|_\infty}{(n+1)!}\left\|\frac{d^{n+1}}{dx^{n+1}}\left(x^{n/2}f(x)\right)\right\|_\infty \\ &\leq \frac{(\sqrt{b}-\sqrt{a})^{n+1}\sqrt{b}}{2^n(n+1)!}\left\|\frac{d^{n+1}}{dx^{n+1}}\left(x^{n/2}f(x)\right)\right\|_\infty \\ &:= \left\|\frac{d^{n+1}}{dx^{n+1}}\left(x^{n/2}f(x)\right)\right\|_\infty \tilde{R}_{S_n}. \end{aligned}$$

The proof of this result is analogous to the polynomial case.

Let us now compare the usual Lagrange interpolation by algebraic polynomials of degree not exceeding $n$, with the above interpolation by $S_n$-polynomials.

Let $\hat{\xi}_1 < \ldots < \hat{\xi}_{n+1}$ be the zeros of the Chebyshev polynomial $\hat{T}_{n+1}(x) = T_{n+1}\left(\frac{2x-(a+b)}{b-a}\right)$. The Lagrange interpolating polynomial of $f(x)$ at $\hat{\xi}_1, \ldots, \hat{\xi}_{n+1}$ is then given by

$$L_n(x) := L_n(f;x) = \sum_{k=1}^{n+1}\hat{f}_k\hat{l}_k(x),$$

where $\hat{l}_k(x) = \frac{\hat{\omega}(x)}{(x-\hat{\xi}_k)\hat{\omega}'(\hat{\xi}_k)}$, $k = 1, \ldots, n+1$, $\hat{\omega}(x) = (x-\hat{\xi}_1)(x-\hat{\xi}_2)\ldots(x-\hat{\xi}_{n+1})$ and

$$\begin{aligned} r_n(f;x) &:= \|f - L_n(x)\|_\infty \\ &\leq \frac{\|\hat{\omega}(x)\|_\infty}{(n+1)!}\|f^{(n+1)}\|_\infty \end{aligned}$$

$$\begin{aligned}&\leq \frac{(b-a)^{n+1}}{2^{2n+1}(n+1)!}\|f^{(n+1)}\|_{[a,b]}\\ &:= \|f^{(n+1)}\|_\infty \tilde{R}_n.\end{aligned}$$

It is interesting that

$$\frac{\tilde{R}_{S_n}}{\tilde{R}_n} = \frac{2^{n+1}\sqrt{b}}{(\sqrt{b}+\sqrt{a})^{n+1}},$$

i.e., the error estimate we have obtained for the interpolation by $S_n$-polynomials is better than the error estimate for the interpolation by algebraic polynomials for all sufficiently large $n$ if $\sqrt{a}+\sqrt{b} > 2$.

We shall compare also the error of interpolation with the best approximation by $S_n$-polynomials.

THEOREM 3 Let $n$ be a positive even integer, $f \in C[a,b]$ and, let $R_n(f;x)$ and $E_{\bar{S}_n}(f;x)$ be defined by (3.5) and (3.2), respectively. Then the inequality

$$R_n(f;x) \leq E_{\bar{S}_n}(f)(1 + \max_{a\leq x\leq b} \sum_{j=1}^{n+1} |\ell_j(x)|)$$

holds. Moreover, if the nodes of interpolation coincide with the zeros $\{\tilde{\xi}_i\}_{i=1}^{n+1}$ of $\tau_{n+1}$ then

$$R_n(f;x) \leq \left(1 + \frac{b}{a}(1 + \frac{2}{\pi}\log n)\right) E_{\bar{S}_n}(f).$$

Proof: Since $n$ is an even integer, we have $\bar{S}_n = S_n$. Let $s_n^*$ be the best approximant of $f$ from $S_n$. This means that

$$\|s_n^* - f\|_\infty = E_{\bar{S}_n}(f).$$

But,

$$|f(x) - \mathcal{L}_n(f;x)| \leq |f(x) - s_n^*(x)| + |s_n^*(x) - \mathcal{L}_n(f;x)|.$$

From the unicity of $\mathcal{L}_k(f;x)$, we have $s_n^*(x) = \mathcal{L}_n(s_n^*;x)$, and then $s_n^*(x) - \mathcal{L}_n(f;x) = \mathcal{L}_n(s_n^* - f;x)$. Thus,

$$|f(x) - \mathcal{L}_n(f;x)| \leq E_{\bar{S}_n}(f) + |\mathcal{L}_n(s_n^* - f;x)|. \tag{3.6}$$

Now,

$$\begin{aligned}|\mathcal{L}_n(s_n^* - f)| &= |\sum_{j=1}^{n+1} \ell_j(x)(s_n^*(\tilde{\xi}_j) - f(\tilde{\xi}_j))| \\ &\le \max_{a\le x\le b} |s_n^*(x) - f(x)| \max_{a\le x\le b} \sum_{j=1}^{n+1} |\ell_j(x)| \qquad (3.7) \\ &= E_{\bar{S}_n} \max_{a\le x\le b} \sum_{j=1}^{n+1} |\ell_j(x)|.\end{aligned}$$

The first statement of the theorem follows immediately from (3.6) and (3.7).
In order to prove the second one, we shall estimate the size of the constant $\max_{a\le x\le b} \sum_{j=1}^{n+1} |\ell_j(x)|$.

Consider the interpolating polynomials

$$\mathcal{L}_n(f;x) = \sum_{j=1}^{n+1} \ell_j(x) f(\tilde{\xi}_j) \in S_n \quad \text{and} \quad L_n(f;x) = \sum_{j=1}^{n+1} l_j(x) f(\xi_j) \in I\!P_n$$

over the zeros of $\tau_{n+1}(x)$ and $T_{n+1}(x)$, respectively. Then we have

$$\ell_j(x) = \frac{\tau_{n+1}(x)}{(x-\tilde{\xi}_j)\tau'_{n+1}(\tilde{\xi}_j)} \quad \text{and} \quad l_j(x) = \frac{T_{n+1}(x)}{(x-\xi_j)T'_{n+1}(\xi_j)}\,, \; j = 1, \ldots, n+1.$$

Equations (1.4) and (2.1) imply

$$|\ell_j(x)| = \frac{|T_{n+1}(u(x))|}{|T'_{n+1}(\xi_j)||x-\xi_j|} \left( \frac{2(\sqrt{b}-\sqrt{a})x\sqrt{x}}{x+\sqrt{ab}} \right) \frac{|x-\xi_j|}{|x-\tilde{\xi}_j|}.$$

On the other hand, the inequality

$$\left| \frac{x-\xi_j}{x-\tilde{\xi}_j} \right| = \frac{1}{\sqrt{b}-\sqrt{a}} \left| \frac{\sqrt{\tilde{\xi}_j x} + \sqrt{ab}}{\sqrt{\tilde{\xi}_j x}(\sqrt{x}+\sqrt{\tilde{\xi}_j})} \right| \le \frac{1}{\sqrt{b}-\sqrt{a}} \left( \frac{\sqrt{a}+\sqrt{b}}{2a} \right)$$

holds because the function on the left-hand side of this inequality is a decreasing one in $[a,b]$.

It is easy to see that $\left( \dfrac{2(\sqrt{b}-\sqrt{a})x\sqrt{x}}{x+\sqrt{ab}} \right)$ is an increasing function in $[a,b]$. This fact immediately yields

$$|\ell_j(x)| \le \frac{b}{a}\, |l_j(u(x))|\,, \; j = 1, \ldots, n+1$$

and then

$$\max_{a\le x\le b}\sum_{j=1}^{n+1}|\ell_j(x)|\le\frac{b}{a}\max_{-1\le u\le 1}\sum_{j=1}^{n+1}|l_j(u(x))|.$$

Now we employ the well-known estimate

$$\max_{-1\le u\le 1}\sum_{j=1}^{n+1}|l_j(u)|\le 1+\frac{2}{\pi}\log n$$

for the Lebesgue's function in order to complete the proof. □

## 4 EXTREMAL PROPERTIES OF THE CHEBYSHEV $S_n$-POLYNOMIALS

Consider the following extremal problem:

PROBLEM 3 Let $1\le q\le\infty$ and $n\in\mathbb{N}$ be given and

$$\|f\|_{q,n}:=\int_a^b |x^{-n}f(x)|^q\,((b-x)(x-a))^{-1/2}dx.$$

Find the polynomials $r_{2n-1}^*\in\mathbb{P}_{2n-1}$ for which

$$\|x^{2n}-r_{2n-1}^*\|_{q,n}=\inf_{r_{2n-1}\in\mathbb{P}_{2n-1}}\|x^{2n}-r_{2n-1}\|_q.$$

THEOREM 4 For any $q$, $1\le q\le\infty$, and $n\in\mathbb{N}$ PROBLEM 3 has a unique solution $r_{2n-1}^*$ and it is

$$r_{2n-1}^*(x)=x^{2n}-2^{-2n+1}(\sqrt{b}-\sqrt{a})^{2n}x^n\tau_{2n}(x).$$

The extremal polynomial $x^{2n}-r_{2n-1}^*(x)$ may be written also in the following equivalent form:

$$\begin{aligned}x^{2n}-r_{2n-1}^*(x) &= \frac{(\sqrt{b}-\sqrt{a})^{2n}}{2^{2n-1}}\Big((x-\sqrt{ab}+\sqrt{(x-a)(x-b)})^{2n}\\ &\quad+(x-\sqrt{ab}-\sqrt{(x-a)(x-b)})^{2n}\Big).\end{aligned}\tag{4.1}$$

Proof: Given a polynomial $Q\in\mathbb{P}_m$, such that $Q(x)>0$ for $x\in[a,b]$ and an integer $\nu\ge 2m$, let $r_{\nu-1}^*\in\mathbb{P}_{\nu-1}$ be the unique solution of the extremal problem

$$\|(x^\nu-r_{\nu-1}^*(x))/\sqrt{Q(x)}\|_\infty=\inf_{r_{\nu-1}\in\mathbb{P}_{\nu-1}}\|(x^\nu-r_{\nu-1}(x))/\sqrt{Q(x)}\|_\infty.$$

Bernstein in [2, 3] proved that for any $q$, $1 \le q \le \infty$, $r^*_{\nu-1}$ is also the unique solution of the extremal problem

$$\int_a^b |(x^\nu - r^*_{\nu-1}(x))/\sqrt{Q(x)}|^q((b-x)(x-a))^{-1/2}dx =$$

$$= \inf_{r_{\nu-1}\in \mathbb{P}_{\nu-1}} \int_a^b |(x^\nu - r_{\nu-1}(x))/\sqrt{Q(x)}|^q((b-x)(x-a))^{-1/2}dx.$$

The statement of the theorem is an immediate consequence of Bernstein's result for $Q(x) = x^{2n}$, $\nu = 2n = 2m$, and the fact that $\tau_{2n}$ is the general Chebyshev polynomial associated with the Chebyshev space $S_n$ in $(a,b)$. The representation (4.2) follows from (2.5). □

We have already proved a Jackson type theorem for approximation by $S_n$-polynomials. An important question in Approximation Theory is to establish inverse type theorems. Basic tools for proving such theorems for approximation by algebraic polynomials are the inequalities of Markov [11] and Bernstein [1] (see also [12, 14]). It is known that Markov's and Bernstein's inequalities are sharp in the sense that there exist extremal polynomials for which equalities are attained. In fact, in both cases the extremal polynomial is the Chebyshev one. Chebyshev polynomials are solutions of some other interesting extremal problems concerning algebraic polynomials [14, Chapter 2B], [12, Chapter 6]. It is natural to expect that the Chebyshev $S_n$-polynomial $\tau_n$ plays a similar role in the corresponding problems concerning $S_n$-polynomials. We state our expectations in the following conjectures:

CONJECTURE 1 [Markov's inequality] For every $s \in S_n$

$$||s'||_\infty \le \frac{\sqrt{b}+\sqrt{a}}{2a(\sqrt{b}-\sqrt{a})} n^2 ||s||_\infty. \tag{4.2}$$

Equality is attained if, and only if, $s = \tau_n$ and for $x = a$.

CONJECTURE 2 [Bernstein's inequality] For every $s \in S_n$ and any $x \in (a,b)$

$$|s'(x)|_\infty \le \frac{x+\sqrt{ab}}{2x\sqrt{(b-x)(x-a)}} n ||s||_\infty, \tag{4.3}$$

where equality is attained only for $s = \tau_n$ at the zeros of $\tau_n$.

CONJECTURE 3 [Szegö's inequality] For every $s \in S_n$, for which $||s||_\infty \le 1$

$$\frac{4(b-x)(x-a)x^2}{(x+\sqrt{ab})^2}(y')^2 + n^2y^2 \le n^2, \quad \text{for } x \in [a,b]. \tag{4.4}$$

Equality is attained if, and only if, $s = \tau_n$ when (4.4) becomes the identity (2.5).

CONJECTURE 4 [Bojanov-Kristiansen's inequality] For any $p, 1 \leq p, \leq \infty$ and every $s \in S_n$

$$||s'||_p \leq ||\tau_n'||_p ||s||_\infty, \tag{4.5}$$

where $||\cdot||_\infty$ and $||\cdot||_p$ are the uniform and $L_p$-norms in $(a, b)$, respectively. Equality is attained if, and only if, $s = \tau_n$.

CONJECTURE 5 [The longest polynomial] Among all the $S_n$-polynomials $s$, for which $|s(x)| \leq 1$ for $x \in [a, b]$, $\tau_n$ has the maximum length of arc.

Inequalities (4.2) and (4.3) are analogs of the celebrated Markov's [11] and Bernstein's [1] inequalities, respectively. The polynomial version of inequality (4.4) was proved first by Szegö [16] and rediscovered later by van der Corput and Schaake [8]. Inequality (4.5) for algebraic polynomials was discovered independently by Bojanov [4] and Kristiansen [10]. Note that for $p = \infty$ (4.5) reduces to (4.2) and for $p = 1$ (4.5) can be verified just like in the polynomial case (see [4] and [14, p.149]). Bojanov [5] and Kristiansen [10] proved also that the Chebyshev polynomial $T_n$ is "the longest" one among all algebraic polynomials whose modulus is bounded by one in $[-1, 1]$, thus verifying a long-standing conjecture of Erdös. For these and many other interesting extremal problems for algebraic polynomials we refer to [12] and [14].

## REFERENCES


1. S.N.Bernstein, On best approximation of continuous functions by polynomials of given degree, Soobšč. Har'kov. Mat. Obšč. (2) 13: 47-194, (1912); reprinted in his Collected works, Vol.1, pp. 11-104, Izdat. Akad. Nauk. SSSR, Moskow, (1952); abriged French transl., Acad. Roy. Belg. Cl. Sci. Mèm. (2) 4: 1-103, fasc. 1, 1912.

2. S.N.Bernstein, On a class of orthogonal polynomials, Collected Papers, Vol.1, pp.452-465, Izdat. Akad. Nauk. SSSR, Moskow, 1952, (in Russian).

3. S.N.Berstein, Appendix to the paper " On a class of orthogonal polynomials ", Collected Papers, Vol.1, pp. 466-467, Izdat. Akad. Nauk. SSSR, Moskow, 1952, (in Russian).

4. B.D.Bojanov, An extension of the Markov inequality, J.Approx.Theory 35: 181-190, 1982.

5. B.D.Bojanov, Proof of a conjecture of Erdös about the longest polynomial, Proc.Amer.Math.Soc. 84: 99-103, 1982.

6. P.Borwein, Zeros of Chebyshev polynomials in Markov systems, J. Approx. Theory 63: 56-64, 1990.

7. P.Borwein and E.B.Saff, On the denseness of weighted incomplete approximations, Progress in Approximation Theory: An International Perspective (A.A.Gonchar and E.B.Saff, eds.): 419-429, Springer-Verlag, New York, 1992.

8. J.G. van der Corput and G. Schaake, Ungleichungen für Polynome und trigonometrishe Polynome, Composito Math. 2: 321-361 (1935); Correction, Composito Math. 3: 128, 1936.

9. W.B.Jones, W.J.Thron and H.Waadeland, A strong Stieltjes moment problem, Trans. Amer. Math. Soc. 261: 503-528, 1980.

10. G.K.Kristiansen, Some inequalities for algebraic and trigonometric polynomials, J. London Math. Soc.(2) 20: 300-314, 1979.

11. A.A.Markov, On a problem of Mendeleev, Zap. Imp. Acad. Nauk., St. Petersburg 62: 1-24, 1889. (in Russian)

12. G.V.Milovanović, D.S. Mitrinović and Th.R. Rassias, Topics in Polynomials: Extremal Problems, Inequalities, Zeros, World Scientific, Singapore, 1994.

13. P.J.O'Hara and R.S.Rodriguez, Some properies of self-inverse polynomials, Proc.Amer.Math.Soc. 44: 331-335, 1974.

14. T.J.Rivlin, Chebyshev Polynomials: From Approximation Theory to Algebra and Number Theory, 2nd ed., Wiley, New York, 1990.

15. A. Sri Ranga, Symmetric orthogonal polynomials and the associated orthogonal L-polynomials, Proc. Amer. Math. Soc. 123: 3135-3141, 1995.

16. G. Szegö, Über einen Satz der Herrn Serge Bernstein, Schriften Königsberg Gel. Ges., Naturw. Klasse 5: 59-70, 1928. Reprinted in: Gabor Szegö: Collected Papers, Vol.2, pp. 207-219, (R.Askey, ed.), Birkhäuser, Boston, 1982.

17. G. Szegö, Orthogonal polynomials, 4th ed., Amer. Math. Soc. Coll. Publ., Vol. 23, Providence, RI, 1975.

18. V.Totik, Weighted Approximation with Varying Weight, Lecture Notes in Mathematics, Vol. 1569, Springer-Verlag, Berlin, 1994.

# Natural Solutions of Indeterminate Strong Stieltjes Moment Problems Derived from PC-Fractions

CATHERINE M. BONAN-HAMADA* Department of Computer Science, Mathematics and Statistics, Mesa State College, Grand Junction, Colorado

WILLIAM B. JONES* Department of Mathematics, University of Colorado, Boulder, Colorado

OLAV NJÅSTAD Department of Mathematical Sciences, Norwegian University of Science and Technology, Trondheim, Norway

**Research supported in part by the National Science Foundation under Grant DMS-9302584.*

## 1 INTRODUCTION

Let $\{c_k\}_{k=-\infty}^{\infty}$ be a bisequence of real numbers such that the strong Stieltjes moment problem (SSMP) for $\{c_k\}_{k=-\infty}^{\infty}$ has a solution. That is, there exists a distribution function, $\psi$, (bounded, nondecreasing with infinitely many points of increase on $(0,\infty)$) such that

$$c_k = \int_0^\infty (-t)^k d\psi(t), \quad k = 0, \pm 1, \pm 2, \ldots .$$

A solvable moment problem is said to be determinate if there exists one and only one solution; otherwise it is called indeterminate.

In [4] it is shown that if $\psi$ is a solution to the SSMP for $\{c_k\}_{k=-\infty}^{\infty}$, then the quadratic forms

$$
\begin{aligned}
Q(s,n) &:= \sum_{i=-n}^{n}\sum_{j=-n}^{n}(-1)^s c_{i+j+s}u_i u_j \\
&= \int_0^\infty \left(t^s \sum_{i=-n}^{n}\sum_{j=-n}^{n}(-t)^i u_i (-t)^j u_j\right) d\psi(t) \\
&= \int_0^\infty t^s \left(\sum_{j=-n}^{n}(-t)^j u_j\right)^2 d\psi(t), \quad n=0,1,2,\ldots,\ s=0,\pm1,\pm2,\ldots,
\end{aligned}
$$

are positive definite. This in turn, is equivalent to the condition that for $n = 0,1,2,\ldots$, $s=0,\pm1,\pm2,\ldots$, and $k=0,1,\ldots,2n$, the determinants

$$
\begin{vmatrix}
(-1)^s c_{-2n+s} & (-1)^s c_{-2n+1+s} & \cdots & (-1)^s c_{-2n+k+s} \\
(-1)^s c_{-2n+1+s} & (-1)^s c_{-2n+2+s} & \cdots & (-1)^s c_{-2n+k+1+s} \\
\vdots & \vdots & \ddots & \vdots \\
(-1)^s c_{-2n+k+s} & (-1)^s c_{-2n+k+1+s} & \cdots & (-1)^s c_{-2n+2k+s}
\end{vmatrix}
$$

are all positive. Hence we have the following theorem.

THEOREM 1.1 Let $\psi$ be a solution of the SSMP for a bisequence $\{c_k\}_{k=-\infty}^{\infty}$. Then for $n=0,1,2,\ldots$, $s=0,\pm1,\pm2,\ldots$, and $k=0,1,\ldots,2n$

$$
(-1)^{s(k+1)}H_{k+1}^{(-2n+s)} > 0, \tag{1.1a}
$$

where $H_0^{(m)} := 1$ and

$$
H_l^{(m)} := \begin{vmatrix}
c_m & c_{m+1} & \cdots & c_{m+l-1} \\
c_{m+1} & c_{m+2} & \cdots & c_{m+l} \\
\vdots & \vdots & \ddots & \vdots \\
c_{m+l-1} & c_{m+l} & \cdots & c_{m+2l-2}
\end{vmatrix} \tag{1.1b}
$$

for $m = 0 \pm 1, \pm 2, \ldots$, $l = 1,2,\ldots$.

In the remainder of this paper we assume that the SSMP for $\{c_k\}_{k=-\infty}^{\infty}$ is solvable and hence the inequalities (1.1) are valid throughout. It should be noted that the conditions $H_{n+1}^{(-n)} > 0$, $n = 0,1,2,\ldots$, and $(-1)^n H_n^{(-n)} > 0$, $n = 1,2,\ldots$, are necessary and sufficient for the SSMP for $\{c_k\}_{k=-\infty}^{\infty}$ to have a solution [4].

Let $(L_0, L_\infty)$ be the pair of formal power series (fps)

$$
L_0(z) := \sum_{k=1}^{\infty} -c_{-k}z^k \tag{1.2a}
$$

and

$$
L_\infty(z) := \sum_{k=0}^{\infty} c_k z^{-k}. \tag{1.2b}
$$

Define

$$\alpha_1 := H_1^{(0)} = c_0, \tag{1.3a}$$

$$\alpha_{2n+1} := \frac{-H_{n+1}^{(-n)} H_{n-1}^{(-n+2)}}{\left(H_n^{(-n+1)}\right)^2}, \quad n = 1, 2, \dots, \tag{1.3b}$$

$$\beta_0 := 0, \tag{1.3c}$$

$$\beta_{2n} := \frac{(-1)^n H_n^{(-n)}}{H_n^{(-n+1)}}, \quad n = 1, 2, \dots, \tag{1.3d}$$

and

$$\beta_{2n+1} := \frac{(-1)^n H_n^{(-n+2)}}{H_n^{(-n+1)}}, \quad n = 0, 1, 2, \dots. \tag{1.3e}$$

It is a straight forward application of (1.1) to determine that $\alpha_1 > 0$, $\alpha_{2n+1} \neq 0$, $\beta_0 = 0$, $\beta_{2n} \neq 0$ and $\beta_{2n+1} \neq 0$.

THEOREM 1.2 Let $\{c_k\}_{k=-\infty}^{\infty}$ be a bisequence of real numbers for which the SSMP is solvable. Let $(L_0, L_\infty)$ be the pair of fps defined by (1.2) and let the sequences $\{\alpha_{2n+1}\}_{n=0}^{\infty}$ and $\{\beta_n\}_{n=0}^{\infty}$ be defined by (1.3). Then since $H_{n+1}^{(-n)} \neq 0$ for $n = 0, 1, 2, \dots$, and $\beta_m \neq 0$ for $m = 1, 2, \dots$, there exists a unique modified PC-fraction

$$\frac{\alpha_1}{1} + \frac{1/z}{\beta_2} + \frac{\alpha_3 z}{\beta_3 z} + \frac{1}{\beta_4} + \frac{\alpha_5}{\beta_5} + \frac{1/z}{\beta_6} + \frac{\alpha_7 z}{\beta_7 z} + \frac{1}{\beta_8} + \frac{\alpha_9}{\beta_9} + \cdots, \tag{1.4}$$

corresponding to $(L_0, L_\infty)$ in the sense that for $n = 0, 1, 2, \dots$,

$$L_0(z) - \frac{A_{2n}(z)}{B_{2n}(z)} = \beta_{2n+2} \prod_{j=0}^{n} \alpha_{2j+1} z^{n+1} + O(z^{n+2}), \tag{1.5a}$$

$$L_0(z) - \frac{A_{2n+1}(z)}{B_{2n+1}(z)} = -\frac{\prod_{j=0}^{n} \alpha_{2j+1}}{\beta_{2n+1}} z^n + O(z^{n+1}), \tag{1.5b}$$

$$L_\infty(z) - \frac{A_0(z)}{B_0(z)} = \alpha_1 + O\left(\frac{1}{z}\right), \tag{1.5c}$$

$$L_\infty(z) - \frac{A_{2n}(z)}{B_{2n}(z)} = \frac{\prod_{j=0}^{n} \alpha_{2j+1}}{\beta_{2n} z^n} + O\left(\left(\frac{1}{z}\right)^{n+1}\right), \tag{1.5d}$$

$$L_\infty(z) - \frac{A_{2n+1}(z)}{B_{2n+1}(z)} = -\beta_{2n+3} \frac{\prod_{j=0}^{n} \alpha_{2j+1}}{z^{n+1}} + O\left(\left(\frac{1}{z}\right)^{n+2}\right), \tag{1.5e}$$

where

$$A_0(z) := 0, \quad A_1(z) := \alpha_1, \quad B_0(z) := 1, \quad B_1(z) := 1,$$

$$\begin{bmatrix} A_{4n+1}(z) \\ B_{4n+1}(z) \end{bmatrix} := \beta_{4n+1} \begin{bmatrix} A_{4n}(z) \\ B_{4n}(z) \end{bmatrix} + \alpha_{4n+1} \begin{bmatrix} A_{4n-1}(z) \\ B_{4n-1}(z) \end{bmatrix}, \quad n = 1, 2, 3, \ldots, \tag{1.6a}$$

$$\begin{bmatrix} A_{4n+2}(z) \\ B_{4n+2}(z) \end{bmatrix} := \beta_{4n+2} \begin{bmatrix} A_{4n+1}(z) \\ B_{4n+1}(z) \end{bmatrix} + \frac{1}{z} \begin{bmatrix} A_{4n}(z) \\ B_{4n}(z) \end{bmatrix}, \quad n = 0, 1, 2, \ldots, \tag{1.6b}$$

$$\begin{bmatrix} A_{4n+3}(z) \\ B_{4n+3}(z) \end{bmatrix} := \beta_{4n+3} z \begin{bmatrix} A_{4n+2}(z) \\ B_{4n+2}(z) \end{bmatrix} + \alpha_{4n+3} z \begin{bmatrix} A_{4n+1}(z) \\ B_{4n+1}(z) \end{bmatrix}, \quad n = 0, 1, 2, \ldots, \tag{1.6c}$$

and

$$\begin{bmatrix} A_{4n+4}(z) \\ B_{4n+4}(z) \end{bmatrix} := \beta_{4n+4} \begin{bmatrix} A_{4n+3}(z) \\ B_{4n+3}(z) \end{bmatrix} + \begin{bmatrix} A_{4n+2}(z) \\ B_{4n+2}(z) \end{bmatrix}, \quad n = 0, 1, 2, \ldots. \tag{1.6d}$$

REMARK 1: $A_n(z)$ and $B_n(z)$ are called the nth numerator and denominator, respectively, of the modified PC-fraction (1.4).

REMARK 2: In Theorem 1.2, $O(z^n)$ denotes a fps of increasing powers of $z$ starting with a power greater than or equal to $n$. Similarly, $O((1/z)^n)$ denotes a formal Laurent series (fLs) of increasing posers of $1/z$ starting with a power greater than or equal to $n$.

Proof: In [2] it is shown that the pair of formal power series $(L_0, L_\infty)$ correspond to the PC-fraction

$$\frac{\alpha_1}{1} + \frac{1}{\beta_2 z} + \frac{\alpha_3 z}{\beta_3} + \frac{1}{\beta_4 z} + \frac{\alpha_5 z}{\beta_5} + \frac{1}{\beta_6 z} + \cdots. \tag{1.7}$$

Using an equivalence transformation [3, Theorem 2.6] with

$$r_0 = 1, \quad r_{4k+1} = 1, \quad r_{4k+2} = \frac{1}{z}, \quad r_{4k+3} = z \quad \text{and} \quad r_{4k+4} = \frac{1}{z}, \quad k = 0, 1, 2, \dots,$$

we see that the modified PC-fraction (1.4) is equivalent to the PC-fraction (1.7) and hence Theorem 2.1 follows from Theorems 2.1, 2.2B and 4.1 in [2].

The even canonical contraction [5] of the modified PC-fraction (1.4) is a modified T-fraction

$$\frac{F_1}{\frac{1}{z} + G_1} + \frac{F_2}{1 + G_2 z} + \frac{F_3}{\frac{1}{z} + G_3} + \frac{F_4}{1 + G_4 z} + \cdots, \tag{1.8a}$$

where

$$F_1 := \alpha_1 \beta_2 = -H_1^{(-1)} > 0, \tag{1.8b}$$

$$F_n := -\frac{\alpha_{2n-1}\beta_{2n}}{\beta_{2n-2}} = -\frac{H_{n-2}^{(-n+3)} H_n^{(-n)}}{H_{n-1}^{(-n+2)} H_{n-1}^{(-n+1)}} > 0, \quad n = 2, 3, \dots, \tag{1.8c}$$

$$G_1 := \beta_2 = -\frac{H_1^{(-1)}}{H_1^{(0)}} > 0, \tag{1.8d}$$

and

$$G_n := \frac{\beta_{2n}}{\beta_{2n-2}} = -\frac{H_n^{(-n)} H_{n-1}^{(-n+2)}}{H_n^{(-n+1)} H_{n-1}^{(-n+1)}} > 0, \quad n = 2, 3, \dots. \tag{1.8e}$$

If $P_n(z)$ denotes the nth numerator of (1.8) and $Q_n(z)$ denotes the nth denominator of (1.8) then $\{P_n(z)\}$ and $\{Q_n(z)\}$ are defined by

$$P_{-1}(z) := 1, \quad P_0(z) := 0, \quad Q_{-1}(z) := 0, \quad Q_0(z) := 1, \tag{1.9a}$$

$$\begin{bmatrix} P_{2n+1}(z) \\ Q_{2n+1}(z) \end{bmatrix} := \left(\frac{1}{z} + G_{2n+1}\right) \begin{bmatrix} P_{2n}(z) \\ Q_{2n}(z) \end{bmatrix} + F_{2n+1} \begin{bmatrix} P_{2n-1}(z) \\ Q_{2n-1}(z) \end{bmatrix}, \quad n = 0, 1, 2, \dots, \tag{1.9b}$$

and

$$\begin{bmatrix} P_{2n}(z) \\ Q_{2n}(z) \end{bmatrix} := (1+G_{2n}z)\begin{bmatrix} P_{2n-1}(z) \\ Q_{2n-1}(z) \end{bmatrix} + F_{2n}\begin{bmatrix} P_{2n-2}(z) \\ Q_{2n-2}(z) \end{bmatrix}, \quad n=1,2,3,\dots. \tag{1.9c}$$

Since the modified positive T-fraction is the even canonical contraction of the modified PC-fraction (1.4), the numerators $P_n(z)$ and denominators $Q_n(z)$ of (1.8) are related to the numerators $A_n(z)$ and denominators $B_n(z)$ of (1.4) by

$$P_n(z) = A_{2n}(z) \tag{1.10a}$$

and

$$Q_n(z) = B_{2n}(z). \tag{1.10b}$$

The odd canonical contraction of the modified PC-fraction (1.4) is a modified M-fraction

$$\alpha_1 + \frac{U_1}{z+V_1} + \frac{U_2}{1+\frac{V_2}{z}} + \frac{U_3}{z+V_3} + \frac{U_4}{1+\frac{V_4}{z}} + \cdots, \tag{1.11a}$$

where

$$U_1 := -\alpha_1\beta_3 = H_1^{(1)} < 0, \tag{1.11b}$$

$$U_n := -\frac{\alpha_{2n-1}\beta_{2n+1}}{\beta_{2n-1}} = -\frac{H_{n-2}^{(-n+3)}H_n^{(-n+2)}}{H_{n-1}^{(-n+2)}H_{n-1}^{(-n+3)}} > 0, \quad n=2,3,\dots, \tag{1.11c}$$

$$V_1 := \beta_3 = -\frac{H_1^{(1)}}{H_1^{(0)}} > 0, \tag{1.11d}$$

and

$$V_n := \frac{\beta_{2n+1}}{\beta_{2n-1}} = -\frac{H_{n-1}^{(-n+2)}H_n^{(-n+2)}}{H_n^{(-n+1)}H_{n-1}^{(-n+3)}} > 0, \quad n=2,3,\dots. \tag{1.11e}$$

If $R_n(z)$ and $S_n(z)$ denote the nth numerator and denominator, respectively, of (1.11) then $\{R_n(z)\}$ and $\{S_n(z)\}$ are defined by

$$R_{-1}(z) := 1, \quad R_0(z) := \alpha_1, \quad S_{-1}(z) := 0, \quad S_0(z) := 1, \tag{1.12a}$$

$$\begin{bmatrix} R_{2n+1}(z) \\ S_{2n+1}(z) \end{bmatrix} := (z+V_{2n+1})\begin{bmatrix} R_{2n}(z) \\ S_{2n}(z) \end{bmatrix} + U_{2n+1}\begin{bmatrix} R_{2n-1}(z) \\ S_{2n-1}(z) \end{bmatrix}, \quad n=0,1,2,\dots, \tag{1.12b}$$

and

$$\begin{bmatrix} R_{2n}(z) \\ S_{2n}(z) \end{bmatrix} := \left(1+\frac{V_{2n}}{z}\right)\begin{bmatrix} R_{2n-1}(z) \\ S_{2n-1}(z) \end{bmatrix} + U_{2n}\begin{bmatrix} R_{2n-2}(z) \\ S_{2n-2}(z) \end{bmatrix}, \quad n=1,2,3,\dots. \tag{1.12c}$$

Since the modified M-fraction (1.11) is the odd canonical contraction of the modified PC-fraction (1.4), the numerators and denominators $R_n(z)$ and $S_n(z)$, respectively, of (1.11) are related to the numerators $A_n(z)$ and denominators $B_n(z)$ of (1.4) by

$$R_n(z) = A_{2n+1}(z) \tag{1.13a}$$

and

$$S_n(z) = B_{2n+1}(z). \tag{1.13b}$$

## 2 THE MAIN RESULT

THEOREM 2.1 Let $\{c_k\}_{k=-\infty}^{\infty}$ be a bisequence of real numbers such that the SSMP for $\{c_k\}_{k=-\infty}^{\infty}$ is indeterminate. Let $A_n(z)$ and $B_n(z)$ be the nth numerator and denominator, respectively, of the modified PC-fraction (1.4). Let $P_n(z)$ and $Q_n(z)$ denote the nth numerator and denominator, respectively, of the modified positive T-fraction (1.8) which is the even canonical contraction of (1.4) and let $R_n(z)$ and $S_n(z)$ denote the nth numerator and denominator, respectively, of the modified M-fraction (1.11) which is the odd canonical contraction of (1.4). Then

A. There exist (natural) solutions $\mu^{(0)}, \mu^{(\infty)}, \nu^{(0)}$, and $\nu^{(\infty)}$ to the SSMP for $\{c_k\}_{k=-\infty}^{\infty}$ such that

$$\lim_{n\to\infty}\frac{P_{2n}(z)}{Q_{2n}(z)} = \int_0^\infty \frac{d\mu^{(0)}(t)}{z+t}, \tag{2.1a}$$

$$\lim_{n\to\infty}\frac{P_{2n+1}(z)}{Q_{2n+1}(z)} = \int_0^\infty \frac{d\mu^{(\infty)}(t)}{z+t}, \tag{2.1b}$$

$$\lim_{n\to\infty}\frac{R_{2n}(z)}{S_{2n}(z)} = \int_0^\infty \frac{d\nu^{(0)}(t)}{z+t}, \tag{2.1c}$$

$$\lim_{n\to\infty}\frac{R_{2n+1}(z)}{S_{2n+1}(z)} = \int_0^\infty \frac{d\nu^{(\infty)}(t)}{z+t}. \tag{2.1d}$$

B. $\{Q_n(-z)\}_{n=0}^{\infty}$ and $\{S_n(-z)\}_{n=0}^{\infty}$ are orthogonal Laurent polynomial sequences (OLPS) with respect to any solution of the SSMP, in particular to $\mu^{(0)}, \mu^{(\infty)}, \nu^{(0)}$, and $\nu^{(\infty)}$, such that $\{Q_n(z)\}_{n=1}^{\infty}$ is associated with the base $\{1, z^{-1}, z, z^{-2}, z^2, \dots\}$

and $\{S_n(z)\}_{n=1}^{\infty}$ is associated with the base $\{1, z, z^{-1}, z^2, z^{-2}, \ldots\}$. That is,

$$\begin{aligned}
&Q_0(z) = 1,\\
&Q_{2n}(z) \in \mathrm{Span}\{1, z^{-1}, z, \ldots, z^{-n}, z^n\} - \mathrm{Span}\{1, z^{-1}, z, \ldots, z^{-n}\},\\
&Q_{2n+1}(z) \in \mathrm{Span}\{1, z^{-1}, z, \ldots, z^n, z^{-(n+1)}\} - \mathrm{Span}\{1, z^{-1}, z, \ldots, z^n\},\\
&S_0(z) = 1,\\
&S_{2n}(z) \in \mathrm{Span}\{1, z, z^{-1}, \ldots, z^n, z^{-n}\} - \mathrm{Span}\{1, z, z^{-1}, \ldots, z^n\},\\
&S_{2n+1}(z) \in \mathrm{Span}\{1, z, z^{-1}, \ldots, z^{-n}, z^{n+1}\} - \mathrm{Span}\{1, z, z^{-1}, \ldots, z^{-n}\}.
\end{aligned}$$

C. The sequences $\{W_{2n}(z)\}_{n=1}^{\infty}$, $\{W_{2n+1}(z)\}_{n=1}^{\infty}$, $\{T_{2n}(z)\}_{n=1}^{\infty}$, $\{T_{2n+1}(z)\}_{n=1}^{\infty}$, $\{X_{2n}(z)\}_{n=1}^{\infty}$, $\{X_{2n+1}(z)\}_{n=1}^{\infty}$, $\{Y_{2n}(z)\}_{n=1}^{\infty}$, and $\{Y_{2n+1}(z)\}_{n=1}^{\infty}$, of Laurent polynomials (L-polynomials) defined by

$$W_m(z) := \left(\prod_{k=1}^{m} e_k\right) P_m(z), \tag{2.2a}$$

$$T_m(z) := \left(\prod_{k=1}^{m} e_k\right) Q_m(z), \tag{2.2b}$$

$$X_m(z) := \left(\prod_{k=1}^{m} \epsilon_k\right) R_m(z), \tag{2.2c}$$

$$Y_m(z) := \left(\prod_{k=1}^{m} \epsilon_k\right) S_m(z), \tag{2.2d}$$

where

$$e_1 := \frac{1}{F_1}, \quad e_{2n-1} := \frac{\prod_{k=1}^{n-1} F_{2k}}{\prod_{k=1}^{n} F_{2k-1}}, \quad e_{2n} := \frac{\prod_{k=1}^{n} F_{2k-1}}{\prod_{k=1}^{n} F_{2k}}, \tag{2.2e}$$

and

$$\epsilon_1 := \frac{1}{|U_1|}, \quad \epsilon_{2n-1} := \frac{\prod_{k=1}^{n-1} U_{2k}}{\prod_{k=1}^{n} |U_{2k-1}|}, \quad \epsilon_{2n} := \frac{\prod_{k=1}^{n} |U_{2k-1}|}{\prod_{k=1}^{n} U_{2k}}, \tag{2.2f}$$

converge uniformly on compact subsets of $\mathcal{C} - \{0\}$ to holomorphic functions $W^{(0)}(z)$, $W^{(\infty)}(z)$, $T^{(0)}(z)$, $T^{(\infty)}(z)$, $X^{(0)}(z)$, $X^{(\infty)}(z)$, $Y^{(0)}(z)$, and $Y^{(\infty)}(z)$, respectively.

D.

$$\lim_{n\to\infty} \frac{P_{2n}(z)}{Q_{2n}(z)} = \frac{W^{(0)}(z)}{T^{(0)}(z)} = \frac{X^{(\infty)}(z)}{Y^{(\infty)}(z)} = \lim_{n\to\infty} \frac{R_{2n+1}(z)}{S_{2n+1}(z)} \tag{2.3a}$$

and

$$\lim_{n\to\infty} \frac{P_{2n+1}(z)}{Q_{2n+1}(z)} = \frac{W^{(\infty)}(z)}{T^{(\infty)}(z)} = \frac{X^{(0)}(z)}{Y^{(0)}(z)} = \lim_{n\to\infty} \frac{R_{2n}(z)}{S_{2n}(z)}. \tag{2.3b}$$

E.

$$\mu^{(0)} = \nu^{(\infty)} \quad \text{and} \quad \mu^{(\infty)} = \nu^{(0)}. \tag{2.4}$$

Proof A: The modified positive T-fraction (1.8) is equivalent to a positive T-fraction. Hence existence of solutions $\mu^{(0)}$ and $\mu^{(\infty)}$ satisfying (2.1a) and (2.1b) is shown in [4]. The modified M-fraction (1.11) is equivalent to the continued fraction

$$\alpha_1 + \frac{\left(\frac{U_1}{V_1}\right)}{1+\frac{1}{V_1}z} + \frac{\left(\frac{U_2}{V_1V_2}\right)z}{1+\frac{1}{V_2}z} + \frac{\left(\frac{U_3}{V_2V_3}\right)z}{1+\frac{1}{V_3}z} + \cdots. \tag{2.5}$$

If we let $\hat{R}_n(z)$ and $\hat{S}_n(z)$ denote the nth numerator and denominator, respectively, of (2.5) then

$$\hat{R}_n(z) = \left(\prod_{k=0}^{n} r_k\right) R_n(z) \quad \text{and} \quad \hat{S}_n(z) = \left(\prod_{k=0}^{n} r_k\right) S_n(z) \tag{2.6a}$$

where

$$r_0 := 1, \quad r_{2n} = \frac{1}{V_{2n}}z, \; n = 1, 2, \ldots, \text{ and } r_{2n+1} = \frac{1}{V_{2n+1}}, \; n = 0, 1, 2, \ldots. \tag{2.6b}$$

Recall from (1.11) that $U_1 < 0, U_n > 0$ for $n = 2, 3, \ldots$, and $V_n > 0$ for $n = 1, 2, \ldots$. Thus, the continued fraction (2.5) can be written as

$$\alpha_1 - \frac{1}{z}\left[\frac{\left(\frac{|U_1|}{V_1}\right)z}{1+\frac{1}{V_1}z} + \frac{\left(\frac{U_2}{V_1V_2}\right)z}{1+\frac{1}{V_2}z} + \frac{\left(\frac{U_3}{V_2V_3}\right)z}{1+\frac{1}{V_3}z} + \cdots\right]. \tag{2.7}$$

If we let $C_n(z)$ and $D_n(z)$ denote the nth numerator and denominator, respectively, of the positive T-fraction

$$\frac{\left(\frac{|U_1|}{V_1}\right)z}{1+\frac{1}{V_1}z} + \frac{\left(\frac{U_2}{V_1V_2}\right)z}{1+\frac{1}{V_2}z} + \frac{\left(\frac{U_3}{V_2V_3}\right)z}{1+\frac{1}{V_3}z} + \cdots, \tag{2.8}$$

then

$$\frac{\hat{R}_n(z)}{\hat{S}_n(z)} = \alpha_1 - \frac{1}{z}\left(\frac{C_n(z)}{D_n(z)}\right), \tag{2.9a}$$

and hence

$$\hat{R}_n(z) = \alpha_1 D_n(z) - \frac{1}{z}C_n(z), \tag{2.9b}$$

and

$$\hat{S}_n(z) = D_n(z). \tag{2.9c}$$

Finally, from (2.6) and (2.9) we have

$$R_{2n}(z)=\left(\prod_{k=0}^{2n}V_k\right)\frac{1}{z^n}\left(\alpha_1 D_{2n}(z)-\frac{1}{z}C_{2n}(z)\right), \tag{2.10a}$$

$$R_{2n+1}(z)=\left(\prod_{k=0}^{2n+1}V_k\right)\frac{1}{z^n}\left(\alpha_1 D_{2n+1}(z)-\frac{1}{z}C_{2n+1}(z)\right), \tag{2.10b}$$

$$S_{2n}(z)=\left(\prod_{k=0}^{2n}V_k\right)\frac{1}{z^n}D_{2n}(z), \tag{2.10c}$$

and

$$S_{2n+1}(z)=\left(\prod_{k=0}^{2n+1}V_k\right)\frac{1}{z^n}D_{2n+1}(z). \tag{2.10d}$$

Since $C_n(z)$ and $D_n(z)$ are the nth numerator and denominator, respectively, of the positive T-fraction (2.8), then $C_n(z)$ and $D_n(z)$ are polynomials in $z$ of degree $n$. The $n$ zeros of $D_n(z)$ are simple and negative. The zeros of $C_n(z)$ separate the zeros of $D_n(z)$. And the sequences $\left\{\frac{C_{2n}(z)}{D_{2n}(z)}\right\}$ and $\left\{\frac{C_{2n+1}(z)}{D_{2n+1}(z)}\right\}$ converge uniformly on compact subsets of $C-(-\infty,0]$ [4]. It thus follows from (2.10) that the zeros of $S_n(z)$ are simple and negative and are distinct from the zeros of $R_n(z)$. Also, we see from (2.6) and (2.9a) that the sequences $\left\{\frac{R_{2n}(z)}{S_{2n}(z)}\right\}$ and $\left\{\frac{R_{2n+1}(z)}{S_{2n+1}(z)}\right\}$ converge uniformly on compact subsets of $C-(-\infty,0]$. A proof of existence of solutions $\nu^{(0)}$ and $\nu^{(\infty)}$ to the SSMP for $\{c_k\}_{k=-\infty}^{\infty}$ such that (2.1c) and (2.1d) hold now parallels the proofs of Theorems 3.2, 3.4 and 4.1 in [4].

Proof B: Using induction on the recurrence relations (1.9) one sees that

$$Q_0(z)=1 \quad\text{and}\quad Q_1(z)=\frac{1}{z}+G_1, \tag{2.11a}$$

$$Q_{2n}(z)=\frac{1}{z^n}+\cdots+\left(\prod_{k=1}^{2n}G_k\right)z^n,\quad n=1,2,\dots, \tag{2.11b}$$

and

$$Q_{2n+1}(z)=\frac{1}{z^{n+1}}+\cdots+\left(\prod_{k=1}^{2n+1}G_k\right)z^n,\quad n=0,1,\dots. \tag{2.11c}$$

Similarly, using induction on (1.12) we have

$$S_0(z) = 1 \quad \text{and} \quad S_1(z) = V_1 + z, \tag{2.12a}$$

$$S_{2n}(z) = \frac{\left(\prod_{k=1}^{2n} V_k\right)}{z^n} + \cdots + z^n, \quad n = 1, 2, \ldots, \tag{2.12b}$$

and

$$S_{2n+1}(z) = \frac{\left(\prod_{k=1}^{2n+1} V_k\right)}{z^n} + \cdots + z^{n+1}, \quad n = 0, 1, \ldots. \tag{2.12c}$$

Hence $\{Q_n(z)\}_{n=1}^{\infty}$ is a sequence of L-polynomials associated with the base $\{1, z^{-1}, z, z^{-2}, z^2, \ldots\}$ and $\{S_n(z)\}$ is a sequence of L-polynomials associated with the base $\{1, z, z^{-1}, z^2, z^{-2}, \ldots\}$.
From the correspondence properties (1.5) and from (1.10) we have

$$Q_{2n}(z)L_0(z) - P_{2n}(z) = Q_{2n}(z)\left[\beta_{4n+2}\prod_{j=0}^{2n}\alpha_{2j+1}z^{2n+1} + O(z^{2n+2})\right] \tag{2.13a}$$

and

$$Q_{2n}(z)L_\infty(z) - P_{2n}(z) = Q_{2n}(z)\left[\frac{\prod_{j=0}^{2n}\alpha_{2j+1}}{\beta_{4n}z^{2n}} + O\left(\left(\frac{1}{z}\right)^{2n+1}\right)\right]. \tag{2.13b}$$

Subtracting (2.13b) from (2.13a) and multiplying by $z^n$ we get

$$\begin{aligned} &z^n Q_{2n}(z)\left[L_0(z) - L_\infty(z)\right] = \\ &z^n Q_{2n}(z)\left[O\left(\left(\frac{1}{z}\right)^{2n+1}\right) - \frac{\prod_{j=0}^{2n}\alpha_{2j+1}}{\beta_{4n}z^{2n}} + \beta_{4n+2}\prod_{j=0}^{2n}\alpha_{2j+1}z^{2n+1} + O(z^{2n+2})\right]. \end{aligned} \tag{2.14}$$

On the right hand side of (2.14), the coefficient of $z^0$ is $-\prod_{j=0}^{2n}\alpha_{2j+1}$ and the coefficients of $z, z^2, \ldots, z^{2n}$ are all equal to zero. Let $Q_{2n}(z) = \sum_{k=-n}^{n} q_{2n,k}z^k$. Then comparing coefficients in (2.14) we obtain the system of equations

$$\begin{aligned} q_{2n,-n}(-c_{-2n}) + q_{2n,-n+1}(-c_{-2n+1}) + \cdots + q_{2n,n}(-c_0) &= 0 \\ q_{2n,-n}(-c_{-2n+1}) + q_{2n,-n+1}(-c_{-2n+2}) + \cdots + q_{2n,n}(-c_1) &= 0 \\ &\vdots \\ q_{2n,-n}(-c_{-1}) + q_{2n,-n+1}(-c_0) + \cdots + q_{2n,n}(-c_{2n-1}) &= 0 \\ q_{2n,-n}(-c_0) + q_{2n,-n+1}(-c_1) + \cdots + q_{2n,n}(-c_{2n}) &= -\prod_{j=0}^{2n}\alpha_{2j+1}. \end{aligned}$$

Using Cramer's Rule and the fact that

$$\frac{\prod_{j=0}^{2n}\alpha_{2j+1}}{H_{2n+1}^{(-2n)}}=\frac{1}{H_{2n}^{(-2n+1)}},$$

which is easily verified using equations (1.3), we see that

$$Q_{2n}(z)=\frac{1}{H_{2n}^{(-2n+1)}}\begin{vmatrix} c_{-2n} & c_{-2n+1} & \cdots & c_{-1} & z^{-n} \\ c_{-2n+1} & c_{-2n+2} & \cdots & c_0 & z^{-n+1} \\ \vdots & \vdots & \vdots & \vdots & \vdots \\ c_0 & c_1 & \cdots & c_{2n-1} & z^n \end{vmatrix}.$$

In a similar manner one obtains $Q_0(z)=1$,

$$Q_{2n+1}(z)=\frac{-1}{H_{2n+1}^{(-2n)}}\begin{vmatrix} c_{-2n-1} & c_{-2n} & \cdots & c_{-1} & z^{-n-1} \\ c_{-2n} & c_{-2n+1} & \cdots & c_0 & z^{-n} \\ \vdots & \vdots & \vdots & \vdots & \vdots \\ c_0 & c_1 & \cdots & c_{2n} & z^n \end{vmatrix},$$

$$S_{2n}(z)=\frac{1}{H_{2n}^{(-2n+1)}}\begin{vmatrix} c_{-2n+1} & c_{-2n+2} & \cdots & c_0 & z^{-n} \\ c_{-2n+2} & c_{-2n+3} & \cdots & c_1 & z^{-n+1} \\ \vdots & \vdots & \vdots & \vdots & \vdots \\ c_1 & c_2 & \cdots & c_{2n} & z^n \end{vmatrix},$$

and

$$S_{2n+1}(z)=\frac{1}{H_{2n+1}^{(-2n)}}\begin{vmatrix} c_{-2n} & c_{-2n+1} & \cdots & c_0 & z^{-n} \\ c_{-2n+1} & c_{-2n+2} & \cdots & c_1 & z^{-n+1} \\ \vdots & \vdots & \vdots & \vdots & \vdots \\ c_1 & c_2 & \cdots & c_{2n+1} & z^{n+1} \end{vmatrix}.$$

To show that $\{Q_n(-z)\}_{n=1}^{\infty}$ is orthogonal with respect to any solution,$\psi$, of the SSMP, it is enough to show that $Q_{2n+\sigma}(-z)$ is orthogonal to $z^m$ for $m=-n,\ldots,n-1+\sigma$ where $\sigma=0,1$. Consider the case $\sigma=0$. Let $m\in\{-n,-n+1,\ldots,n-1\}$. Then

$$\int_0^\infty z^m Q_{2n}(-z)d\psi(z)$$

$$= \frac{1}{H_{2n}^{(-2n+1)}} \begin{vmatrix} c_{-2n} & c_{-2n+1} & \cdots & c_{-1} & (-1)^m \int_0^\infty (-z)^{m-n} d\psi(z) \\ c_{-2n+1} & c_{-2n+2} & \cdots & c_0 & (-1)^m \int_0^\infty (-z)^{m-n+1} d\psi(z) \\ \vdots & \vdots & \vdots & \vdots & \vdots \\ c_0 & c_1 & \cdots & c_{2n-1} & (-1)^m \int_0^\infty (-z)^{m+n} d\psi(z) \end{vmatrix}$$

$$= \frac{(-1)^m}{H_{2n}^{(-2n+1)}} \begin{vmatrix} c_{-2n} & c_{-2n+1} & \cdots & c_{-1} & c_{m-n} \\ c_{-2n+1} & c_{-2n+2} & \cdots & c_0 & c_{m-n+1} \\ \vdots & \vdots & \vdots & \vdots & \\ c_0 & c_1 & \cdots & c_{2n-1} & c_{m+n} \end{vmatrix} = 0$$

since for $m \in \{-n, -n+1, \dots, n-1\}$, the last column of the determinant will be the same as a previous column. One proves the case $\sigma = 1$ similarly. Thus $\{Q_n(-z)\}_{n=1}^\infty$ is orthogonal with respect to any solution of the SSMP and in particular the solutions $\mu^{(0)}$, $\mu^{(\infty)}$, $\nu^{(0)}$ and $\nu^{(\infty)}$. The proof that $\{S_n(-z)\}_{n=1}^\infty$ is orthogonal with respect to any solution, and in particular the solutions $\mu^{(0)}$, $\mu^{(\infty)}$, $\nu^{(0)}$ and $\nu^{(\infty)}$, of the SSMP is analogous to the proof for $\{Q_n(-z)\}_{n=1}^\infty$ and is hence omitted.

Proof C: Using an equivalence transformation with $r_0 = 1$ and $r_n = e_n$, $n = 1, 2, \dots$, where the $e_n$ are defined by (2.2e), one sees that the modified positive T-fraction (1.8) is equivalent to the continued fraction

$$\frac{1}{\frac{e_1}{z} + d_1} + \frac{1}{e_2 + d_2 z} + \frac{1}{\frac{e_3}{z} + d_3} + \frac{1}{e_4 + d_4 z} + \cdots, \tag{2.15}$$

where

$$d_n := G_n e_n.$$

The *mth* numerator $W_m(z)$ and denominator, $T_m(z)$, respectively, of (2.15) are given by equations (2.2a) and (2.2b). Since the SSMP for $\{c_k\}_{k=-\infty}^\infty$ is indeterminate and $\sum_{n=1}^\infty e_n < \infty$ [see for example, 4], it is shown in [1] that the sequences $\{W_{2n}(z)\}_{n=1}^\infty$, $\{W_{2n+1}(z)\}_{n=1}^\infty$, $\{T_{2n}(z)\}_{n=1}^\infty$, and $\{T_{2n+1}(z)\}_{n=1}^\infty$, converge uniformly on compact subsets of $C-\{0\}$ to holomorphic functions $W^{(0)}(z)$, $W^{(\infty)}(z)$, $T^{(0)}(z)$, and $T^{(\infty)}(z)$, respectively, where

$$W^{(\infty)}(z)T^{(0)}(z) - W^{(0)}(z)T^{(\infty)}(z) = 1. \tag{2.16}$$

Using an equivalence transformation with $r_0 = 1$ and $r_n = \epsilon_n$, $n = 1, 2, \dots$, where the $\epsilon_n$ are defined by (2.2f), one sees that the modified M-fraction (1.11) is equivalent to the continued fraction

$$\alpha_1 - \frac{1}{\epsilon_1 z + \delta_1} + \frac{1}{\epsilon_2 + \frac{\delta_2}{z}} + \frac{1}{\epsilon_3 z + \delta_3} + \frac{1}{\epsilon_4 + \frac{\delta_4}{z}} + \cdots, \tag{2.17}$$

where

$$\delta_n := V_n \epsilon_n.$$

The mth numerator $X_m(z)$ and denominator $Y_m(z)$ of (2.17) are given by equations (2.2c) and (2.2d). It has been shown (2.1c,d) that $\left\{\frac{R_{2n}(z)}{S_{2n}(z)}\right\}$ and $\left\{\frac{R_{2n+1}(z)}{S_{2n+1}(z)}\right\}$ converge. Hence by (2.2c,d), $\left\{\frac{X_{2n}(z)}{Y_{2n}(z)}\right\}$ and $\left\{\frac{X_{2n+1}(z)}{Y_{2n+1}(z)}\right\}$ converge. If it can be shown that $\lim_{n\to\infty}\frac{X_{2n}(z)}{Y_{2n}(z)} = \lim_{n\to\infty}\frac{R_{2n}(z)}{S_{2n}(z)} \neq \lim_{n\to\infty}\frac{R_{2n+1}(z)}{S_{2n+1}(z)} = \lim_{n\to\infty}\frac{X_{2n+1}(z)}{Y_{2n+1}(z)}$ then by Van Vleck's Criteria we will have $\sum_{n=1}^{\infty}\epsilon_n < \infty$ and $\sum_{n=1}^{\infty}\delta_n < \infty$. Proofs that $\{X_{2n}(z)\}_{n=1}^{\infty}$, $\{X_{2n+1}(z)\}_{n=1}^{\infty}$, $\{Y_{2n}(z)\}_{n=1}^{\infty}$, and $\{Y_{2n+1}(z)\}_{n=1}^{\infty}$, converge uniformly on compact subsets of $C-\{0\}$ to holomorphic functions $X^{(0)}(z)$, $X^{(\infty)}(z)$, $Y^{(0)}(z)$, and $Y^{(\infty)}(z)$, respectively, then parallel those found in [1] for convergence of $\{W_{2n}(z)\}_{n=1}^{\infty}$, $\{W_{2n+1}(z)\}_{n=1}^{\infty}$, $\{T_{2n}(z)\}_{n=1}^{\infty}$, and $\{T_{2n+1}(z)\}_{n=1}^{\infty}$. By using the recurrence relations (1.6) for the modified PC-fraction (1.4), the facts that $P_n(z) = A_{2n}(z)$, $Q_n(z) = B_{2n}(z)$, $R_n(z) = A_{2n+1}(z)$ and $S_n(z) = B_{2n+1}(z)$, and equations (2.2a,b), we have

$$\frac{R_{2n}(z)}{S_{2n}(z)} = \frac{A_{4n+1}(z)}{B_{4n+1}(z)}$$

$$= \frac{(\beta_{4n+2})^{-1}[A_{4n+2}(z) - \frac{1}{z}A_{4n}(z)]}{(\beta_{4n+2})^{-1}[B_{4n+2}(z) - \frac{1}{z}B_{4n}(z)]}$$

$$= \frac{P_{2n+1}(z) - \frac{1}{z}P_{2n}(z)}{Q_{2n+1}(z) - \frac{1}{z}Q_{2n}(z)}$$

$$= \frac{\left(\prod_{k=1}^{2n+1} e_k\right)^{-1} W_{2n+1}(z) - \frac{1}{z}\left(\prod_{k=1}^{2n} e_k\right)^{-1} W_{2n}(z)}{\left(\prod_{k=1}^{2n+1} e_k\right)^{-1} T_{2n+1}(z) - \frac{1}{z}\left(\prod_{k=1}^{2n} e_k\right)^{-1} T_{2n}(z)}$$

$$= \frac{W_{2n+1}(z) - e_{2n+1}\frac{1}{z}W_{2n}(z)}{T_{2n+1}(z) - e_{2n+1}\frac{1}{z}T_{2n}(z)}.$$

Letting $n \to \infty$ we have

$$\lim_{n\to\infty}\frac{R_{2n}(z)}{S_{2n}(z)} = \lim_{n\to\infty}\frac{W_{2n+1}(z) - e_{2n+1}\frac{1}{z}W_{2n}(z)}{T_{2n+1}(z) - e_{2n+1}\frac{1}{z}T_{2n}(z)} = \frac{W^{(\infty)}(z)}{T^{(\infty)}(z)} \tag{2.18}$$

since $\sum_{k=1}^{\infty} e_k < \infty$ implies that $\lim_{k\to\infty} e_k = 0$. Similarly,

$$\frac{R_{2n+1}(z)}{S_{2n+1}(z)} = \frac{A_{4n+3}(z)}{B_{4n+3}(z)}$$

$$= \frac{(\beta_{4n+4})^{-1}[A_{4n+4}(z) - A_{4n+2}(z)]}{(\beta_{4n+4})^{-1}[B_{4n+4}(z) - B_{4n+2}(z)]}$$

$$= \frac{P_{2n+2}(z) - P_{2n+1}(z)}{Q_{2n+2}(z) - Q_{2n+1}(z)}$$

$$= \frac{\left(\prod_{k=1}^{2n+2} e_k\right)^{-1} W_{2n+2}(z) - \left(\prod_{k=1}^{2n+1} e_k\right)^{-1} W_{2n+1}(z)}{\left(\prod_{k=1}^{2n+2} e_k\right)^{-1} T_{2n+2}(z) - \left(\prod_{k=1}^{2n+1} e_k\right)^{-1} T_{2n+1}(z)}$$

$$= \frac{W_{2n+2}(z) - e_{2n+2}W_{2n+1}(z)}{T_{2n+2}(z) - e_{2n+2}T_{2n+1}(z)}.$$

Thus

$$\lim_{n\to\infty} \frac{R_{2n+1}(z)}{S_{2n+1}(z)} = \lim_{n\to\infty} \frac{W_{2n+2}(z) - e_{2n+2}W_{2n+1}(z)}{T_{2n+2}(z) - e_{2n+2}T_{2n+1}(z)} = \frac{W^{(0)}(z)}{T^{(0)}(z)}. \tag{2.19}$$

But from (2.16) we see that

$$W^{(\infty)}(z)T^{(0)}(z) - W^{(0)}(z)T^{(\infty)}(z) = 1$$

and hence

$$\lim_{n\to\infty} \frac{X_{2n}(z)}{Y_{2n}(z)} = \lim_{n\to\infty} \frac{R_{2n}(z)}{S_{2n}(z)} = \frac{W^{(\infty)}(z)}{T^{(\infty)}(z)} \neq \frac{W^{(0)}(z)}{T^{(0)}(z)} = \lim_{n\to\infty} \frac{R_{2n+1}(z)}{S_{2n+1}(z)}$$

$$= \lim_{n\to\infty} \frac{X_{2n+1}(z)}{Y_{2n+1}(z)}.$$

Proof D,E: In (C) it was shown that $W_{2n}(z) \to W^{(0)}(z)$, $W_{2n+1}(z) \to W^{(\infty)}(z)$, $T_{2n}(z) \to T^{(0)}(z)$, $T_{2n+1}(z) \to T^{(\infty)}(z)$, $X_{2n}(z) \to X^{(0)}(z)$, $X_{2n+1}(z) \to X^{(\infty)}(z)$, $Y_{2n}(z) \to Y^{(0)}(z)$, and $Y_{2n+1}(z) \to Y^{(\infty)}(z)$. Using convergence of these sequences together with equations (2.1), (2.18) and (2.19) we have

$$\int_0^\infty \frac{d\nu^{(0)}(t)}{z+t} = \lim_{n\to\infty} \frac{R_{2n}(z)}{S_{2n}(z)} = \lim_{n\to\infty} \frac{X_{2n}(z)}{Y_{2n}(z)} = \frac{X^{(0)}(z)}{Y^{(0)}(z)}$$

$$= \frac{W^{(\infty)}(z)}{T^{(\infty)}(z)} = \lim_{n\to\infty} \frac{W_{2n+1}(z)}{T_{2n+1}(z)} = \lim_{n\to\infty} \frac{P_{2n+1}(z)}{Q_{2n+1}(z)}$$

$$= \int_0^\infty \frac{d\mu^{(\infty)}(t)}{z+t}$$

and

$$\int_0^\infty \frac{d\nu^{(\infty)}(t)}{z+t} = \lim_{n\to\infty} \frac{R_{2n+1}(z)}{S_{2n+1}(z)} = \lim_{n\to\infty} \frac{X_{2n+1}(z)}{Y_{2n+1}(z)} = \frac{X^{(\infty)}(z)}{Y^{(\infty)}(z)}$$

$$= \frac{W^{(0)}(z)}{T^{(0)}(z)} = \lim_{n\to\infty} \frac{W_{2n}(z)}{T_{2n}(z)} = \lim_{n\to\infty} \frac{P_{2n}(z)}{Q_{2n}(z)}$$

$$= \int_0^\infty \frac{d\mu^{(0)}(t)}{z+t}$$

from which (D) and (E) follow.

## REFERENCES


1. Catherine M. Bonan-Hamada, Orthogonal Laurent Polynomials and Indeterminate Strong Stieltjes Moment Problems, dissertation, *Dissertation Abstracts International*, Vol. 56-02 0856 (1994), Catalog 9518603.
2. W. B. Jones, Olav Njåstad and W. J. Thron, Continued Fractions Associated with Trigonometric and Other Strong Moment Problems, *Constr. Approx.* **2** (1986), 197–211.
3. W. B. Jones and W. J. Thron, *Continued Fractions: Analytic Theory and Applications*, in Encyclopedia of Mathematics and Its Applications, Vol. 11, Addison–Wesley, Reading, Mass., 1980. Distributed now by the Cambridge University Press.
4. W. B. Jones, W. J. Thron and H. Waadeland, A strong Stieltjes moment problem, *Trans. Amer. Math. Soc.* **261** (1980), 503–528.
5. L. Lorentzen and H. Waadeland, *Continued Fractions with Applications*, Elsevier Science Publishers B.V., Amsterdam, The Netherlands, 1992.
6. O. Njåstad, *Solutions of the strong Stieltjes moment problem*, Methods and Applications of Analysis, **2** (1995), 309-318.

# A Class of Indeterminate Strong Stieltjes Moment Problems with Discrete Distributions

CATHERINE M. BONAN-HAMADA* Department of Computer Science, Mathematics and Statistics, Mesa State College, Grand Junction, Colorado

WILLIAM B. JONES* and W. J. THRON* Department of Mathematics, University of Colorado, Boulder, Colorado

OLAV NJÅSTAD Department of Mathematical Sciences, Norwegian University of Science and Technology, Trondheim, Norway

## 1 INTRODUCTION

For each pair $(a, b)$ such that $-\infty \leq a < b \leq \infty$ let $\Psi(a, b)$ denote the family of distribution functions $\psi(t)$ (i.e., real-valued, bounded, non-decreasing functions with infinitely many points of increase) on $a < t < b$. The *Stieltjes moment problem* (SMP) *for a sequence* $\{c_n\}_{n=0}^{\infty}$ is to find necessary and sufficient conditions for there to exist a $\psi \in \Psi(0, \infty)$ such that

$$c_n = \int_0^{\infty} (-t)^n d\psi(t), \quad n = 0, 1, 2, \dots . \tag{1.1}$$

A distribution function $\psi(t)$ satisfying (1.1) is called a *solution to the SMP for* $\{c_n\}_{n=0}^{\infty}$. A solvable moment problem is said to be *determinate* if there exists one and only one solution, otherwise it is called *indeterminate*. The SMP was posed and solved by T. J. Stieltjes in 1894 [13] using among others the modified Stieltjes continued fractions (modified S-fractions)

$$\frac{a_1}{1} + \frac{a_2}{z} + \frac{a_3}{1} + \frac{a_4}{z} + \cdots, \quad a_n > 0, \quad n = 1, 2, 3, \dots . \tag{1.2}$$

*Research supported in part by the National Science Foundation under Grants DMS-9302584, DMS-9103141 and INT-9113400.*

He proved that a SMP has a solution if and only if there exists a modified S-fraction corresponding to the formal series

$$L_\infty(z) := c_0 + c_1 z^{-1} + c_2 z^{-2} + \cdots \tag{1.3}$$

in the sense that

$$\frac{A_n(z)}{B_n(z)} - L_\infty(z) = O\left(\left(\frac{1}{z}\right)^n\right), \quad n = 1, 2, 3, \ldots, \tag{1.4}$$

where $A_n(z)$ and $B_n(z)$ denote the (polynomial) $n$th numerator and denominator, respectively, of the modified S-fraction (1.2). Here $O\left(\left(\frac{1}{z}\right)^n\right)$ denotes a formal Laurent series of increasing powers of $\frac{1}{z}$ starting with a power not less than $n > 0$. A SMP is known to be indeterminate if and only if the corresponding modified S-fraction is divergent. In that case the sequences $\{A_{2n+\sigma}(z)\}_{n=1}^{\infty}$ and $\{B_{2n+\sigma}(z)\}_{n=1}^{\infty}$, $\sigma = 0, 1$, converge uniformly on compact subsets of $C$ to entire functions $A^{(\sigma)}(z)$ and $B^{(\sigma)}(z)$, respectively. The zeros $t_k^{(\sigma)}$ of $B^{(\sigma)}(z)$ are all simple and negative and $\{t_k^{(\sigma)}\}_{k=1}^{\infty}$ converges to $-\infty$. Moreover, there exist distinct distribution functions $\psi_0(t)$ and $\psi_1(t)$ that have $\{c_n\}_{n=0}^{\infty}$ as moment sequence (i.e., are solutions to the SMP for $\{c_n\}_{n=0}^{\infty}$) and satisfy

$$F_\sigma(z) := \frac{A^{(\sigma)}(z)}{B^{(\sigma)}(z)} = z\int_0^\infty \frac{d\psi_\sigma(t)}{z+t}, \quad \sigma = 0, 1, \ z \in C - \{t_1^{(\sigma)}, t_2^{(\sigma)}, \ldots\}. \tag{1.5}$$

In fact for $\sigma = 0, 1$, $F_\sigma(z)$ is meromorphic in $C$ with simple poles at the zeros $t_k^{(\sigma)}$ of $B^{(\sigma)}(z)$ and has a partial fraction expansion

$$F_\sigma(z) = \sum_{k=1}^{\infty} \frac{z\lambda_k^{(\sigma)}}{z - t_k^{(\sigma)}}, \tag{1.6}$$

that is uniformly convergent on compact subsets of $C - \{t_1^{(\sigma)}, t_2^{(\sigma)}, \ldots\}$, since $\lambda_k^{(\sigma)} > 0$ and $\sum_{k=1}^{\infty} \lambda_k^{(\sigma)} < \infty$. Each $\psi_\sigma(t)$ is a step function with jump $\lambda_k^{(\sigma)}$ at $t = -t_k^{(\sigma)} > 0$.

Throughout this paper, the superscript $(\sigma)$ is used as an index and not as a derivative notation.

The *strong Stieltjes moment problem* (SSMP) *for a bisequence* $\{c_n\}_{n=-\infty}^{\infty}$ consists of finding necessary and sufficient conditions for there to exist a $\psi \in \Psi(0, \infty)$ such that

$$c_n = \int_0^\infty (-t)^n d\psi(t), \quad n = 0, \pm 1, \pm 2, \ldots. \tag{1.7}$$

A distribution function $\psi(t)$ satisfying (1.7) is called a *solution to the* SSMP *for* $\{c_n\}_{n=-\infty}^{\infty}$. It is also said that $\psi(t)$ has $\{c_n\}_{n=-\infty}^{\infty}$ as moment sequence. The SSMP was posed and solved in 1980 [11] with the help of positive T-fractions

$$\frac{F_1 z}{1 + G_1 z} + \frac{F_2 z}{1 + G_2 z} + \frac{F_3 z}{1 + G_3 z} + \cdots, \quad F_n > 0, \ G_n > 0, \ n = 1, 2, \ldots. \tag{1.8}$$

The situation for the SSMP is more complicated than that for the SMP, but there are a number of similarities. In [11] it was shown that the SSMP for $\{c_n\}_{n=-\infty}^{\infty}$ has a solution if and only if there exists a positive T-fraction corresponding to the pair $(L_0, L_\infty)$ of formal power series (fps)

$$L_0(z) := \sum_{k=1}^{\infty} -c_{-k}z^k \quad \text{and} \quad L_\infty(z) := \sum_{k=0}^{\infty} c_k z^{-k} \tag{1.9}$$

in the sense that, for $n = 1, 2, 3, \ldots,$

$$\frac{P_n(z)}{Q_n(z)} - L_0(z) = O(z^{n+1}) \quad \text{and} \quad \frac{P_n(z)}{Q_n(z)} - L_\infty(z) = O\left(\left(\frac{1}{z}\right)^n\right). \tag{1.10}$$

Here $P_n(z)$ and $Q_n(z)$ denote the $n$th numerator and denominator, respectively, of the positive T-fraction. It was also established in [11] that the SSMP is indeterminate if and only if the positive T-fraction diverges for all $z \in C - \{0\}$. In that case the subsequences $\{P_{2n}(z)/Q_{2n}(z)\}$ and $\{P_{2n+1}(z)/Q_{2n+1}(z)\}$ converge uniformly on compact subsets of the cut plane $S_\pi := \{z : |\arg z| < \pi\}$ to different holomorphic functions $G^{(0)}(z)$ and $G^{(1)}(z)$, respectively. Moreover there exist distinct solutions $\psi^{(0)}(t)$ and $\psi^{(1)}(t)$ to the SSMP for $\{c_n\}_{n=-\infty}^{\infty}$ which satisfy

$$G^{(\sigma)}(z) := \lim_{n\to\infty} \frac{P_{2n+\sigma}(z)}{Q_{2n+\sigma}(z)} = z\int_0^\infty \frac{d\psi^{(\sigma)}(t)}{z+t}, \quad z \in S_\pi,\ \sigma = 0, 1. \tag{1.11}$$

In the present paper we study the class of indeterminate strong Stieltjes moment problems for which the associated positive T-fractions (1.8) satisfy the conditions

$$\sum_{n=1}^{\infty} e_n < \infty \quad \text{where} \quad e_1 := \frac{1}{F_1},\ e_{2n-1} := \frac{\prod_{k=1}^{n-1} F_{2k}}{\prod_{k=1}^{n} F_{2k-1}},\ e_{2n} := \frac{\prod_{k=1}^{n} F_{2k-1}}{\prod_{k=1}^{n} F_{2k}} \tag{1.12a}$$

and

$$G_n = G > 0, \qquad n = 1, 2, 3, \ldots . \tag{1.12b}$$

Condition (1.12a), together with the condition $\sum_{n=1}^{\infty}(e_n G_n) < \infty$, is necessary and sufficient for the SSMP to be indeterminate [11]. Condition (1.12b) ensures that $Q_{2n+1}(-1/G) = 0$ for $n = 0, 1, 2, \ldots,$ a property which is subsequently used to determine convergence properties of sequences of zeros of the $Q_{2n+\sigma}(z)$, $\sigma = 0, 1$. We show (Theorem 5.3) that, for this class of divergent positive T-fractions, the holomorphic functions (1.11) to which the (even or odd order) approximants converge are represented by infinite series of the form

$$G^{(\sigma)}(z) = \sum_{k=-\infty}^{\infty} \frac{z a_k^{(\sigma)}}{z - x_k^{(\sigma)}}, \quad \sigma = 0, 1,\ z \neq x_k^{(\sigma)}, \tag{1.13}$$

where $a_k^{(\sigma)} > 0$, $\sum_{k=-\infty}^{\infty} a_k^{(\sigma)} < \infty$ and the $x_k^{(\sigma)}$ are real, negative, distinct and $\{x_k^{(\sigma)}\}_{k=-\infty}^{\infty}$ has two cluster points, one at 0 and one at $-\infty$. Thus the solutions to the SSMP for $\{c_k\}_{k=-\infty}^{\infty}$, denoted by $\psi^{(\sigma)}(t)$, $\sigma = 0, 1$ are step functions with a jump $a_k^{(\sigma)}$ at each $t = -x_k^{(\sigma)}$ and the sequence $\{-x_k^{(\sigma)}\}_{k=-\infty}^{\infty}$ has cluster points at 0 and at $+\infty$.

Some preliminary results not requiring that condition (1.12) hold are described in Section 2. Condition (1.12a) is imposed beginning in Section 3 and condition (1.12b) in Section 4.

Basic definitions and results summarized in Section 2 are used in Section 3 to prove properties of the zeros of the Laurent polynomials

$$U_{2n+\sigma}(z) := \left(\prod_{k=1}^{2n+\sigma} e_k\right) \frac{P_{2n+\sigma}(z)}{z^{n+\sigma}}, \quad V_{2n+\sigma}(z) := \left(\prod_{k=1}^{2n+\sigma} e_k\right) \frac{Q_{2n+\sigma}(z)}{z^{n+\sigma}}, \quad \sigma = 0, 1. \tag{1.14}$$

Then, subject to condition (1.12a), we show (Theorem 3.3) that if $|G_n| \leq M$, $M$ fixed, $n = 1, 2, 3, \ldots$, the sequences $\{U_{2n+\sigma}(z)\}_{n=1}^{\infty}$ and $\{V_{2n+\sigma}(z)\}_{n=1}^{\infty}$, $\sigma = 0, 1$, converge uniformly on compact subsets of $D := \{z \in C : z \neq 0\}$ to holomorphic functions $U^{(\sigma)}(z)$ and $V^{(\sigma)}(z)$, respectively. In Section 4 further properties of zeros of $V_{2n+\sigma}(z)$ and $V^{(\sigma)}(z)$, $\sigma = 0, 1$, are established assuming the additional condition (1.12b). In particular it is shown that the zeros $\{x_p^{(\sigma)}\}_{p=-\infty}^{\infty}$ of $V^{(\sigma)}(z)$ are negative and distinct and have cluster points at 0 and $-\infty$. Moreover each zero $x_p^{(\sigma)}$ is the limit of a specified sequence of zeros of the $V_{2n+\sigma}(z)$. Properties of the zeros $x_p^{(\sigma)}$ established in Section 4 are used in Section 5 to construct the step functions $\psi^{(0)}(t)$ and $\psi^{(1)}(t)$ which are shown to be solutions to the SSMP. The results of this paper are illustrated in Section 6 by an example involving the moment sequence for log-normal distribution functions. For these moments we are able to give explicit expressions for the location and size of each jump in two discrete distribution functions which generate the log-normal moments. That example is a prototype of the class of SSMP satisfying conditions (1.12) considered in the present paper.

## 2 POSITIVE T-FRACTIONS

We begin by summarizing properties of the positive T-fractions (1.8) that are subsequently used. Throughout the remainder of this paper we use $P_n(z)$ and $Q_n(z)$ to denote the $n$th numerator and denominator, respectively, of a given positive T-fraction (1.8). $P_n(z)$ and $Q_n(z)$ are polynomials defined by the difference equations

$$\begin{aligned} &P_{-1}(z) := 1, \quad P_0(z) := 0, \quad Q_{-1}(z) := 0, \quad Q_0(z) := 1 \\ &P_n(z) := (1 + G_n z) P_{n-1}(z) + F_n z P_{n-2}(z), \quad n = 1, 2, \ldots, \\ &Q_n(z) := (1 + G_n z) Q_{n-1}(z) + F_n z Q_{n-2}(z), \quad n = 1, 2, \ldots . \end{aligned} \tag{2.1}$$

The proofs of the various results contained in the following theorem are only sketched in [11] and are included here for completeness.

THEOREM 2.1

A. For $n = 1, 2, \ldots, P_n(z)$ and $Q_n(z)$ are polynomials in $z$ of degree $n$ of the forms,

$$P_n(z) = F_1 z + \cdots + F_1\left(\prod_{k=2}^{n} G_k\right) z^n, \quad Q_n(z) = 1 + \cdots + \left(\prod_{k=1}^{n} G_k\right) z^n. \quad (2.2)$$

B. For $n = 1, 2, \ldots$, the $n$ zeros $z_{n,k}$, $k = 1, 2, \ldots, n$, of $Q_n(z)$ are simple, negative and separate the zeros of $Q_{n+1}(z)$. We write $z_{n+1,k+1} < z_{n,k} < z_{n+1,k} < 0$ for $k = 1, 2, \ldots, n$.

C. For $n = 2, 3, \ldots$, the $n-1$ zeros of $\frac{P_n(z)}{z}$ separate the zeros of $Q_n(z)$.

D. For $n = 1, 2, 3, \ldots$, the rational function $P_n(z)/Q_n(z)$ has the partial fraction decomposition

$$\frac{P_n(z)}{Q_n(z)} = \sum_{k=1}^{n} \frac{z a_{n,k}}{z - z_{n,k}}, \quad (2.3a)$$

where

$$a_{n,k} := \frac{P_n(z_{n,k})}{z_{n,k} Q_n'(z_{n,k})} > 0, \quad \text{and} \quad \sum_{k=1}^{n} a_{n,k} = \frac{F_1}{G_1} > 0. \quad (2.3b)$$

Proof A: Apply induction to (2.1).

Proof B: From (2.2) we see that $Q_1(z) = 1 + G_1 z$ and so $z_{1,1} = -\frac{1}{G_1} < 0$. Since $Q_2(z) = (1 + G_2 z)Q_1(z) + F_2 z$ and $Q_2(z_{1,1}) = -F_2/G_1 < 0$ while $Q_2(0) = 1 > 0$, there exists a real zero, $z_{2,1}$ of $Q_2(z)$ with $z_{1,1} < z_{2,1} < 0$. Moreover, the parabola $y = Q_2(x)$ is concave up and so the other zero, $z_{2,2}$, is real and lies to the left of $z_{1,1}$; that is, $z_{2,2} < z_{1,1} < z_{2,1} < 0$. Now for a fixed integer $n \geq 2$, suppose that $z_{n,k+1} < z_{n-1,k} < z_{n,k} < 0$ for $k = 1, 2, \ldots, n-1$. From (2.1) we see that

$$Q_{n+1}(z_{n,k}) = F_{n+1} z_{n,k} \left(\prod_{j=1}^{n-1} G_j\right) \prod_{j=1}^{n-1} (z_{n,k} - z_{n-1,j})$$

has sign $(-1)^k$ for $k = 1, 2, \ldots, n$. Consequently there exist zeros $z_{n+1,k+1}$, $k = 1, 2, \ldots, n-1$ of $Q_{n+1}(z)$ so that $z_{n,k+1} < z_{n+1,k+1} < z_{n,k} < 0$ for $k = 1, 2, \ldots, n-1$. Since $Q_{n+1}(z_{n,1}) < 0$ and $Q_{n+1}(0) = 1$ there exists a zero, $z_{n+1,1}$ of $Q_{n+1}(z)$ such that $z_{n,1} < z_{n+1,1} < 0$. The sign of $Q_{n+1}(z_{n,n})$ is the same as the sign of $(-1)^n$ and from (2.2) we see that for the real variable $x$, $Q_{n+1}(x) \to +\infty$ as $x \to -\infty$ if $n$ is odd and $Q_{n+1}(x) \to -\infty$ as $x \to -\infty$ if $n$ is even. Thus there exists a real zero, $z_{n+1,n+1}$, of $Q_{n+1}(z)$ so that $z_{n+1,n+1} < z_{n,n}$. Consequently $z_{n+1,k+1} < z_{n,k} < z_{n+1,k} < 0$ for $k = 1, 2, \ldots, n$.

Proof C: From the determinant formula [10] for continued fractions it follows that

$$\frac{P_n(z)}{z}\,Q_{n-1}(z)-\frac{P_{n-1}(z)}{z}\,Q_n(z)=(-1)^{n-1}z^{n-1}\prod_{j=1}^{n}F_j>0$$

for $z<0$. Since $Q_n(z_{n,k})=0$, it follows that $P_n(z_{n,k})Q_{n-1}(z_{n,k})/z_{n,k}>0$, $k=1,2,\ldots,n$, and hence $Q_{n-1}(z_{n,k})$ and $P_n(z_{n,k})/z_{n,k}$ have the same sign. For each $k=1,2,\ldots,n-1$ there exists exactly one zero $z_{n-1,k}$ of $Q_{n-1}(z)$ so that $z_{n,k+1}<z_{n-1,k}<z_{n,k}$. Thus $Q_{n-1}(x)$ changes sign exactly once in the interval $(z_{n,k+1},z_{n,k})$, namely at $x=z_{n-1,k}$; therefore for each $k=1,2,\ldots,n-1$, $P_n(x)/x$ changes sign an odd numer of times for $x$ between $z_{n,k+1}$ and $z_{n,k}$. But $Q_n(z)$ has $n$ zeros while $P_n(z)/z$ has $n-1$ zeros. Hence there exists exactly one zero of $P_n(z)/z$ between $z_{n,k+1}$ and $z_{n,k}$, $k=1,2,\ldots,n-1$.

Proof D: Since the zeros of $P_n(z)/z$ separate the zeros of $Q_n(z)$ and the zeros of $Q_n(z)$ are simple, there exists a partial fraction decomposition

$$\frac{P_n(z)}{zQ_n(z)}=\frac{P_n(z)/z}{\left(\prod_{k=1}^{n}G_k\right)\left(\prod_{k=1}^{n}z-z_{n,k}\right)}=\frac{a_{n,1}}{z-z_{n,1}}+\frac{a_{n,2}}{z-z_{n,2}}+\cdots+\frac{a_{n,n}}{z-z_{n,n}}. \quad (2.4)$$

From (2.4) we see that

$$\frac{P_n(z)}{z}=\frac{a_{n,1}Q_n(z)}{z-z_{n,1}}+\frac{a_{n,2}Q_n(z)}{z-z_{n,2}}+\cdots+\frac{a_{n,n}Q_n(z)}{z-z_{n,n}}.$$

Consequently for each $k=1,2,\ldots,n$,

$$\frac{P_n(z_{n,k})}{z_{n,k}}=\lim_{z\to z_{n,k}}\frac{a_{n,k}Q_n(z)}{z-z_{n,k}}=a_{n,k}Q_n'(z_{n,k})$$

from which (2.3) follows.

It will be shown that $a_{n,k}>0$, $k=1,2,\ldots,n$ by proving that $P_n(z_{n,k})/z_{n,k}$ and $Q_n'(z_{n,k})$ have the same sign for $k=1,2,\ldots,n$. $Q_n'(z_{n,k})\neq 0$, $k=1,2,\ldots,n$ since $z_{n,k}$ is a simple zero of $Q_n(z)$. Moreover $Q_n(0)=1$ and so it follows that the sign of $Q_n'(z_{n,k})$ is $(-1)^{k-1}$, $k=1,2,\ldots,n$. Now $\lim\limits_{z\to 0}\dfrac{P_n(z)}{z}=F_1>0$ and the largest zero of $P_n(z)/z$ lies in $(z_{n,2},z_{n,1})$. Thus $P_n(z_{n,1})/z_{n,1}>0$ and it follows by (C) that $P_n(z_{n,k})/z_{n,k}$ has the same sign as $(-1)^{k-1}$ for $k=1,2,\ldots,n$. Finally from (2.2) and (2.4) we see that

$$\sum_{k=1}^{n}a_{n,k}=\lim_{z\to\infty}z\sum_{k=1}^{n}\frac{a_{n,k}}{z-z_{n,k}}=\lim_{z\to\infty}\frac{P_n(z)}{Q_n(z)}=\frac{F_1}{G_1}>0.$$

## 3 PROPERTIES OF THE LAURENT POLYNOMIALS $U_n(z)$ AND $V_n(z)$

Using an equivalence transformation [10, Theorem 2.6] with

$$r_0 = 1, \quad r_{2n-1} = \frac{e_{2n-1}}{z} \quad \text{and} \quad r_{2n} = e_{2n}, \quad n = 1, 2, \dots, \tag{3.1}$$

we see that the positive T-fraction (1.8) is equivalent to the continued fraction

$$\frac{1}{\frac{e_1}{z} + d_1} + \frac{1}{e_2 + d_2 z} + \frac{1}{\frac{e_3}{z} + d_3} + \frac{1}{e_4 + d_4 z} + \cdots, \tag{3.2a}$$

where $e_1 := 1/F_1$ and for $n = 1, 2, 3, \dots$,

$$e_{2n-1} := \frac{\prod_{k=1}^{n-1} F_{2k}}{\prod_{k=1}^{n} F_{2k-1}}, \quad e_{2n} := \frac{\prod_{k=1}^{n} F_{2k-1}}{\prod_{k=1}^{n} F_{2k}}, \quad d_n := G_n e_n. \tag{3.2b}$$

The $n$th numerator $U_n(z)$ and denominator $V_n(z)$ of (3.2a) are related to $P_n(z)$ and $Q_n(z)$ by the equations

$$U_n(z) = \left(\prod_{k=1}^{n} r_k\right) P_n(z), \quad n = 1, 2, \dots \tag{3.3a}$$

and

$$V_n(z) = \left(\prod_{k=1}^{n} r_k\right) Q_n(z), \quad n = 1, 2, \dots. \tag{3.3b}$$

From (3.1) and (3.3) we obtain for $\sigma = 0, 1$, and $n = 1-\sigma, 2-\sigma, 3-\sigma, \dots$,

$$U_{2n+\sigma}(z) = \left(\prod_{k=1}^{2n+\sigma} e_k\right) \frac{P_{2n+\sigma}(z)}{z^{n+\sigma}}, \tag{3.4a}$$

and

$$V_{2n+\sigma}(z) = \left(\prod_{k=1}^{2n+\sigma} e_k\right) \frac{Q_{2n+\sigma}(z)}{z^{n+\sigma}}. \tag{3.4b}$$

THEOREM 3.1 The $n$th numerator $U_n(z)$ and denominator $V_n(z)$ of the continued fraction (3.2) are Laurent polynomials satisfying the following properties:

A. $U_1(z) = 1$ and for $\sigma = 0, 1$,

$$U_{2n+\sigma}(z) = \left(\prod_{k=2}^{2n+\sigma} e_k\right) z^{-n+1-\sigma} + \cdots + \left(\prod_{k=2}^{2n+\sigma} d_k\right) z^n, \; n = 1, 2, \dots, \tag{3.5a}$$

$$V_{2n+\sigma}(z) = \left(\prod_{k=1}^{2n+\sigma} e_k\right) z^{-n-\sigma} + \cdots + \left(\prod_{k=1}^{2n+\sigma} d_k\right) z^n, \; n = 1-\sigma, 2-\sigma, \ldots . \quad (3.5b)$$

B. For $n = 1, 2, \ldots$, the $n$ zeros of $V_n(z)$ are the same as the $n$ zeros, $z_{n,k}$, $k = 1, 2, \ldots, n$, of $Q_n(z)$ and hence they are simple, negative and separate the zeros of $V_{n+1}(z)$.

C. For $n = 2, 3, \ldots$, the $n-1$ zeros of $U_n(z)$ are the same as the zeros of $P_n(z)/z$ and hence they are simple, negative and separate the zeros of $V_n(z)$.

D. For each $n = 1, 2, \ldots$, $\dfrac{U_n(z)}{V_n(z)} = \dfrac{P_n(z)}{Q_n(x)}$ has the partial fraction decomposition

$$\frac{U_n(z)}{V_n(z)} = \sum_{k=1}^{n} \frac{z a_{n,k}}{z - z_{n,k}}, \quad (3.6a)$$

where

$$a_{n,k} = \frac{P_n(z_{n,k})}{z_{n,k} Q_n'(z_{n,k})} = \frac{U_n(z_{n,k})}{z_{n,k} V_n'(z_{n,k})} > 0, \quad (3.6b)$$

and

$$\sum_{k=1}^{n} a_{n,k} = \frac{F_1}{G_1} > 0. \quad (3.6c)$$

Proof A: Apply (2.2) and (3.4).

Proofs B,C,D: These are immediate consequences of (3.4) and Theorem 2.1.

The next theorem was proved in [7] using ideas of von Koch; the authors of [7] were able to make stronger statements by assuming additional information and structure.

THEOREM 3.2 Let $\mathop{K}_{n=1}^{\infty} \left(\dfrac{1}{\varepsilon_n f_n(z)}\right)$ be a continued fraction such that $\{\varepsilon_n\}_{n=1}^{\infty}$ is a sequence of complex constants satisfying $\sum_{n=1}^{\infty} |\varepsilon_n| < \infty$ and $\{f_n(z)\}$ is a sequence of functions holomorphic in a region $D$ and is uniformly bounded on every compact subset of $D$. Let $I_n(z)$ and $J_n(z)$ denote the $n$th numerator and denominator, respectively, of $\mathop{K}_{n=1}^{\infty} \left(\dfrac{1}{\varepsilon_n f_n(z)}\right)$. Then:

A. The sequences $\{I_{2n+\sigma}(z)\}_{n=1}^{\infty}$ and $\{J_{2n+\sigma}(z)\}_{n=1}^{\infty}$, $\sigma = 0, 1$, converge uniformly on compact subsets of $D$ to holomorphic functions $I^{(\sigma)}(z)$ and $J^{(\sigma)}(z)$, respectively.

B. For all $z \in D$, $I^{(1)}(z)J^{(0)}(z) - I^{(0)}(z)J^{(1)}(z) = 1$ and hence the functions $I^{(\sigma)}(z)/J^{(\sigma)}(z)$, $\sigma = 0, 1$, are meromorphic in $D$ and satisfy

$$\frac{I^{(0)}(z)}{J^{(0)}(z)} = \lim_{n\to\infty} \frac{I_{2n}(z)}{J_{2n}(z)} \neq \lim_{n\to\infty} \frac{I_{2n+1}(z)}{J_{2n+1}(z)} = \frac{I^{(1)}(z)}{J^{(1)}(z)}.$$

THEOREM 3.3 Let $U_n(z)$ and $V_n(z)$ denote the $n$th numerator and denominator, respectively, of the continued fraction (3.2) such that

$$\sum_{n=1}^{\infty} e_n < \infty \quad \text{and} \quad |G_n| \leq M, \quad n = 1, 2, \ldots . \tag{3.7}$$

Then for $\sigma = 0, 1$, the sequences $\{U_{2n+\sigma}(z)\}_{n=1}^{\infty}$ and $\{V_{2n+\sigma}(z)\}_{n=1}^{\infty}$ converge uniformly on compact subsets of the punctured plane $D := \{z \in C : z \neq 0\}$ to holomorphic functions $U^{(\sigma)}(z)$ and $V^{(\sigma)}(z)$, respectively. Moreover

$$U^{(1)}(z)V^{(0)}(z) - U^{(0)}(z)V^{(1)}(z) = 1 \tag{3.8}$$

and hence the functions $U^{(\sigma)}(z)/V^{(\sigma)}(z)$, $\sigma = 0, 1$, are meromorphic in $D$ and satisfy

$$\frac{U^{(0)}(z)}{V^{(0)}(z)} = \lim_{n\to\infty} \frac{U_{2n}(z)}{V_{2n}(z)} \neq \lim_{n\to\infty} \frac{U_{2n+1}(z)}{V_{2n+1}(z)} = \frac{U^{(1)}(z)}{V^{(1)}(z)}. \tag{3.9}$$

Proof: We recall from (3.2) that $d_n := G_n e_n$, $n = 1, 2, \ldots$. Thus (3.2) can be written as $\mathop{K}_{n=1}^{\infty}\left(\frac{1}{\varepsilon_n f_n(z)}\right)$ where $f_{2n-1}(z) := \frac{1}{z} + G_{2n-1}$, $f_{2n}(z) := 1 + G_{2n}z$, and $\varepsilon_n = e_n$ for $n = 1, 2, \ldots$. Our proof is now an application of Theorem 3.2.

## 4 ADDITIONAL PROPERTIES OF THE ZEROS OF $V_{2n+\sigma}(z)$

In this section we establish some additional properties of the zeros of $V_{2n+\sigma}(z)$. After Lemma 4.1 we restrict ourselves to positive T-fractions (1.8) which satisfy condition (1.12b). The additional restriction (1.12a) is made following Corollary 4.5.

The continued fraction (3.2) can be written as

$$\frac{1}{e_1\left(\frac{1}{z} + G_1\right)} + \frac{1}{e_2(1 + G_2 z)} + \frac{1}{e_3\left(\frac{1}{z} + G_3\right)} + \frac{1}{e_4(1 + G_4 z)} + \cdots . \tag{4.1}$$

Hereafter we refer to (4.1) as a modified positive T-fraction. Since $U_n(z)$ and $V_n(z)$ are the $n$th numerator and denominator, respectively, of (4.1) they satisfy the recurrence relations

$$U_{-1}(z) := 1, \quad U_0(z) := 0, \quad V_{-1}(z) := 0, \quad V_0(z) := 1 \tag{4.2a}$$

$$U_{2n}(z) = (1 + G_{2n}z)e_{2n}U_{2n-1}(z) + U_{2n-2}(z), \quad n = 1, 2, \dots, \tag{4.2b}$$

$$U_{2n+1}(z) = \left(\frac{1}{z} + G_{2n+1}\right) e_{2n+1}U_{2n}(z) + U_{2n-1}(z), \quad n = 0, 1, \dots, \tag{4.2c}$$

$$V_{2n}(z) = (1 + G_{2n}z)e_{2n}V_{2n-1}(z) + V_{2n-2}(z), \quad n = 1, 2, \dots, \tag{4.2d}$$

$$V_{2n+1}(z) = \left(\frac{1}{z} + G_{2n+1}\right) e_{2n+1}V_{2n}(z) + V_{2n-1}(z), \quad n = 0, 1, \dots. \tag{4.2e}$$

LEMMA 4.1 Define $G_0 := 0$. Then for $n = 1, 2, \dots$ and $z \neq -\frac{1}{G_m}$, $m = 1, 2, \dots$, the sequences $\{V_{2n}(z)\}$ and $\{V_{2n-1}(z)\}$ satisfy the recurrence relations

$$V_{-2}(z) := 1, \quad V_{-1}(z) := 0, \quad V_0(z) := 1, \quad V_1(z) := \left(\frac{1}{z} + G_1\right) e_1 \tag{4.3a}$$

$$\begin{aligned} V_{2n}(z) = &\left[e_{2n}e_{2n-1}(1 + G_{2n}z)\left(\frac{1}{z} + G_{2n-1}\right) + \frac{(1 + G_{2n}z)e_{2n}}{(1 + G_{2n-2}z)e_{2n-2}} + 1\right] V_{2n-2}(z) \\ &- \frac{(1 + G_{2n}z)e_{2n}}{(1 + G_{2n-2}z)e_{2n-2}} V_{2n-4}(z), \end{aligned} \tag{4.3b}$$

$$\begin{aligned} V_{2n+1}(z) = &\\ &\left[e_{2n}e_{2n+1}(1 + G_{2n}z)\left(\frac{1}{z} + G_{2n+1}\right) + \frac{(1 + G_{2n+1}z)e_{2n+1}}{(1 + G_{2n-1}z)e_{2n-1}} + 1\right] V_{2n-1}(z) \\ &- \frac{(1 + G_{2n+1}z)e_{2n+1}}{(1 + G_{2n-1}z)e_{2n-1}} V_{2n-3}(z). \end{aligned} \tag{4.3c}$$

Proof: Apply (4.2).

From this point on we restrict ourselves to the special case where the continued fraction (4.1) satisfies (1.12b). That is, $G_n = G > 0$ for $n = 1, 2, \dots$. We recall from Theorem 3.1(B) that the $n$ zeros of $V_n(z)$ are the same as the $n$ zeros, $z_{n,k}$, $k = 1, 2, \dots, n$, of $Q_n(z)$ and hence they are simple, negative and separate the zeros of $V_{n+1}(z)$. In the following lemma we relabel the zeros of $V_n(z)$ for reasons that are apparent in (4.4) and (4.5).

LEMMA 4.2 Let $G_n = G > 0$ for $n = 1, 2, \dots$. Then:

A. $V_{2n+1}(-1/G) = 0$, $n = 0, 1, \dots$, and $V_{2n}(-1/G) = 1$, $n = 1, 2, \dots$.

B. For each $n = 1, 2, \dots$, let $x_{2n,p}$, $-n + 1 \leq p \leq n$, denote the zeros (all simple) of $V_{2n}(z)$ arranged so that $x_{2n,p} < x_{2n,p+1}$ for $p = -n + 1, \dots, n - 1$; then

$$x_{2n,-n+1} < x_{2n,-n+2} < \cdots < x_{2n,0} < \frac{-1}{G} < x_{2n,1} < \cdots < x_{2n,n} < 0. \tag{4.4}$$

C. Let $x_{1,0}$ denote the zero $x_{1,0} = -1/G$ of $V_1(z)$ and for each $n = 1, 2, \ldots$, let $x_{2n+1,p}$, $-n \le p \le n$, denote the zeros (all simple) of $V_{2n+1}(z)$ arranged so that $x_{2n+1,p} < x_{2n+1,p+1}$ for $p = -n, \ldots, n$; then

$$x_{2n+1,-n} < x_{2n+1,-n+1} < \cdots < x_{2n+1,0} = \frac{-1}{G} < x_{2n+1,1} < \cdots < x_{2n+1,n} < 0. \tag{4.5}$$

D. For each $n = 1, 2, \ldots$,

$$x_{2n+1,p-1} < x_{2n,p} < x_{2n+1,p}, \quad p = -n+1, \ldots, n \tag{4.6a}$$

$$x_{2n,p-1} < x_{2n-1,p-1} < x_{2n,p}, \quad p = -n+2, \ldots, n. \tag{4.6b}$$

Proof A: follows from induction on (4.2e) and (4.2d), with $G_n = G > 0$, $n = 1, 2, \ldots$.

Proofs B, C, D: These are immediate consequences of (A) and the fact that the zeros of $V_n(z)$ separate the zeros of $V_{n+1}(z)$.

LEMMA 4.3 Let $x \in I\!R$. Then for $\sigma = 0, 1$:

A.

$$V_1(x) \to -\infty \text{ as } x \to 0^- \quad \text{and} \quad V_1(x) \to d_1 = e_1 G_1 \text{ as } x \to -\infty$$

$$V_{2n+\sigma}(x) \to (-1)^n \infty \quad \text{as} \quad x \to -\infty, \quad n = 1, 2, \ldots,$$

$$V_{2n+\sigma}(x) \to (-1)^{n+\sigma} \infty \quad \text{as} \quad x \to 0^-, \quad n = 1, 2, \ldots .$$

B. For $n = 1, 2, \ldots$,

$$\text{sign} V_{2n}(x) = \begin{cases} \text{sign}(-1)^n & \text{on } (-\infty, x_{2n,-n+1}) \cup (x_{2n,n}, 0), \\ \text{sign}(-1)^{p-1} & \text{on } (x_{2n,p-1}, x_{2n,p}), \quad p = -n+2, \ldots, n, \end{cases}$$

and consequently $V_{2n}(-1/G) > 0$ since $x_{2n,0} < -1/G < x_{2n,1}$.

C. $V_1(x) > 0$ for $x \in (-\infty, -1/G)$ and $V_1(x) < 0$ for $x \in (-1/G, 0)$. For $n = 1, 2, \ldots$,

$$\text{sign}\, V_{2n+1}(x) = \begin{cases} \text{sign}(-1)^{n+1} & \text{on } (x_{2n+1,n}, 0), \\ \text{sign}(-1)^n & \text{on } (-\infty, x_{2n+1,-n}), \\ \text{sign}(-1)^p & \text{on } (x_{2n+1,p-1}, x_{2n+1,p}), \quad p = -n+1, \ldots, n. \end{cases}$$

Proof A: Follows from (3.5b) and the fact that $\prod_{k=1}^{m} d_k > 0$ for $m = 1, 2, \dots$.

Proofs B,C: Apply (4.4), (4.5), Lemma 4.3 and Theorem 3.1(B).

In Section 3 (Theorem 3.1) it was shown that the zeros of $V_n(z)$ separate the zeros of $V_{n+1}(z)$. We now prove that when (1.12b) holds, the zeros of $V_{2n+\sigma}(z)$ separate the zeros of $V_{2(n+1)+\sigma}(z)$, $\sigma = 0, 1$.

THEOREM 4.4 Let the modified positive T-fraction (4.1) satisfy (1.12b). Then

A. For $n \geq 2$, $-1/G < x_{2n,1} < x_{2n-2,1}$ and $x_{2n-2,n-1} < x_{2n,n} < 0$ and for each $n \geq 3$,

$$-1/G < x_{2n-2,p-1} < x_{2n,p} < x_{2n-2,p} < 0$$

for $p = 2, 3, \dots, n-1$.

B. For $n \geq 2$, $x_{2n,-n+1} < x_{2n-2,-n+2}$ and $x_{2n-2,0} < x_{2n,0} < -1/G$ and for each $n \geq 3$,

$$x_{2n-2,p} < x_{2n,p} < x_{2n-2,p+1} < -1/G$$

for $p = -n+2, \dots, -1$.

C. For $n \geq 2$, $-1/G < x_{2n+1,1} < x_{2n-1,1}$ and $x_{2n-1,n-1} < x_{2n+1,n} < 0$ and for each $n \geq 3$,

$$-1/G < x_{2n-1,p-1} < x_{2n+1,p} < x_{2n-1,p} < 0$$

for $p = 2, 3, \dots, n-1$.

D. For $x_{2n+1,-n} < x_{2n-1,-n+1}$ and $x_{2n-1,-1} < x_{2n+1,-1} < -1/G$ and for each $n \geq 3$,

$$x_{2n-1,p} < x_{2n+1,p} < x_{2n-1,p+1} < -1/G$$

for $p = -n+1, \dots, -2$.

Proof A: We first consider $n = 2$. From Lemma 4.3, $V_4(-1/G) > 0$, $V_4(x) \to +\infty$ as $x \to 0^-$ and from (4.3b) with $G_m = G$ for $m = 1, 2, \dots$, we see that $V_4(x_{2,1}) = -e_4/e_2 < 0$. Since $-1/G < x_{2,1} < 0$ and $V_4(z)$ has two zeros greater than $-1/G$ we have $-1/G < x_{4,1} < x_{2,1} < x_{4,2} < 0$. We know that $-1/G < x_{4,1} < x_{4,2} < 0$ and that $V_6(-1/G) > 0$. It follows from (4.3b) with $G_m = G$ for all $m$ that $V_6(x_{4,k}) = (-e_6/e_4)\, V_2(x_{4,k})$, $k = 1, 2$. But from Lemma 4.3(B), $V_2(x_{4,1}) > 0$ and $V_2(x_{4,2}) < 0$ since $x_{2,0} < x_{4,1} < x_{2,1}$ and $x_{2,1} < x_{4,2} < 0$. Hence $V_6(x_{4,1}) < 0$ while $V_6(x_{4,2}) > 0$. Moreover $V_6(x) \to -\infty$ as $x \to 0^-$ and $V_6(z)$ has three zeros greater than $-1/G$. Thus arranging the zeros as in (4.4) we see that $-1/G < x_{6,1} < x_{4,1} < x_{6,2} < x_{4,2} < x_{6,3} < 0$ and so (A) is true for $n = 3$. As induction hypothesis we assume that for some fixed $n > 3$, $-1/G < x_{2n-2,1} < x_{2n-4,1}$, $x_{2n-4,n-2} < x_{2n-2,n-1} < 0$

and $x_{2n-4,p-1} < x_{2n-2,p} < x_{2n-4,p} < 0$ for $p = 2, 3, \ldots, n-2$. Then from (4.3b) with $G_m = G$ for all $m$ and from Lemma 4.3 it follows that for $k = 1, 2, \ldots, n-1$,

$$\operatorname{sign} V_{2n}(x_{2n-2,k}) = \operatorname{sign} \frac{-e_{2n}}{e_{2n-2}} V_{2n-4}(x_{2n-2,k}) = \operatorname{sign}(-1)^k, \tag{4.7}$$

since $x_{2n-4,k-1} < x_{2n-2,k} < x_{2n-4,k}$ and $V_{2n-4}(x)$ has the same sign as $(-1)^{k-1}$ on the interval $(x_{2n-4,k-1}, x_{2n-4,k})$. Hence there exists at least one zero of $V_{2n}(x)$ between $x = x_{2n-2,k}$ and $x = x_{2n-2,k+1}$ for each $k = 1, 2, \ldots, n-2$. $V_{2n}(-1/G) > 0$ while we see from (4.7) that $V_{2n}(x_{2n-2,1}) < 0$ and so there exists at least one zero of $V_{2n}(x)$ between $x = -1/G$ and $x = x_{2n-2,1}$. From (4.7) and Lemma 4.3(A) there exists at least one zero of $V_{2n}(x)$ between $x = x_{2n-2,n-1}$ and $x = 0$ since $V_{2n}(x_{2n-2,n-1})$ has the same sign as $(-1)^{n-1}$ while $V_{2n}(x) \to (-1)^n\infty$ as $x \to 0^-$. But $V_{2n}(x)$ has exactly $n$ zeros between $x = -1/G$ and $x = 0$. Hence for each $k = 1, 2, \ldots, n-2$, there exists exactly one zero of $V_{2n}(x)$ between $x = x_{2n-2,k}$ and $x = x_{2n-2,k+1}$, exactly one zero of $V_{2n}(x)$ between $x = -1/G$ and $x = x_{2n-2,1}$, and exactly one zero of $V_{2n}(x)$ between $x = x_{2n-2,n-1}$ and $x = 0$. Ordering the zeros of $V_{2n}(z)$ as in (4.4) we obtain (A) by induction.

Proof B: An argument for (B) is similar to that for (A) and hence is omitted.

Proof C: The proofs that $x_{2n-1,n-1} < x_{2n+1,n} < 0$ for $n \geq 2$ and for $n \geq 3$, $-1/G < x_{2n-1,p-1} < x_{2n+1,p} < x_{2n-1,p} < 0$, $p = 2, 3, \ldots, n-1$ are similar to the proof of (A) and are thus omitted. To show that $-1/G < x_{2n+1,1} < x_{2n-1,1}$ one uses the fact (4.6) that $-1/G < x_{2n,1} < x_{2n-1,1}$ and shows, using (4.2e) with $G_m = G$ for all $m$ and using Lemma 4.3, that $V_{2n+1}(x_{2n-1,1}) = (1/x_{2n-1,1} + G)e_{2n+1}V_{2n}(x_{2n-1,1}) > 0$ and $V_{2n+1}(x_{2n,1}) = V_{2n-1}(x_{2n,1}) < 0$.

Proof D: A proof for (D) similar to that given for (C) is omitted.

COROLLARY 4.5 Let the modified positive T-fraction (4.1) satisfy condition (1.12b). Then for $\sigma = 0, 1$ and for each $p \in \mathcal{Z}$ the sequence $\{x_{2n+\sigma,p}\}_{n=|p|+1}^{\infty}$ converges and

$$\lim_{n\to\infty} x_{2n+\sigma,p} =: x_p^{(\sigma)} < 0. \tag{4.8}$$

Proof: For fixed $p \leq 0$ we see from Theorem 4.4 that for $\sigma = 0, 1$, the sequence $\{x_{2n+\sigma,p}\}_{n=|p|+1}^{\infty}$ is monotone increasing and bounded above by $-1/G$. For fixed $p \geq 1$, $\{x_{2n+\sigma,p}\}_{n=|p|+1}^{\infty}$ is a monotone decreasing sequence of negative numbers bounded below by $-1/G$.

Hereafter we restrict ourselves to continued fractions (4.1) which satisfy both (1.12a) and (1.12b). For completeness we state a version of Hurwitz' theorem found in [14] that is used to prove several of the following results.

THEOREM 4.6 (Hurwitz) For each $n = 1, 2, 3, \ldots$, let $f_n(z)$ be holomorphic and single-valued in an open region $D$ and let $\{f_n(z)\}$ converge uniformly to a non-constant function $f(z)$ in every compact subset of $D$. Let $a \in D$. Then $f(a) = 0$ if and only if there exists a sequence $\{z_n\}$ with $\lim z_n = a$ and there exists $n_0$ such that $f_n(z_n) = 0$ for each $n > n_0$.

THEOREM 4.7 Let the modified positive T-fraction (4.1) satisfy conditions (1.12). For $\sigma = 0, 1$, let $V^{(\sigma)}(z) := \lim_{n\to\infty} V_{2n+\sigma}(z)$ (see Theorem 3.3) and let $x_p^{(\sigma)} := \lim_{n\to\infty} x_{2n+\sigma,p}$, $p = 0, \pm1, \pm2, \ldots$, (see (4.8)). Then $V^{(\sigma)}(x_p^{(\sigma)}) = 0$ for all $p \in \mathcal{Z}$ and $V^{(\sigma)}(z) \not\equiv 0$, $\sigma = 0, 1$.

Proof: Recall that for $\sigma = 0, 1$, $\{V_{2n+\sigma}(z)\}_{n=1}^{\infty}$ is a sequence of functions holomorphic in $R := \{z \in C : z \neq 0\}$and $\{V_{2n+\sigma}(z)\}_{n=1}^{\infty}$ converges uniformly to $V^{(\sigma)}(z)$ in every compact set in $R$. Moreover for each $p \in \mathcal{Z}$, $\lim_{n\to\infty} x_{2n+\sigma,p} = x_p^{(\sigma)}$ and $V_{2n+\sigma}(x_{2n+\sigma,p}) = 0$ for each $n \geq |p|+1$. If it can be shown that for $\sigma = 0, 1$, $V^{(\sigma)}(z)$ is not identically constant then the assertion follows by Hurwitz' Theorem. It is enough to show that for $\sigma = 0, 1$, $V^{(\sigma)}(z) \not\equiv 0$ since $\{V_{2n+\sigma}(z)\}$ converges uniformly to $V^{(\sigma)}(z)$ on every compact subset $K$ of $R$ and there exist compact subsets $K$ of $R$ where $V_{2n+\sigma}(z) = 0$. It follows from Lemma 4.2(A) that $V^{(0)}(-1/G) = 1$ since $V_{2n}(-1/G) = 1$ for each $n = 1, 2, \ldots$. Consequently $V^{(0)}(z) \not\equiv 0$ and so by Hurwitz' Theorem $V^{(0)}(x_p^{(0)}) = 0$ for all $p \in \mathcal{Z}$. Suppose $V^{(1)}(z) \equiv 0$. Then $V^{(1)}(x_p^{(0)}) = 0$ for each $p \in \mathcal{Z}$. But this contradicts (3.8) and the fact that $V^{(0)}(x_p^{(0)}) = 0$. Therefore $V^{(1)}(z) \not\equiv 0$.

LEMMA 4.8 Let the modified positive T-fraction (4.1) satisfy (1.12). Then for each $p \in \mathcal{Z}$,

$$x_p^{(\sigma)} < x_{p+1}^{(\sigma)} < 0.$$

Proof: Since $x_{2n+\sigma,p} < x_{2n+\sigma,p+1}$ for all $n > |p| + 1$, we have $x_p^{(\sigma)} \leq x_{p+1}^{(\sigma)}$. The zeros of $U_{2n+\sigma}(z)$ separate the zeros of $V_{2n+\sigma}(z)$ and so there exists a sequence $\{y_{2n+\sigma,p}\}_{n=|p|+1}^{\infty}$ of zeros of $U_{2n+\sigma}(z)$ such that $x_{2n+\sigma,p} < y_{2n+\sigma,p+\sigma} < x_{2n+\sigma,p+1}$. Suppose $x_p^{(\sigma)} = x_{p+1}^{(\sigma)}$. Then $\lim_{n\to\infty} x_{2n+\sigma,p} = \lim_{n\to\infty} y_{2n+\sigma,p+\sigma} = \lim_{n\to\infty} x_{2n+\sigma,p+1} = x_p^{(\sigma)}$. From Theorem 4.7 we have $V^{(\sigma)}(x_p^{(\sigma)}) = 0$ and using Theorem 4.6 one can show that $U^{(\sigma)}(x_p^{(\sigma)}) = 0$. This contradicts the determinant formula (3.8) with $z = x_p^{(\sigma)}$. Hence $x_p^{(\sigma)} < x_{p+1}^{(\sigma)}$.

LEMMA 4.9 Let the modified positive T-fraction (4.1) satisfy conditions (1.12) and for $\sigma = 0, 1$, let $x_p^{(\sigma)}$, $p = 0, \pm1, \pm2, \ldots$ be defined as in (4.8). Then

A. $\lim_{p\to\infty} x_p^{(\sigma)} = 0.$

B. $\lim_{p\to-\infty} x_p^{(\sigma)} = -\infty.$

Proof A: From Lemma 4.8 we see that $\{x_p^{(\sigma)}\}_{p=1}^{\infty}$ is a monotone increasing sequence bounded above by 0. Suppose that $\lim_{p\to\infty} x_p^{(\sigma)} = c < 0$. Then since $V^{(\sigma)}(z)$ is continuous on $\mathcal{C} - \{0\}$, we have $V^{(\sigma)}(c) = V^{(\sigma)}(\lim_{p\to\infty} x_p^{(\sigma)}) = \lim_{p\to\infty} V^{(\sigma)}(x_p^{(\sigma)}) = 0$ which implies by the Identity Theorem [8] that $V^{(\sigma)}(z) \equiv 0$ giving a contradiction to Theorem 4.7.

Proof B: To prove (B) use the fact that $\{x_{-p}^{(\sigma)}\}_{p=1}^{\infty}$ is a monotone decreasing sequence and obtain a proof similar to that given for (A).

THEOREM 4.10 If the modified positive T-fraction (4.1) satisfies (1.12), then, for $\sigma = 0, 1$, the zeros of $V^{(\sigma)}(z)$ are exactly $x_p^{(\sigma)}$, $p = 0, \pm 1, \pm 2, \ldots$, and they are simple.

Proof: Suppose $V^{(\sigma)}(b) = 0$ and $b \neq x_p^{(\sigma)}$ for all $p \in \mathcal{Z}$. It follows from Theorem 4.6 (Hurwitz) that $-\infty < b < 0$ since for each $n \in \mathcal{Z}$, the zeros of $V_{2n+\sigma}(z)$ lie in $(-\infty, 0)$. Consequently from Lemma 4.9 there exists $p^{(\sigma)} \in \mathcal{Z}$ so that $x_{p^{(\sigma)}}^{(\sigma)} < b < x_{p^{(\sigma)}+1}^{(\sigma)}$. Fix $\varepsilon < \frac{1}{4}\min\{x_{p^{(\sigma)}+1}^{(\sigma)} - b, b - x_{p^{(\sigma)}}^{(\sigma)}, |x_{p^{(\sigma)}+1}^{(\sigma)}|\}$. Since for each $p \in \mathcal{Z}$, the sequence $\{x_{2n+\sigma,p}\}_{n=|p|+1}^{\infty}$ converges to $x_p^{(\sigma)}$, there exists an integer $N_1^{(\sigma)}$ so large that $|x_{2n+\sigma,p^{(\sigma)}} - x_{p^{(\sigma)}}^{(\sigma)}| < \varepsilon$ for $n \geq N_1^{(\sigma)}$ and $|x_{2n+\sigma,p^{(\sigma)}+1} - x_{p^{(\sigma)}+1}^{(\sigma)}| < \varepsilon$ for $n \geq N_1^{(\sigma)}$. Now $V^{(\sigma)}(b) = 0$ and so by Theorem 4.6 (Hurwitz) there exists an integer $N_2^{(\sigma)}$ so that for $n \geq N_2^{(\sigma)}$, $V_{2n+\sigma}(z)$ has a zero in the neighborhood $B(b, \varepsilon) := \{z : |z - b| < \varepsilon\}$. Let $N^{(\sigma)} = \max\{N_1^{(\sigma)}, N_2^{(\sigma)}\}$. Then $V_{2N^{(\sigma)}+\sigma}(z)$ has a zero in $B(b, \varepsilon)$, $|x_{2N^{(\sigma)}+\sigma,p^{(\sigma)}+1} - x_{p^{(\sigma)}+1}^{(\sigma)}| < \varepsilon$ and $|x_{2N^{(\sigma)}+\sigma,p^{(\sigma)}} - x_{p^{(\sigma)}}^{(\sigma)}| < \varepsilon$. This implies that $V_{2N^{(\sigma)}+\sigma}(z)$ has a zero $x_{2N^{(\sigma)}+\sigma,*}$ with $x_{2N^{(\sigma)},p^{(\sigma)}} < x_{2N^{(\sigma)}+\sigma,*} < x_{2N^{(\sigma)}+\sigma,p^{(\sigma)}+1}$, which contradicts (4.4) and (4.5). Hence we have $b = x_p^{(\sigma)}$ for some $p \in \mathcal{Z}$.

We now show that for each $p \in \mathcal{Z}$, $x_p^{(\sigma)}$ is a simple zero of $V^{(\sigma)}(z)$. Fix $p \in \mathcal{Z}$. By Lemma 4.8 we have $x_{p-1}^{(\sigma)} < x_p^{(\sigma)} < x_{p+1}^{(\sigma)}$. Choose $\varepsilon < \frac{1}{4}\min\{x_{p+1}^{(\sigma)} - x_p^{(\sigma)}, x_p^{(\sigma)} - x_{p-1}^{(\sigma)}, |x_{p+1}^{(\sigma)}|\}$, define $\Gamma := \{z \in \mathcal{C} : |z - x_p^{(\sigma)}| = \varepsilon\}$ and set $\delta = \inf\{|V^{(\sigma)}(z)| : z \in \Gamma\} > 0$. Fix $N_p^{(\sigma)} \in \mathcal{Z}$ so large that $|x_{2n+\sigma,p-1} - x_{p-1}^{(\sigma)}| < \varepsilon$ for $n \geq N_p^{(\sigma)}$, $|x_{2n+\sigma,p} - x_p^{(\sigma)}| < \varepsilon$ for $n \geq N_p^{(\sigma)}$, $|x_{2n+\sigma,p+1} - x_{p+1}^{(\sigma)}| < \varepsilon$ for $n \geq N_p^{(\sigma)}$ and $|V^{(\sigma)}(z) - V_{2n+\sigma}(z)| < \frac{1}{2}\,\delta$ for $n \geq N_p^{(\sigma)}$, $z \in \Gamma$. Then $V_{2N_p^{(\sigma)}+\sigma}(z)$ has exactly one zero, $x_{2N_p^{(\sigma)}+\sigma,p}$ in $B(x_p^{(\sigma)}, \varepsilon)$ and it is simple. Now $V^{(\sigma)}(z)$ and $V_{2N_p^{(\sigma)}+\sigma}(z)$ are holomorphic on $\bar{B}(x_p^{(\sigma)}, \varepsilon)$ with no zeros or poles on $\Gamma$ and $|V^{(\sigma)}(z) - V_{2N_p^{(\sigma)}+\sigma}(z)| < \frac{1}{2}\,\delta < |V^{(\sigma)}(z)| \leq |V^{(\sigma)}(z)| + |V_{2N_p^{(\sigma)}+\sigma}(z)|$. Hence by Rouche's Theorem [4], $V^{(\sigma)}(z)$ has exactly one zero (counted according to multiplicity) in $B(x_p^{(\sigma)}, \varepsilon)$. Thus $x_p^{(\sigma)}$ is a simple zero of $V^{(\sigma)}(z)$.

THEOREM 4.11 For $\sigma = 0, 1$, $V^{(\sigma)}(z)$ has essential singularities at $z = 0$ and $z = \infty$.

Proof: We have shown (Theorem 3.3) that $V^{(\sigma)}(z)$ is analytic in $\mathcal{C} - \{0\}$. If $V^{(\sigma)}(z)$ had a pole at $z = 0$, then we would have $\lim_{z\to 0}|V^{(\sigma)}(z)| = \infty$. This is impossible since $V^{(\sigma)}(x_p^{(\sigma)}) = 0$ for each $p \in \mathcal{Z}$ and $\lim_{p\to\infty} x_p^{(\sigma)} = 0$. If $V^{(\sigma)}(z)$ had a removable singularity at $z = 0$, then we would have $\lim_{z\to 0} zV^{(\sigma)}(z) = 0$. Consequently the function $g^{(\sigma)}(z) := zV^{(\sigma)}(z)$ would have zeros at $\{x_p^{(\sigma)}\}_{p=-\infty}^{\infty} \cup \{0\}$. However, $\lim_{p\to\infty} x_p^{(\sigma)} = 0$ and so $g^{(\sigma)}(z) \equiv 0$ which implies $V^{(\sigma)}(z) \equiv 0$ and contradicts Theorem 4.7. Hence $z = 0$ is an essential singularity of $V^{(\sigma)}(z)$, $\sigma = 0, 1$. One shows that $V^{(\sigma)}(z)$ has an essential singularity at $\infty$ by showing that $V^{(\sigma)}\left(\frac{1}{z}\right)$ has an essential singularity at $z = 0$. This can be done in a manner similar to that used above using $V^{(\sigma)}\left(\frac{1}{z}\right) = 0$ at $z = \frac{1}{x_p^{(\sigma)}}$ and $\lim_{p\to-\infty} \frac{1}{x_p^{(\sigma)}} = 0$.

## 5 CONSTRUCTION OF STEP-FUNCTION SOLUTIONS OF SSMPs

Suppose $\{c_k\}_{k=-\infty}^{\infty}$ is a bisequence that leads to an indeterminate SSMP and such that the associated positive T-fraction (1.8) satisfies (1.12). The properties developed in previous sections are now used to construct two distribution functions $\psi^{(0)}(t), \psi^{(1)}(t) \in \Psi(0,\infty)$ that are step functions and that generate the sequence $\{c_k\}_{k=-\infty}^{\infty}$. Every convex linear combination ($\psi(\alpha, t) := (1-\alpha)\psi^{(0)}(t) + \alpha\psi^{(1)}(t)$, $0 \le \alpha \le 1$) of these distribution functions is a distribution function which generates the same sequence of moments.

We recall that the zeros of $V_{2n+\sigma}(z)$ denoted by $z_{2n+\sigma,k}$, $k = 1, 2, \ldots, 2n+\sigma$, in Theorem 3.1 were relabeled as $x_{2n+\sigma,p}$, $p = -n+1-\sigma, \ldots, n$ in Lemma 4.2.

Since we now wish to apply Theorem 3.1(D), we rewrite its assertions in terms of the $x_{2n+\sigma,p}$ notation. For $\sigma = 0, 1$ and $n = 1, 2, \ldots$, we have by Theorem 3.1(D),

$$\frac{U_{2n+\sigma}(z)}{V_{2n+\sigma}(z)} = \sum_{p=-n+1-\sigma}^{n} \frac{z a_{2n+\sigma,p}}{z - x_{2n+\sigma,p}} \tag{5.1a}$$

where

$$a_{2n+\sigma,p} := \frac{U_{2n+\sigma}(x_{2n+\sigma,p})}{x_{2n+\sigma,p}V'_{2n+\sigma}(x_{2n+\sigma,p})} > 0 \quad \text{and} \quad \sum_{p=-n+1-\sigma}^{n} a_{2n+\sigma,p} = \frac{F_1}{G}. \tag{5.1b}$$

Since for $\sigma = 0, 1$, $\{V_{2n+\sigma}(z)\}_{n=1}^{\infty}$ and $\{U_{2n+\sigma}(z)\}_{n=1}^{\infty}$ converge uniformly on compact subsets of $\mathcal{C} - \{0\}$ to holomorphic functions $V^{(\sigma)}(z)$ and $U^{(\sigma)}(z)$, respectively, and since $\{x_{2n+\sigma,p}\}_{n=1}^{\infty}$ converges to $x_p^{(\sigma)}$ for each $p \in \mathcal{Z}$, the bisequences $\{a_p^{(\sigma)}\}_{p=-\infty}^{\infty}$, $\sigma = 0, 1$, where $a_p^{(\sigma)}$ is given by

$$a_p^{(\sigma)} := \lim_{n\to\infty} a_{2n+\sigma,p} = \lim_{n\to\infty} \frac{U_{2n+\sigma}(x_{2n+\sigma,p})}{x_{2n+\sigma,p}V'_{2n+\sigma}(x_{2n+\sigma,p})} = \frac{U^{(\sigma)}(x_p^{(\sigma)})}{x_p^{(\sigma)}V^{(\sigma)\prime}(x_p^{(\sigma)})} \tag{5.2}$$

are well defined.

THEOREM 5.1 For $\sigma = 0, 1$ and $p \in \mathcal{Z}$, the numbers $a_p^{(\sigma)}$ defined as in (5.2) satisfy

A. For all $p \in \mathcal{Z}$, $a_p^{(\sigma)} > 0$.

B. $\displaystyle\sum_{p=-\infty}^{\infty} a_p^{(\sigma)} \le \frac{F_1}{G}$.

REMARK: In Corollary 5.4 we show that $\le$ in (B) can be replaced by $=$.

Proof A: Let integers $\sigma, n$ and $p$ satisfy $\sigma = 0$ or $1$, $n \ge 1$, and $-n+1-\sigma \le p \le n$. From (5.1b) we know that $a_{2n+\sigma,p} > 0$ and hence $a_p^{(\sigma)} \ge 0$. Since $V^{(\sigma)}(x_p^{(\sigma)}) = 0$, we see from the determinant formula (3.8) that $U^{(\sigma)}(x_p^{(\sigma)}) \ne 0$. Hence $a_p^{(\sigma)} \ne 0$.

Proof B: Fix $k \in \mathcal{N}$. Then for all $n \ge k+1-\sigma$ we have by (5.1b) that

$$\sum_{p=-k}^{k} a_{2n+\sigma,p} \le \sum_{p=-n+1-\sigma}^{n} a_{2n+\sigma,p} = \frac{F_1}{G}.$$

Thus

$$\sum_{p=-k}^{k} a_p^{(\sigma)} = \sum_{p=-k}^{k} \lim_{n\to\infty} a_{2n+\sigma,p} = \lim_{n\to\infty} \sum_{p=-k}^{k} a_{2n+\sigma,p} \le \frac{F_1}{G},$$

and hence

$$\sum_{p=-\infty}^{\infty} a_p^{(\sigma)} = \lim_{k\to\infty} \sum_{p=-k}^{k} a_p^{(\sigma)} \le \frac{F_1}{G}.$$

THEOREM 5.2 For $\sigma = 0, 1$ and $n = 1, 2, \ldots$, let $\psi_{2n+\sigma}(t)$ denote the real-valued step function on $[0, \infty)$ with jump $a_{2n+\sigma,p} > 0$ at $t = -x_{2n+\sigma,p} > 0$, $p = -n+1-\sigma, \ldots, n$, defined by

$$\psi_{2n+\sigma}(t) := \begin{cases} -\displaystyle\sum_{\nu=1}^{n} a_{2n+\sigma,\nu} & \text{for} \quad 0 \le t < -x_{2n+\sigma,n}, \\ -\displaystyle\sum_{\nu=1}^{p} a_{2n+\sigma,\nu} & \text{for} \quad -x_{2n+\sigma,p+1} \le t < -x_{2n+\sigma,p}, \ p=1,\ldots,n-1, \\ 0 & \text{for} \quad -x_{2n+\sigma,1} \le t < -x_{2n+\sigma,0}, \\ \displaystyle\sum_{\nu=0}^{p-1} a_{2n+\sigma,-\nu} & \text{for} \quad -x_{2n+\sigma,-p+1} \le t < -x_{2n+\sigma,-p}, \ p=1,\ldots,n-1+\sigma, \\ \displaystyle\sum_{\nu=0}^{n-1+\sigma} a_{2n+\sigma,-\nu} & \text{for} \quad -x_{2n+\sigma,-n+1-\sigma} \le t < \infty. \end{cases} \tag{5.3}$$

Then for each $\sigma = 0, 1$ the sequence $\{\psi_{2n+\sigma}(t)\}_{n=1}^{\infty}$ converges pointwise on $[0, \infty)$ to the step function $\psi^{(\sigma)}(t)$ with jump $a_p^{(\sigma)} > 0$ at $t = -x_p^{(\sigma)} > 0$, $p = 0, \pm 1, \pm 2, \dots$, defined by

$$\psi^{(\sigma)}(t) := \begin{cases} -\sum_{\nu=1}^{p} a_\nu^{(\sigma)} & \text{for} \quad -x_{p+1}^{(\sigma)} \le t < -x_p^{(\sigma)},\ p = 1, 2, \dots, \\ 0 & \text{for} \quad -x_1^{(\sigma)} \le t \le -x_0^{(\sigma)}, \\ \sum_{\nu=0}^{p-1} a_{-\nu}^{(\sigma)} & \text{for} \quad -x_{-p+1}^{(\sigma)} < t \le -x_{-p}^{(\sigma)}. \end{cases} \tag{5.4}$$

Proof: Fix $t$ so that

$$-x_1^{(\sigma)} \le t \le -x_0^{(\sigma)}.$$

Recall that for each $k = 0, -1, -2, \dots$, the sequence $\{-x_{2n+\sigma,k}\}_{n=1}^{\infty}$ decreases monotonically to $-x_k^{(\sigma)}$, for each $k = 1, 2, \dots$, the sequence $\{-x_{2n+\sigma,k}\}_{n=1}^{\infty}$ increases monotonically to $-x_k^{(\sigma)}$ and

$$0 < \dots < -x_2^{(\sigma)} < -x_1^{(\sigma)} < -x_0^{(\sigma)} < -x_{-1}^{(\sigma)} < -x_{-2}^{(\sigma)} < \dots.$$

Thus

$$-x_{2n+\sigma,1} < -x_1^{(\sigma)} \le t \le -x_0^{(\sigma)} \le -x_{2n+\sigma,0}$$

for $n = 1, 2, \dots$, and hence $\psi_{2n+\sigma}(t) = 0$ for $n = 1, 2, \dots$. Consequently,

$$\lim_{n\to\infty} \psi_{2n+\sigma}(t) = 0 = \psi^{(\sigma)}(t), \quad \text{for} \quad -x_1^{(\sigma)} \le t \le -x_0^{(\sigma)}.$$

Fix $t$ so that $-x_0^{(\sigma)} < t$. Since $lim_{p\to-\infty} - x_p^{(\sigma)} = \infty$, there exists a positive integer $p_t^{(\sigma)}$ so that

$$-x^{(\sigma)}_{-p_t^{(\sigma)}+1} < t \le -x^{(\sigma)}_{-p_t^{(\sigma)}}.$$

Since for each $k = -p_t^{(\sigma)}, -p_t^{(\sigma)} + 1, \dots$, the sequence $\{-x_{2n+\sigma,k}\}_{n=1}^{\infty}$ decreases monotonically to $-x_k^{(\sigma)}$, there exists a positive integer $N_t^{(\sigma)}$ so that

$$-x^{(\sigma)}_{-p_t^{(\sigma)}+1} < -x_{2n+\sigma,-p_t^{(\sigma)}+1} < t \le -x^{(\sigma)}_{-p_t^{(\sigma)}} < -x_{2n+\sigma,-p_t^{(\sigma)}}$$

for all $n \ge N_t^{(\sigma)}$. Consequently, from (5.3)

$$\psi_{2n+\sigma}(t) = \sum_{\nu=0}^{p_t^{(\sigma)}-1} a_{2n+\sigma,-\nu}$$

for all $n > N_t^{(\sigma)}$ and we have for $-x_0^{(\sigma)} < t$

$$\begin{aligned}\lim_{n\to\infty} \psi_{2n+\sigma}(t) &= \lim_{n\to\infty} \sum_{\nu=0}^{p_t^{(\sigma)}-1} a_{2n+\sigma,-\nu} \\ &= \sum_{\nu=0}^{p_t^{(\sigma)}-1} \lim_{n\to\infty} a_{2n+\sigma,-\nu} \\ &= \sum_{\nu=0}^{p_t^{(\sigma)}-1} a_{-\nu}^{(\sigma)} = \psi^{(\sigma)}(t).\end{aligned}$$

Fix $t$ so that $0 < t < -x_1^{(\sigma)}$. Since $lim_{p\to+\infty} - x_p^{(\sigma)} = 0$, there exists a positive integer $p_t^{(\sigma)}$ so that

$$-x_{p_t^{(\sigma)}+1}^{(\sigma)} \le t < -x_{p_t^{(\sigma)}}^{(\sigma)}.$$

Since for each $k = p_t^{(\sigma)}, p_t^{(\sigma)}+1, \dots$, the sequence $\{-x_{2n+\sigma,k}\}_{n=1}^{\infty}$ increases monotonically to $-x_k^{(\sigma)}$, there exists a positive integer $N_t^{(\sigma)}$ so that

$$-x_{2n+\sigma,p_t^{(\sigma)}+1} < -x_{p_t^{(\sigma)}+1}^{(\sigma)} \le t < -x_{2n+\sigma,p_t^{(\sigma)}} < -x_{p_t^{(\sigma)}}^{(\sigma)}$$

for all $n \ge N_t^{(\sigma)}$. Consequently, from (5.3)

$$\psi_{2n+\sigma}(t) = -\sum_{\nu=1}^{p_t^{(\sigma)}} a_{2n+\sigma,\nu}$$

for all $n > N_t^{(\sigma)}$ and we have for $0 < t < -x_1^{(\sigma)}$

$$\begin{aligned}\lim_{n\to\infty} \psi_{2n+\sigma}(t) &= \lim_{n\to\infty} -\sum_{\nu=1}^{p_t^{(\sigma)}} a_{2n+\sigma,\nu} \\ &= -\sum_{\nu=1}^{p_t^{(\sigma)}} \lim_{n\to\infty} a_{2n+\sigma,\nu} \\ &= -\sum_{\nu=1}^{p_t^{(\sigma)}} a_\nu^{(\sigma)} = \psi^{(\sigma)}(t).\end{aligned}$$

THEOREM 5.3 Let $\{c_k\}_{k=-\infty}^{\infty}$ be a bisequence of real numbers that has a corresponding (see (1.10)) divergent positive T-fraction (1.8) satisfying conditions (1.12) and hence that leads to an indeterminant SSMP. Then for each $\sigma = 0, 1$, the function $\psi^{(\sigma)}(t)$ given in (5.4) satisfies:

A.

$$\frac{U^{(\sigma)}(z)}{V^{(\sigma)}(z)} = \lim_{n\to\infty} \frac{U_{2n+\sigma}(z)}{V_{2n+\sigma}(z)} = z\int_0^{\infty} \frac{d\psi^{(\sigma)}(t)}{z+t} = z\sum_{p=-\infty}^{\infty} \frac{a_p^{(\sigma)}}{z - x_p^{(\sigma)}} \tag{5.5}$$

where the series converges uniformly on compact subsets of $C - \{x_p^{(\sigma)}\}_{p=-\infty}^{\infty} - \{0\}$.

B. $\psi^{(0)}(t)$ and $\psi^{(1)}(t)$ are distinct solutions to the SSMP for $\{c_k\}_{k=-\infty}^{\infty}$. That is, for $\sigma = 0, 1$,

$$c_k = \int_0^{\infty} (-t)^k d\psi^{(\sigma)}(t), \quad k = 0, \pm 1, \pm 2, \ldots . \tag{5.6}$$

Proof A: Since $\lim_{n\to\infty} U_{2n+\sigma}(z) = U^{(\sigma)}(z)$ and $\lim_{n\to\infty} V_{2n+\sigma}(z) = V^{(\sigma)}(z)$ we have

$$\frac{U^{(\sigma)}(z)}{V^{(\sigma)}(z)} = \lim_{n\to\infty} \frac{U_{2n+\sigma}(z)}{V_{2n+\sigma}(z)}.$$

It follows from (5.1a), (5.3) and Stieltjes integration that

$$\frac{U_{2n+\sigma}(z)}{V_{2n+\sigma}(z)} = \sum_{p=-n+1-\sigma}^{n} \frac{z a_{2n+\sigma,p}}{z - x_{2n+\sigma,p}} = \int_0^{\infty} \frac{z}{z+t}\, d\psi_{2n+\sigma}(t).$$

Fix $z \in C - (-\infty, 0]$. Then $g(t) := \dfrac{z}{z+t}$ is a continuous complex valued function of the real variable $t$ on $(0,\infty)$ such that $\lim_{t\to\infty} g(t) = 0$. By (3.6c) and Lemma 5.2 we have

$$0 \leq \psi_{2n+\sigma}(t) \leq \frac{F_1}{G_1} \quad \text{for} \quad 0 < t < \infty, \quad n = 1, 2, 3, \ldots \quad \text{and} \quad \sigma = 0, 1$$

and

$$\lim_{n\to\infty} \psi_{2n+\sigma}(t) = \psi^{(\sigma)}(t) \leq \frac{F_1}{G_1} \quad \text{for} \quad 0 < t < \infty, \quad \sigma = 0, 1.$$

Thus Helly's second theorem extended to an infinite interval [12, Theorem 6, pg 240] can be applied to yield

$$\lim_{n\to\infty} z \int_0^{\infty} \frac{d\psi_{2n+\sigma}(t)}{z+t} = z \int_0^{\infty} \frac{d\psi^{(\sigma)}(t)}{z+t}, \quad z \in C - (-\infty, 0], \quad \sigma = 0, 1.$$

Equations (5.5) follow from this and the definition of $\psi^{(\sigma)}(t)$. The series converges uniformly on compact subsets of $C - \{x_p^{(\sigma)}\}_{p=-\infty}^{\infty} - \{0\}$ since $|z - x_p^{(\sigma)}|$ is bounded away from zero and since from Theorem 5.1 we have

$$\sum_{p=-\infty}^{\infty} |a_p^{(\sigma)}| = \sum_{p=-\infty}^{\infty} a_p^{(\sigma)} \leq \frac{F_1}{G}.$$

Proof B: That $\psi^{(0)}(t)$ and $\psi^{(1)}(t)$ are solutions to the SSMP for $\{c_k\}_{k=-\infty}^{\infty}$ follows from part (A) and [11, Theorem 4.1].

COROLLARY 5.4 For $\sigma = 0, 1$ and $p = 0, \pm 1, \pm 2, \ldots$, let $a_p^{(\sigma)}$ be defined as in (5.2). Then

$$\sum_{p=-\infty}^{\infty} a_p^{(\sigma)} = \frac{F_1}{G}. \tag{5.7}$$

Proof: From (5.5) and (5.6) we have

$$\sum_{p=-\infty}^{\infty} \frac{a_p^{(\sigma)}}{1 - \frac{x_p^{(\sigma)}}{z}} = z \sum_{p=-\infty}^{\infty} \frac{a_p^{(\sigma)}}{z - x_p^{(\sigma)}} = z \int_0^{\infty} \frac{d\psi^{(\sigma)}(t)}{z+t} = \int_0^{\infty} \left(1 - \frac{t}{z+t}\right) d\psi^{(\sigma)}(t)$$
$$= \int_0^{\infty} d\psi^{(\sigma)}(t) - \frac{1}{z}\int_0^{\infty} \frac{t}{1+\frac{t}{z}}\, d\psi^{(\sigma)}(t) = c_0 - \frac{1}{z}\int_0^{\infty} \frac{t}{1+\frac{t}{z}}\, d\psi^{(\sigma)}(t)$$

where the series converges uniformly on compact subsets of $C - \{x_p^{(\sigma)}\}_{p=-\infty}^{\infty} - \{0\}$. Letting $z \to \infty$ with $|\arg z| < \pi$ we have $\sum_{p=-\infty}^{\infty} a_p^{(\sigma)} = c_0$. It remains to be seen that $c_0 = F_1/G$. From (1.9), (1.10) and (2.2) we have $\frac{P_1(z)}{Q_1(z)} - L_\infty(z) = O\left(\frac{1}{z}\right)$ and so

$$\frac{F_1 z}{1+Gz} - \sum_{k=0}^{\infty} c_k z^{-k} = \left(\frac{F_1}{G} - \frac{F_1}{G^2 z} - \frac{F_1}{G^3 z^2} - \cdots\right) - \left(c_0 + \frac{c_1}{z} + \frac{c_2}{z} + \cdots\right)$$
$$= O\left(\frac{1}{z}\right).$$

Consequently $c_0 = F_1/G$.

COROLLARY 5.5 Let $\{c_k\}_{k=-\infty}^{\infty}$ be a bisequence that leads to an indeterminate SSMP and such that the associated positive T-fraction (1.8) satisfies (1.12). Let $\psi^{(0)}(t)$ and $\psi^{(1)}(t)$ be the solutions to the SSMP for $\{c_k\}_{k=-\infty}^{\infty}$ which are given in (5.4). Then if $\psi(t)$ is any solution to the SSMP for $\{c_k\}_{k=-\infty}^{\infty}$ we have for $z > 0$

$$\int_0^{\infty} \frac{z}{z+t}\, d\psi^{(0)}(t) \le \int_0^{\infty} \frac{z}{z+t}\, d\psi(t) \le \int_0^{\infty} \frac{z}{z+t}\, d\psi^{(1)}(t). \tag{5.8}$$

Proof: Our proof of Corollary 5.5 is based on a result obtained in the proof of [11, Theorem 6.2]. Recall that the polynomials $P_n(z)$ and $Q_n(z)$ defined in (2.1) are the $n$th numerator and denominator respectively of (1.8). Let $\psi(t)$ be a solution to the SSMP for $\{c_k\}_{k=-\infty}^{\infty}$. In the proof of [11, Theorem 6.2] it is shown that for $z > 0$ and $n = 1, 2, 3, \ldots$,

$$\frac{P_{2n}(z)}{Q_{2n}(z)} < \int_0^{\infty} \frac{z}{z+t}\, d\psi(t) < \frac{P_{2n+1}(z)}{Q_{2n+1}(z)}.$$

Since $U_n(z)$ and $V_n(z)$ are the $n$th numerator and denominator, respectively, of the continued fraction (3.2) which is equivalent to the continued fraction (1.8), we have using (3.3)

$$\frac{U_{2n}(z)}{V_{2n}(z)} < \int_0^\infty \frac{z}{z+t}\, d\psi(t) < \frac{U_{2n+1}(z)}{V_{2n+1}(z)}$$

for $z > 0$, $n = 1, 2, \ldots$. Letting $n \to \infty$ and using (5.5) gives the desired result.

## 6 AN EXAMPLE INVOLVING LOG-NORMAL MOMENTS

We use as an example a result obtained in [2]. The *classical log-normal distribution function* $\varphi(t) \in \Psi(0, \infty)$ defined by

$$\varphi'(t) := \frac{q^{\frac{1}{2}} e^{-\left(\frac{\log t}{2\kappa}\right)^2}}{2\kappa\sqrt{\pi}}, \quad 0 < t < \infty,\ 0 < q = e^{-2\kappa^2} < 1, \tag{6.1}$$

generates the *sequence of log-normal moments* $\{c_k\}_{k=-\infty}^{\infty}$ given by

$$c_k := \int_0^\infty (-t)^k \varphi'(t) dt = (-1)^k q^{-\frac{k^2}{2}-k}, \quad k = 0, \pm 1, \pm 2, \ldots . \tag{6.2}$$

Other distribution functions generating the same moment sequence (6.2) are known (see for instance [1, 3]). In this example we construct two distribution functions $\psi^{(0)}(t), \psi^{(1)}(t) \in \Psi(0, \infty)$ which also generate $\{c_k\}_{k=-\infty}^{\infty}$. They are both step functions and the location of the steps and the size of the jumps are given explicitly. We base our work on the earlier sections of this paper and on [9, 5 and 2].

In [5] it was shown that the positive T-fraction

$$\mathop{K}_{n=1}^{\infty} \frac{F_n z}{1 + G_n z}, \ F_1 = G_1 := q^{\frac{1}{2}},\ F_n := q^{-n+\frac{3}{2}}(1 - q^{n-1}),\ G_n := q^{\frac{1}{2}},\ n = 2, 3, 4, \ldots . \tag{6.3}$$

corresponds to the pair $(L_0, L_\infty)$ of fps

$$L_0(z) := \sum_{k=1}^{\infty} (-1)^{k-1} q^{-\frac{k^2}{2}+k} z^k, \quad L_\infty(z) := \sum_{k=0}^{\infty} (-1)^k q^{-\frac{k^2}{2}-k} z^{-k}. \tag{6.4}$$

We now consider the sequence $\{e_n\}_{n=1}^{\infty}$ where the $e_n$ are defined as in (1.12a) and where the $F_n$ and $G_n$ are given in (6.3). It is shown in [5] that $e_1 = q^{-\frac{1}{2}}$, $e_2 = \dfrac{1}{1-q}$, and

$$e_{2n-1} = q^{n-\frac{3}{2}} \frac{\prod_{k=1}^{n-1} (1 - q^{2k-1})}{\prod_{k=1}^{n-1} (1 - q^{2k})}, \quad e_{2n} = q^n \frac{\prod_{k=1}^{n-1} (1 - q^{2k})}{\prod_{k=1}^{n} (1 - q^{2k-1})}, \quad n = 2, 3, \ldots, \tag{6.5a}$$

and

$$\sum_{n=1}^{\infty} e_n < \infty. \tag{6.5b}$$

For each $n = 1, 2, 3, \ldots$, let $V_n(z)$ and $U_n(z)$ be defined as in (3.4) where $P_n(z)$ and $Q_n(z)$ are the $n$th numerator and denominator, respectively, of (6.3) and where the $e_n$ are given in (6.5). Since $\sum_{n=1}^{\infty} e_n < \infty$ and $G_n = G = q^{\frac{1}{2}}$ for $n \geq 1$, we see from Theorem 3.3 that for $\sigma = 0, 1$, the sequences $\{U_{2n+\sigma}(z)\}_{n=1}^{\infty}$ and $\{V_{2n+\sigma}(z)\}_{n=1}^{\infty}$ converge uniformly on compact subsets of the punctured plane $D := \{z \in C : z \neq 0\}$ to holomorphic functions $U^{(\sigma)}(z)$ and $V^{(\sigma)}(z)$, respectively.

In [9] it was shown that for $\sigma = 0, 1$ the sequences $\{\widetilde{V}_{2n+\sigma}(z)\}$ and $\{\widetilde{U}_{2n+\sigma}(z)\}$, defined for $n = 1, 2, \ldots$, by

$$\widetilde{V}_{2n+\sigma}(z) := \frac{\gamma_{2n+\sigma} Q_{2n+\sigma}(-z)}{z^{n+\sigma}} \quad \text{and} \quad \widetilde{U}_{2n+\sigma}(z) := \frac{\gamma_{2n+\sigma} P_{2n+\sigma}(-z)}{z^{n+\sigma}}, \tag{6.6}$$

where $\gamma_{2n+\sigma} := (-1)^{n+\sigma} q^{n^2 - \frac{n}{2} + \sigma(n - \frac{1}{2})} \prod_{j=1}^{\infty}(1 - q^j)$, $n = 1, 2, \ldots$, $\sigma = 0, 1$, converge uniformly on compact subsets of $\frac{1}{R} < |z| < R$, $R > 1$ to $\widetilde{V}^{(\sigma)}(z)$ and $\widetilde{U}^{(\sigma)}(z)$, respectively, and that

$$\widetilde{V}^{(\sigma)}(z) := \lim_{n\to\infty} \widetilde{V}_{2n+\sigma}(z) = \sum_{j=-\infty}^{\infty} q^{j^2}(-q^{\sigma+\frac{1}{2}} z)^j \tag{6.7a}$$

and

$$\widetilde{U}^{(\sigma)}(z) := \lim_{n\to\infty} \widetilde{U}_{2n+\sigma}(z) = \sum_{j=-\infty}^{\infty} q^{j^2} \left( \sum_{k=0}^{\infty} (-1)^k q^{\frac{k^2}{2} + (2j+\sigma-\frac{1}{2})k} \right) (-q^{\sigma+\frac{1}{2}} z)^j. \tag{6.7b}$$

The proof of the following theorem can be found in [2].

THEOREM 6.1 For $\sigma = 0, 1$, the zeros of $\widetilde{V}^{(\sigma)}(z)$ are $q^{2p - \frac{3}{2} + \sigma}$, $p = 0, \pm 1, \pm 2, \ldots$.

Combining equations (3.4) and (6.6) we see that for $\sigma = 0, 1$ and $n = 1, 2, \ldots$,

$$U_{2n+\sigma}(z) = (-1)^{n+\sigma} \frac{\left(\prod_{k=1}^{2n+\sigma} e_k\right) \widetilde{U}_{2n+\sigma}(-z)}{\gamma_{2n+\sigma}}, \tag{6.8a}$$

$$V_{2n+\sigma}(z) = (-1)^{n+\sigma} \frac{\left(\prod_{k=1}^{2n+\sigma} e_k\right) \widetilde{V}_{2n+\sigma}(-z)}{\gamma_{2n+\sigma}}, \quad n = 1, 2, \ldots \tag{6.8b}$$

Moreover, it follows from (6.5) and (6.6) that

$$(-1)^{n+\sigma} \frac{\prod_{k=1}^{2n+\sigma} e_k}{\gamma_{2n+\sigma}} = \frac{1}{\prod_{k=1}^{n}(1 - q^{2k-1+\sigma}) \prod_{j=1}^{\infty}(1 - q^j)}, \quad \sigma = 0, 1. \tag{6.9}$$

But since $0 < q < 1$ we have from (6.8) and (6.9)

$$U^{(\sigma)}(z) = \lim_{n\to\infty} (-1)^{n+\sigma} \frac{\left(\prod_{k=1}^{2n+\sigma} e_k\right)}{\gamma_{2n+\sigma}} \widetilde{U}_{2n+\sigma}(-z) = \frac{\widetilde{U}^{(\sigma)}(-z)}{\prod_{k=1}^{\infty}(1-q^{2k-1+\sigma})\prod_{j=1}^{\infty}(1-q^j)} \tag{6.10a}$$

and

$$V^{(\sigma)}(z) = \lim_{n\to\infty} (-1)^{n+\sigma} \frac{\left(\prod_{k=1}^{2n+\sigma} e_k\right)}{\gamma_{2n+\sigma}} \widetilde{V}_{2n+\sigma}(-z) = \frac{\widetilde{V}^{(\sigma)}(-z)}{\prod_{k=1}^{\infty}(1-q^{2k-1+\sigma})\prod_{j=1}^{\infty}(1-q^j)} \tag{6.10b}$$

Hence by Theorem 6.1 and equations (6.10) we have the following result:

THEOREM 6.2 For $\sigma = 0, 1$, the zeros of $V^{(\sigma)}(z)$ are $-q^{2p-\frac{3}{2}+\sigma}$ where $p = 0, \pm1, \pm2, \ldots$. That is, using the notation of this paper, $x_p^{(\sigma)} = -q^{2p-\frac{3}{2}+\sigma}$, $p = 0, \pm1, \pm2, \ldots$.

THEOREM 6.3 Let $\psi^{(\sigma)}(t)$, $\sigma = 0, 1$, be defined by (5.4) where $a_p^{(\sigma)}$ is given by (5.2), $U^{(\sigma)}(z)$ and $V^{(\sigma)}(z)$ are given by (6.10) and $x_p^{(\sigma)} = -q^{2p-\frac{3}{2}+\sigma}$, $p = 0, \pm1, \pm2, \ldots$. Then $\psi^{(\sigma)}(t)$ is a solution to the SSMP for the log-normal moments (6.2) and $\psi^{(\sigma)}(t)$ is a step function with jump

$$a_p^{(\sigma)} = \frac{U^{(\sigma)}(-q^{2p-\frac{3}{2}+\sigma})}{-q^{2p-\frac{3}{2}+\sigma}V^{(\sigma)\prime}(-q^{2p-\frac{3}{2}+\sigma})} = \frac{\widetilde{U}^{(\sigma)}(q^{2p-\frac{3}{2}+\sigma})}{q^{2p-\frac{3}{2}+\sigma}\widetilde{V}^{(\sigma)\prime}(q^{2p-\frac{3}{2}+\sigma})}$$
$$= \frac{\sum_{j=-\infty}^{\infty}\left(\sum_{k=0}^{\infty}(-1)^k q^{\frac{k^2}{2}+(2j+\sigma-\frac{1}{2})k}\right)(-q^{2\sigma+2p-1+j})^j}{\sum_{j=-\infty}^{\infty} j(-q^{2\sigma+2p-1+j})^j}$$

at $t = q^{2p-\frac{3}{2}+\sigma}$.

Proof: Apply Theorem 5.3 and use equations (6.10) and (6.7).

REMARK: For each $r \in \mathcal{Z}$, results analogous to Theorem 6.3 are given in [2] for the moments $c_k^{(r)}$, $k = 0, \pm1, \pm2, \ldots$, associated with $\varphi_r'(t) := t^r\varphi'(t)$ and defined by

$$c_k^{(r)} := \int_0^\infty (-t)^k \varphi_r'(t)dt = \int_0^\infty (-1)^r(-t)^{k+r}\varphi'(t)dt = (-1)^r c_{k+r}.$$

## REFERENCES


1. R. Askey, Beta integrals and $q$-extensions, in *Proceedings, Ramanujan Centennial International Conference*, Anuamalainagar, December 15–18, 1987.
2. C. Bonan-Hamada, W. B. Jones, A. Magnus and W. J. Thron, Discrete distribution functions for log-normal moments, *Continued Fractions and Orthogonal Functions: Theory and Applications*, (S. Clement Cooper and W. J. Thron, eds.), Marcel Dekker, New York, 1994, 1–21.
3. T. S. Chihara, *Introduction to Orthogonal Polynomials*, Gordon, New York, 1978.
4. John B. Conway, *Functions of One Complex Variable*, Springer–Verlag, New York, 1973.
5. S. Clement Cooper, W. B. Jones and W. J. Thron, Orthogonal Laurent polynomials and continued fractions associated with log-normal distributions, *J. Comput. Appl. Math.* **32** (1990), 39–46.
6. S. Clement Cooper, W. B. Jones and W. J. Thron, Asymptotics of orthogonal Laurent polynomials for log-normal distributions, *Constr. Approx.* **8** (1992), 59–67.
7. S. Clement Cooper, W. B. Jones and W. J. Thron, Separate convergence for log-normal modified S-fractions, *Continued Fractions and Orthogonal Functions: Theory and Applications*, (S. Clement Cooper and W. J. Thron, eds.), Marcel Dekker, New York, 1994, 101–114.
8. Einar Hille, *Analytic Function Theory*, vol. I, Blaisdell, New York, 1959.
9. W. B. Jones, A. Magnus and W. J. Thron, PC-fractions and orthogonal Laurent polynomials for log-normal distribution, *J. Math. Anal. Appl.* **170** (1992), 225–244.
10. W. B. Jones and W. J. Thron, *Continued Fractions: Analytic Theory and Applications*, in Encyclopedia of Mathematics and Its Applications, Vol. 11, Addison–Wesley, Reading, Mass., 1980. Distributed now by the Cambridge University Press.
11. W. B. Jones, W. J. Thron and H. Waadeland, A strong Stieltjes moment problem, *Trans. Amer. Math. Soc.* **261** (1980), 503–528.
12. I. P. Natanson, *Theory of Functions of a Real Variable.* I, Ungar, New York, 1955.
13. T. J. Stieltjes, Recherches sur les fractions continues, *Ann. Fac. Sci. Toulouse* **8** (1894), J, 1–122; **9** (1894), A, 1–47; *Oeuvres*, Vol. 2, pp. 402–566. Also published in *Mémoires présentés par divers savants a l'Académie de sciences de l'Institut National de France*, Vol. 33, pp. 1–196.
14. W. J. Thron, *Introduction to the Theory of Functions of a Complex Variable*, Wiley, New York, 1953.

# Symmetric Orthogonal L-Polynomials in the Complex Plane

C. F. BRACCIALI Departamento de Ciências de Computação e Estatística, IBILCE, Universidade Estadual Paulista (UNESP), 15054-000 São José do Rio Preto, SP, Brazil.

J. M. V. CAPELA Departamento de Físico-Química, IQ, Universidade Estadual Paulista (UNESP), 14801-970 Araraquara, SP, Brazil.

A. SRI RANGA Departamento de Ciências de Computação e Estatística, IBILCE, Universidade Estadual Paulista (UNESP), 15054-000 São José do Rio Preto, SP, Brazil.

## 1 INTRODUCTION

In [7], we have shown the relation that exists between real symmetric orthogonal polynomials and certain inverse symmetric orthogonal L-polynomials. We obtain this relation through the transformation

$$x(u) = \left(\sqrt{u} - \beta/\sqrt{u}\right)/\left(2\sqrt{\alpha}\right), \quad \alpha > 0, \quad \beta > 0, \tag{1.1}$$

which represents a one to one correspondence between $\Gamma_1 = \{x:\ -\infty < x < \infty\}$ and $\Lambda_1 = \{u:\ 0 < u < \infty\}$. The principal results given in [7] can be summarized as follow:

For any $const > 0$, let

$$\xi(u) = const\ u^{-1/2}\,\zeta(x(u)).$$

Then $\zeta(x)$ is a symmetric weight function over $(-d, d)$ if, and only if, $\xi(u)$ is a strong weight function over $(\beta^2/b, b)$ such that

$$\sqrt{u}\ \xi(u) = \sqrt{\beta^2/u}\ \xi(\beta^2/u). \tag{1.2}$$

---

This research was supported in part by CNPq and FAPESP of Brazil.

The relation between $b$ and $d$ is $\sqrt{b} = \sqrt{\alpha d^2 + \beta} + \sqrt{\alpha}\, d$. Further, in this case,

$$\tilde{B}_n(\xi; u) = \left\{2\sqrt{\alpha u}\right\}^n Q_n(\zeta; x), \qquad n \geq 0.$$

Here, $Q_n(\zeta; x)$ is the monic symmetric orthogonal polynomial of degree $n$ associated with $\zeta(x)$ and $\tilde{B}_n(\xi; u)$ is the monic polynomial of degree $n$ defined by

$$\int_{\beta^2/b}^{b} u^{-n+s}\, \tilde{B}_n(\xi; u)\, \xi(u) du = 0, \qquad 0 \leq s \leq n-1. \tag{1.3}$$

Since $\xi(u)$ satisfies (1.2), the polynomials $\tilde{B}_n(\xi; u)$, $n \geq 0$, satisfies the inverse symmetric property (see [6])

$$\tilde{B}_n(\xi; u) = u^n\, \tilde{B}_n(\xi; \beta^2/u)/(-\beta)^n, \qquad n \geq 0.$$

Polynomials defined as in (1.3) are closely related to orthogonal Laurent polynomials. For more information about polynomials defined this way and the associated orthogonal Laurent polynomials see for example [4], [5], [8] and [9]. Here, we will refer to the polynomials $\tilde{B}_n(\xi; u)$ simply as orthogonal L-polynomials.

In this article we extend the transformation (1.1) into the complex domains $Z^{(\alpha,\beta)}$ and $W$, where

$$Z^{(\alpha,\beta)} = \left\{z\text{: } z \in \mathbb{C}, \text{ where if } \operatorname{Re}(z) = 0 \text{ then } |\operatorname{Im}(z)| < \sqrt{\beta/\alpha}\right\},$$

and

$$W = \left\{w\text{: } w \in \mathbb{C}, \text{ where } |\arg(w)| < \pi \text{ and } w \neq 0\right\},$$

and obtain some information regarding orthogonal L-polynomials over certain interesting contours in these domains. Throughout this article we assume $\beta > 0$ and $\alpha > 0$.

We write the extended transformation by

$$z(w) = \frac{1}{2\sqrt{\alpha}}\left\{\sqrt{w} - \beta/\sqrt{w}\right\}. \tag{1.4}$$

The inverse of this transformation can be given as

$$w(z) = \alpha\left\{\sqrt{z^2 + \beta/\alpha} + z\right\}^2.$$

Observation: For $f(z)$, not lying on the negative real axis, we take $\sqrt{f(z)}$ as the root that satisfy $|\arg\{\sqrt{f(z)}\}| < \pi/2$.

## 2 SYMMETRIC CONTOURS

Let $w = u + iv = re^{i\theta} \in W$ and $z = x + iy \in Z^{(\alpha,\beta)}$. Then from (1.4)

$$x = \frac{1}{2\sqrt{\alpha}}\left\{\sqrt{r} - \beta/\sqrt{r}\right\}\cos(\theta/2), \qquad y = \frac{1}{2\sqrt{\alpha}}\left\{\sqrt{r} + \beta/\sqrt{r}\right\}\sin(\theta/2). \tag{2.1}$$

Since $-\pi < \theta < \pi$, we obtain from (2.1) that $y = 0$ *iff* $\theta = 0$. This situation represents the transformation (1.1), where $\Gamma_1$ in $Z^{(\alpha,\beta)}$ is mapped onto $\Lambda_1$ in $W$.

Since $-\pi < \theta < \pi$, we also have that $\cos(\theta/2) > 0$. Hence from (2.1) it follows that $x = 0$ *iff* $r = \beta$, and that for $r = \beta$, $y = \sqrt{\beta/\alpha}\,\sin(\theta/2)$. Here, as $\theta$ varies from $-\pi$ to $\pi$, $y$ varies from $-\sqrt{\beta/\alpha}$ to $\sqrt{\beta/\alpha}$. Hence, if we define

$$\Gamma_2 = \left\{z\colon\ z = iy,\ \text{ where } -\sqrt{\beta/\alpha} < y < \sqrt{\beta/\alpha}\right\}$$

and

$$\Lambda_2 = \left\{w\colon\ w = \beta e^{i\theta},\ \text{ where } -\pi < \theta < \pi\right\},$$

then we have

THEOREM 2.1 The transformation (1.4) gives a one to one correspondence between the "linear" contour $\Gamma_2$ in $Z^{(\alpha,\beta)}$ and the "circular" contour $\Lambda_2$ in $W$.

The contour $\Gamma_2$, like $\Gamma_1$, is symmetric about the origin. Any contour $\Gamma$ is said to be symmetric about the origin, or simply symmetric, if $z \in \Gamma$ then so is $-z \in \Gamma$.

We say that any contour $\Lambda$ is $\beta$-inverse symmetric, if $w \in \Lambda$ implies $\beta^2/w \in \Lambda$. The contours $\Lambda_1$ and $\Lambda_2$ are therefore $\beta$-inverse symmetric contours.

It is clear that the transformation (1.4) maps a symmetric contour $\Gamma$ in $Z^{(\alpha,\beta)}$ onto a $\beta$-inverse symmetric contour $\Lambda$ in $W$.

For $0 \le \tau < \pi/2$, we now consider the symmetric contours in $Z^{(\alpha,\beta)}$, given by

$$\Gamma(\tau, d) = \left\{z\colon\ z = e^{i\tau}\rho,\ \text{ where } -d < \rho < d \le \infty\right\}.$$

If $d \le \sqrt{\beta/\alpha}$ then we may also choose $\tau = \pi/2$.

Observation: If we write $z = \rho e^{i\tau}$ then by convention $\rho$ must be taken to be positive. To avoid this, we write here $z = e^{i\tau}\rho$.

The contour $\Gamma(\tau, \infty)$ represents the line that passes through the origin with an angle $\tau$. For $z \in \Gamma(\tau, d)$, we have from (2.1) that $\tan(\tau) = y/x = \{(r+\beta)/(r-\beta)\}\tan(\theta/2)$. If we write $\Lambda(\tau, \beta, d\sqrt{\alpha/\beta})$ as the image of $\Gamma(\tau, d)$ through the transformation (1.4) then the contour $\Lambda(\tau, \beta, \infty)$ in $W$ is given by the equation

$$\tan(\theta/2) = \frac{r-\beta}{r+\beta}\tan(\tau). \tag{2.2}$$

If $w = u + iv \in \Lambda(\tau, \beta, \infty)$, then (2.2) can be given parametrically in terms of $r$ as

$$\left.\begin{aligned} u = u(r) &= r\frac{(r+\beta)^2 - \tan^2(\tau)(r-\beta)^2}{(r+\beta)^2 + \tan^2(\tau)(r-\beta)^2}, \\ v = v(r) &= r\frac{2(r+\beta)(r-\beta)\tan(\tau)}{(r+\beta)^2 + \tan^2(\tau)(r-\beta)^2}, \end{aligned}\right\} \quad 0 < r < \infty. \tag{2.3}$$

To obtain the contour $\Lambda(\tau, \beta, d\sqrt{\alpha/\beta})$, we must also include

$$\frac{1}{4\alpha}(\sqrt{r} - \beta/\sqrt{r})^2\cos^2(\theta/2) + \frac{1}{4\alpha}(\sqrt{r} + \beta/\sqrt{r})^2\sin^2(\theta/2) < d^2, \tag{2.4}$$

and hence

THEOREM 2.2 The transformation (1.4) gives a one to one correspondence between the contour $\Gamma(\tau, d)$ in $Z^{(\alpha,\beta)}$ and the contour $\Lambda(\tau, \beta, d\sqrt{\alpha/\beta})$ in $W$ given by (2.3) with $r_1(\tau, d) < r < r_2(\tau, d)$, where $r_1(\tau, d)$ and $r_2(\tau, d)$ are the two positive solutions, in terms of $r$, of

$$\frac{(r+\beta)^2\,(r-\beta)^2}{(r+\beta)^2 + (r-\beta)^2\tan^2(\tau)} = \frac{4\alpha d^2 r}{1+\tan^2(\tau)}.$$

If we take $d = \sqrt{\beta/\alpha}$ then (2.4) reduces to $(r-\beta)^2 < 4\beta r\cos^2(\theta/2)$. We thus obtain

COROLLARY 2.1 The transformation (1.4) gives a one to one correspondence between the contour $\Gamma(\tau, \sqrt{\beta/\alpha})$ in $Z^{(\alpha,\beta)}$ and the contour $\Lambda(\tau, \beta, 1)$ in $W$ given by (2.3) with $r_1(\tau, \sqrt{\beta/\alpha}) < r < r_2(\tau, \sqrt{\beta/\alpha})$, where $r_1(\tau, \sqrt{\beta/\alpha})$ and $r_2(\tau, \sqrt{\beta/\alpha})$ are the two positive solutions of

$$\frac{(r+\beta)^2}{(r+\beta)^2 + (r-\beta)^2\tan^2(\tau)} = \frac{(r-\beta)^2}{4\beta r}.$$

Note that $\Lambda(\tau, \beta, 1)$ is independent of $\alpha$. For example, the length of the linear contour $\Lambda(0, \beta, 1)$ is $\left\{(\sqrt{2}+1)^2\beta\right\}-\left\{(\sqrt{2}-1)^2\beta\right\}=4\sqrt{2}\beta$. Note also that, $\Gamma_2$ and $\Lambda_2$ can be considered as the respective limits of $\Gamma(\tau, \sqrt{\beta/\alpha})$ and $\Lambda(\tau, \beta, 1)$, as $\tau \to \pi/2$. We simply write $\Lambda(\pi/2, \beta, 1) = \Lambda_2$.

The contours $\Lambda(\tau, 1, d\sqrt{\alpha/\beta})$ associated with four values of $d$, specifically, $d = \infty$, $d = 1.01\sqrt{\beta/\alpha}$, $d = \sqrt{\beta/\alpha}$ and $d = 0.99\sqrt{\beta/\alpha}$, are shown in figures 1, 2, 3 and 4, respectively. In each of these figures, the contours are sketched for $\tau = \tan^{-1}(0.5)$, $\tau = \tan^{-1}(-6)$ and $\tau = \tan^{-1}(40)$.

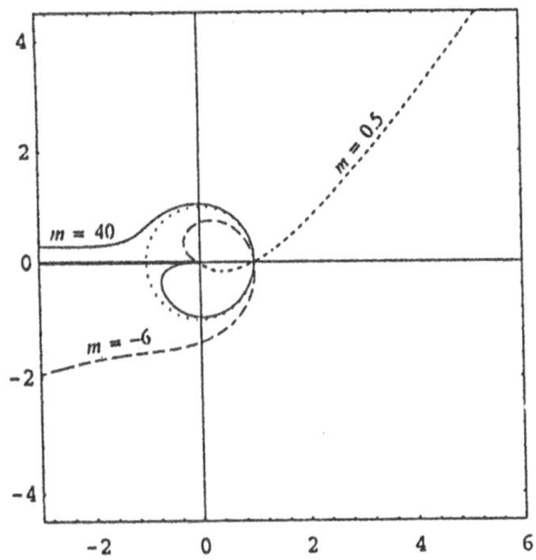


Figure 1. The contours $\Lambda(\tau, 1, \infty)$ for three values of $m = \tan(\tau)$.

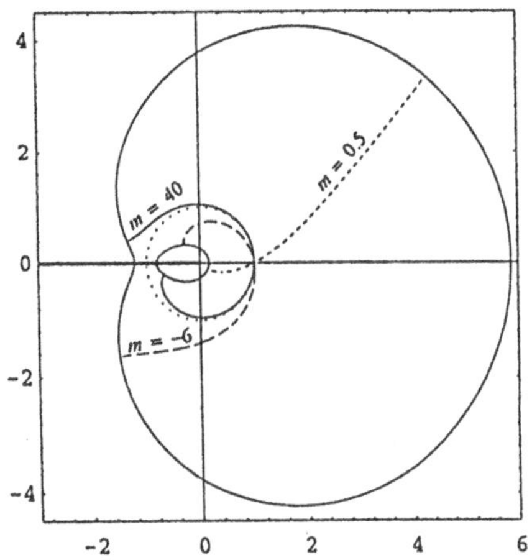


Figure 2. The contours $\Lambda(\tau, 1, 1.01)$ for three values of $m = \tan(\tau)$.

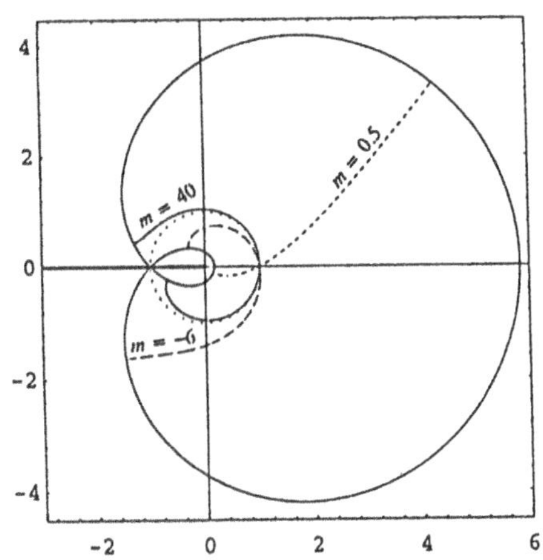


Figure 3. The contours $\Lambda(\tau, 1, 1)$ for three values of $m = \tan(\tau)$.

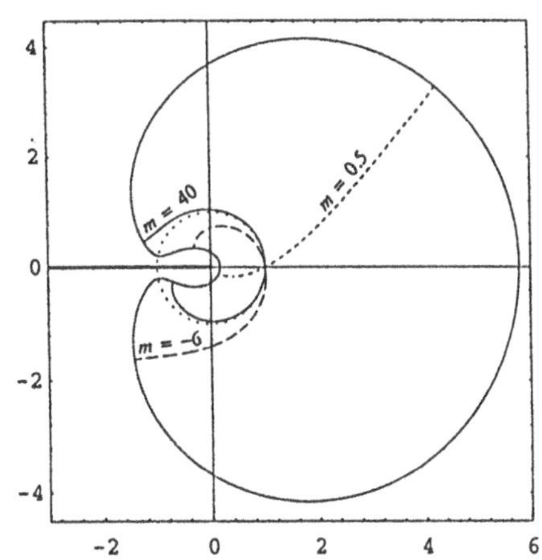


Figure 4. The contours $\Lambda(\tau, 1, 0.99)$ for three values of $m = \tan(\tau)$.

## 3 ORTHOGONAL POLYNOMIALS AND L-POLYNOMIALS

Let $\Gamma$ be a symmetric contour in $Z^{(\alpha,\beta)}$. Let the complex valued function $\zeta(z)$, defined on $\Gamma$, be such that

- $$\zeta(z) = \zeta(-z) \quad \text{for} \quad z \in \Gamma. \tag{3.1}$$

- The moments $\mu_m^\zeta = \int_\Gamma z^m \zeta(z)dz \in \mathbb{C}$, $m = 0,1,2,\cdots$, all exist.

- The sequence of monic polynomials $\{Q_n(\zeta; z)\}$, defined by
$$\int_\Gamma z^s Q_n(\zeta; z)\zeta(z)dz = 0, \qquad 0 \le s \le n-1,$$
where $Q_n(\zeta; z)$ is of degree *n*, exists.

Under these conditions we refer to $\zeta(z)$ as a symmetric weight function in $\Gamma$. Following Chihara[1], we refer to the polynomials $Q_n(\zeta; z)$, $n \ge 0$, as orthogonal polynomials. However, this concept of orthogonal polynomials is some what different from the standard concept of orthogonal polynomials in a complex variable (see for example [10]). For $\zeta(z)$ the following results hold:

If $f(z)$, a function defined over $\Gamma$ and integrable with respect to $\zeta(z)$, is such that $f(-z) = -f(z)$, then $\int_\Gamma f(z)\zeta(z)dz = 0$. Consequently, $\mu_{2m+1}^\zeta = 0$, $m = 0,1,2,\cdots$. Furthermore, the orthogonal polynomials satisfy

$$Q_n(\zeta; -z) = (-1)^n Q_n(\zeta; z), \quad n \ge 0$$

and

$$Q_{n+1}(\zeta; z) = zQ_n(\zeta; z) - \alpha_{n+1}^\zeta Q_{n-1}(\zeta; z), \quad n \ge 1, \tag{3.2}$$

where
$$\alpha_{n+1}^\zeta = \frac{\int_\Gamma z^n Q_n(\zeta; z)\zeta(z)dz}{\int_\Gamma z^{n-1} Q_{n-1}(\zeta; z)\zeta(z)dz} \in \mathbb{C}, \quad n \ge 1.$$

We also have the following result, which will be used later.

THEOREM 3.1 Let $\zeta(z)$ be a symmetric weight function defined over a symmetric contour $\Gamma$ in $Z^{(\alpha,\beta)}$. For $n \ge 1$, let $P_n(z)$ be a monic polynomial of degree *n* such that $P_n(-z) = (-1)^n P_n(z)$. Then $P_n(z) = Q_n(\zeta; z)$ if, and only if

$$\int_\Gamma \left\{\sqrt{\alpha z^2 + \beta} + \sqrt{\alpha}\, z\right\}^{-(n-1)+2s} \frac{P_n(z)}{\sqrt{\alpha z^2 + \beta}} \zeta(z)dz = 0, \qquad 0 \le s \le n-1.$$

The proof of this result (see [7] for the real case) follows from the Binomial expansion of $\{\sqrt{\alpha z^2+\beta}+\sqrt{\alpha}z\}^{-(n-1)+2s}$ and the symmetry of $P_n(z)$.

Now, let $\Lambda$ be a $\beta$-inverse symmetric contour in $W$. Let the complex valued function $\xi(w)$, defined on $\Lambda$, be such that

- $$\sqrt{w}\,\xi(w) = \sqrt{\beta^2/w}\,\zeta(\beta^2/w) \quad \text{for} \quad w\in\Lambda. \tag{3.3}$$

- The moments $\mu_m^\xi = \int_\Lambda w^m \xi(w)dw \in \mathbb{C}$, $m = 0, \pm 1, \pm 2, \cdots$, all exist.

- The monic polynomials $\tilde{B}_n(\xi; w)$, $n \geq 0$, defined by

$$\int_\Lambda w^{-n+s}\tilde{B}_n(\xi; w)\xi(w)dw = 0, \qquad 0 \leq s \leq n-1,$$

where $\tilde{B}_n(\xi; w)$ is of degree $n$, all exist.

We refer to $\xi(w)$ as a strong inverse symmetric weight function about the point $\beta$. For this weight function the following results hold:

$$\mu_m^\xi = \beta^{2m+1}\mu_{-m-1}^\xi.$$

In general, if $f(w)$ is integrable with respect to $\xi(w)$ on $\Lambda$, then

$$\int_\Lambda f(w)\xi(w)dw = \int_\Lambda (\beta/w)f(\beta^2/w)\xi(w)dw.$$

The polynomials $\tilde{B}_n(\xi; w)$, $n \geq 0$, satisfy

$$\tilde{B}_n(\xi; w) = w^n\,\tilde{B}_n(\xi; \beta^2/w)\,/(-\beta)^n, \quad n \geq 0$$

and

$$\tilde{B}_{n+1}(\xi; w) = (w-\beta)\,\tilde{B}_n(\xi; w) - \tilde{\alpha}_{n+1}^\xi\, w\,\tilde{B}_{n-1}(\xi; w), \quad n \geq 1, \tag{3.4}$$

where $$\tilde{\alpha}_{n+1}^\xi = \frac{\int_\Lambda \tilde{B}_n(\xi; w)\xi(w)dw}{\int_\Lambda \tilde{B}_{n-1}(\xi; w)\xi(w)dw} \in \mathbb{C}, \quad n \geq 1.$$

The Laurent polynomials given by

$$R_{2m}(w) = w^{-m}\tilde{B}_{2m}(\xi; w), \quad R_{2m+1}(w) = w^{-m-1}\tilde{B}_{2m+1}(\xi; w), \quad m \geq 0,$$

form a sequence of orthogonal Laurent polynomials on $\Lambda$ in relation to the weight function $\xi(w)$. Such complex orthogonal Laurent polynomials are treated, for example, in [2] and [3].

THEOREM 3.2 Let the symmetric contour $\Gamma$ in $Z^{(\alpha,\beta)}$ and the β-inverse symmetric contour $\Lambda$ in $W$ be images of each other through the transformation (1.4). For any $const \in \mathbb{C}$, let

$$\xi(w) = const\ w^{-1/2}\zeta(z(w)).$$

Then $\zeta(z)$ is a symmetric weight function on $\Gamma$ if, and only if, $\xi(w)$ is a strong inverse symmetric weight function about the point $\beta$ on $\Lambda$. Furthermore,

$$\widetilde{B}_n(\xi; w) = \left\{2\sqrt{\alpha w}\right\}^n Q_n(\zeta; z(w)), \quad n \geq 0. \tag{3.5}$$

and

$$\widetilde{\alpha}_{n+1}^{\xi} = 4\alpha\, \alpha_{n+1}^{\zeta}, \quad n \geq 1.$$

Proof: It is clear that $\zeta(z)$ is defined on $\Gamma$ *iff* $\xi(w)$ is defined on $\Lambda$. Now, one needs to show the following two results to establish the first result of the theorem.
(1) If $\zeta(z)$ satisfies (3.1) then $\xi(w)$ satisfies (3.3), and vice versa.
(2) If the moments of $\zeta(z)$ exist then so do the moments (including negative ones) of $\xi(w)$, and vice versa.
These results can be obtained in the same way as in [7] for the real case.
We now consider the relation $B_n(w) = \{2\sqrt{\alpha w}\}^n Q_n(z(w))$. Here, one has that $Q_n(z)$ is a monic polynomial of degree *n* such that $Q_n(z) = (-1)^n Q_n(-z)$ *iff* $B_n(w)$ is a monic polynomial of degree *n* such that $B_n(w) = w^n B_n(\beta^2/w)/(-\beta)^n$. Hence, the result (3.5) of the theorem follows from theorem 3.1 with the substitution (1.4).
The last result of this theorem is obtained from (3.5) and the two recurrence relations (3.2) and (3.4).

## 4 A SPECIAL CASE

Let $\zeta^{(0)}(z)$ be a symmetric weight function over the symmetric contour $\Gamma(0,d) = \{z\colon z = x,\ -d < x < d \leq \infty\}$ in $Z^{(\alpha,\beta)}$. Let $\zeta^{(0)}(z)$ also be positive (i.e., non-negative, but not identically zero) for $z \in \Gamma(0,d)$. Then it is well known (see [1], [10]) that the associated orthogonal polynomials $Q_n(\zeta^{(0)}; z)$, $n \geq 0$, exist and satisfy

$$Q_{n+1}(\zeta^{(0)}; z) = z Q_n(\zeta^{(0)}; z) - \alpha_{n+1}^{\zeta^{(0)}} Q_{n-1}(\zeta^{(0)}; z), \quad n \geq 1, \tag{4.1}$$

with $Q_0(\zeta^{(0)}; z) = 1$, $Q_1(\zeta^{(0)}; z) = z$ and $\alpha_{n+1}^{\zeta^{(0)}} > 0$, $n \geq 1$.

Substituting $z$ by $ze^{-i\tau}$, where we assume $-\pi/2 < \tau < \pi/2$, we obtain for the polynomials $Q_n(\zeta^{(\tau)}; z) = e^{in\tau} Q_n(\zeta^{(0)}; ze^{-i\tau})$, $n \geq 0$, the following results.

$$\int_{\Gamma(\tau,d)} z^s Q_n(\zeta^{(\tau)};z)\,\zeta^{(\tau)}(z)dz = 0, \qquad 0 \le s \le n-1,$$

where $\zeta^{(\tau)}(z) = \zeta^{(0)}(ze^{-i\tau})$, and

$$Q_{n+1}(\zeta^{(\tau)};z) = zQ_n(\zeta^{(\tau)};z) - \alpha_{n+1}^{\zeta^{(\tau)}} Q_{n-1}(\zeta^{(\tau)};z), \quad n \ge 1, \tag{4.2}$$

with $Q_0(\zeta^{(\tau)};z)=1$, $Q_1(\zeta^{(\tau)};z)=z$ and $\alpha_{n+1}^{\zeta^{(\tau)}} = \alpha_{n+1}^{\zeta^{(0)}} e^{i2\tau}$, $n \ge 1$.

Now, if we let

$$\xi^{(\tau)}(w) = const\ w^{-1/2}\zeta^{(\tau)}(z(w)), \tag{4.3}$$

then we get from the theorems 2.2 and 3.2 that $\xi^{(\tau)}(w)$ is a strong inverse symmetric weight function about the point $\beta$ defined over $\Lambda(\tau,\beta,d\sqrt{\alpha/\beta})$ and that

$$\tilde{B}_n(\xi^{(\tau)};w) = \left\{2\sqrt{\alpha w}\right\}^n Q_n(\zeta^{(\tau)};z(w)), \quad n \ge 0$$

and

$$\tilde{B}_{n+1}(\xi^{(\tau)};w) = (w-\beta)\tilde{B}_n(\xi^{(\tau)};w) - \tilde{\alpha}_{n+1}^{\xi^{(\tau)}}\, w\tilde{B}_{n-1}(\xi^{(\tau)};w), \quad n \ge 1, \tag{4.4}$$

where $\tilde{\alpha}_{n+1}^{\xi^{(\tau)}} = 4\alpha_{n+1}^{\zeta^{(\tau)}}\alpha$, $n \ge 1$. In the case of $d < \infty$, we may choose $\alpha$ such that $d \le \sqrt{\beta/\alpha}$ and then $\tau$ can also take the value $\pi/2$.

From (4.2) and (4.4) we have that $\tilde{\alpha}_{n+1}^{\zeta^{(\tau)}} = \tilde{\alpha}_{n+1}^{\zeta^{(0)}} e^{i2\tau}$. From this result one can make the following observation. Given the recurrence relation

$$B_{n+1}(w) = (w-\beta)B_n(w) - \tilde{\alpha}_{n+1}\, wB_{n-1}(w), \quad n \ge 1,$$

with $B_0(w)=1$ and $B_1(w)=w-\beta$, if $\tilde{\alpha}_{n+1} > 0$ for $n \ge 1$ then it is known that the zeros of $B_n(w)$, $n \ge 1$, are inside an interval $(\beta^2/b, b) \subseteq (0,\infty)$. We can represent this interval by the contour $\Lambda(0,\beta,d\sqrt{\alpha/\beta})$, where $d = \left\{\sqrt{b} - \beta/\sqrt{b}\right\}/(2\sqrt{\alpha})$. Now if we substitute all the $\tilde{\alpha}_{n+1}$ by $\tilde{\alpha}_{n+1}e^{i2\tau}$, where $-\pi/2 < \tau < \pi/2$, then the zeros of $B_n(w)$, $n \ge 1$, lie in the contour $\Lambda(\tau,\beta,(\sqrt{b}-\beta/\sqrt{b})/(2\sqrt{\beta}))$. Here, $\tau$ can also take the value $\pi/2$ if $b \le (\sqrt{2}+1)^2\beta$.

**Example:** With $\lambda > -1$, we consider $\zeta^{(0)}(z) = (1-z^2)^\lambda$ defined over $\Gamma(0,1)$ in $Z^{(\beta,\beta)}$. The orthogonal polynomials $Q_n(\zeta^{(0)};z)$, $n \ge 0$, are the monic symmetric Jacobi polynomials. If $\lambda > -1/2$, then they are the Gegenbauer or ultraspherical polynomials. These polynomials satisfy the recurrence relation (4.1) with

$$\alpha_{n+1}^{\zeta^{(0)}} = \frac{n(n+2\lambda)}{(2n+2\lambda-1)(2n+2\lambda+1)}, \quad n \geq 1.$$

We immediately see that for $\zeta^{(\tau)}(z) = (1-z^2e^{-i2\tau})^{\lambda}$ on $\Gamma(\tau, 1)$, the associated monic orthogonal polynomials $Q_n(\zeta^{(\tau)}; z)$, $n \geq 0$, satisfy the recurrence relation (4.2) with $\alpha_{n+1}^{\zeta^{(\tau)}} = \alpha_{n+1}^{\zeta^{(0)}} e^{i2\tau}$, $n \geq 1$. Furthermore, from (4.3) and (4.4), for the strong weight function

$$\xi^{(\tau)}(w) = w^{-\lambda}\left\{e^{-i2\tau}(b^{(\tau)}-w)(w-\beta^2/b^{(\tau)})\right\}^{\lambda-1/2},$$

on $\Lambda(\tau, \beta, 1)$, where $b^{(\tau)} = \beta\left\{(1+2e^{i2\tau})+2e^{i2\tau}\sqrt{1+e^{-i2\tau}}\right\}$, the monic polynomials $\tilde{B}_n(\xi^{(\tau)}; w)$, $n \geq 0$, defined by

$$\int_{\Lambda(\tau,\beta,1)} w^{-n+s}\, \tilde{B}_n(\xi^{(\tau)}; w)\, \xi^{(\tau)}(w)dw = 0, \quad 0 \leq s \leq n-1,$$

satisfy the recurrence relation

$$\tilde{B}_{n+1}(\xi^{(\tau)}; w) = (w-\beta)\tilde{B}_n(\xi^{(\tau)}; w) - \tilde{\alpha}_{n+1}^{\xi^{(\tau)}}\, w\tilde{B}_{n-1}(\xi^{(\tau)}; w), \quad n \geq 1,$$

where

$$\tilde{\alpha}_{n+1}^{\xi^{(\tau)}} = 4\beta\frac{n(n+2\lambda)}{(2n+2\lambda-1)(2n+2\lambda+1)}e^{i2\tau}, \quad n \geq 1.$$

Since $\alpha = \beta$, (i.e., $\alpha$ is chosen such that $d = 1 = \sqrt{\beta/\alpha}$ ), $\Lambda(\pi/2, \beta, 1)$ is the circular contour $\Lambda_2$. The results associated with $\xi^{(\pi/2)}(w)$ on $\Lambda_2$ also follows from [3].

## REFERENCES


1. T.S. Chihara, An Introduction to Orthogonal Polynomials, *Mathematics and its Applications Series*, Gordon and Breach, New York (1978).
2. E. Hendriksen, A weight function for the associated Jacobi-Laurent polynomials, *J. Comput. Appl. Math.*, *33*: 171-180 (1990).
3. E. Hendriksen and H. van Rossum, Orthogonal Laurent polynomials, *Nederl. Akad. Wetensch. Proc. Ser. A*, *89*: 17-36, (1986).
4. W.B. Jones, O. Njåstad and W.J. Thron, Two-point Padé expansions for a family of analytic functions, *J. Comput. Appl. Math.*, *9*: 105-123, (1983).
5. W.B. Jones, W.J. Thron and H. Waadeland, A strong Stieltjes moment problem, *Trans. Amer. Math. Soc.*, *261*: 503-528, (1980).
6. A. Sri Ranga, Another quadrature rule of highest algebraic degree of precision, *Numerische Math.*, *68*: 283-294, (1994).

7. A. Sri Ranga, Symmetric orthogonal polynomials and the associated orthogonal L-polynomials, *Proc. Amer. Math. Soc.*, *123*: 3135-3141, (1995).
8. A. Sri Ranga, E.X.L. de Andrade, Zeros of polynomials which satisfy a certain three term recurrence relation, *Comm. in the Anal. Theory of Continued Fractions*, *1*: 61-65, (1992).
9. A. Sri Ranga, E.X.L. de Andrade and J.H. McCabe, Some consequences of symmetry in strong distributions, *J. Math. Anal. and Appl.*, *193*: 158-168, (1995).
10. G. Szegö, Orthogonal Polynomials, *Amer. Math. Soc. Colloq. Publ.* Vol.23, Amer. Math. Soc. New York, (1939), 3rd Edition (1967).

# Continued Fractions and Orthogonal Rational Functions

ADHEMAR BULTHEEL Departement Computerwetenschappen, Katholieke Universiteit Leuven, Heverlee, Belgium

PABLO GONZÁLEZ-VERA Departamento Analisis Matematico, Universidad de La Laguna, Canary Islands, Spain

ERIK HENDRIKSEN Faculteit Wiskunde en Informatica, Universiteit van Amsterdam, Amsterdam, The Netherlands

OLAV NJÅSTAD Department of Mathematical Sciences, Norwegian University of Science and Technology, Trondheim, Norway

Abstract: To a given modified $PC$-fraction with approximants $A_n/B_n$ there exists a unique pair $(L_0, L_\infty)$ of formal power series at 0 and $\infty$ such that $A_{2m}/B_{2m}$ is the weak $(m,m)$ two-point Padé approximant to $L_0$ of order $m+1$ and to $L_\infty$ of order $m$, while $A_{2m+1}/B_{2m+1}$ is the weak $(m,m)$ two-point Padé approximant to $L_0$ of order $m$ and to $L_\infty$ of order $m+1$. The canonical denominators of the even contraction of the modified $PC$-fraction (which is a modified $T$-fraction) are orthogonal Laurent polynomials obtained from the basis $\{1, z^{-1}, z, z^{-2}, z^2, \dots\}$, while the canonical denominators of the odd contractions of the modified $PC$-fraction (which is a modified $M$-fraction) are orthogonal Laurent polynomials obtained from the basis $\{1, z, z^{-1}, z^2, z^{-2}, \dots\}$.

An analogous situation arises when the pair of power series $(L_0, L_\infty)$ is replaced by Newton series determined by a general interpolation table of points on the real line, the modified $PC$-fraction and its contractions are replaced by appropriate analogous continued fractions, and orthogonal Laurent polynomials are replaced by orthogonal rational functions with poles in the set of interpolation points. In this paper an investigation of these relationships is carried out.

## 1 INTRODUCTION

By a *modified PC-fraction (modified Perron-Carathéodory fraction)* we mean a continued fraction

$$\kappa_0 + K_{n=1}^{\infty} \frac{a_n}{b_n} \tag{1.1}$$

where

$$a_{4m} = 1, \quad b_{4m} = \mu_{4m}, \quad m = 1, 2, \dots \tag{1.2}$$

$$a_{4m+1} = \lambda_{4m+1}, \quad b_{4m+1} = \mu_{4m+1}, \quad m = 0, 1, 2, \dots \tag{1.3}$$

$$a_{4m+2} = \frac{1}{z}, \quad b_{4m+2} = \mu_{4m+2}, \quad m = 0, 1, 2, \ldots \tag{1.4}$$

$$a_{4m+3} = \lambda_{4m+3} z, \quad b_{4m+3} = \mu_{4m+3} z, \quad m = 0, 1, 2, \ldots \tag{1.5}$$

with $\lambda_n, \mu_n$ constants, $\mu_1 = 1$. The constants $\lambda_n, \mu_n$ are required to satisfy the relation

$$\lambda_{2n+1} = 1 - \mu_{2n}\mu_{2n+1}, \quad n = 1, 2, \ldots . \tag{1.6}$$

(A modified *PC*-fraction is related to a standard *PC*-fraction through a simple equivalence transformation. The coefficients $\lambda_n, \mu_n$ have usually been called $\alpha_n, \beta_n$, but we need especially the notation $\alpha_n$ for other purposes.)

Let $A_n$ and $B_n$ denote the canonical numerators and denominators of the modified *PC*-fraction (1.1). These functions are of the form

$$A_{4m} = \frac{\tilde{p}_{2m}}{z^m}, \quad B_{4m} = \frac{p_{2m}}{z^m}, \quad m = 1, 2, \ldots \tag{1.7}$$

$$A_{4m+1} = \frac{\tilde{q}_{2m}}{z^m}, \quad B_{4m+1} = \frac{q_{2m}}{z^m}, \quad m = 0, 1, 2, \ldots \tag{1.8}$$

$$A_{4m+2} = \frac{\tilde{r}_{2m+1}}{z^{m+1}}, \quad B_{4m+2} = \frac{r_{2m+1}}{z^{m+1}}, \quad m = 0, 1, 2, \ldots \tag{1.9}$$

$$A_{4m+3} = \frac{\tilde{s}_{2m+1}}{z^{m+1}}, \quad B_{4m+3} = \frac{s_{2m+1}}{z^{m+1}}, \quad m = 0, 1, 2, \ldots \tag{1.10}$$

where $p_n, q_n, r_n, s_n, \tilde{p}_n, \tilde{q}_n, \tilde{r}_n, \tilde{s}_n$ are polynomials of degree $n$.

The modified *PC*-fraction determines two formal power series

$$L_0 = c_0^{(0)} + \sum_{k=1}^{\infty} c_k z^k \tag{1.11}$$

$$L_\infty = -c_0^{(\infty)} - \sum_{k=1}^{\infty} c_{-k} z^{-k} \tag{1.12}$$

which are related to the continued fraction in the following way: The even approximants (partial fractions) $\frac{A_{2m}}{B_{2m}}$ are weak $(m, m)$ two-point Padé approximants for the pair $(L_0, L_\infty)$ of order $(m+1, m)$, and the odd approximants $\frac{A_{2m+1}}{B_{2m+1}}$ are weak $(m, m)$ two-point Padé approximants for the pair $(L_0, L_\infty)$ of order $(m, m+1)$. This means that

$$A_{2m} - L_0 B_{2m} = O(z^{m+1}) \quad \text{as} \quad z \to 0, \quad m = 1, 2, \ldots \tag{1.13}$$

$$A_{2m} - L_\infty B_{2m} = O(1) \quad \text{as} \quad z \to \infty, \quad m = 1, 2, \dots \tag{1.14}$$

$$A_{2m+1} - L_0 B_{2m+1} = O(z^m) \quad \text{as} \quad z \to 0, \quad m = 0, 1, 2, \dots \tag{1.15}$$

$$A_{2m+1} - L_\infty B_{2m+1} = O\left(\frac{1}{z}\right) \quad \text{as} \quad z \to \infty, \quad m = 0, 1, 2, \dots . \tag{1.16}$$

When $\mu_{2m} \neq 0$ for all $m$, the even contraction of (1.1) exists (with $\dfrac{A_{2m}}{B_{2m}}$ as approximants). This continued fraction

$$\kappa_0 + K_{n=1}^{\infty} \frac{\theta_n}{\kappa_n} \tag{1.17}$$

is a *modified T-fraction* with elements

$$\theta_{2m} = -F_{2m}, \quad \kappa_{2m} = 1 - G_{2m} z, \quad m = 1, 2, \dots \tag{1.18}$$

$$\theta_{2m+1} = -F_{2m+1}, \quad \kappa_{2m+1} = \frac{1}{z} - G_{2m+1}, \quad m = 0, 1, 2, \dots \tag{1.19}$$

with $F_n, G_n$ constants. The canonical denominators (which are the functions $B_{2m}$) are orthogonal Laurent polynomials obtained by orthogonalization of the basis $\{1, z^{-1}, z, z^{-2}, z^2, \dots\}$ with respect to a given measure or linear functional. (Cf. Section 2).

When $\mu_{2m+1} \neq 0$ for all $m$, the odd contraction of (1.1) exists (with $\dfrac{A_{2m+1}}{B_{2m+1}}$ as approximants). This continued fraction

$$\kappa_0' + K_{n=1}^{\infty} \frac{\rho_n}{\sigma_n} \tag{1.20}$$

is a *modified M-fraction* with elements

$$\rho_{2m} = -U_{2m}, \quad \sigma_{2m} = (1 - V_{2m} \cdot \frac{1}{z}), \quad m = 1, 2, \dots \tag{1.21}$$

$$\rho_{2m+1} = -U_{2m+1}, \quad \sigma_{2m+1} = (z - V_{2m+1}), \quad m = 0, 1, 2, \dots \tag{1.22}$$

with $U_n, V_n$ constants. The canonical denominators (which are the functions $B_{2m+1}$) are orthogonal Laurent polynomials obtained by orthogonalization of the basis $\{1, z, z^{-1}, z^2, z^{-2}, \dots\}$.

We refer to [1,2,3,16,17,18,19,20,22,23] for more details on two-point Padé approximants and related continued fractions. Especially relevant for the discussion in this section are [2,18,19].

An analogous situation arises when interpolation at $\{0, \infty\}$ is replaced by interpolation tables determined by a general sequence of basic points $\{\alpha_1, \alpha_2, \alpha_3, \ldots\}$ on the real line. It is known that regular orthogonal rational functions with poles in the set $\{\alpha_1, \alpha_2, \alpha_3, \ldots\}$ are canonical denominators of a continued fraction called an *MP*-fraction. And vice versa: The canonical denominators of an *MP*-fraction are orthogonal rational functions with respect to a functional. The theory of such orthogonal rational functions and the associated continued fractions is briefly reviewed in Section 2.

The main content of this paper is the following:

**I.** We show how to extend an *MP*-fraction to a continued fraction of a new type, an *EMP*-fraction, which has the *MP*-fraction as its even contraction. (Section 3).

**II.** We formally introduce *EMP*-fractions and show that the canonical numerators and denominators of an *EMP*-fraction has a structure analogous to that of the numerators and denominators of a modified *PC*-fraction. (Section 4).

**III.** We show that an *EMP*-fraction determines a Newton series $\Gamma$ associated with the interpolation table $\{\infty, 0, \alpha_1, \alpha_1, \alpha_2, \alpha_2, \alpha_3, \alpha_3, \ldots\}$ and a Newton series $\Delta$ associated with the interpolation table $\{\infty, 0, \alpha_2, \alpha_2, \alpha_1, \alpha_1, \alpha_4, \alpha_4, \ldots\}$, such that the even approximants of the *EMP*-fraction are multi-point Padé approximants to $\Gamma$ and the odd approximants are multi-point Padé approximants to $\Delta$. (Section 5.)

**IV.** We show that the odd contraction (when it exists) of an *EMP*-fraction is equivalent to an *MP*-fraction, whose canonical denominators are orthogonal rational functions associated with the sequence of basic points $\{\alpha_2, \alpha_1, \alpha_4, \alpha_3, \ldots\}$. (Section 6.)

## 2 MP-FRACTIONS

Let $\{\alpha_n : n = 1, 2, \ldots\}$ be a sequence of not necessarily distinct points on the (finite) real line $\boldsymbol{R}$. We call these points *basic points*.

We define the functions $\omega_n$ by

$$\omega_0 = 1, \quad \omega_n(z) = \prod_{k=1}^{n} (z - \alpha_k), \quad n = 1, 2, \ldots \tag{2.1}$$

and the spaces $\mathcal{L}_n$ and $\mathcal{L}$ by

$$\mathcal{L}_n = \text{Span}\left\{\frac{1}{\omega_0}, \frac{1}{\omega_1}, \ldots, \frac{1}{\omega_n}\right\}, \quad \mathcal{L} = \bigcup_{n=0}^{\infty} \mathcal{L}_n. \tag{2.2}$$

The elements of $\mathcal{L}_n$ are exactly the functions that can be written in the form

$$L(z) = \frac{p(z)}{\omega_n(z)}, \quad p \in \mathcal{P}_n \tag{2.3}$$

where $\mathcal{P}_n$ denotes the space of polynomials of degree at most $n$.

Let $M$ be a given linear functional on $\mathcal{L} \cdot \mathcal{L}$. We define the form $\langle , \rangle$ on $\mathcal{L} \times \mathcal{L}$ by

$$\langle K, L \rangle = M[K \cdot \bar{L}] \tag{2.4}$$

We assume that $M$ is real for all functions in $\mathcal{L} \cdot \mathcal{L}$ that are real on $\boldsymbol{R}$, and also that $M$ is definite, i.e., that

$$\langle L, L \rangle \neq 0 \quad \text{for} \quad L \in \mathcal{L}, \quad L \not\equiv 0. \tag{2.5}$$

Let $\{\Phi_n : n = 0, 1, 2, \ldots\}$ be the essentially unique orthogonal sequence obtained by the Gram-Schmidt process from the sequence $\left\{\frac{1}{\omega_n} : n = 0, 1, 2, \ldots\right\}$. Each $\Phi_n$ can be written in the form

$$\Phi_n(z) = \frac{K_n(z)}{\omega_n(z)} \tag{2.6}$$

where $K_n \in \mathcal{P}_n$. It follows from the construction that

$$K_n(\alpha_n) \neq 0, \quad n = 0, 1, 2, \ldots \tag{2.7}$$

The function $\Phi_n$ is called *regular* if $K_n(\alpha_{n-1}) \neq 0$.

*We shall in this and the next section assume that the sequence* $\{\Phi_n\}$ *is regular, i.e.,*

$$K_n(\alpha_{n-1}) \neq 0, \quad n = 1, 2, \ldots . \tag{2.8}$$

To avoid considering exceptional cases in some formulas, we shall use the following conventions:

$$z - \alpha_{-1} \quad \text{means} \quad z \tag{2.9}$$

$$z - \alpha_0 \quad \text{means} \quad 1. \tag{2.10}$$

The sequence $\{\Phi_n\}$ satisfies a recurrence relation of the form

$$\Phi_n(z) = \left(h_n + \frac{g_n}{z - \alpha_n}\right)\Phi_{n-1}(z) + F_n \frac{z - \alpha_{n-2}}{z - \alpha_n}\Phi_{n-2}(z), \quad n = 1, 2, \dots . \tag{2.11}$$

with initial conditions

$$\Phi_0 = 1, \quad \Phi_{-1} = 0. \tag{2.12}$$

The coefficients satisfy the inequalities

$$F_n \neq 0, \quad g_n + h_n(\alpha_{n-1} - \alpha_n) \neq 0, \quad n = 1, 2, \dots . \tag{2.13}$$

We note that the sum $h_n + \dfrac{g_n}{z - \alpha_n}$ may be written in various ways:

$$h_n + \frac{g_n}{z - \alpha_n} = \frac{x_n \zeta_n(z) + y_n \tau_n(z)}{z - \alpha_n}, \tag{2.14}$$

where $\{\zeta_n, \tau_n\}$ is an arbitrary basis for the space $\mathcal{P}_1$.

The functions $\Phi_n$ are canonical denominators of a continued fraction

$$\kappa_0 + K_{n=1}^{\infty} \frac{\theta_n}{\kappa_n} \tag{2.15}$$

where

$$\theta_n = F_n \frac{z - \alpha_{n-2}}{z - \alpha_n}, \quad n = 1, 2, \dots \tag{2.16}$$

$$\kappa_n = h_n + \frac{g_n}{z - \alpha_n}, \quad n = 1, 2, \dots \tag{2.17}$$

We denote by $\Psi_n$ the canonical numerators of the continued fraction, i.e., the solution of the three-term recurrence relation with elements $\theta_n$, $\kappa_n$ and initial values

$$\Psi_0 = \kappa_0, \quad \Psi_{-1} = 1, \quad \kappa_0 = M[1]^{-\frac{1}{2}}. \tag{2.18}$$

The functions $\Psi_n, \Phi_n$ thus satisfy the relation

$$\begin{bmatrix} \Psi_n \\ \Phi_n \end{bmatrix} = \kappa_n \begin{bmatrix} \Psi_{n-1} \\ \Phi_{n-1} \end{bmatrix} + \theta_n \begin{bmatrix} \Psi_{n-2} \\ \Phi_{n-2} \end{bmatrix}, \quad n = 1, 2, \ldots \tag{2.19}$$

with initial conditions

$$\begin{bmatrix} \Psi_0 \\ \Phi_0 \end{bmatrix} = \begin{bmatrix} \kappa_0 \\ 1 \end{bmatrix}, \quad \begin{bmatrix} \Psi_{-1} \\ \Phi_{-1} \end{bmatrix} = \begin{bmatrix} 1 \\ 0 \end{bmatrix}. \tag{2.20}$$

Continued fractions of the kind desribed above are called *MP-fractions (Multi-point Padé continued fractions).*

We have seen that a regular sequence of orthogonal functions in $\mathcal{L}$ determines an *MP*-fraction, the orthogonal functions being the canonical denominators of the continued fraction. On the other hand, the sequence of canonical denominators of an *MP*-fraction is a regular sequence of orthogonal functions in $\mathcal{L}$ determined by a suitable functional. Thus there is a one-to-one correspondence between functionals $M$ giving rise to regular sequences of orthogonal functions and *MP*-fractions. Properties tied to such a functional $M$ may then be considered as properties tied to the corresponding continued fraction, and vice versa.

The sequence of approximants $\left\{\frac{\Psi_n}{\Phi_n}\right\}$ determines a sequence of interpolation values with respect to the interpolation table $\{\infty, 0, \alpha_1, \alpha_1, \alpha_2, \alpha_2, \alpha_3, \alpha_3,$
$\ldots\}$. We shall give a detailed discussion of this relationship in the course of the discussions in Section 5.

For more information on orthogonal functions in the spaces $\mathcal{L}$, corresponding continued fractions, and related topics, we refer to [4, 5, 6, 7, 8, 9, 10, 13, 14,
15, 24, 25].

## 3 EXTENSION OF MP-FRACTIONS

We shall construct a continued fraction

$$\kappa_0 + K_{n=1}^{\infty} \frac{a_n}{b_n} \tag{3.1}$$

which has (2.15)-(2.17) as its even contraction and with the property that the approximants (partial fractions) represent multipoint Padé approximants to Newton series determined by the functional $M$ and the sequence $\{\alpha_n : n = 1, 2, \dots\}$ (or alternatively, by the continued fraction) according to certain patterns.

*We shall in this section assume that* $\alpha_{2m} \neq \alpha_{2m-1}, \quad m = 1, 2, \dots$ (3.2)

We note that this condition is in particular satisfied when $\{\alpha_n\}$ is obtained by cyclic repetition of a finite number of distinct points $\alpha_1, \dots, \alpha_p \quad (p > 1)$.

We make use of the following observation: When (3.2) is satisfied, $\{z - \alpha_{2m-1}, z - \alpha_{2m}\}$ is a base for $\mathcal{P}_1$. The elements (2.17) of (2.15) may therefore be represented in the form

$$\kappa_{2m} = H_{2m} + G_{2m} \frac{z - \alpha_{2m-1}}{z - \alpha_{2m}}, \quad m = 1, 2, \dots \tag{3.3}$$

$$\kappa_{2m+1} = H_{2m+1} \frac{z - \alpha_{2m}}{z - \alpha_{2m+1}} + G_{2m+1} \frac{z - \alpha_{2m-1}}{z - \alpha_{2m+1}}, \quad m = 0, 1, 2, \dots . \tag{3.4}$$

(If (3.2) is not satisfied, the continued fraction (2.15) with elements (2.16), (3.3)-(3.4) is not associated with a sequence of orthogonal rational functions.) Expressions for the new coefficients are

$$G_{2m} = \frac{g_{2m}}{\alpha_{2m} - \alpha_{2m-1}}, \quad m = 1, 2, \dots \tag{3.5}$$

$$H_{2m} = h_{2m} - \frac{g_{2m}}{\alpha_{2m} - \alpha_{2m-1}}, \quad m = 1, 2, \dots \tag{3.6}$$

$$G_{2m+1} = \frac{g_{2m+1}}{\alpha_{2m} - \alpha_{2m-1}} - h_{2m+1}\frac{\alpha_{2m+1} - \alpha_{2m}}{\alpha_{2m} - \alpha_{2m-1}}, \quad m = 1, 2, \ldots \tag{3.7}$$

$$H_{2m+1} = \frac{-g_{2m+1}}{\alpha_{2m} - \alpha_{2m-1}} + h_{2m+1}\frac{\alpha_{2m+1} - \alpha_{2m-1}}{\alpha_{2m} - \alpha_{2m-1}}, \quad m = 1, 2, \ldots \tag{3.8}$$

$$G_1 = h_1, \quad H_1 = g_1 - \alpha_1 h_1. \tag{3.9}$$

It follows from (2.13) that

$$G_{2m+1} \neq 0, \ m = 0, 1, 2, , \ldots, \quad H_{2m} \neq 0, \ m = 1, 2, \ldots \tag{3.10}$$

*We shall in this section assume that the sequence* $\{\Phi_n\}$ *is strongly regular, i.e., that also*
$$G_{2m} \neq 0, \ m = 1, 2, \ldots, \quad H_{2m+1} \neq 0, \ m = 0, 1, 2, \ldots \tag{3.11}$$

(This is always the case if $\alpha_{2m+1} = \alpha_{2m-1}$ for all $m$, cf. the situation where $\alpha_{2m} = \infty$, $\alpha_{2m-1} = 0$ for all $m$.)

*We shall assume that the functions* $\Phi_n$ *are normalized such that*
$$H_n = 1, \quad n = 1, 2, \ldots . \tag{3.12}$$

We may then write

$$\kappa_{2m} = 1 + G_{2m}\frac{z - \alpha_{2m-1}}{z - \alpha_{2m}}, \quad m = 1, 2, \ldots \tag{3.13}$$

$$\kappa_{2m+1} = \frac{z - \alpha_{2m}}{z - \alpha_{2m+1}} + G_{2m+1}\frac{z - \alpha_{2m-1}}{z - \alpha_{2m+1}}, \quad m = 0, 1, 2, \ldots \tag{3.14}$$

From general formulas for the even contraction of a continued fraction (see e.g. [3, 16, 22, 26]) we find that $\{a_n\}$, $\{b_n\}$ in the extension (3.1) must satisfy the equations

$$b_2 a_1 = \theta_1 \tag{3.15}$$

$$\frac{b_{2n}a_{2n-1}a_{2n-2}}{b_{2n-2}} = -\theta_n, \quad n = 2, 3, \ldots \tag{3.16}$$

$$a_2 + b_1 b_2 = \kappa_1 \tag{3.17}$$

$$a_{2n} + b_{2n}b_{2n-1} + a_{2n-1}\frac{b_{2n}}{b_{2n-2}} = \kappa_n, \quad n = 2, 3, \ldots \tag{3.18}$$

By taking into account (2.16), (3.12)-(3.13) we may write these equations in the following form:

$$b_2 a_1 = \frac{F_1 z}{z - \alpha_1} \tag{3.19}$$

$$\frac{b_{4m}a_{4m-1}a_{4m-2}}{b_{4m-2}} = -F_{2m}\frac{z - \alpha_{2m-2}}{z - \alpha_{2m}}, \quad m = 1, 2, \ldots \tag{3.20}$$

$$\frac{b_{4m+2}a_{4m+1}a_{4m}}{b_{4m}} = -F_{2m+1}\frac{z - \alpha_{2m-1}}{z - \alpha_{2m+1}}, \quad m = 1, 2, \ldots \tag{3.21}$$

$$a_2 + b_1 b_2 = \frac{1}{z - \alpha_1} + G_1\frac{z}{z - \alpha_1} \tag{3.22}$$

$$a_{4m} + b_{4m}b_{4m-1} + a_{4m-1}\frac{b_{4m}}{b_{4m-2}} = 1 + G_{2m}\frac{z - \alpha_{2m-1}}{z - \alpha_{2m}}, \quad m = 1, 2, \ldots \tag{3.23}$$

$$a_{4m+2} + b_{4m+2}b_{4m+1} + a_{4m+1}\frac{b_{4m+2}}{b_{4m}} = \frac{z - \alpha_{2m}}{z - \alpha_{2m+1}} + G_{2m+1}\frac{z - \alpha_{2m-1}}{z - \alpha_{2m+1}}, \quad m = 1, 2, \ldots \tag{3.24}$$

Our first task is to find an appropriate solution of these functional equations. We introduce the notation

$$\lambda_1 = \frac{F_1}{G_1}, \quad \mu_1 = 1 \tag{3.25}$$

$$\mu_{2n} = G_1 G_2 \cdots G_{n-1} G_n, \quad n = 1, 2, \ldots \tag{3.26}$$

and define

$$a_1 = \frac{\lambda_1 z}{z - \alpha_1}, \quad b_1 = \frac{\mu_1 z}{z - \alpha_1} = \frac{z}{z - \alpha_1} \tag{3.27}$$

$$a_{4m} = 1, \quad m = 1, 2, \ldots \tag{3.28}$$

$$a_{4m+2} = \frac{z - \alpha_{2m}}{z - \alpha_{2m+1}}, \quad m = 0, 1, 2, \ldots \tag{3.29}$$

$$b_{2n} = \mu_{2n}, \quad n = 1, 2, \ldots \tag{3.30}$$

We observe that with these definitions, (3.19) and (3.22) are satisfied.

It follows from (3.20)-(3.21) and the definitions above that

$$a_{4m-1} = \lambda_{4m-1} \frac{z - \alpha_{2m-1}}{z - \alpha_{2m}}, \quad m = 1, 2, \ldots \tag{3.31}$$

$$a_{4m+1} = \lambda_{4m+1} \frac{z - \alpha_{2m-1}}{z - \alpha_{2m+1}}, \quad m = 1, 2, \ldots \tag{3.32}$$

where

$$\lambda_{2n+1} = -\frac{F_{n+1}}{G_{n+1}}, \quad n = 1, 2, \ldots \tag{3.33}$$

We note that by (2.13),

$$\lambda_{2n+1} \neq 0, \quad n = 0, 1, 2, \ldots \tag{3.34}$$

Similarly it follows from (3.23)-(3.24) and the definitions above that

$$b_{4m-1} = \mu_{4m-1} \frac{z - \alpha_{2m-1}}{z - \alpha_{2m}}, \quad m = 1, 2, \ldots \tag{3.35}$$

$$b_{4m+1} = \mu_{4m+1}\frac{z-\alpha_{2m-1}}{z-\alpha_{2m+1}}, \quad m=1,2,\ldots \tag{3.36}$$

where

$$\mu_{2n+1} = \frac{G_{n+1}}{\mu_{2n+2}} - \frac{\lambda_{2n+1}}{\mu_{2n}} = \frac{G_{n+1}+F_{n+1}}{\mu_{2n+2}}, \quad n=1,2,\ldots \tag{3.37}$$

In this way we have determined a continued fraction (3.1) whose even contraction is the continued fraction (2.15)-(2.16), (3.13)-(3.14). For convenience we repeat the form of the elements:

$$a_{4m-1} = \lambda_{4m-1}\frac{z-\alpha_{2m-1}}{z-\alpha_{2m}}, \quad b_{4m-1} = \mu_{4m-1}\frac{z-\alpha_{2m-1}}{z-\alpha_{2m}}, \quad m=1,2,\ldots \tag{3.38}$$

$$a_{4m} = 1, \quad b_{4m} = \mu_{4m}, \quad m=1,2,\ldots \tag{3.39}$$

$$a_{4m+1} = \lambda_{4m+1}\frac{z-\alpha_{2m-1}}{z-\alpha_{2m+1}}, \quad b_{4m+1} = \mu_{4m+1}\frac{z-\alpha_{2m-1}}{z-\alpha_{2m+1}}, \quad m=0,1,2,\ldots \tag{3.40}$$

$$a_{4m+2} = \frac{z-\alpha_{2m}}{z-\alpha_{2m+1}}, \quad b_{4m+2} = \mu_{4m+2}, \quad m=0,1,2,\ldots \tag{3.41}$$

where the constants $\lambda_n$, $\mu_n$ are given by formulas (3.25)-(3.26), (3.33), (3.37).

We find that the following relationship holds:

$$\lambda_{2n+1} + \mu_{2n}\mu_{2n+1} = 1, \quad n=1,2,\ldots \tag{3.42}$$

These continued fractions belong to a class of continued fractions that we shall call *EMP-fractions (Extended multi-point Padé continued fractions)*. We shall give a formal definition and general discussion of *EMP*-fractions in Section 4.

We formulate the main extension result as a theorem.

THEOREM 1. *The continued fraction* $\kappa_0 + K_{n=1}^{\infty}\frac{a_n}{b_n}$ *with* $a_n, b_n$ *given by formulas (3.25)-(3.26), (3.33), (3.37)-(3.41), has as its even contraction the MP-fraction* $\kappa_0 + K_{n=1}^{\infty} \frac{\theta_n}{\kappa_n}$, *with* $\theta_n$, $\kappa_n$ *given by formulas (2.16), (3.13)-(3.14).*

## 4 EMP-FRACTIONS

We define an *EMP-fraction (Extended multi-point Padé continued fraction)* to be a continued fraction

$$\kappa_0 + K_{n=1}^{\infty} \frac{a_n}{b_n} \tag{4.1}$$

where the elements are given by (cf. (3.38)-(3.41))

$$a_{4m-1} = \lambda_{4m-1} \frac{z - \alpha_{2m-1}}{z - \alpha_{2m}}, \quad b_{4m-1} = \mu_{4m-1} \frac{z - \alpha_{2m-1}}{z - \alpha_{2m}}, \quad m = 1, 2, \ldots \tag{4.2}$$

$$a_{4m} = 1, \quad b_{4m} = \mu_{4m}, \quad m = 1, 2, \ldots \tag{4.3}$$

$$a_{4m+1} = \lambda_{4m+1} \frac{z - \alpha_{2m-1}}{z - \alpha_{2m+1}}, \quad b_{4m+1} = \mu_{4m+1} \frac{z - \alpha_{2m-1}}{z - \alpha_{2m+1}}, \quad m = 0, 1, 2, \ldots \tag{4.4}$$

$$a_{4m+2} = \frac{z - \alpha_{2m}}{z - \alpha_{2m+1}}, \quad b_{4m+2} = \mu_{4m+2}, \quad m = 0, 1, 2, \ldots \tag{4.5}$$

where $\lambda_n$, $\mu_n$ are constants,

$$\mu_1 = 1 \tag{4.6}$$

$$\lambda_{2n+1} \neq 0, \quad n = 0, 1, 2, \ldots \tag{4.7}$$

and where

$$\lambda_{2n+1} + \mu_{2n}\mu_{2n+1} = 1, \quad n = 1, 2, \ldots \tag{4.8}$$

Let $A_n$ and $B_n$ denote the canonical numerators and denominators of the *EMP*-fraction (4.1). We write out explicitely the recurrence formulas for $\{A_n\}$, $\{B_n\}$:

$$\begin{bmatrix} A_{4m} \\ B_{4m} \end{bmatrix} = \mu_{4m} \begin{bmatrix} A_{4m-1} \\ B_{4m-1} \end{bmatrix} + \begin{bmatrix} A_{4m-2} \\ B_{4m-2} \end{bmatrix}, \quad m = 1, 2, \ldots \tag{4.9}$$

$$\begin{bmatrix} A_{4m+1} \\ B_{4m+1} \end{bmatrix} = \mu_{4m+1} \frac{z - \alpha_{2m-1}}{z - \alpha_{2m+1}} \begin{bmatrix} A_{4m} \\ B_{4m} \end{bmatrix} + \lambda_{4m+1} \frac{z - \alpha_{2m-1}}{z - \alpha_{2m+1}} \begin{bmatrix} A_{4m-1} \\ B_{4m-1} \end{bmatrix}, \quad m = 0, 1, 2, \ldots \tag{4.10}$$

$$\begin{bmatrix} A_{4m+2} \\ B_{4m+2} \end{bmatrix} = \mu_{4m+2} \begin{bmatrix} A_{4m+1} \\ B_{4m+1} \end{bmatrix} + \frac{z - \alpha_{2m}}{z - \alpha_{2m+1}} \begin{bmatrix} A_{4m} \\ B_{4m} \end{bmatrix}, \quad m = 0, 1, 2, \ldots \tag{4.11}$$

$$\begin{bmatrix} A_{4m+3} \\ B_{4m+3} \end{bmatrix} = \mu_{4m+3} \frac{z - \alpha_{2m+1}}{z - \alpha_{2m+2}} \begin{bmatrix} A_{4m+2} \\ B_{4m+2} \end{bmatrix} + \lambda_{4m+3} \frac{z - \alpha_{2m+1}}{z - \alpha_{2m+2}} \begin{bmatrix} A_{4m+1} \\ B_{4m+1} \end{bmatrix}, \quad m = 0, 1, 2, \ldots \tag{4.12}$$

$$\begin{bmatrix} A_0 \\ B_0 \end{bmatrix} = \begin{bmatrix} \kappa_0 \\ 1 \end{bmatrix}, \quad \begin{bmatrix} A_{-1} \\ B_{-1} \end{bmatrix} = \begin{bmatrix} 1 \\ 0 \end{bmatrix}. \tag{4.13}$$

We shall describe the structure of $A_n$ and $B_n$. In the following, $p_n, \tilde{p}_n, q_n, \tilde{q}_n, r_n, \tilde{r}_n, s_n, \tilde{s}_n$ denote polynomials in $\mathcal{P}_n$.

We find from (4.6), (4.10) that

$$A_1 = (\kappa_0 + \lambda_1) \frac{z}{z - \alpha_1} = \frac{\tilde{q}_1}{\omega_1}, \tag{4.14}$$

$$B_1 = \frac{z}{z - \alpha_1} = \frac{q_1}{\omega_1}. \tag{4.15}$$

Similarly we find from (4.11) that

$$A_2 = \mu_2 \frac{\tilde{q}_1}{\omega_1} + \frac{1}{z-\alpha_1} \cdot \kappa_0 = \frac{\tilde{r}_1}{\omega_1}, \tag{4.16}$$

$$B_2 = \mu_2 \frac{q_1}{\omega_1} + \frac{1}{z-\alpha_1} \cdot 1 = \frac{r_1}{\omega_1}, \tag{4.17}$$

and from (4.12) that

$$A_3 = \mu_3 \frac{z-\alpha_1}{z-\alpha_2} \cdot \frac{\tilde{r}_1}{z-\alpha_1} + \lambda_3 \frac{z-\alpha_1}{z-\alpha_2} \cdot \frac{\tilde{q}_1}{z-\alpha_1} = \frac{\tilde{s}_1}{z-\alpha_2}, \tag{4.18}$$

$$B_3 = \mu_3 \frac{z-\alpha_1}{z-\alpha_2} \cdot \frac{r_1}{z-\alpha_1} + \lambda_3 \frac{z-\alpha_1}{z-\alpha_2} \cdot \frac{q_1}{z-\alpha_1} = \frac{s_1}{z-\alpha_2}. \tag{4.19}$$

From here it follows by straightforward induction that

$$A_{4m}(z) = \frac{\tilde{p}_{2m}(z)}{\omega_{2m}(z)}, \quad B_{4m}(z) = \frac{p_{2m}(z)}{\omega_{2m}(z)}, \quad m = 1, 2, \ldots \tag{4.20}$$

$$\begin{aligned} A_{4m+1}(z) &= \frac{\tilde{q}_{2m}(z)}{(z-\alpha_{2m+1})(z-\alpha_{2m})\omega_{2m-2}(z)}, \\ B_{4m+1}(z) &= \frac{q_{2m}(z)}{(z-\alpha_{2m+1})(z-\alpha_{2m})\omega_{2m-2}(z)}, \quad m = 1, 2, \ldots \end{aligned} \tag{4.21}$$

$$A_{4m+2}(z) = \frac{\tilde{r}_{2m+1}(z)}{\omega_{2m+1}(z)}, \quad B_{2m+2}(z) = \frac{r_{2m+1}(z)}{\omega_{2m+1}(z)}, \quad m = 0, 1, 2, \ldots \tag{4.22}$$

$$A_{4m+3}(z) = \frac{\tilde{s}_{2m+1}(z)}{(z-\alpha_{2m+2})\omega_{2m}(z)}, \quad B_{4m+3}(z) = \frac{s_{2m+1}(z)}{(z-\alpha_{2m+2})\omega_{2m}(z)}, \quad m = 0, 1, 2, \ldots \tag{4.23}$$

We shall also make use of the following notation for the numerator polynomials in the canonical numerators and denominators:

$$P_0 = \kappa_0, \quad Q_0 = 1, \quad P_1 = \tilde{q}_1, \quad Q_1 = q_1 \tag{4.24}$$

$$P_{4m} = \tilde{p}_{2m}, \quad Q_{4m} = p_{2m}, \quad m = 1, 2, \ldots \tag{4.25}$$

$$P_{4m+1} = \tilde{q}_{2m}, \quad Q_{4m+1} = q_{2m}, \quad m = 1, 2, \ldots \tag{4.26}$$

$$P_{4m+2} = \tilde{r}_{2m+1}, \quad Q_{4m+2} = r_{2m+1}, \quad m = 0, 1, 2, \ldots \tag{4.27}$$

$$P_{4m+3} = \tilde{s}_{2m+1}, \quad Q_{4m+3} = s_{2m+1}, \quad m = 0, 1, 2, \ldots \tag{4.28}$$

We formulate the main result on the structure of the canonical numerators and denominators in a theorem.

THEOREM 2. *The canonical numerators and denominators* $A_n$, $B_n$ *of the EMP-fraction* $\kappa_0 + K_{n=1}^{\infty} \frac{a_n}{b_n}$ *determined by (4.2)-(4.8) are of the form given by formulas (4.14)-(4.15), (4.20)-(4.23).*

## 5 INTERPOLATION

We shall now show how the approximants of the *EMP*-fraction (4.1)-(4.8) determine Newton series for which these approximants are multi-point Padé approximants. (Thus in particular a functional $M$ which gives rise to an *MP*-fraction (2.15)-(2.16), (3.13)-(3.14) and hence to an *EMP*-fraction by extension, also may be said to determine such a series). For basic information on Newton series and multi-point Padé approximation, see [11,12,20,21,26,27]. We briefly recall that a Newton series with respect to a sequence $\{\beta_n : n = 1, 2, \ldots\}$ of (not necessarily distinct) interpolation points is a formal series

$$c_0 + c_1(z - \beta_1) + c_2(z - \beta_1)(z - \beta_2) + \cdots + c_n(z - \beta_1)\cdots(z - \beta_n) + \cdots \tag{5.1}$$

and that the (formal) Newton series expansion of the function $\varphi$ is the unique Newton series (5.1) with the property

$$\begin{aligned}\varphi(z) - [c_0 + c_1(z-\beta_1) + \cdots + c_n(z-\beta_1)\cdots(z-\beta_n)] \\ = O[(z-\beta_1)\cdots(z-\beta_{n+1})], \quad n = 1,2,\ldots\end{aligned} \tag{5.2}$$

This means that in the formal Newton series expansion of $\varphi(z) - [c_0 + c_1(z-\beta_1) + \cdots + c_n(z-\beta_1)\cdots(z-\beta_n)]$, the coefficients of $1,\ (z-\beta_1), \cdots (z-\beta_1)\cdots(z-\beta_n)$ are zero. We refer especially to [7, 9] for related discussions of the situation when the functional $M$ is positive definite.

We shall call the continued fraction (4.1)–(4.8) *normal* if the following conditions are satisfied: The polynomials $p_{2m}, q_{2m}, r_{2m+1}$ and $s_{2m+1}$ are of full degree and do not vanish at the origin, $p_{2m}$ and $q_{2m}$ do not vanish at any of the points $\alpha_1, \ldots, \alpha_{2m}, r_{2m+1}$ does not vanish at any of the points $\alpha_1, \ldots, \alpha_{2m+1}$, and $s_{2m+1}$ does not vanish at any of the points $\alpha_1, \ldots, \alpha_{2m}, \alpha_{2m+2}$. (When $\mu_n \neq 0$ for all $n$, so that the even and the odd contraction of the continued fraction exists, cf. Section 6, this requirement means that 0 is a regular value for all the orthogonal functions $B_{2m}$ and $B_{2m+1}$. For this concept of a regular value, see [6].)

*We shall in this section assume that the EMP-fraction (4.1)–(4.8) is normal.*

For later use we need expressions for the products $a_1a_2a_3\cdots a_n$. We find from (3.38)-(3.41):

$$a_1a_2\cdots a_{4m} = \frac{\Lambda_{4m}z}{(z-\alpha_{2m-1})(z-\alpha_{2m})}, \quad m = 1,2,\ldots \tag{5.3}$$

$$a_1a_2\cdots a_{4m+1} = \frac{\Lambda_{4m+1}z}{(z-\alpha_{2m})(z-\alpha_{2m+1})}, \quad m = 0,1,2\ldots \tag{5.4}$$

$$a_1a_2\cdots a_{4m+2} = \frac{\Lambda_{4m+2}z}{(z-\alpha_{2m+1})^2}, \quad m = 0,1,2\ldots \tag{5.5}$$

$$a_1 a_2 \cdots a_{4m+3} = \frac{\Lambda_{4m+3} z}{(z-\alpha_{2m+1})(z-\alpha_{2m+2})}, \quad m = 0,1,2\ldots \tag{5.6}$$

where the $\Lambda_n$ are constants.

We shall make use of the following formulas connecting different numerators and denominators (see e.g. [3, 16, 22, 26]):

$$A_n B_{n-1} - A_{n-1} B_n = (-1)^{n-1} a_1 a_2 \cdots a_n, \quad n = 1,2,\ldots \tag{5.7}$$

$$A_{n+1} B_{n-1} - A_{n-1} B_{n+1} = (-1)^{n-1} a_1 a_2 \cdots a_n b_{n+1}, \quad n = 1,2,\ldots \tag{5.8}$$

By substituting from (4.2)-(4.5) and (4.20)-(4.27) in (5.7)-(5.8) we obtain the following formulas ($e_n$ and $\varepsilon_n$ are constants):

$$\frac{P_{4m+1}}{Q_{4m+1}} - \frac{P_{4m}}{Q_{4m}} = e_{4m} \frac{z\omega_{2m-2}\omega_{2m}}{p_{2m} q_{2m}}, \quad m = 1,2,\ldots \tag{5.9}$$

$$\frac{P_{4m+2}}{Q_{4m+2}} - \frac{P_{4m+1}}{Q_{4m+1}} = e_{4m+1} \frac{z(z-\alpha_{2m})\omega_{2m-2}\omega_{2m}}{q_{2m} r_{2m+1}}, \quad m = 1,2,\ldots \tag{5.10}$$

$$\frac{P_{4m+3}}{Q_{4m+3}} - \frac{P_{4m+2}}{Q_{4m+2}} = e_{4m+2} \frac{z\omega_{2m}\omega_{2m}}{r_{2m+1} s_{2m+1}}, \quad m = 0,1,2,\ldots \tag{5.11}$$

$$\frac{P_{4m+4}}{Q_{4m+4}} - \frac{P_{4m+3}}{Q_{4m+3}} = e_{4m+3} \frac{z(z-\alpha_{2m+2})\omega_{2m}\omega_{2m}}{s_{2m+1} p_{2m+2}}, \quad m = 0,1,2,\ldots \tag{5.12}$$

$$\frac{P_{4m+1}}{Q_{4m+1}} - \frac{P_{4m-1}}{Q_{4m-1}} = \varepsilon_{4m-1} \frac{z(z-\alpha_{2m})\omega_{2m-2}\omega_{2m-2}}{s_{2m-1} q_{2m}}, \quad m = 1,2,\ldots \tag{5.13}$$

$$\frac{P_{4m+2}}{Q_{4m+2}} - \frac{P_{4m}}{Q_{4m}} = \varepsilon_{4m} \frac{z\omega_{2m-1}\omega_{2m}}{p_{2m} r_{2m+1}}, \quad m = 1,2,\ldots \tag{5.14}$$

$$\frac{P_{4m+3}}{Q_{4m+3}} - \frac{P_{4m+1}}{Q_{4m+1}} = \varepsilon_{4m+1} \frac{z(z-\alpha_{2m})\omega_{2m-2}\omega_{2m}}{q_{2m}s_{2m+1}}, \quad m = 1, 2, \ldots \tag{5.15}$$

$$\frac{P_{4m+4}}{Q_{4m+4}} - \frac{P_{4m+2}}{Q_{4m+2}} = \varepsilon_{4m+2} \frac{z\omega_{2m}\omega_{2m+1}}{r_{2m+1}p_{2m+2}}, \quad m = 1, 2, \ldots \tag{5.16}$$

We note that

$$\frac{P_{n+1}}{Q_{n+1}} - \frac{P_n}{Q_n} = O\left(\frac{1}{z}\right) \quad \text{as} \quad z \to \infty, \quad n = 1, 2, \ldots \tag{5.17}$$

We shall consider Newton series with respect to the interpolation polynomial sequences

$$\begin{aligned} \{1, z, z(z-\alpha_1), \ldots, \quad z\omega_{2m-2}^2, \quad z(z-\alpha_{2m-1})\omega_{2m-2}^2, \\ z\omega_{2m-1}^2, z(z-\alpha_{2m})\omega_{2m-1}^2, \quad z\omega_{2m}^2, \ldots\} \end{aligned} \tag{5.18}$$

and

$$\begin{aligned} \{1, z, z(z-\alpha_2), \ldots, \quad z\omega_{2m-2}^2, \quad z(z-\alpha_{2m})\omega_{2m-2}^2, \\ z(z-\alpha_{2m})^2\omega_{2m-2}^2, \quad z(z-\alpha_{2m})^2(z-\alpha_{2m-1})\omega_{2m-2}^2, \quad z\omega_{2m}^2 \ldots\} \end{aligned} \tag{5.19}$$

i.e., with respect to the interpolation tables

$$\{\infty, 0, \alpha_1, \alpha_1, \alpha_2, \alpha_2, \alpha_3, \alpha_3, \alpha_4, \alpha_4, \ldots\} \tag{5.20}$$

and

$$\{\infty, 0, \alpha_2, \alpha_2, \alpha_1, \alpha_1, \alpha_4, \alpha_4, \alpha_3, \alpha_3, \ldots\} \tag{5.21}$$

where the interpolation at $\infty$ will be written out separately.

These series have the form

$$\begin{aligned} \gamma_0 + \gamma_1 z + \gamma_2 z(z-\alpha_1) + \gamma_3 z\omega_1^2 + \gamma_4(z-\alpha_2)\omega_1^2 + \cdots \\ \cdots + \gamma_{4m-3} z\omega_{2m-2}^2 + \gamma_{4m-2} z(z-\alpha_{2m-1})\omega_{2m-2}^2 + \\ \gamma_{4m-1} z\omega_{2m-1}^2 + \gamma_{4m} z(z-\alpha_{2m})\omega_{2m-1}^2 + \cdots \end{aligned} \tag{5.22}$$

and

$$\delta_0 + \delta_1 z + \delta_2 z(z-\alpha_2) + \delta_3 z(z-\alpha_2)^2 + \delta_4 z(z-\alpha_1)(z-\alpha_2)^2 + \cdots$$
$$\cdots + \delta_{4m-3} z\omega_{2m-2}^2 + \delta_{4m-2} z(z-\alpha_{2m})\omega_{2m-2}^2$$
$$+\delta_{4m-1} z(z-\alpha_{2m})^2\omega_{2m-2}^2$$
$$+\delta_{4m} z(z-\alpha_{2m-1})(z-\alpha_{2m})^2\omega_{2m-2}^2 + \cdots \tag{5.23}$$

Let

$$\gamma_0^{(4m)} + \gamma_1^{(4m)} z + \gamma_2^{(4m)} z(z-\alpha_1) + \gamma_3^{(4m)} z(z-\alpha_1)^2 + \cdots$$
$$+\gamma_{4m-3}^{(4m)} z\omega_{2m-2}^2 + \gamma_{4m-2}^{(4m)} z(z-\alpha_{2m-1})\omega_{2m-2}^2$$
$$+\gamma_{4m-1}^{(4m)} z\omega_{2m-1}^2 + \gamma_{4m}^{(4m)} z(z-\alpha_{2m})\omega_{2m-1}^2 + \cdots \tag{5.24}$$

and

$$\gamma_0^{(4m+2)} + \gamma_1^{(4m+2)} z + \gamma_2^{(4m+2)} z(z-\alpha_1) + \gamma_3^{(4m+2)} z(z-\alpha_1)^2 + \cdots$$
$$+\gamma_{4m-3}^{(4m+2)} z\omega_{2m-2}^2 + \gamma_{4m-2}^{(4m+2)} z(z-\alpha_{2m-1})\omega_{2m-2}^2$$
$$+\gamma_{4m-1}^{(4m+2)} z\omega_{2m-1}^2 + \gamma_{4m}^{(4m+2)} z(z-\alpha_{2m})\omega_{2m-1}^2$$
$$+\gamma_{4m+1}^{(4m+2)} z\omega_{2m}^2 + \gamma_{4m+2}^{(4m+2)} z(z-\alpha_{2m+1})\omega_{2m}^2 + \cdots \tag{5.25}$$

be the formal Newton series expansions of $\dfrac{P_{4m}}{Q_{4m}}$ and $\dfrac{P_{4m+2}}{Q_{4m+2}}$ with respect to the sequence (5.18). It follows from (5.14) and (5.16) that

$$\gamma_k^{(4m)} = \gamma_k^{(4m+2)}, \quad k = 0, 1, \ldots, \quad 4m-1, \tag{5.26}$$

$$\gamma_k^{(4m+2)} = \gamma_k^{(4m+4)}, \quad k = 0, 1, \ldots, \quad 4m+1. \tag{5.27}$$

The formulas (5.26)-(5.27) imply that $\gamma_k^{(2n)}$ does not depend on $n$ for $k < 2n$. Hence if $n$ is large enough, then the numbers $\gamma_k = \gamma_k^{(2n)}$ are well defined and independent of $n$:

$$\gamma_k = \gamma_k^{(2n)}, \quad 2n \geq k+1. \tag{5.28}$$

Thus the *EMP*-fraction (4.1)-(4.5) determines a unique formal Newton series (5.22) via the Newton series expansions of the even approximants. (Hence the functional $M$ also determines such a formal Newton series through the corresponding *MP*-fraction and then the *EMP*-fraction obtained by extension.)

Similarly let

$$\begin{aligned}\delta_0^{(4m+1)} + \delta_1^{(4m+1)} z + \delta_2^{(4m+1)} z(z-\alpha_2) + \delta_3^{(4m+1)} z(z-\alpha_2)^2 \\ + \cdots + \delta_{4m-3}^{(4m+1)} z\omega_{2m-2}^2 + \delta_{4m-2}^{(4m+1)} z(z-\alpha_{2m})\omega_{2m-2}^2 \\ + \delta_{4m-1}^{(4m+1)} z(z-\alpha_{2m})^2\omega_{2m-2}^2 + \delta_{4m}^{(4m+1)} z(z-\alpha_{2m})\omega_{2m-2}\omega_{2m} + \cdots\end{aligned} \tag{5.29}$$

and

$$\begin{aligned}\delta_0^{(4m+3)} + \delta_1^{(4m+3)} z + \delta_2^{(4m+3)} z(z-\alpha_2) + \delta_3^{(4m+3)} z(z-\alpha_2)^2 \\ + \cdots + \delta_{4m-3}^{(4m+3)} z\omega_{2m-2}^2 + \delta_{4m-2}^{(4m+3)} z(z-\alpha_{2m})\omega_{2m-2}^2 \\ + \delta_{4m-1}^{(4m+3)} z(z-\alpha_{2m})^2\omega_{2m-2}^2 + \delta_{4m}^{(4m+3)} z(z-\alpha_{2m-1})(z-\alpha_{2m})^2\omega_{2m-2}^2 \\ + \delta_{4m+1}^{(4m+3)} z\omega_{2m}^2 + \delta_{4m+2}^{(4m+3)} z(z-\alpha_{2m+2})\omega_{2m}^2 + \cdots\end{aligned} \tag{5.30}$$

be the formal Newton series expansions of $\dfrac{P_{4m+1}}{Q_{4m+1}}$ and of $\dfrac{P_{4m+3}}{Q_{4m+3}}$ with respect to the sequence (5.19). It follows from (5.13) and (5.15) that

$$\delta_k^{(4m+1)} = \delta_k^{(4m+3)}, \quad k = 0, 1, \ldots \quad 4m-1, \tag{5.31}$$

$$\delta_k^{(4m+3)} = \delta_k^{(4m+5)}, \quad k = 0, 1, \ldots\ldots \quad 4m+1. \tag{5.32}$$

Thus we see as above that the *EMP*-fraction (4.1)-(4.5) determines a unique formal Newton series (5.23) via the Newton series expansions of the odd approximants:

$$\delta_k = \delta_k^{(2n+1)}, \quad 2n \geq k+1. \tag{5.33}$$

Clearly $\delta_0 = \gamma_0 = \kappa_0$. From (5.17) we see that there is a unique well defined value $\gamma_\infty$ given by

$$\gamma_\infty = \lim_{z\to\infty} \frac{P_n(z)}{Q_n(z)} = \lambda_1 + \kappa_0, \quad n = 1, 2, \ldots \tag{5.34}$$

Let $\Gamma$ and $\Delta$ denote the formal Newton series (5.22) and (5.23) determined by (5.28) and (5.33), and let $P_n$, $Q_n$, $P_n/Q_n$ in addition to having their basic meaning also denote Newton series expansions of these functions with respect to the sequences (5.18) and (5.19).

It follows from the preceeding arguments that

$$\frac{P_{4m}}{Q_{4m}} = \gamma_0 + \gamma_1 z + \gamma_2 z(z-\alpha_1) + \cdots$$
$$+\gamma_{4m-1} z\omega_{2m-1}^2 + \gamma_{4m}^{(4m)} z(z-\alpha_{2m})\omega_{2m-1}^2 + \cdots, \quad m = 1,2,\ldots \tag{5.35}$$

$$\frac{P_{4m+2}}{Q_{4m+2}} = \gamma_0 + \gamma_1 z + \gamma_2 z(z-\alpha_1) + \cdots$$
$$+\gamma_{4m+1} z\omega_{2m}^2 + \gamma_{4m+2}^{(4m+2)} z(z-\alpha_{2m+1})\omega_{2m}^2 + \cdots, \quad m = 0,1,2,\ldots \tag{5.36}$$

$$\frac{P_{4m+1}}{Q_{4m+1}} = \gamma_0 + \delta_1 z + \delta_2 z(z-\alpha_2) + \cdots$$
$$+\delta_{4m-1} z(z-\alpha_{2m})^2\omega_{2m-2}^2 + \delta_{4m}^{(4m+1)}(z-\alpha_{2m-1})(z-\alpha_{2m})^2\omega_{2m-2}^2 + \cdots,$$
$$m = 1,2,\ldots \tag{5.37}$$

$$\frac{P_{4m+3}}{Q_{4m+3}} = \gamma_0 + \delta_1 z + \delta_2 z(z-\alpha_2) + \cdots$$
$$+\delta_{4m+1} z\omega_{2m}^2 + \delta_{4m+2}^{(4m+3)} z(z-\alpha_{2m+2})\omega_{2m}^2 + \cdots, \quad m = 0,1,2,\ldots \tag{5.38}$$

$$\frac{P_n}{Q_n} - \gamma_\infty = O\left(\frac{1}{z}\right) \quad \text{as} \quad z \to \infty, \quad n = 1,2,\ldots \tag{5.39}$$

Furthermore (with $x_n$ and $y_n$ constants)

$$P_{4m} - \Gamma Q_{4m} = x_{4m} z\omega_{2m-1}\omega_{2m} + \cdots, \quad m = 1,2,\ldots \tag{5.40}$$

$$P_{4m+2} - \Gamma Q_{4m+2} = x_{4m+2} z \omega_{2m} \omega_{2m+1} + \cdots, \quad m = 0, 1, 2, \ldots \tag{5.41}$$

$$P_{4m+1} - \Delta Q_{4m+1} = x_{4m+1} z (z - \alpha_{2m}) \omega_{2m-2} \omega_{2m} + \cdots, \quad m = 1, 2, \ldots \tag{5.42}$$

$$P_{4m+3} - \Delta Q_{4m+3} = x_{4m+3} z (z - \alpha_{2m+2}) \omega_{2m}^2 + \cdots, \quad m = 0, 1, 2, \ldots . \tag{5.43}$$

By (5.9) and (5.42) it follows that $P_{4m} - \Delta Q_{4m}$ is of the form (5.23) with $\delta_k = 0$ for $k = 0, 1, \ldots, 4m-1$. Hence

$$P_{4m} - \Delta Q_{4m} = y_{4m} z (z - \alpha_{2m}) \omega_{2m-2}^2 + \cdots, \quad m = 1, 2, \ldots \tag{5.44}$$

Similarly we find by using (5.11) and (5.43), (5.10) and (5.41), and (5.12) and (5.40) that

$$P_{4m+2} - \Delta Q_{4m+2} = y_{4m+2} z \omega_{2m}^2 + \cdots, \quad m = 0, 1, 2, \ldots \tag{5.45}$$

$$P_{4m+1} - \Gamma Q_{4m+1} = y_{4m+1} z \omega_{2m-2} \omega_{2m-1} + \cdots, \quad m = 1, 2, \ldots \tag{5.46}$$

$$P_{4m+3} - \Gamma Q_{4m+3} = y_{4m+3} z \omega_{2m}^2 + \cdots, \quad m = 0, 1, 2, \ldots \tag{5.47}$$

Since $P_{4m}$ and $Q_{4m}$ are polynomials of degree $2m$, the quotient $\frac{P_{4m}}{Q_{4m}}$ has $4m+1$ free parameters. It follows from (5.35) (or (5.40)) and (5.39) that $\frac{P_{4m}}{Q_{4m}}$ satisfies $4m+1$ interpolation conditions (at $\infty, 0, \alpha_1, \alpha_1, \ldots \alpha_{2m-1}, \alpha_{2m-1}, \alpha_{2m}$) determining the coefficients $\gamma_\infty, \gamma_0, \ldots, \gamma_{4m-1}$. We conclude that $\frac{P_{4m}}{Q_{4m}}$ has maximal degree of interpolation to $\gamma_\infty$ at $\infty$ and the series $\Gamma$. Thus $\frac{P_{4m}}{Q_{4m}}$ is the $(2m, 2m)$ multi-point Padé appoximant to $\gamma_\infty$ and

$\Gamma$. Similarly we find that $\frac{P_{4m+2}}{Q_{4m+2}}$ is the $(2m+1, 2m+1)$ multi-point Padé approximant to $\gamma_\infty$ and $\Gamma$, that $\frac{P_{4m+1}}{Q_{4m+1}}$ is the $(2m, 2m)$ multi-point Padé approximant to $\gamma_\infty$ and $\Delta$, and that $\frac{P_{4m+3}}{Q_{4m+3}}$ is the $(2m+1, 2m+1)$ multi-point Padé approximant to $\gamma_\infty$ and $\Delta$. — Multi-point Padé approximants to a Newton series, i.e. to interpolation data given with respect to a fixed sequence of interpolation points, are also called Newton-Padé approximants.

We also note that $\frac{P_{4m}}{Q_{4m}}$ and $\frac{P_{4m+2}}{Q_{4m+2}}$ interpolate $\gamma_\infty$ and $\Delta$ with order 2 and 1 less than maximal, and that $\frac{P_{4m+1}}{Q_{4m+1}}$ and $\frac{P_{4m+3}}{Q_{4m+3}}$ interpolate $\gamma_\infty$ and $\Gamma$ with order 2 and 1 less than maximal.

We formulate the main interpolation result in a theorem.

THEOREM 3. *The EMP-fraction (4.1)-(4.8) determines a Newton series $\Gamma$ with respect to the sequence (5.18) and a Newton series $\Delta$ with respect to the sequence (5.19), and a value $\gamma_\infty$ at $\infty$. The even approximants $\frac{P_{2n}}{Q_{2n}}$ of the EMP-fraction are $(n, n)$ multi-point Padé approximants to $\gamma_\infty$ and $\Gamma$, the odd approximants $\frac{P_{2n+1}}{Q_{2n+1}}$ of the EMP-fraction are $(n, n)$ multi-point Padé approximants to $\gamma_\infty$ and $\Delta$. The even approximants interpolate $\gamma_\infty$ and $\Delta$ with order less than maximal, and the odd approximants interpolate $\gamma_\infty$ and $\Gamma$ with order less than maximal.*

REMARK. The interpolation results above are proved with reference to the *EMP*-fraction (4.1)-(4.8). When $\alpha_{2m} \neq \alpha_{2m-1}$ for all $m$, the *MP*-fraction (2.15)-(2.16), (3.13)-(3.14) is the even contraction of an *EMP*-fraction, and thus interpolation to $\gamma_\infty$ and $\Gamma$ of approximants of this *MP*-fraction are also proved. Even if $\alpha_{2m} \neq \alpha_{2m-1}$ is not satisfied for all $m$, the formulas (5.14) and (5.16) can still be proved directly from the recurrence relation of the *MP*-fraction. Hence interpolation to $\gamma_\infty$ and $\Gamma$ of the approximants of the *MP*-fraction can be proved in the general case (as long as the sequence $\{\Phi_n\}$ of orthogonal functions is regular). — Similar considerations concerning interpolation properties connected with the odd approximants of the *EMP*-fraction and $\gamma_\infty, \Delta$ can be made when the nature of these approximants has been studied in Section 6.

## 6 CONTRACTION OF EMP-FRACTIONS

The *EMP*-fraction discussed in Section 3 has by the very construction the *MP*-fraction (2.15)-(2.16), (3.13)-(3.14) as its even contraction. And it follows from (3.25)-(3.26) and (3.33) that

$$F_1 = \lambda_1\mu_2, \quad G_1 = \mu_2 \tag{6.1}$$

$$F_n = -\lambda_{2n-1}\frac{\mu_{2n}}{\mu_{2n-2}}, \quad n = 2, 3, \ldots \tag{6.2}$$

$$G_n = \frac{\mu_{2n}}{\mu_{2n-2}}, \quad n = 2, 3, \ldots . \tag{6.3}$$

Conversely, assume that

$$\mu_{2n} \neq 0, \quad n = 1, 2, \ldots \tag{6.4}$$

in the *EMP*-fraction (4.1)-(4.8). Then this *EMP*-fraction has an even contraction

$$\kappa_0 + K_{n=1}^{\infty}\frac{\theta_n}{\kappa_n} \tag{6.5}$$

determined by the equations (3.15)-(3.18). Substituting from (4.2)-(4.5) and taking into account (4.7)-(4.8) we get

$$\theta_n = F_n\frac{z-\alpha_{n-2}}{z-\alpha_n}, \quad n = 1, 2, \ldots \tag{6.6}$$

$$\kappa_{2m} = 1 + G_{2m}\frac{z-\alpha_{2m-1}}{z-\alpha_{2m}}, \quad m = 1, 2, \ldots \tag{6.7}$$

$$\kappa_{2m+1} = \frac{z-\alpha_{2m}}{z-\alpha_{2m+1}} + G_{2m+1}\frac{z-\alpha_{2m-1}}{z-\alpha_{2m+1}}, \quad m=0,1,2,\ldots \tag{6.8}$$

where $F_n$ and $G_n$ are defined by(6.1)-(6.3). Thus the even contraction of the *EMP*-fraction (which exists when (6.4) is satisfied) is an *MP*-fraction of the form (2.15)-(2.16), (3.13)-(3.14).

We note that if the *EMP*-fraction (4.1)–(4.8) is normal then the canonical denominators $K_{2m} = B_{4m}$, $K_{2m+1} = B_{4m+2}$, are completely regular in the sense that the numerator polynomials $p_{2m}, r_{2m+1}$ have no common factors with the corresponding denominator polynomials $w_{2m}, w_{2m+1}$.

We shall now investigate the odd contraction of the *EMP*-fraction (4.1)-(4.8). This odd contraction is by definition (when it exists) the continued fraction

$$\kappa_0' + K_{n=1}^{\infty}\frac{\xi_n}{\eta_n} \tag{6.9}$$

whose $n$th canonical numerator and denominator are the $(2n+1)$th canonical numerator and denominator of the *EMP*-fraction. Thus if $C_n$ and $D_n$ denote the canonical numerator and denominator of (6.9) we have

$$C_n = A_{2n+1}, \quad D_n = B_{2n+1}, \quad n=0,1,2,\ldots \tag{6.10}$$

*We shall in the rest of this section assume that*

$$\mu_{2n+1} \neq 0, \quad n=0,1,2,\ldots \tag{6.11}$$

From general formulas for the odd contraction of a continued fraction (see e.g. [3, 16, 22, 26]) we find that $\kappa_0'$, $\{\xi_n\}$, $\{\eta_n\}$ are given by

$$\kappa_0' = \frac{A_1}{B_1} = \kappa_0 + \lambda_1 \tag{6.12}$$

$$\xi_n = -a_{2n-1}a_{2n}\frac{b_{2n+1}}{b_{2n-1}}, \quad n=1,2,\ldots \tag{6.13}$$

$$\eta_n = a_{2n+1} + b_{2n}b_{2n+1} + a_{2n}\frac{b_{2n+1}}{b_{2n-1}}, \quad n = 1, 2, \dots . \tag{6.14}$$

Substituting from (4.1)-(4.5), taking into account (4.7)-(4.8), we get

$$\xi_{2m} = -\lambda_{4m-1}\frac{\mu_{4m+1}}{\mu_{4m-1}} \cdot \frac{z-\alpha_{2m-1}}{z-\alpha_{2m+1}}, \quad m = 1, 2, \dots \tag{6.15}$$

$$\eta_{2m} = \frac{z-\alpha_{2m-1}}{z-\alpha_{2m+1}} + \frac{\mu_{4m+1}}{\mu_{4m-1}} \cdot \frac{z-\alpha_{2m}}{z-\alpha_{2m+1}}, \quad m = 1, 2, \dots \tag{6.16}$$

$$\xi_{2m+1} = -\lambda_{4m+1}\frac{\mu_{4m+3}}{\mu_{4m+1}} \cdot \frac{z-\alpha_{2m}}{z-\alpha_{2m+2}}, \quad m = 0, 1, 2, \dots \tag{6.17}$$

$$\eta_{2m+1} = \frac{z-\alpha_{2m+1}}{z-\alpha_{2m+2}} + \frac{\mu_{4m+3}}{\mu_{4m+1}} \cdot \frac{(z-\alpha_{2m})(z-\alpha_{2m+1})}{(z-\alpha_{2m-1})(z-\alpha_{2m+2})}, \quad m = 0, 1, 2, \dots \tag{6.18}$$

By an equivalence transformation we obtain a continued fraction

$$\kappa_0' + K_{n=1}^{\infty}\frac{\rho_n}{\sigma_n} \tag{6.19}$$

whose canonical numerators $\Pi_n$ and denominators $\Omega_n$ are given by

$$\Pi_{2m} = \frac{z-\alpha_{2m+1}}{z-\alpha_{2m-1}}C_{2m} = \frac{z-\alpha_{2m+1}}{z-\alpha_{2m-1}}A_{4m+1}, \quad m = 0, 1, 2, \dots \tag{6.20}$$

$$\Omega_{2m} = \frac{z-\alpha_{2m+1}}{z-\alpha_{2m-1}}D_{2m} = \frac{z-\alpha_{2m+1}}{z-\alpha_{2m-1}}B_{4m+1}, \quad m = 1, 2, \dots \tag{6.21}$$

$$\Pi_{2m+1} = C_{2m+1} = A_{4m+3}, \quad m = 0, 1, 2, \dots \tag{6.22}$$

$$\Omega_{2m+1} = D_{2m+1} = B_{4m+3}, \quad m = 0, 1, 2, \ldots \tag{6.23}$$

We observe that $\Pi_{2m}$, $\Omega_{2m}$ belong to
$\operatorname{Span}\left\{1, \ldots, \frac{1}{\omega_{2m-2}}, \frac{1}{(z-\alpha_{2m})\omega_{2m-2}}, \frac{1}{\omega_{2m}}\right\} = \mathcal{L}_{2m}$ and
$\Pi_{2m+1}$, $\Omega_{2m+1}$ belong to
$\operatorname{Span}\left\{1, \ldots, \frac{1}{\omega_{2m-2}}, \frac{1}{(z-\alpha_{2m})\omega_{2m-2}}, \frac{1}{\omega_{2m}}, \frac{1}{(z-\alpha_{2m+1})\omega_{2m}}\right\}$.
We also note that the continued fractions (6.9) and (6.19) are identical if $\alpha_{2m-1} = \alpha_{2m+1}$ for all $m$.

By substituting from (6.20)-(6.23) in the recurrence relation

$$\begin{bmatrix} C_n \\ D_n \end{bmatrix} = \eta_n \begin{bmatrix} C_{n-1} \\ D_{n-1} \end{bmatrix} + \xi_n \begin{bmatrix} C_{n-2} \\ D_{n-2} \end{bmatrix}, \quad n = 1, 2, \ldots \tag{6.24}$$

or by using general formulas expressing relationships between equivalent continued fractions (see e.g. [3, 16, 22, 26]) we find that

$$\rho_{2m} = U_{2m} \frac{z - \alpha_{2m-3}}{z - \alpha_{2m-1}}, \quad m = 1, 2, \ldots \tag{6.25}$$

$$\sigma_{2m} = 1 + V_{2m} \frac{z - \alpha_{2m}}{z - \alpha_{2m-1}}, \quad m = 1, 2, \ldots \tag{6.26}$$

$$\rho_{2m+1} = U_{2m+1} \frac{z - \alpha_{2m}}{z - \alpha_{2m+2}}, \quad m = 0, 1, 2, \ldots \tag{6.27}$$

$$\sigma_{2m+1} = \frac{z - \alpha_{2m-1}}{z - \alpha_{2m+2}} + V_{2m+1} \frac{z - \alpha_{2m}}{z - \alpha_{2m+2}}, \quad m = 0, 1, 2, \ldots . \tag{6.28}$$

where

$$U_{2m} = -\lambda_{4m-1} \frac{\mu_{4m+1}}{\mu_{4m-1}}, \quad m = 1, 2, \ldots \tag{6.29}$$

$$V_{2m} = \frac{\mu_{4m+1}}{\mu_{4m-1}}, \quad m = 1, 2, \ldots \tag{6.30}$$

$$U_{2m+1} = -\lambda_{4m+1}\frac{\mu_{4m+3}}{\mu_{4m+1}}, \quad m = 0, 1, 2, \ldots \tag{6.31}$$

$$V_{2m+1} = \frac{\mu_{4m+3}}{\mu_{4m+1}}, \quad m = 0, 1, 2, \ldots \tag{6.32}$$

This shows that the continued fraction (6.19) is a continued fraction of the special type (2.15)-(2.16), (3.13)-(3.14) determined by the basic sequence $\{\beta_n : n = 1, 2, \ldots\}$, where

$$\beta_{2m} = \alpha_{2m-1}, \quad m = 1, 2, \ldots, \quad \beta_{2m+1} = \alpha_{2m+2}, \quad m = 0, 1, 2, \ldots \tag{6.33}$$

Thus if (3.2) is satisfied, the canonical denominators $\Omega_n$ are orthogonal rational functions and the canonical numerators are associated functions of the second kind corresponding to the basic sequence $\{\alpha_2, \alpha_1, \alpha_4, \alpha_3, \ldots\}$. Because of the one-to-one correspondence between (certain) functionals and *EMP*-fractions satisfying (6.4) and (6.11), it follows that when these conditions are satisfied, the canonical denominators of the even contraction of the *EMP*-fraction and the canonical denominators of the odd contraction of the *EMP*-fraction are orthogonal functions with respect to the same functional $M$ (but with different orderings of the basic points).

We note that if the *EMP*-fraction (4.1)–(4.8) is normal, then the canonical denominators $\Omega_{2m} = \dfrac{z-\alpha_{2m+1}}{z-\alpha_{2m-1}}B_{4m+1}$, $\Omega_{2m+1} = B_{4m+3}$, are completely regular in the sense that the numerator polynomials $q_{2m}, s_{2m+1}$ have no common factors with the corresponding denominator polynomials $w_{2m}, w_{2m+1}$.

We formulate the main result on the odd contraction of an *EMP*-fraction as a theorem.

THEOREM 4. *Let the EMP-fraction* $\kappa_0 + K_{n=1}^{\infty}\dfrac{a_n}{b_n}$ *(associated with a basic sequence* $\{\alpha_1, \alpha_2, \alpha_3, \alpha_4, \ldots\}$*) have an odd contraction* $\kappa_0' + K_{n=1}^{\infty}\dfrac{\xi_n}{\eta_n}$. *Then this odd contraction is equivalent to an MP-fraction* $\kappa_0' + K_{n=1}^{\infty}\dfrac{\rho_n}{\sigma_n}$ *associated with the basic sequence*

$\{\alpha_2, \alpha_1, \alpha_4, \alpha_3, \ldots\}$. *When the EMP-fraction has both an even and an odd contraction, the canonical denominators of both MP-fractions are orthogonal functions with respect to the same functional.*

## REFERENCES


1. G.A. Baker, Jr. and P.R. Graves-Morris: Padé Approximants I,II, *Encyclopedia of Mathematics and its Applications 13, 14*. Addison-Wesley (1980).

2. C. Bonan-Hamada, W.B. Jones and O. Njåstad: Natural solutions of strong Stieltjes moment problems derived from PC-fractions, In W.B. Jones and A. Sri Ranga, editors, *Orthogonal Functions, Moment Theory and Continued Fractions: Theory and Applications*, Marcel Dekker (1997).

3. A. Bultheel: Laurent series and their Padé approximants. *Operator Theory: Advances and Applications, OT-27*, Birkhäuser Verlag (1987).

4. A. Bultheel, P. Gonzalez-Vera, E. Hendriksen and O. Njåstad: Moment problems and orthogonal functions. *J. Comp. Appl. Math. 48* (1993) 49–68.

5. A. Bultheel, P. Gonzalez-Vera, E. Hendriksen and O. Njåstad: Recurrence relations for orthogonal functions. In S.C. Cooper and W.J. Thron, editors, *Continued fractions and orthogonal functions*, Marcel Dekker (1994) 23–46.

6. A. Bultheel, P. Gonzalez-Vera, E. Hendriksen and O. Njåstad: Orthogonal rational functions with poles on the unit circle. *J. Math. Anal. Appl. 182* (1994) 221–243.

7. A. Bultheel, P. Gonzalez-Vera, E. Hendriksen and O. Njåstad: Orthogonality and boundary interpolation. In A.Cuyt, editor, *Nonlinear Numerical Methods and Rational Approximation II*, Kluwer (1994) 37–47.

8. A. Bultheel, P. Gonzalez-Vera, E. Hendriksen and O. Njåstad: A Favard theorem for rational functions with poles on the unit circle. Submitted.

9. A. Bultheel, P. Gonzalez-Vera, E. Hendriksen and O. Njåstad: Interpolation of Nevanlinna functions by rationals with poles on the real line. Submitted.

10. A. Bultheel, P. Gonzalez-Vera, E. Hendriksen and O. Njåstad: Orthogonal rational functions. Submitted.

11. M.A. Gallucci and W.B. Jones: Rational approximations corresponding to Newton series (Newton-Padé Approximants). *J. Approx. Theory 17* (1976) 366–392.

12. A.A. Gonchar and G. Lopez: On Markov's theorem for multipoint Padé approximants for functions of Stieltjes type. *Mat. Sb. 105* (1978) 512–524. English translation, *Math. USSR-Sb. 34* (1978) 449-459.

13. E. Hendriksen and O. Njåstad: A Favard theorem for rational functions. *J. Math. Anal. Appl. 142* (1989) 508–520.

14. E. Hendriksen and O. Njåstad: Positive multipoint Padé continued fractions. *Proc. Edinburgh Math. Soc. 32* (1989) 261–269.

15. M.E.H. Ismail and D.R. Masson: Generalized orthogonality and continued fractions. *J. Approx. Theory 83* (1995) 1–40.

16. W.B. Jones and W.J. Thron: Continued Fractions: Analytic Theory and Applications. *Encyclopedia of Mathematics and its Applications 11.* Addison-Wesley (1980).

17. W.B. Jones, O. Njåstad and W.J. Thron: Two-point Padé expansions of a family of analytic functions. *J. Comp. Appl. Math. 9* (1983) 105–123.

18. W.B. Jones, O. Njåstad and W.J. Thron: Continued fractions associated with trigonometric and other strong moment problems. *Constructive Approximation 2* (1986) 197-211.

19. W.B. Jones, O. Njåstad and W.J. Thron: Perron-Carathéodory continued fractions. In: J. Gilewicz, M. Pindor and W. Siemaszko, editors, Rational Approximation and its Applications in Mathematics and Physics, *Springer Lecture Notes in Mathematics 1237*, Springer Verlag (1987) 188–206.

20. J. Karlsson: Rational interpolation and best rational approximation. *J. Math. Anal. Appl. 52* (1976) 38–52.

21. G. Lopez: Conditions for convergence of multipoint Padé approximants for functions of Stieltjes type. *Math. USSR-Sb. 35* (1979) 363–375.

22. L. Lorentzen and H. Waadeland: Continued Fractions with Applications. *Studies in Computational Mathematics 3.* North-Holland (1992).

23. J.H. McCabe and J.A. Murphy: Continued fractions which correspond to power series expansions at two points. *J. Inst. Math. Appl. 17* ( 1976) 233–247.

24. O. Njåstad: Multipoint Padé approximation and orthogonal rational functions. In: A. Cuyt, editor, *Nonlinear Numerical Methods and Rational Approximation*, Reidel (1988) 259–270.

25. O. Perron: Die Lehre von den Kettenbrüchen, Vol. 2 (3rd edition), Teubner (1957).

26. E.B. Saff: An extension of Montessus de Ballore's theorem on the convergence of interpolating rational functions. *J. Approx. Theory 6* (1972) 63–67.

27. H. Wallin: Rational interpolation to meromorphic functions. In: M.G. de Bruin and H. van Rossum, editors, Padé Approximation and its Applications, *Springer Lecture Notes in Mathematics 88*, Springer Verlag (1981) 371–382.

# Interpolation of Nevanlinna Functions by Rationals with Poles on the Ral Line

ADHEMAR BULTHEEL Department of Computer Science, K.U.Leuven, Belgium
PABLO GONZÁLEZ-VERA Department Análisis Math., Univ. La Laguna, Tenerife, Spain
ERIK HENDRIKSEN Department of Math., Univ. of Amsterdam, The Netherlands
OLAV NJÅSTAD Department of Math. Sciences, Norwegian Univ. of Science and Technology, Trondheim, Norway

Abstract: Let $\mu$ be a measure on the extended real line $\hat{\mathbb{R}} = \mathbb{R} \cup \{\infty\}$. Its Nevanlinna transform is defined as $\Omega_\mu(z) = \int D(t,z)d\mu(t)$ with $D(t,z) = i(1+tz)/(z-t)$. It is a holomorphic function in the upper half plane with nonnegative real part. In this paper we construct rational approximants $\Omega_n(z)$ of degree $n$, which belong to the same class and are such that for an arbitrary sequence of points $\alpha_k \in \hat{\mathbb{R}}$, $k = 1, 2, \ldots$, we have Hermite interpolation at these points for the boundary functions of $\Omega_\mu$ and $\Omega_n$. This means that the nontangential limits of $\Omega_\mu - \Omega_n$ and an appropriate number of its derivatives vanish in the given points $\alpha_k$.

Keywords: orthogonal rational functions, rational interpolation.
AMS Classification: 30E05, 30E15, 41A20

## 1 HAMBURGER MOMENT PROBLEMS AND RATIONAL INTERPOLATION

Let $\mu$ be a finite positive measure on the extended real line $\hat{\mathbb{R}} = \mathbb{R} \cup \{\infty\}$. Consider a sequence $A = \{\alpha_k\}_{k\geq 1}$ with $\alpha_k \in \hat{\mathbb{R}} \setminus \{0\}$. We define $\mathcal{L}_n$ to be the space of rational functions of degree $n$ whose poles are in $A$, i.e., if $\Pi_n$ is the set of polynomials of degree $n$ at most, then

$$\mathcal{L}_n = \left\{ \frac{p_n(z)}{\pi_n(z)} : p_n \in \Pi_n, \pi_n = \prod_{k=1}^{n} \left(1 - \frac{z}{\alpha_k}\right) \right\}.$$

This research was performed as part of the European project ROLLS under contract CHRX-CT93-0416.

Note that if all $\alpha_k = \infty$, then $\mathcal{L}_n = \Pi_n$. To highlight this fact, we choose as a natural set of basis functions

$$b_0 = 1, \quad b_n(z) = Z_1(z)Z_2(z)\cdots Z_n(z), \quad n = 1, 2, \ldots$$

with

$$Z_k(z) = \frac{z}{1 - z/\alpha_k}, \quad k = 0, 1, \ldots$$

This is indeed a natural generalization since these basis functions reduce to the powers of $z$ when all $\alpha_k = \infty$. Obviously

$$\mathcal{L}_n = \text{span}\{b_0, b_1, \ldots, b_n\}.$$

Suppose we know the generalized moments

$$m_k = \int b_k(t)d\mu(t), \quad k = 0, 1, \ldots$$

(here and in the sequel, all integrals are over $\hat{\mathbb{R}}$). To solve the moment problem means to find conditions for existence and uniqueness of a measure having these moments and if solutions exist to construct them. In the special case where all $\alpha_k = \infty$, this is the classical Hamburger moment problem [1, 11, 2, 16]. In the general case, this is almost the extended Hamburger moment problem as studied in [12, 13, 15]. The difference is that there $b_k$ is replaced by $\pi_k^{-1}$ and that there are only a finite number of different $\alpha_k \in \mathbb{R}$, say $\alpha_1, \ldots, \alpha_p$ which are cyclically repeated to give the sequence $A$. We refer to this as the cyclic case. Generalizations to the noncyclic case were considered in [9, 10, 5] which treat the trigonometric analog, i.e., where the real line is replaced by the unit circle. See also [14] where the poles are in a finite subset of $\mathbb{R}$. Some aspects of the problem in the case where the cyclic sequence of $\alpha_k$'s is replaced by an arbitrary sequence of real numbers are discussed in [4, 7, 6].

We remark that if we assume all the moments to be finite, then, since $b_n$ has poles at all the points in $A_n = \{\alpha_1, \ldots, \alpha_n\}$, $\mu$ can not have a mass point in any of the $\alpha_k$. In the classical polynomial case, all the poles are at $\infty$ and therefore it is then natural to take integrals over $\mathbb{R}$ rather than over $\hat{\mathbb{R}}$ because $\infty$ can not be a mass point. In general however, if none of the poles is infinite, a function in $\mathcal{L}_n$ will be finite at $\infty$ and a mass point at $\infty$ may well be possible, that is why we take integrals over $\hat{\mathbb{R}}$.

We preferred to use the $b_n$ to define the moments instead of the $\pi_n^{-1}$ because this allows us to treat $\alpha_k = \infty$ in the same way as the other finite $\alpha_k$. The polynomial situation of the classical Hamburger moment problem drops out as a special case.

For completeness, we mention that also a strong version of the Hamburger (and other) moment problem has been studied. In that case the power moments $\int t^k d\mu(t)$ are prescribed for $k \in \mathbb{Z}$. Neither this problem nor its rational generalization will be considered in this paper. Also the problem where all the $\alpha_k$ are in the upper half plane instead of on its boundary (i.e. the real line) has been considered in a series of papers, mainly in its trigonometric form. See for example [3, 4] for an introduction. Here we only discuss the boundary case.

What we do want to discuss in this paper is one aspect which shows up in the analysis of this kind of moment problem, namely the construction of rational interpolants for the Nevanlinna transform [1, p. 92]

$$\Omega_\mu(z) = \int D(t,z)d\mu(t), \quad D(t,z) = -i\frac{1+tz}{t-z}.$$

This function is holomorphic in the upper half plane (in fact everywhere off the real line) and it maps the open upper half plane to the closed right half plane. We shall call the class of such functions the class $\mathcal{P}$ of positive real functions (they have a nonnegative real part in the upper half plane). These functions have a pseudo-analytic continuation to the lower half plane by the definition

$$f(z) = \overline{f(\overline{z})}.$$

The boundary values exist a.e. and one has

$$\lim_{y\downarrow 0} f(x+iy) = \lim_{h\downarrow 0}\frac{\mu(x-h,x+h)}{2h}$$

which can be used in an inversion formula for the Nevanlinna transform.

In the polynomial case, the Nevanlinna kernel $D(t,z)$ is often replaced by the simpler kernel $-i(t-z)^{-1}$. However, this is only possible if [1, p. 93]

$$\sup_{y\geq 1} |y\Omega(iy)| < \infty.$$

If $\mu$ has a mass point at $\infty$, this condition may not be satisfied and that is why we use here the more complicated kernel $D(t,z)$.

The rational interpolants $\Omega_n$ of degree $n$ we are looking for will be obtained as the Nevanlinna transform of a discrete measure $\mu_n$ with $n$ mass points

$$\Omega_n(z) = \int D(t,z)d\mu_n(t) = \sum_{k=1}^{n}\lambda_{nk}D(\xi_{nk},z)$$

with $\xi_{nk} \in \hat{\mathbb{R}}$ and $\lambda_{nk} > 0$. The construction of such a measure will be based on quasi-orthogonal functions (just as in the polynomial or the cyclic case). However, if we want to exploit orthogonality, we need an inner product in $\mathcal{L} = \cup_{n=0}^{\infty}\mathcal{L}_n$. Now in the polynomial case or the cyclic case we have $\mathcal{R} = \mathcal{L}\cdot\mathcal{L} = \mathcal{L}$. Therefore we can define an inner product in $\mathcal{L}$ given the moments $m_k$, $k = 0, 1, \ldots$ However when we are in the noncyclic case, then $\mathcal{R} = \mathcal{L}\cdot\mathcal{L} \neq \mathcal{L}$. Thus to define an inner product in $\mathcal{L}$, we need to assume that all the integrals $\int f(t)d\mu(t)$ exist for $f \in \mathcal{R}$. This means that all the moments $m_{k,l} = \int b_k(t)b_l(t)d\mu_n(t)$, $k, l = 0, 1, \ldots$ should be defined. We call these the moments in $\mathcal{R}$ while the previous moments are the moments in $\mathcal{L}$. The measure $\mu_n$ that will be constructed shall fit a number of the moments in $\mathcal{R}$, namely

$$\int b_k(t)b_l(t)d\mu_n(t) = m_{kl} = \int b_k(t)b_l(t)d\mu(t), \quad k,l = 0,1,\ldots,n-1.$$

Note that this implies that it also fits the moments in $\mathcal{L}$

$$\int b_k(t)d\mu_n(t) = m_k = \int b_k(t)d\mu(t), \quad k = 0, 1, \ldots, n-1.$$

Although we do not want to solve the moment problem here, we mention that essentially by letting $n$ tend to $\infty$ and by using Helly's theorems, it follows that an accumulation point of the sequence $\{\mu_n\}$ will be a solution of the moment problem in $\mathcal{R}$ and thus also of the moment problem in $\mathcal{L}$ [16, 8].

The interpolation properties that we shall give below generalize the polynomial case of Akhiezer [1, p. 22] and for the cyclic case in [14, 10]. Analogous results were obtained in the trigonometric case for the noncyclic case (using a different approach) in [6].

## 2 Orthogonal rational functions and quadrature

In order to construct the quadrature formulas we gave at the end of the previous section, we first have to introduce orthogonal rational functions. Suppose $\mu$ is a finite positive measure on $\hat{\mathbb{R}}$. We suppose it to be normalized by $m_0 = \int d\mu(t) = 1$. Let $\alpha_1, \alpha_2, \ldots$ be a sequence of points in $\hat{\mathbb{R}} \setminus \{0\}$ and define the basis functions $b_k$ and $\mathcal{L}_n = \text{span}\{b_0, \ldots, b_n\}$ as before. We also introduce $\mathcal{R}_n = \mathcal{L}_n \cdot \mathcal{L}_n = \{fg : f, g \in \mathcal{L}_n\}$. Furthermore, let $\mathcal{L} = \cup_{n=0}^{\infty}\mathcal{L}_n$, $\Pi_n = \text{span}\{1, z, \ldots, z^n\}$, $\Pi = \cup_{n=0}^{\infty}\Pi_n$ and $\mathcal{R} = \cup_{b=0}^{\infty}\mathcal{R}_n$. Note that if all $\alpha_k = \infty$, then $\mathcal{L}_n = \Pi_n$ and $\mathcal{R}_n = \Pi_{2n}$. Assume also that the inner product

$$\langle f, g\rangle = \int f(t)\overline{g(t)}d\mu(t)$$

is well defined for $f, g \in \mathcal{L}$. This means that we should have

$$\int |b_n(t)|^2 d\mu(t) < \infty, \quad \forall n = 0, 1, 2, \ldots$$

Let $\phi_0, \phi_1, \ldots$ be an orthonormal basis obtained by applying the Gram-Schmidt procedure on the basis $b_0, b_1, b_2, \ldots$. If

$$\phi_n(z) = \phi(0) + \cdots + \kappa_n' b_{n-1}(z) + \kappa_n b_n(z)$$

then we normalise $\phi_n$ by taking $\kappa_n > 0$. With these orthonormal functions we associate the functions of the second kind defined by

$$\psi_0 = iz, \quad \psi_n(z) = \int D(t, z)[\phi_n(t) - \phi_n(z)]d\mu(t), \quad n \geq 1.$$

Note that $\psi_n \in \mathcal{L}_n$ for $n \geq 1$.

Such an orthonormal function $\phi_n(z) = p_n(z)/\pi_n(z)$ is called regular if $p_n(\alpha_{n-1}) \neq 0$. If $\alpha_{n-1} = \infty$, then $p_n(\alpha_{n-1}) \neq 0$ means that the polynomial $p_n$ has degree $n$ (and not lower). We call the sequence $\{\phi_n\}$ regular if all $\phi_n$ in the sequence are regular.

Quasi-orthogonal functions are defined for $\tau \in \hat{\mathbb{R}}$ as

$$Q_n(z,\tau) = \phi_n(z) + \tau \frac{Z_n(z)}{Z_{n-1}(z)} \phi_{n-1}(z), \quad k \geq 1.$$

If $\tau = \infty$, then this should be interpreted as

$$Q_n(z,\infty) = \frac{Z_n(z)}{Z_{n-1}(z)} \phi_{n-1}(z).$$

The associated functions are given by

$$P_n(z,\tau) = \psi_n(z) + \tau \frac{Z_n(z)}{Z_{n-1}(z)} \psi_{n-1}(z), \quad \tau \in \hat{\mathbb{R}}, \quad n \geq 1$$

with a similar interpretation for $\tau = \infty$.

Along the same lines as in the proof of [5] for the case of the unit circle, or as in [12] for the cyclic situation of the real line, one can prove

THEOREM 2.1 *Let the sequence $\{\phi_n\}$ of orthonormal functions be regular. Then for each $n$ there are infinitely many so called regular values $\tau \in \hat{\mathbb{R}}$ for which the quasi-orthogonal functions $Q_n(z,\tau)$ have precisely $n$ simple zeros $\xi_{nk}(\tau)$, $k = 1, \ldots, n$ in $\hat{\mathbb{R}} \setminus \{0, \alpha_1, \ldots, \alpha_n\}$. Furthermore, if we define*

$$\lambda_{nk}(\tau) = \left[ \sum_{i=0}^{n-1} |\phi_i(\xi_{nk}(\tau))|^2 \right]^{-1}, \quad k = 1, \ldots, n$$

*and if $\mu_n(\cdot,\tau)$ is the discrete measure wich takes masses $\lambda_{nk}(\tau)$ at the points $\xi_{nk}(\tau)$, then*

$$\int f(t) d\mu(t) = \int f(t) d\mu_n(t,\tau)$$

*for all $f \in \mathcal{R}_{n-1} = \mathcal{L}_{n-1} \cdot \mathcal{L}_{n-1}$. If $\tau = 0$ is a regular value, then the formula holds for all $f \in \mathcal{L}_n \cdot \mathcal{L}_{n-1}$.*

The theorem says that for general $\tau \in \hat{\mathbb{R}}$, the quadrature formula

$$\int f(t) d\mu_n(t,\tau) = \sum_{k=1}^{n} \lambda_{nk}(\tau) f(\xi_{nk}(\tau))$$

is exact in $\mathcal{R}_{n-1}$ or equivalently that the inner product in $\mathcal{L}_{n-1}$ with respect to $\mu$ and with respect to $\mu_n$ is the same.

## 3 MEASURES AND INTERPOLATION

In this section we prove that equality of integrals with respect to two measures implies interpolation properties for their Nevanlinna transforms. This interpolation will hold at the points $\alpha_k \in \hat{\mathbb{R}}$. Since a Nevanlinna transform $\Omega_\mu(z)$ is defined by the integral $\int D(t,z) d\mu(t)$ for $z$ off $\hat{\mathbb{R}}$, we have to consider boundary values for $\Omega_\mu$ in these points $\alpha_k$. These are obtained by taking nontangential limits. For simplicity, we shall assume we take limits along vertical lines and call them orthogonal limits.

We set by definition for $t \in \mathbb{R}$ (note the special arrow $\mapsto$ used in the limit)

$$\lim_{z \mapsto t} f(z) = \lim_{y \downarrow t} f(t + iy)$$

while for $t = \infty$, we set

$$\lim_{z \mapsto \infty} f(z) = \lim_{z \mapsto 0} f(1/z).$$

Furthermore we introduce the notation $A_n$ to denote the set $A_n = \{\alpha_1, \alpha_2, \ldots, \alpha_n\}$ and $\tilde{A}_{n-1}$ for the sequence (including repetitions) $\tilde{A}_{n-1} = (\alpha_1, \alpha_1, \ldots, \alpha_{n-1}, \alpha_{n-1})$ while $\tilde{A}'_n$ adds the single point $\alpha_n$ to it: $\tilde{A}'_n = (\alpha_1, \alpha_1, \alpha_2, \alpha_2, \ldots, \alpha_{n-1}, \alpha_{n-1}, \alpha_n)$. By $\alpha^{\#}$ we indicate the multiplicity of $\alpha \in A_{n-1}$ in the sequence $\tilde{A}_{n-1}$, that is the number of times $\alpha$ occurs in the sequence $\tilde{A}_{n-1}$. At other places, the same notation is also used to indicate the multiplicity of $\alpha \in A_n$ in the sequence $\tilde{A}'_n$

Let $\mu$ be such that $\mathcal{L}_n \cdot \mathcal{L}_{n-1} \subset L^1(\mu)$ and denote its Nevanlinna transform as

$$\Omega_\mu(z) = \int D(t, z) d\mu(t).$$

Then we have

LEMMA 3.1 *The integrals for functions in $\mathcal{L}_n \cdot \mathcal{L}_{n-1}$ are completely defined by the values*

$$\Omega_\mu(i) \quad \textit{and} \quad \lim_{w \mapsto \alpha} \Omega_\mu^{(k)}(w), \quad \alpha \in A_n, \quad k = 0, 1, \ldots, \alpha^{\#} - 1$$

*where the superscript $^{(k)}$ denotes the kth derivative, $A_n = \{\alpha_1, \alpha_1, \ldots, \alpha_n\}$ and $\alpha^{\#}$ is the multiplicity of $\alpha$ in the sequence $\tilde{A}'_n = (\alpha_1, \alpha_1, \alpha_2, \alpha_2, \ldots, \alpha_{n-1}, \alpha_{n-1}, \alpha_n)$.*

Proof. These integrals involve the computation of integrals of the form $\int b_k b_l d\mu$. By partial fraction decomposition of the integrand, it is seen that besides $\Omega_\mu(i) = \int d\mu(t)$ these will be completely defined in terms of integrals of the form

$$\int \frac{d\mu(t)}{(t-\alpha)^{k+1}}, \quad k = 0, 1, \ldots, \alpha^{\#} - 1 \tag{3.1}$$

where $\alpha$ ranges over all finite elements of $A_n$ and if $\alpha = \infty$ is also in $A_n$, then we should add integrals of the form

$$\int t^{k+1} d\mu(t), \quad k = 0, 1, \ldots, \infty^{\#} - 1. \tag{3.2}$$

For $\alpha$ finite, it is easily seen that taking the $k$th derivative with respect to the variable $z$, we have

$$\partial_z^k \Omega_\mu(z) = \int \partial_z^k D(t, z) d\mu(t) = -i(-1)^k k! \int \frac{d\mu(t)}{(t-z)^{k+1}}. \tag{3.3}$$

If $\alpha \in \mathbb{R}$ and $\epsilon > 0$, then

$$\frac{1}{|t-\alpha|} - \frac{1}{|t-\alpha-i\epsilon|} \leq \left| \frac{1}{t-\alpha} - \frac{1}{t-\alpha-i\epsilon} \right| = \frac{\epsilon}{|t-\alpha||t-\alpha-i\epsilon|}.$$

For $t \in \hat{\mathbb{R}}$, it holds that $|t - \alpha - i\epsilon| \geq \epsilon$. Hence the previous expression is bounded by $|t - \alpha|^{-1}$ and therefore we have for $z$ in the upper half plane and $t \in \hat{\mathbb{R}}$ that $|t - z|^{-1} \leq 2|t - \alpha|^{-1}$. Now we can apply the dominated convergence theorem when taking the orthogonal limit $z \mapsto \alpha$ in (3.3). So we see that there is a one-to-one correspondence between the generalized moments (3.1) and the values $\lim_{z \mapsto \alpha} \Omega_\mu^{(k)}(\alpha)$, $k = 0, 1, \ldots, \alpha^\# - 1$.

By a similar argument and recalling that $\lim_{z \mapsto \infty} \Omega_\mu(z) = \lim_{z \mapsto 0} \Omega_\mu(1/z)$, it is found that the moments (3.2) are defined by the values $\lim_{z \mapsto \infty} \Omega_\mu^{(k)}(z)$, $k = 0, 1, \ldots, \infty^\# - 1$. □

Using this lemma in combination with Theorem 2.1, immediately gives the following theorem.

THEOREM 3.2 *With the conditions and notation of Theorem 2.1, we define*

$$\Omega_\mu(z) = \int D(t, z) d\mu(t) \quad \textit{and} \quad \Omega_n(z, \tau) = \int D(t, z) d\mu_n(t, \tau).$$

*Then*

$$\Omega_\mu(i) = \Omega_n(i, \tau) \quad \textit{and} \quad \lim_{w \mapsto \alpha} [\Omega_\mu^{(k)}(w) - \Omega_n^{(k)}(w, \tau)] = 0,$$
$$\textit{for} \quad \alpha \in A_{n-1}, \quad k = 0, 1, \ldots, \alpha^\# - 1$$

*with* $A_{n-1} = \{\alpha_1, \alpha_2, \ldots, \alpha_{n-1}\}$ *and* $\alpha^\#$ *the multiplicity of* $\alpha$ *in the sequence* $\tilde{A}_{n-1} = (\alpha_1, \alpha_1, \alpha_2, \alpha_2, \ldots, \alpha_{n-1}, \alpha_{n-1})$.
*If* $\tau = 0$ *is a regular value, then with* $\Omega_n(z) = \Omega_n(z, 0)$

$$\Omega_\mu(i) = \Omega_n(i) \quad \textit{and} \quad \lim_{w \mapsto \alpha} [\Omega_\mu^{(k)}(w) - \Omega_n^{(k)}(w)] = 0, \quad \alpha \in A_n, \quad k = 0, 1, \ldots, \alpha^\# - 1$$

*with* $A_n = \{\alpha_1, \alpha_2, \ldots, \alpha_n\}$ *and* $\alpha^\#$ *the multiplicity of* $\alpha$ *in the sequence* $\tilde{A}'_n = (\alpha_1, \alpha_1, \alpha_2, \alpha_2, \ldots, \alpha_{n-1}, \alpha_{n-1}, \alpha_n)$.

Proof. Because by Theorem 2.1, the inner product in $\mathcal{L}_{n-1}$ is the same when using the measure $\mu$ or the measure $\mu_n$. Hence by the previous lemma, we have the interpolation properties as stated.

When $\tau = 0$ is a regular value, then the quadrature formulas are exact in $\mathcal{L}_{n-1} \cdot \mathcal{L}_n$ and we have the extra interpolation as described above. □

## 4 AN ALTERNATIVE EXPRESSION FOR $\Omega_n(z, \tau)$

By its definition we have the following expression for the Nevanlinna function $\Omega_n(z, \tau)$

$$\Omega_n(z, \tau) = \sum_{k=1}^{n} \lambda_{nk}(\tau) D(\xi_{nk}(\tau), z) = i \sum_{k=1}^{n} \lambda_{nk}(\tau) \frac{1 + \xi_{nk}(\tau) z}{z - \xi_{nk}(\tau)}.$$

It is however possible to give a simple expression in terms of the quasi-orthogonal functions and their associates.

To reach this goal, we first give the following lemma.

LEMMA 4.1 *Let $\phi_n$ be the orthonormal functions and $\psi_n$ the associated functions of the second kind. Then for $n > 0$ and for any function $f$ such that (as a function of $t$), $D(t,z)[f(t) - f(z)] \in \mathcal{L}_{n-1}$ we have*

$$\psi_n(z)f(z) = \int D(t,z)[\phi_n(t)f(t) - \phi_n(z)f(z)]d\mu(t).$$

Proof. This expression for $\psi_n$ corresponds to the original definition if it holds that

$$\int D(t,z)\phi_n(t)[f(t) - f(z)]d\mu(t) = 0.$$

If $D(t,z)[f(t) - f(z)] \in \mathcal{L}_{n-1}$, then this is zero because $\phi_n \perp \mathcal{L}_{n-1}$. □

We have as a corollary the following result.

LEMMA 4.2 *Let $Q_n(z,\tau)$ be the quasi-orthogonal functions and $P_n(z,\tau)$ the associated functions. Then for $n \geq 2$*

$$\frac{Z_{n-1}(z)}{Z_n(z)}P_n(z,\tau) = \int D(t,z)\left[\frac{Z_{n-1}(t)}{Z_n(t)}Q_n(t,\tau) - \frac{Z_{n-1}(z)}{Z_n(z)}Q_n(z,\tau)\right]d\mu(t).$$

Proof. The right hand side is

$$\int D(t,z)\left[\frac{Z_{n-1}(t)}{Z_n(t)}\phi_n(t) - \frac{Z_{n-1}(z)}{Z_n(z)}\phi_n(z)\right]d\mu(t)$$
$$+\tau\int D(t,z)[\phi_{n-1}(t) - \phi_{n-1}(z)]d\mu(t).$$

The second term equals $\psi_{n-1}(z)$ by definition. The first term turns out to be

$$\frac{Z_{n-1}(z)}{Z_n(z)}\psi_n(z)$$

by the previous lemma when we use $f(z) = Z_{n-1}(z)/Z_n(z)$. Thus the lemma is proved. □

We can now prove the expression for $\Omega_\mu(z,\tau)$.

THEOREM 4.3 *Let $\mu_n(\cdot,\tau)$ be the discrete measure of the quadrature formula, as it was constructed in the previous sections. Denote by $Q_n(z,\tau)$ and $P_n(z,\tau)$ the quasi-orthogonal functions and the associated functions of the second kind. Then for $n \geq 2$*

$$\Omega_n(z,\tau) = \int D(t,z)d\mu(t) = -\frac{P_n(z,\tau)}{Q_n(z,\tau)}.$$

Proof. By the previous lemma

$$\frac{Z_{n-1}(z)}{Z_n(z)}P_n(z,\tau) = \int D(t,z)\left[\frac{Z_{n-1}(t)}{Z_n(t)}Q_n(t,\tau) - \frac{Z_{n-1}(z)}{Z_n(z)}Q_n(z,\tau)\right]d\mu(t).$$

It is immediately checked that the integrand is in $\mathcal{R}_{n-1}$ and therefore we can replace $d\mu$ by $d\mu_n$. Furthermore, recalling that $Q_n(\xi_{nk}(\tau),\tau) = 0$ by definition, we get

$$\frac{Z_{n-1}(z)}{Z_n(z)}P_n(z,\tau) = -\left[\sum_{k=1}^{n} D(\xi_{nk}(\tau),z)\lambda_{nk}(\tau)\right]\frac{Z_{n-1}(z)}{Z_n(z)}Q_n(z,\tau).$$

In the square brackets, we recognize $\Omega_n(z,\tau)$ so that the required equality follows. □

Note that for the case when $\tau = 0$ is a regular value, we have

$$\Omega_n(z,0) = -\frac{\psi_n(z)}{\phi_n(z)}, \quad n \geq 2.$$

To conclude, we check how the polynomial case can be recovered from our general formulation. So assume that all $\alpha_k = \infty$. Theorem 3.2 says that in that case

$$\Omega_\mu(i) = \Omega_n(i,\tau) \quad \text{and} \quad \lim_{z\mapsto\infty}[\Omega_\mu^{(k)}(z) - \Omega_n^{(k)}(z,\tau)] = 0, \quad k = 0,1,\ldots,2n-3.$$

Therefore we have

$$\left\{-\frac{P_n(z,\tau)}{Q_n(z,\tau)} - i\left[s_0 + \frac{s_1}{z} + \frac{s_2}{z^2} + \cdots + \frac{s_{2n-3}}{z^{2n-3}}\right]\right\}z^{2n-3} \to 0, \quad z \to \infty \tag{4.1}$$

where

$$s_0 = m_1, \quad s_k = m_{k-1} + m_{k+1} \quad k \geq 1, \quad m_k = \int t^k d\mu(t).$$

This can be seen as follows.

$$D(t,z) = -i\frac{1+tz}{t-z} = i\left[t + \frac{1+t^2}{z} + \frac{(1+t^2)t}{z^2} + \cdots + \frac{(1+t^2)t^{l-1}}{z^l} + \frac{1+t^2)t^l}{(z-t)z^l}\right].$$

Thus,

$$\Omega_\mu(t) = \int D(t,z)d\mu(t) = i\left[s_0 + \frac{s_1}{z} + \frac{s_2}{z^2} + \cdots + \frac{s_{2n-3}}{z^{2n-3}} + \int \frac{(1+t^2)t^{2n-3}}{(z-t)z^{2n-3}}d\mu(t)\right].$$

However, because the quadrature formula is exact for all polynomials in $\Pi_{2n-2}$, it follows that the first $2n-2$ coefficients in these expansions for $\Omega_n$ and $\Omega_\mu$ at $\infty$ are the same. Since the integral of the last term in the expansion for $\Omega_\mu$ vanishes when it is multiplied by $z^{2n-3}$ and when $z \mapsto \infty$, the result (4.1) follows.

We should note that in Akhiezer [1, p. 22], not the kernel $D(t,z)$ is used but the simpler kernel $(t-z)^{-1}$. Therefore (and this can also be seen from the previous formulas) one has that $\int(t-z)^{-1}d\mu_n(t,\tau)$ is a rational function where the coefficients of the asymptotic expansion at $\infty$ matches the coefficients in the expansion of $\int(t-z)^{-1}d\mu(t)$ up to the coefficient in $z^{-(2n-2)}$. Note that in the polynomial case $\tau = 0$ is a regular value for each $n$. Thus we can choose $\tau = 0$ and we get in that case correspondence up to the coefficient in $z^{-(2n-1)}$. With our choice of the kernel $D(t,z)$, this extra interpolation property at $\infty$ is replaced by an interpolation property at $i$.

## 5 References


[1] N.I. Akhiezer [Achieser]. *The classical moment problem.* Oliver and Boyd, Edinburgh, 1969. Originally published Moscow, 1961.

[2] E. Alden. A survey of weak and strong moment problems with generalizations. Technical report, Dept. Math. Umeå, 1985.

[3] A. Bultheel, P. González-Vera, E. Hendriksen, and O. Njåstad. The computation of orthogonal rational functions and their interpolating properties. *Numer. Algorithms*, 2(1):85–114, 1992.

[4] A. Bultheel, P. González-Vera, E. Hendriksen, and O. Njåstad. Moment problems and orthogonal functions. *J. Comput. Appl. Math.*, 48:49–68, 1993.

[5] A. Bultheel, P. González-Vera, E. Hendriksen, and O. Njåstad. Orthogonal rational functions with poles on the unit circle. *J. Math. Anal. Appl.*, 182:221–243, 1994.

[6] A. Bultheel, P. González-Vera, E. Hendriksen, and O. Njåstad. Orthogonality and boundary interpolation. In A.M. Cuyt, editor, *Nonlinear Numerical Methods and Rational Approximation II*, pages 37–48. Kluwer, 1994.

[7] A. Bultheel, P. González-Vera, E. Hendriksen, and O. Njåstad. Recurrence relations for orthogonal functions. In S.C. Cooper and W.J. Thron, editors, *Continued Fractions and Orthogonal Functions*, volume 154 of *Lecture Notes in Pure and Appl. Math.*, pages 24–46. Marcel Dekker, 1994.

[8] A. Bultheel, P. González-Vera, E. Hendriksen, and O. Njåstad. A rational moment problem on the unit circle. *Methods Appl. Anal.*, 1997. Accepted.

[9] P. González-Vera and O. Njåstad. Szegő functions and multipoint padé approximation. *J. Comput. Appl. Math.*, 32:107–116, 1990.

[10] E. Hendriksen and O. Njåstad. A Favard theorem for rational functions. *J. Math. Anal. Appl.*, 142(2):508–520, 1989.

[11] H.J. Landau, editor. *Moments in mathematics*, volume 37 of *Proc. Sympos. Appl. Math.* Amer. Math. Soc., Providence, R.I., 1987.

[12] O. Njåstad. An extended Hamburger moment problem. *Proc. Edinburgh Math. Soc.*, 28:167–183, 1985.

[13] O. Njåstad. Unique solvability of an extended Hamburger moment problem. *J. Math. Anal. Appl.*, 124:502–519, 1987.

[14] O. Njåstad. Multipoint Padé approximation and orthogonal rational functions. In A. Cuyt, editor, *Nonlinear numerical methods and rational approximation*, pages 258–270, Dordrecht, 1988. D. Reidel Publ. Comp.

[15] O. Njåstad. A modified Schur algorithm and an extended Hamburger moment problem. *Trans. Amer. Math. Soc.*, 327(1):283–311, 1991.

[16] O. Njåstad. Classical and strong moment problems. *Comm. Anal. Th. Continued Fractions*, 4:4–38, 1995.

# Symmetric Orthogonal Laurent Polynomials

LYLE COCHRAN, Department of Mathematics and Computer Science, Whitworth College, Spokane, WA 99251-3702, USA

S. CLEMENT COOPER, Department of Pure and Applied Mathematics, Washington State University, Pullman, WA 99164-3113, USA

## 1 INTRODUCTION

In this paper, we investigate a class of functions called *symmetric orthogonal Laurent polynomials.* These Laurent polynomials originate from strong moment functionals whose odd-indexed moments vanish. Chebyshev Laurent polynomials [3] provide an example of these polynomials. In this section we give some basic definitions, followed by a brief review of results found in [2].

In the next section we establish a relationship between a SMF and a symmetric SMF which will be used to derive results about symmetric SMF's. The results include information about the support, positive-definiteness, three term recurrence relations, coefficients in the three term recurrence relations, conditions on the Hankel determinants, a Favard-type theorem and separation of zeros.

In section 3, associated symmetric orthogonal Laurent polynomials and related continued fractions are introduced. A separation of zeros property is given for the associated symmetric orthogonal Laurent polynomials.

The paper concludes with the derivation of integral representations for positive-definite symmetric SMF's. Step functions are constructed using the associated symmetric orthogonal Laurent polynomials which are then used to produce integral representations with the help of Grommer's and Helly's Selection Theorems.

A **Laurent polynomial** or **L-polynomial** is a function of a real variable $x$ of the form $R(x) = \sum_{j=m}^{n} r_j x^j$, where $m, n \in \mathbb{Z}$ and $r_j \in \mathbb{C}$ for $j = m, \ldots, n$. If $n \geq 0$, $m = -n$ and $r_n \neq 0$, then $R(x)$ is said to be of **L-degree** $2n$. In this case, $r_n$ is called the **L-leading coefficient** and $r_{-n}$ is called the **L-trailing coefficient**. If $m < 0$, $m = -n-1$ and $r_{-n-1} \neq 0$, then $R(x)$ is said to be of **L-degree** $2n+1$ and $r_{-n-1}$ is called the **L-leading coefficient** of $R(x)$ while $r_n$ is called the **L-trailing coefficient**. Note that in any case, the L-leading coefficient is always defined to be nonzero. An L-polynomial is said to be **monic** if the L-leading coefficient is equal to one. An L-polynomial $R(x)$ of **L-degree** $n$ is said to be **regular** if the L-trailing coefficient is also nonzero. A sequence $\{R_n(x)\}_{n=0}^{\infty}$ is said to a **regular sequence** of L-polynomials if for every $k \in \mathbb{Z}_0^+$, $R_k(x)$ is a regular L-polynomial of L-degree $k$.

Suppose $\{\mu_n\}_{-\infty}^{\infty}$ is a bisequence of complex numbers and let $\mathcal{L}$ be a complex-valued functional satisfying

$$\mathcal{L}[R(x)] = \sum_{j=m}^{n} r_j \mu_j, \quad m, n \in \mathbb{Z}$$

for any L-polynomial $R(x) = \sum_{j=m}^{n} r_j x^j$. The linear functional $\mathcal{L}$ is called the **strong moment functional** (abbreviated SMF) with respect to $\{\mu_n\}_{-\infty}^{\infty}$. Note that $\mathcal{L}[x^n] = \mu_n$ for all $n \in \mathbb{Z}_0^+$. In this paper we focus on the special case where $\mathcal{L}[x^{2m+1}] = 0$ for all $m \in \mathbb{Z}_0^+$. A SMF $\mathcal{L}$ is said to be **symmetric** if

$$\mathcal{L}[z^{2n+1}] = 0 \quad \text{for all } n \in \mathbb{Z}.$$

(In [4, p. 23], the symmetric case is called the *lacunary* case.)

A sequence $\{R_n(x)\}_{n=0}^{\infty}$ is called an **orthogonal Laurent polynomial sequence** (abbreviated OLPS) with respect to the SMF $\mathcal{L}$ if $R_k(x)$ is of L-degree $k$ for all $k \in \mathbb{Z}_0^+$ and

$$\mathcal{L}[R_m(x)R_n(x)] = K_n \delta_{m,n}$$

for all $m, n \in \mathbb{Z}_0^+$ were $\delta_{m,n}$ is Kronecker's delta function and $K_n \neq 0$ for all $n \in \mathbb{Z}_0^+$. If $K_n = 1$ for all $n \in \mathbb{Z}_0^+$, then we call $\{R_n(x)\}_{n=0}^{\infty}$ an **orthonormal** Laurent polynomial sequence. An OLPS corresponding to a symmetric SMF is defined to be a **symmetric OLPS**.

An important example of a symmetric OLPS is the strong Chebyshev Laurent polynomial sequence, found in [3]. In this case,

$$\mathcal{L}(x^n) = \int_B x^n \, d\Psi(x)$$

for all $n \in \mathbb{Z}$ where $B = [-b, a] \cup [a, b]$, $0 < a < b < \infty$ and

$$\Psi'(x) = \begin{cases} \frac{|x|}{\sqrt{b^2 - x^2}\sqrt{x^2 - a^2}}, & x \in B \\ 0 & x \notin B \end{cases}.$$

Explicit formulae for the corresponding OLPS can be found in [3].

As illustrated in [2], for a given SMF, a corresponding OLPS may not exist. A SMF, $\mathcal{L}$, for which an OLPS does exist is said to be **quasi-definite**. In order to discuss necessary and sufficient conditions for the existence of an OLPS we must introduce Hankel determinants. A **Hankel** determinant $H_k^{(n)}$ associated with a given bisequence $\{\mu_n\}_{-\infty}^{\infty}$ is defined by

$$H_0^{(n)} = 1, H_k^{(n)} = \begin{vmatrix} \mu_n & \mu_{n+1} & \cdots & \mu_{n+k-1} \\ \mu_{n+1} & \mu_{n+2} & \cdots & \mu_{n+k} \\ \vdots & \vdots & \ddots & \vdots \\ \mu_{n+k-1} & \mu_{n+k} & \cdots & \mu_{n+2k-2} \end{vmatrix}$$

for all $n \in \mathbb{Z}_0^+$ and $k \in \mathbb{Z}^+$.

The theorems that follow summarize some basic results concerning orthogonal Laurent polynomials and the proofs can be found in [2]. Henceforth we will be working with sequences of Laurent polynomials, so in order to distinguish between the coefficients of the different Laurent polynomials, we will adopt the double subscript convention $R_n(x) = \sum_{j=r}^{s} r_{n,j}x^j$.

**THEOREM 1.1** *Suppose $\mathcal{L}$ is a SMF with the moment sequence $\{\mu_n\}_{-\infty}^{\infty}$.*
*(a) The SMF $\mathcal{L}$ is quasi-definite if and only if*

$$H_{2m}^{(-2m)} \neq 0 \quad and \quad H_{2m+1}^{(-2m)} \neq 0 \tag{1}$$

*for all $m \in \mathbb{Z}_0^+$.*
*(b) If $\{R_n(x)\}_{n=0}^{\infty}$ is an OLPS for $\mathcal{L}$, then each $R_n(x)$ is uniquely determined up to an arbitrary nonzero factor. That is, if $\{Q_n(x)\}_{n=0}^{\infty}$ is also an OLPS for $\mathcal{L}$, then there exist constants $c_n \neq 0$ such that*

$$Q_n(x) = c_n R_n(x),$$

*for all $n \in \mathbb{Z}_0^+$.*
*(c) If $\mathcal{L}$ is a quasi-definite SMF and $\{R_n(x)\}_{n=0}^{\infty}$ is the corresponding monic OLPS, then*

$$\mathcal{L}[x^{-(m+1)}R_{2m}(x)] = -r_{2m+1,m}\mathcal{L}[R_{2m}^2(x)] = \frac{H_{2m+1}^{-(2m+1)}}{H_{2m}^{(-2m)}} \tag{2}$$

*and*

$$\mathcal{L}[x^{(m+1)}R_{2m+1}(x)] = -r_{2m+2,-(m+1)}\mathcal{L}[R_{2m+1}^2(x)] = \frac{-H_{2m+2}^{-(2m+1)}}{H_{2m+1}^{(-2m)}} \tag{3}$$

*for all $m \in \mathbb{Z}_0^+$.*
*(d) If $\mathcal{L}$ is quasi-definite, then the corresponding OLPS $\{R_n(x)\}_{n=0}^{\infty}$ is regular if and only if $H_{2m}^{-2m+1} \neq 0$ and $H_{2m+1}^{(-2m-1)} \neq 0$ for all $m \in \mathbb{Z}_0^+$.*
*(e) Suppose $\mathcal{L}$ is quasi-definite with monic OLPS. For each $m \in \mathbb{Z}_0^+$,*

$$R_{2m}(x) = \frac{1}{H_{2m}^{(-2m)}} \begin{vmatrix} \mu_{-2m} & \cdots & \mu_{-1} & x^{-m} \\ \vdots & \ddots & \vdots & \vdots \\ \mu_{-1} & \cdots & \mu_{2m-2} & x^{m-1} \\ \mu_0 & \cdots & \mu_{2m-1} & x^m \end{vmatrix}$$

*and*

$$R_{2m+1}(x) = \frac{-1}{H_{2m+1}^{(-2m)}} \begin{vmatrix} \mu_{-(2m+1)} & \cdots & \mu_{-1} & x^{-(m+1)} \\ \mu_{(-2m)} & \cdots & \mu_0 & x^{-m} \\ \vdots & \ddots & \vdots & \vdots \\ \mu_0 & \cdots & \mu_{2m} & x^m \end{vmatrix}.$$

*(f) If* $\{R_n(x)\}_{n=0}^{\infty}$ *is a regular OLPS corresponding to* $\mathcal{L}$ *where*

$$R_{2m}(x) = \sum_{j=-m}^{m} r_{2m,j}x^j \quad \text{and} \quad R_{2m+1}(x) = \sum_{j=-(m+1)}^{m} r_{2m+1,j}x^j$$

*for all* $m \in \mathbb{Z}_0^+$, *then*

$$R_{2m+1}(x) = \left(\frac{x^{-1}}{\alpha_{2m+1}} + \beta_{2m+1}\right) R_{2m}(x) + \lambda_{2m+1} R_{2m-1}(x) \tag{4}$$

*and*

$$R_{2m+2}(x) = \left(\frac{x}{\alpha_{2m+2}} + \beta_{2m+2}\right) R_{2m+1}(x) + \lambda_{2m+2} R_{2m}(x) \tag{5}$$

*where*

$$\alpha_{2m+1} = \frac{H_{2m}^{-(2m-1)}}{H_{2m}^{-(2m)}}, \quad \alpha_{2m+2} = -\frac{H_{2m+1}^{-(2m+1)}}{H_{2m+1}^{-(2m)}}, \tag{6}$$

$$\beta_{m+1} = \alpha_{m+2} \tag{7}$$

*and* $\lambda_1$ *is arbitrary,*

$$\lambda_{2m+2} = \frac{-H_{2m+2}^{-(2m+1)} H_{2m}^{-(2m)}}{H_{2m+1}^{-(2m)} H_{2m+1}^{-(2m+1)}}, \quad \lambda_{2m+3} = \frac{-H_{2m+3}^{-(2m+3)} H_{2m+1}^{-(2m)}}{H_{2m+2}^{-(2m+2)} H_{2m+2}^{-(2m+1)}} \tag{8}$$

*for all* $m \in \mathbb{Z}_0^+$ *with* $R_{-1}(x) \equiv 0$.

A SMF $\mathcal{L}$ is said to be **positive-definite** on a set $E$ if $\mathcal{L}[R(x)] > 0$ for any L-polynomial $R(x)$ such that $R(x) \geq 0$ for all $x \in E$ and $R(x)$ is not identically equal to 0 on $E$. We say $\mathcal{L}$ is positive-definite if it is positive-definite on the set of all real numbers. The following theorem summarizes some important results concerning positive-definite strong moment functionals.

**THEOREM 1.2** *Suppose* $\mathcal{L}$ *is a quasi-definite SMF with respect to the bisequence* $\{\mu_n\}_{-\infty}^{\infty}$ *and let* $\{R_n(x)\}_{n=0}^{\infty}$ *be the corresponding OLPS whose elements satisfy equations (4) and (5) for all* $n \in \mathbb{Z}_0^+$ *with* $\lambda_1 = \mu_0 r_{1,0}$. *Then each of the following statements is equivalent to* $\mathcal{L}$ *being positive-definite.*

*(a) All the elements in* $\{\mu_n\}_{-\infty}^{\infty}$ *are real and*

$$H_{2m}^{-(2m)} > 0, \quad \text{and} \quad H_{2m+1}^{-(2m)} > 0 \ \ \textit{for all } m \in \mathbb{Z}_0^+. \tag{9}$$

*(b) The quotient*

$$\frac{\lambda_n \alpha_n}{\beta_n} > 0 \ \ \textit{for all } n \in \mathbb{Z}^+ \tag{10}$$

*and* $\{R_n(x)\}_{n=0}^{\infty}$ *is a real OLPS.*

The following theorem concerns the zeros of a OLPS corresponding to a SMF which is positive-definite on $(0, \infty)$.

**THEOREM 1.3** *Assume $\mathcal{L}$ is a SMF which is positive-definite on $(0, \infty)$ and let $\{R_n(x)\}_{n=0}^{\infty}$ be the monic OLPS corresponding to $\mathcal{L}$. Then each of the following hold:*

*(a) If $\lambda_1 = \mu_0 r_{1,0}$, then the elements of $\{R_n(x)\}_{n=0}^{\infty}$ satisfy equations (4) and (5) for all $n \in \mathbb{Z}_0^+$ with $\lambda_n < 0$ for all $n \in \mathbb{Z}^+$.*

*(b) The OLPS $\{R_n(x)\}_{n=0}^{\infty}$ is a regular sequence of L-polynomials and for each $n \in \mathbb{Z}^+$, $R_n(x)$ has $n$ simple zeros contained in $(0, \infty)$.*

*(c) For each $n \in \mathbb{Z}^+$, let the zeros of $R_n(x)$ be denoted by $x_{n,i}$, $i = 1, 2, \ldots, n$ such that $x_{n,1} < x_{n,2} < \cdots < x_{n,n}$. Then the zeros of $R_n(x)$ separate the zeros of $R_{n+1}(x)$ for each $n \in \mathbb{Z}^+$. I.e.,*

$$x_{n+1,i} < x_{n,i} < x_{n+1,i+1}$$

*for $i = 1, 2, \ldots, n$ where $n$ is any element of $\mathbb{Z}^+$.*

The following theorem is known as Favard's theorem. A similar theorem for the symmetric case will be developed later on in this paper.

**THEOREM 1.4** *Let $\{\alpha_n\}_{n=1}^{\infty}$ and $\{\beta_n\}_{n=1}^{\infty}$ be sequences of nonzero complex numbers related by $\beta_n = \alpha_{n+1}$ for $n \in \mathbb{Z}^+$ (where $\alpha_1 = 1$) and let $\{\lambda_n\}_{n=1}^{\infty}$ be a sequence of complex numbers. Define $\{R_n(x)\}_{n=1}^{\infty}$ by letting $R_{-1}(x) \equiv 0$, $R_0(x) \equiv 1$,*

$$R_{2m+1}(x) = (\frac{x^{-1}}{\alpha_{2m+1}} + \beta_{2m+1})R_{2m}(x) + \lambda_{2m+1}R_{2m-1}(x) \tag{11}$$

*and*

$$R_{2m+2}(x) = (\frac{x}{\alpha_{2m+2}} + \beta_{2m+2})R_{2m+1}(x) + \lambda_{2m+2}R_{2m}(x) \tag{12}$$

*for all $m \in \mathbb{Z}_0^+$. Then $\{R_n(x)\}_{n=0}^{\infty}$ is regular with the L-trailing coefficients of $R_{2m+1}(x)$ and $R_{2m+2}(x)$, given by $\beta_{2m+1}$ and $\beta_{2m+2}$, respectively, while the respective L-leading coefficients are both equal to one for all $m \in \mathbb{Z}_0^+$. There also exists a unique SMF $\mathcal{L}$ for which $\{R_n(x)\}_{n=0}^{\infty}$ is the corresponding monic OLPS if and only if $\lambda_n \neq 0$ for all $n \in \mathbb{Z}^+$. The SMF $\mathcal{L}$ is positive-definite if and only if*

$$\frac{\lambda_n \alpha_n}{\beta_n} > 0 \text{ for all } n \in \mathbb{Z}^+ \tag{13}$$

*and $\{R_n(x)\}_{n=0}^{\infty}$ is a real OLPS for $\mathcal{L}$.*

# 2 CONSEQUENCES OF A RELATIONSHIP BETWEEN A SMF AND A SYMMETRIC SMF

In this section, we derive results about a symmetric strong moment functional using a relationship between symmetric SMF's and SMF's. These results include information about the support, positive-definiteness, three-term recurrence relations, coefficients in three term recurrence relations, conditions on the Hankel determinants, a Favard-type theorem, and a separation of zeros theorem. Brief comments about the symmetric case can be found in [4, 7].

The following theorem will be useful in establishing both the connection between a symmetric SMF and a SMF and the fact that a symmetric OLPS satisfies a three term recurrence relation. This theorem was shown to hold for a specific example in [7, pp. 538, 539], but not in general.

**THEOREM 2.1** *Let* $\{S_n(x)\}_{n=0}^{\infty}$ *denote the corresponding monic OLPS for a quasi-definite SMF* $\mathcal{M}$. *Then* $\mathcal{M}$ *is symmetric if and only if*

$$S_{2m}(x) = (-1)^m S_{2m}(-x) \quad \text{and} \quad S_{2m+1}(x) = (-1)^{m+1} S_{2m+1}(-x) \tag{14}$$

*for all* $m \in \mathbb{Z}_0^+$.

**Comment:** From this theorem, it follows that $S_{4m}(x)$ and $S_{4m+3}(x)$ consist entirely of even powers of $x$ while $S_{4m+1}(x)$ and $S_{4m+2}(x)$ possess only odd powers $x$, for all $m \in \mathbb{Z}_0^+$.

**Proof:** ($\Rightarrow$) Since $\mathcal{M}$ is symmetric,

$$\mathcal{M}[S_m(-x)S_n(-x)] = \mathcal{M}[S_m(x)S_n(x)] = K_n\delta_{m,n}, \quad \text{where } K_n \neq 0$$

for all $m, n \in \mathbb{Z}_0^+$. Thus $\{S_n(-x)\}_{n=0}^{\infty}$ is an OLPS for $\mathcal{M}$ and by Theorem 1.1,

$$S_{2m}(x) = (-1)^m S_{2m}(-x) \quad \text{and} \quad S_{2m+1}(x) = (-1)^{m+1} S_{2m+1}(-x)$$

for all $m \in \mathbb{Z}_0^+$.

($\Leftarrow$) If (14) holds for all $m \in \mathbb{Z}_0^+$, then $S_{4m}(x)$ and $S_{4m+3}(x)$ each consist only of terms with even powers of $x$, while $S_{4m+1}(x)$ and $S_{4m+2}(x)$ contain only odd powers of $x$ for all $m \in \mathbb{Z}_0^+$. Therefore,

$$S_{4m+1}(x) = \sum_{j=-m}^{m} s_{4m+1,2j-1}x^{2j-1}, \quad s_{4m+1,-2m-1} = 1 \tag{15}$$

and

$$S_{4m+2}(x) = \sum_{j=-m}^{m+1} s_{4m+2,2j-1}x^{2j-1}, \quad s_{4m+2,2m+1} = 1 \tag{16}$$

for all $m \in \mathbb{Z}_0^+$. Then

$$S_1(x) = \frac{1}{x} \quad \text{and} \quad S_2(x) = \frac{s_{2,-1}}{x} + x$$

and by induction on the successive elements of the sequence and by using the fact that $\mathcal{M}[S_n(x)] = 0$ for all $n \in \mathbb{Z}^+$, it follows that $\mu_{2n+1} = 0$ for all $n \in \mathbb{Z}$. ∎

The following theorem relates a SMF to a symmetric SMF.

**THEOREM 2.2** *For a given bisequence* $\{\mu_n\}_{-\infty}^{\infty}$, *define the strong moment functionals* $\mathcal{L}$ *and* $\mathcal{M}$ *such that*

$$\mathcal{L}(x^n) = \mathcal{M}(x^{2n}) = \mu_n \tag{17}$$

*and*

$$\mathcal{M}(x^{2n+1}) = 0 \tag{18}$$

*for all* $n \in \mathbb{Z}$. *Let* $\{R_n(x)\}_{n=0}^{\infty}$ *and* $\{S_n(x)\}_{n=0}^{\infty}$ *be sequences of L-polynomials related by*

$$S_{4m}(x) = R_{2m}(x^2), \quad S_{4m+3}(x) = R_{2m+1}(x^2), \tag{19}$$

$$S_{4m+1}(x) = \frac{R_{2m}(x^2)}{a_{2m}x} \quad \text{and} \quad S_{4m+2}(x) = \frac{xR_{2m+1}(x^2)}{a_{2m+1}} \tag{20}$$

*for nonzero constants* $a_n$, $n \in \mathbb{Z}_0^+$.

*(a) For each L-polynomial* $R$, $\mathcal{L}[R(x)] = \mathcal{M}[R(x^2)]$.
*(b) The sequence* $\{R_n(x)\}_{n=0}^{\infty}$ *is regular if and only if for each* $m \in \mathbb{Z}_0^+$, $S_{2m}(x)$ *is a regular L-polynomial of L-degree* $2m$ *and* $S_{2m+1}(x)$ *is an L-polynomial of L-degree* $2m+1$ *which is not regular.*
*(c) The sequence* $\{R_n(x)\}_{n=0}^{\infty}$ *is a regular OLPS corresponding to* $\mathcal{L}$ *if and only if* $\{S_n(x)\}_{n=0}^{\infty}$ *is an OLPS corresponding to* $\mathcal{M}$ *where* $S_{2n}(x)$ *is regular with L-degree* $2n$, *for all* $n \in \mathbb{Z}_0^+$.
*(d) Finally,* $\{R_n(x)\}_{n=0}^{\infty}$ *is a regular, monic OLPS and* $a_n$ *is the L-trailing coefficient of* $R_n(x)$ *if and only if* $\{S_n(x)\}_{n=0}^{\infty}$ *is a monic OLPS where* $\frac{1}{a_{2m}}$ *is the coefficient of* $x^{2m-1}$ *in* $S_{4m+1}(x)$ *and* $\frac{1}{a_{2m+1}}$ *is the L-trailing coefficient of* $S_{4m+2}(x)$.

**Proof:** (a) This part of the theorem follows by the linearity of a SMF, the symmetry of $\mathcal{M}$ and (19).

(b) ($\Rightarrow$) An immediate consequence of (19) and (20).

($\Leftarrow$) Let $m \in \mathbb{Z}^+$ be given. From the relationship given for $\{R_n(x)\}_{n=0}^{\infty}$ and $\{S_n(x)\}_{n=0}^{\infty}$ in (19) and (20), it follows that $S_{4m}(x)$ and $S_{4m+3}(x)$ consist entirely of even powers of $x$ while $S_{4m+1}(x)$ and $S_{4m+2}(x)$ consist entirely of odd powers of $x$. By the first part of (19), it follows that $R_{2m}(x)$ is a regular L-polynomial of L-degree $2m$. From (20),

$$R_{2m+1}(x^2) = \frac{a_{2m+1}S_{4m+2}(x)}{x}.$$

Since $S_{4m+2}(x)$ is a regular L-polynomial of L-degree $4m+2$, $R_{2m+1}(x^2)$ is an L-polynomial of L-degree $4m+3$ where the coefficients of both $x^{-2m-2}$ and $x^{2m}$ are nonzero. Thus $R_{2m+1}(x)$ is a regular L-polynomial of L-degree $2m+1$.

(c) ($\Rightarrow$) First it will be shown that

$$\mathcal{M}[S_m(x)S_n(x)] = 0 \tag{21}$$

for all $m, n \in \mathbb{Z}_0^+$ such that $m \neq n$. Products $S_{4m}(x)S_{4n+1}(x)$, $S_{4m}(x)S_{4n+2}(x)$, $S_{4m+1}(x)S_{4n+3}(x)$ and $S_{4m+2}(x)S_{4n+3}(x)$ all consist entirely of odd powers of $x$, for all $m, n \in \mathbb{Z}_0^+$, implying that the SMF $\mathcal{M}$ will be equal to 0 when it is evaluated at each of these products. Next, for all $m, n \in \mathbb{Z}_0^+$,

$$\mathcal{M}[S_{4m}(x)S_{4n+3}(x)] = \mathcal{L}[R_{2m}(x)R_{2n+1}(x)] = 0,$$

$$\mathcal{M}[S_{4m+1}(x)S_{4n+2}(x)] = \frac{1}{a_{2m}a_{2n+2}}\mathcal{L}[R_{2m}(x)R_{2n+1}(x)] = 0$$

and if additionally, $m \neq n$,

$$\mathcal{M}[S_{4m}(x)S_{4n}(x)] = \mathcal{L}[R_{2m}(x)R_{2n}(x)] = 0$$

and

$$\mathcal{M}[S_{4m+3}(x)S_{4n+3}(x)] = \mathcal{L}[R_{2m+1}(x)R_{2n+1}(x)] = 0.$$

Let $m, n \in \mathbb{Z}_0^+$ and without loss of generality, let $m < n$. Then

$$\mathcal{M}[S_{4m+1}(x)S_{4n+1}(x)] = \frac{1}{a_{2m}a_{2n}}\mathcal{L}[\frac{1}{x}R_{2m}(x)R_{2n}(x)] = 0$$

since $\frac{1}{x}R_{2m}(x)$ is of L-degree $2n-1$ or less and, similarly,

$$\mathcal{M}[S_{4m+2}(x)S_{4n+2}(x)] = \frac{1}{a_{2m+1}a_{2n+1}}\mathcal{L}[xR_{2m+1}(x)R_{2n+1}(x)] = 0$$

since $xR_{2m+1}(x)$ is of L-degree $2n$ or less. Hence (21) holds for all $m, n \in \mathbb{Z}_0^+$ where $m \neq n$.

It remains to show that

$$\mathcal{M}[S_n^2(x)] \neq 0$$

for all $n \in \mathbb{Z}_0^+$. Let $m \in \mathbb{Z}_0^+$ be given. Then

$$\mathcal{M}[S_{4m}^2(x)] = \mathcal{L}[R_{2m}^2(x)] \neq 0$$

and

$$\mathcal{M}[S_{4m+3}^2(x)] = \mathcal{L}[R_{2m+1}^2(x)] \neq 0.$$

Also,

$$\begin{aligned}\mathcal{M}[S_{4m+1}^2(x)] &= \mathcal{L}[\frac{1}{a_{2m}^2 x}R_{2m}^2(x)] = \frac{1}{a_{2m}^2}\mathcal{L}[\frac{1}{x}R_{2m}(x)R_{2m}(x)] \\ &= \frac{1}{a_{2m}^2}\mathcal{L}[x^{-(m+1)}R_{2m}(x)] = \frac{-r_{2m+1,m}}{a_{2m}^2}\mathcal{L}[R_{2m}^2(x)] \neq 0\end{aligned}$$

and

$$\begin{aligned}\mathcal{M}[S_{4m+2}^2(x)] &= \frac{1}{a_{2m+1}^2}\mathcal{L}[xR_{2m+1}^2(x)] = \frac{1}{a_{2m+1}^2}\mathcal{L}[x^{m+1}R_{2m+1}(x)] \\ &= \frac{-r_{2m+2,-m-1}}{a_{2m+1}^2}\mathcal{L}[R_{2m+1}^2(x)] \neq 0\end{aligned}$$

where the last equality in each expression above follows from Theorem 1.1.

($\Leftarrow$) Since

$$\mathcal{L}[R_{2m}(x)R_{2n}(x)] = \mathcal{M}[S_{4m}(x)S_{4n}(x)],$$
$$\mathcal{L}[R_{2m}(x)R_{2n+1}(x)] = \mathcal{M}[S_{4m}(x)S_{4n+3}(x)]$$

and

$$\mathcal{L}[R_{2m+1}(x)R_{2n+1}(x)] = \mathcal{M}[S_{4m+3}(x)S_{4n+3}(x)],$$

for all $m, n \in \mathbb{Z}_0^+$, it is clear that $\{R_n(x)\}_{n=0}^{\infty}$ is a regular OLPS corresponding to $\mathcal{L}$.

(d) Follows from comparing coefficients in (20). ∎

We now examine the relationship between the support of positive-definite SMF's $\mathcal{L}$ and $\mathcal{M}$ satisfying (17) and (18) for all $n \in \mathbb{Z}$.

**LEMMA 2.3** *Let $a$ and $b$ be fixed constants where $0 < a < b$. Suppose $R(x)$ is a L-polynomial which does not vanish identically on $[-b, -a] \cup [a, b]$ and for which $R(x) \geq 0$ on $[-b, -a] \cup [a, b]$. Then there exist L-polynomials $E(x)$ and $F(x)$, where $F(x)$ consists entirely of odd powers of $x$ such that*

$$R(x) = E(x^2) + F(x). \tag{22}$$

*Furthermore $E(x) \geq 0$ on $[a^2, b^2]$ and $E(x) \not\equiv 0$ on $[a^2, b^2]$.*

**Proof:** By grouping the terms containing odd powers of $x$ together and grouping the terms containing even powers of $x$, (22) is clear.

For all $x \in \mathbb{R}/\{0\}$,

$$R(x) + R(-x) = 2E(x^2). \tag{23}$$

Since $R(x) \not\equiv 0$ on $[-b, -a] \cup [a, b]$, there exists $x_0 \in [-b, -a] \cup [a, b]$ such that $R(x_0) > 0$. By (23), it follows that $E(x^2) \not\equiv 0$ on $[-b, -a] \cup [a, b]$ and $E(x^2) \geq 0$ on $[-b, -a] \cup [a, b]$. Equivalently, $E(x) \geq 0$ on $[a^2, b^2]$ and $E(x) \not\equiv 0$ on $[a^2, b^2]$. ∎

**THEOREM 2.4** *For a given bisequence, $\{\mu_n\}_{-\infty}^{\infty}$, suppose $\mathcal{L}$ and $\mathcal{M}$ are defined as in (17) and (18) for all $n \in \mathbb{Z}$. For fixed constants $a$ and $b$ satisfying $0 < a < b$, $\mathcal{L}$ is positive-definite on $[a^2, b^2]$ if and only if $\mathcal{M}$ is positive-definite on $[-b, -a] \cup [a, b]$.*

**Proof:** ($\Rightarrow$) Suppose $\mathcal{L}$ is positive-definite on $[a^2, b^2]$ and let $R(x)$ be an arbitrary L-polynomial which does not vanish identically on $[-b, -a] \cup [a, b]$ and for which $R(x) \geq 0$ on $[-b, -a] \cup [a, b]$. By the previous lemma,

$$\mathcal{M}[R(x)] = \mathcal{M}[E(x^2)] + \mathcal{M}[F(x)] = \mathcal{M}[E(x^2)] = \mathcal{L}[E(x)] > 0.$$

Hence, $\mathcal{M}$ is positive-definite on $[-b, -a] \cup [a, b]$.

($\Leftarrow$) Suppose $\mathcal{M}$ is positive-definite on $[-b, -a] \cup [a, b]$. Choose a L-polynomial $R(x)$ such that $R(x) \geq 0$ and $R(x) \not\equiv 0$ on $[a^2, b^2]$. Then $R(x^2) \geq 0$ and $R(x^2) \not\equiv 0$ on $[-b, -a] \cup [a, b]$. Therefore

$$0 < \mathcal{M}[R(x^2)] = \mathcal{L}[R(x)]$$

implying that $\mathcal{L}$ is positive-definite on $[a^2, b^2]$. ∎

By replacing occurrences of $[-b,-a]\cup[a,b]$ with $(-\infty,0)\cup(0,\infty)$ and $[a^2,b^2]$ with $(0,\infty)$ in the proofs of Lemma 2.3 and Theorem 2.4, the following theorem results.

**THEOREM 2.5** *For a given bisequence, $\{\mu_n\}_{-\infty}^{\infty}$, suppose $\mathcal{L}$ and $\mathcal{M}$ are defined as in (17) and (18) for all $n \in \mathbb{Z}$. The SMF $\mathcal{L}$ is positive-definite on $(0,\infty)$ if and only if $\mathcal{M}$ is positive-definite on $(-\infty,0)\cup(0,\infty)$.*

The results in the previous section imply that there is a three-term recurrence relation for an OLPS $\{S_n(x)\}_{n=0}^{\infty}$ corresponding to a SMF $\mathcal{M}$ , provided the OLPS is *regular*. However, as noted in [2] and [9], if $\{S_n(x)\}_{n=0}^{\infty}$ is not regular, then, in general, the OLPS can only be guaranteed to satisfy a less useful five-term recurrence relation of the form

$$S_{2n+2}(x) = \gamma_{2n+2,2n+1}S_{2n+1}(x) + (x+\gamma_{2n+2,2n})S_{2n}(x) + \gamma_{2n+2,2n-1}S_{2n-1}(x)$$
$$x + \gamma_{2n+2,2n-2}S_{2n-2}(x) \tag{24}$$

and

$$S_{2n+3}(x) = \gamma_{2n+3,2n+2}S_{2n+2}(x) + (\frac{1}{x}+\gamma_{2n+3,2n+1})S_{2n+1}(x)$$
$$+ \gamma_{2n+3,2n}S_{2n}(x) + \gamma_{2n+3,2n-1}S_{2n-1}(x) \tag{25}$$

for all $n \in \mathbb{Z}_0^+$ where $S_{-2}(x) \equiv S_{-1}(x) \equiv 0$, $S_0(x) \equiv 1$ and $S_1(x) = \frac{1}{x}$.

Here we show that a symmetric OLPS satisfies a non-traditionalthree term recurrence relation. We use the adjective *non-traditional* because the three term recurrence relation for the even indexed L-polynomials involve three successive even-subscripted L-polynomials. Similarly, the three term recurrence relations for the odd-indexed L-polynomials can be written to involve only three consecutive odd-subscripted L-polynomials.

**THEOREM 2.6** *If $\{S_n(x)\}_{n=0}^{\infty}$ is the corresponding monic OLPS for a symmetric SMF $\mathcal{M}$, where $S_{2n}(x)$ is regular for each $n \in \mathbb{Z}_0^+$, then there exist sequences $\{b_n\}_{n=0}^{\infty}$, $\{c_n\}_{n=1}^{\infty}$, and $\{d_n\}_{n=1}^{\infty}$ of nonzero complex numbers, with $b_0 = 1$, such that*

$$S_{2n+1}(x) = \frac{b_n S_{2n}(x)}{x} \tag{26}$$

*and*

$$S_{2n+2}(x) = (x + \frac{c_{n+1}}{x})S_{2n}(x) - d_{n+1}S_{2n-2}(x) \tag{27}$$

*for all $n \in \mathbb{Z}_0^+$, where $S_{-2}(x) \equiv 0$. Further,*

$$b_n = \frac{1}{\prod_{k=1}^{n} c_k} \quad \text{for all } n \in \mathbb{Z}^+. \tag{28}$$

*Finally, if $\mathcal{M}$ is positive-definite, then $c_n < 0$ and $d_{n+1} > 0$ for all $n \in \mathbb{Z}^+$.*

**Comment:** Equation (26) can be found in [7, p. 537] in a more general context and a more general form of equation (27) can be found in [9, p. 67]. New information concerning the recurrence coefficients is provided in the last part of the theorem.

**Proof of Theorem 2.6:** By Theorem 2.2, the monic OLPS $\{S_n(x)\}_{n=0}^{\infty}$ satisfies

$$S_{4m+1}(x) = \frac{S_{4m}(x)}{a_{2m}x} \quad \text{and} \quad S_{4m+3}(x) = \frac{a_{2m+1}S_{4m+2}(x)}{x} \tag{29}$$

for all $m \in \mathbb{Z}_0^+$, where $\{a_n\}_{n=0}^{\infty}$ is a sequence of nonzero constants with $a_0 = 1$. Hence (26) holds if one lets

$$b_{2m} = \frac{1}{a_{2m}} \quad \text{and} \quad b_{2m+1} = a_{2m+1} \tag{30}$$

for all $m \in \mathbb{Z}_0^+$. Note that these assignments yield $b_0 = 1$ since $S_1(x)$ is monic. Let $S_{-2}(x) \equiv R_{-1}(x) \equiv 0$. By Theorem 2.2 there exists a regular, monic OLPS, $\{R_n(x)\}_{n=0}^{\infty}$, satisfying (19) and (20) and so Theorem 1.1 implies

$$R_{2m+1}(x^2) = \left(\beta_{2m+1} + \frac{1}{\alpha_{2m+1}x^2}\right) R_{2m}(x^2) + \lambda_{2m+1}R_{2m-1}(x^2)$$

and

$$R_{2m+2}(x^2) = \left(\frac{x^2}{\alpha_{2m+2}} + \beta_{2m+2}\right) R_{2m+1}(x^2) + \lambda_{2m+2}R_{2m}(x^2)$$

with $\alpha_{m+2} = \beta_{m+1}$ for all $m \in \mathbb{Z}_0^+$. By equations (19), (20), (29) and (30) we can in turn write

$$\frac{b_{2m+1}S_{4m+2}(x)}{x} = \left(\beta_{2m+1} + \frac{1}{\alpha_{2m+1}x^2}\right) S_{4m}(x) + \frac{\lambda_{2m+1}b_{2m-1}S_{4m-2}(x)}{x}$$

and

$$S_{4m+4}(x) = \left(\frac{x^2}{\alpha_{2m+2}} + \beta_{2m+2}\right) \frac{b_{2m+1}}{x} S_{4m+2}(x) + \lambda_{2m+2}S_{4m}(x)$$

for all $m \in \mathbb{Z}_0^+$, where $b_{-1}$ is an arbitrary constant. Hence

$$S_{4m+2}(x) = \left(\frac{\beta_{2m+1}}{b_{2m+1}}x + \frac{1}{b_{2m+1}\alpha_{2m+1}x}\right) S_{4m}(x) + \frac{\lambda_{2m+1}b_{2m-1}}{b_{2m+1}} S_{4m-2}(x)$$

and

$$S_{4m+4}(x) = \left(\frac{b_{2m+1}}{\alpha_{2m+2}}x + \frac{\beta_{2m+2}b_{2m+1}}{x}\right) S_{4m+2}(x) + \lambda_{2m+2}S_{4m-2}(x)$$

for all $m \in \mathbb{Z}_0^+$. Since $S_{4m+2}(x)$ and $S_{4m+4}(x)$ are monic,

$$b_{2m+1} = \alpha_{2m+2} = \beta_{2m+1} \tag{31}$$

for all $m \in \mathbb{Z}_0^+$. Since $\alpha_{2m+1}$ is the L-trailing coefficient of $R_{2m}(x)$, (29) and (30) imply that

$$\frac{1}{b_{2m}} = \alpha_{2m+1} = \beta_{2m} \tag{32}$$

for all $m \in \mathbb{Z}_0^+$. Let

$$c_{2m+1} = \frac{1}{\alpha_{2m+1}\beta_{2m+1}}, \quad c_{2m+2} = \alpha_{2m+2}\beta_{2m+2} \tag{33}$$

$$d_{2m+1} = \frac{-\alpha_{2m}\lambda_{2m+1}}{\alpha_{2m+2}} \text{ and } d_{2m+2} = -\lambda_{2m+2} \tag{34}$$

for all $m \in \mathbb{Z}_0^+$ where $\alpha_0 = b_{-1}$. Then (27) holds for all $n \in \mathbb{Z}_0^+$. Note that $\alpha_1 = \frac{1}{b_0} = 1$ and since $\{R_n(x)\}_{n=0}^{\infty}$ is regular, $\lambda_{n+1} \neq 0$ for all $n \in \mathbb{Z}^+$ which in turn implies that $d_{n+1} \neq 0$ for all $m \in \mathbb{Z}_0^+$. Hence by (33),

$$c_1 = \frac{1}{\alpha_1\beta_1} = \frac{1}{\beta_1}.$$

For each $m \in \mathbb{Z}_0^+$,

$$c_{2m+1}c_{2m+2} = \frac{\alpha_{2m+2}\beta_{2m+2}}{\alpha_{2m+1}\beta_{2m+1}} = \frac{\beta_{2m+2}}{\alpha_{2m+1}} = \frac{b_{2m}}{b_{2m+2}}.$$

Hence

$$\prod_{i=1}^{2m+2} c_i = \frac{b_0}{b_2}\frac{b_2}{b_4}\cdots\frac{b_{2m-2}}{b_{2m}}\frac{b_{2m}}{b_{2m+2}} = \frac{1}{b_{2m+2}}$$

and

$$\prod_{i=1}^{2m+3} c_i = \frac{c_{2m+3}}{b_{2m+2}} = \frac{1}{\beta_{2m+3}} = \frac{1}{b_{2m+3}}$$

for all $m \in \mathbb{Z}_0^+$. Therefore (28) holds for all $n \in \mathbb{Z}^+$. Let

$$\lambda_1 = \mu_0 r_{1,0} = \alpha_2\mu_0.$$

Suppose $\mathcal{M}$ is positive-definite on $(-\infty, 0) \cup (0, \infty)$. Then $\mathcal{L}$ is positive-definite on $(0, \infty)$. Then by Theorem 1.2

$$\lambda_n < 0 \text{ and } \frac{\alpha_n}{\beta_n} < 0 \tag{35}$$

for all $n \in \mathbb{Z}^+$. Hence equation (33) implies $c_n < 0$ for all $n \in \mathbb{Z}^+$. By (34), $d_{2m+2} > 0$ for all $m \in \mathbb{Z}_0^+$ and by (35),

$$\frac{\alpha_{2m}}{\beta_{2m}} < 0 \text{ and } \frac{\alpha_{2m+1}}{\beta_{2m+1}} < 0$$

for all $m \in \mathbb{Z}^+$. Hence $\frac{\alpha_{2m}}{\alpha_{2m+2}} > 0$ for all $m \in \mathbb{Z}^+$. So by (34) and (35), $d_{2m+1} > 0$ for all $m \in \mathbb{Z}^+$ and we conclude that $d_n > 0$ for all $n \geq 2$. ∎

We can now compare the coefficients found in the three-term recurrence relations of a symmetric OLPS with those found in the three-term recurrence relations for a nonsymmetric OLPS. Suppose $\{R_n(x)\}_{n=0}^{\infty}$ is a regular, monic OLPS corresponding to the SMF $\mathcal{L}$ and let $\{S_n(x)\}_{n=0}^{\infty}$ be a monic OLPS corresponding to a symmetric

SMF $\{S_n(x)\}_{n=0}^{\infty}$ where $\mathcal{L}$ and $\mathcal{M}$ are related by (17) and (18). We can then express a single three term recurrence relation for $\{R_n(x)\}_{n=0}^{\infty}$ by combining (4) and (5) to obtain

$$R_n(x) = \left(\frac{x^{[(-1)^n]}}{\alpha_n} + \beta_n\right) R_{n-1}(x) + \lambda_n R_{n-2}(x)$$

for all $n \in \mathbb{Z}^+$ where $R_{-1}(x) \equiv 0$ and $R_0(x) \equiv 1$. Restating the three term recurrence relation for $\{S_n(x)\}_{n=0}^{\infty}$, we have

$$S_{2n+2}(x) = (x + \frac{c_{n+1}}{x})S_{2n}(x) - d_{n+1}S_{2n-2}(x)$$

where

$$S_{2n+1}(x) = \frac{b_n S_{2n}(x)}{x}$$

for all $n \in \mathbb{Z}_0^+$ where $S_{-2}(x) \equiv 0$ and $S_0(x) \equiv 1$. Then

$$b_{2m} = \frac{1}{\alpha_{2m+1}}, \quad b_{2m+1} = \alpha_{2m+2},$$

$$c_{2m+1} = \frac{1}{\alpha_{2m+1}\beta_{2m+1}}, \quad c_{2m+2} = \alpha_{2m+2}\beta_{2m+2},$$

$$d_{2m+1} = \frac{-\alpha_{2m}\lambda_{2m+1}}{\alpha_{2m+2}}, \quad d_{2m+2} = -\lambda_{2m+2}$$

and

$$\beta_m = \alpha_{m+1}$$

for all $m \in \mathbb{Z}_0^+$ where $\alpha_0$ is arbitrary and $\lambda_1 = \alpha_2\mu_0$. From these equations it is easy to find that

$$\alpha_{2m+1} = \frac{1}{b_{2m+1}c_{2m+1}}, \quad \text{and} \quad \alpha_{2m+2} = b_{2m+1}$$

for all $m \in \mathbb{Z}_0^+$.

The next theorem gives more information about the constants appearing in preceding theorem.

**THEOREM 2.7** *With reference to the recurrence formulae in Theorem 2.6 that correspond to a symmetric and positive-definite SMF $\mathcal{M}$, the following are valid provided $d_1 = \mu_0$, for all $n \in \mathbb{Z}^+$.*

*(a)* $d_{n+1} = \dfrac{\mathcal{M}[S_{2n}^2(x)]}{\mathcal{M}[S_{2n-2}^2(x)]} = \dfrac{H_{2n-2}^{-(2n-2)} H_{2n+1}^{(-2n)}}{H_{2n-1}^{-(2n-2)} H_{2n}^{(-2n)}}$.

*(b)* $\mathcal{M}[S_{2n-2}^2(x)] = \prod_{k=1}^{n} d_k$.

*(c) The coefficient of $x^{-n}$ in $S_{2n}(x)$ is $\prod_{k=1}^{n} a_k = \dfrac{1}{b_n}$ and $b_n$ is the coefficient of $x^{n-1}$ in $S_{2n+1}(x)$.*

(d) $b_n = \dfrac{\mathcal{M}[x^{n+1}S_{2n+1}(x)]}{\mathcal{M}[S_{2n}^2(x)]} = \dfrac{-H_{2n}^{(-2n)}H_{2n+2}^{-(2n+1)}}{[H_{2n+1}^{(-2n)}]^2} = \dfrac{-\mathcal{M}[S_{2n-1}^2(x)]}{b_{n-1}\mathcal{M}[S_{2n-2}^2(x)]}$.

(e) $a_n = \dfrac{b_{n-1}}{b_n} = -\dfrac{b_{n-1}^2\mathcal{M}[S_{2n-2}^2(x)]}{\mathcal{M}[S_{2n-1}^2(x)]}$.

The following theorem gives results for certain Hankel determinants in the symmetric case.

**THEOREM 2.8** *Suppose $\mathcal{M}$ is a positive-definite symmetric SMF with a corresponding monic OLPS $\{S_n(x)\}_{n=0}^{\infty}$. Then*

$$H_{2n}^{-2n} > 0,\ H_{2n+1}^{-2n} > 0,\ sign\{H_{2n}^{(-2n+1)}\} = (-1)^n \text{ and } H_{2n+1}^{-(2n+1)} = 0 \qquad (36)$$

*for all $n \in \mathbb{Z}_0^+$.*

**Proof:** By (26), $\frac{1}{b_n}$ is the coefficient of $x^{-n}$ in $S_{2n}(x)$. Therefore, by Theorem 1.1,

$$\frac{1}{b_n} = \frac{H_{2n}^{(-2n+1)}}{H_{2n}^{(-2n)}} \quad \text{for all } n \in \mathbb{Z}_0^+.$$

By Theorem 1.2,

$$H_{2n}^{(-2n)} > 0 \text{ and } H_{2n+1}^{(-2n)} > 0 \text{ for all } n \in \mathbb{Z}_0^+.$$

Therefore $b_n$ and $H_{2n}^{(-2n+1)}$ are of the same sign. Since $c_n < 0$ and $b_n = \frac{1}{\prod_{k=1}^{n} c_k}$, the sign of $b_n$ and $H_{2n}^{(-2n+1)}$ are both $(-1)^n$. According to Theorem 1.1, the coefficient of $x^n$ in $S_{2n+1}(x)$ is $\frac{H_{2n+1}^{(-2n-1)}}{H_{2n+1}^{(-2n)}}$ and since $S_{2n+1}(x)$ is not regular, $H_{2n+1}^{(-2n-1)} = 0$. ∎

We are now ready to state and prove a Favard-type theorem for the symmetric case.

**THEOREM 2.9** *Let $\{b_n\}_{n=0}^{\infty}$ and $\{c_n\}_{n=1}^{\infty}$ be given sequences of nonzero complex numbers with $b_0 = 1$, and let $\{d_n\}_{n=1}^{\infty}$ be a sequence of complex numbers. Define the sequence of L-polynomials, $\{S_n(x)\}_{n=0}^{\infty}$, by letting*

$$S_{2n+1}(x) = \frac{b_n S_{2n}(x)}{x} \qquad (37)$$

*and*

$$S_{2n+2}(x) = (x + \frac{c_{n+1}}{x})S_{2n}(x) - d_{n+1}S_{2n-2}(x) \qquad (38)$$

*for all $n \in \mathbb{Z}_0^+$, where $S_{-2}(x) \equiv 0$. Then $\{S_n(x)\}_{n=0}^{\infty}$ is a monic OLPS corresponding to a symmetric SMF $\mathcal{M}$ if and only if $d_{n+1} \neq 0$ and*

$$b_n = \frac{1}{\prod_{k=1}^{n} c_k} \qquad (39)$$

*for all $n \in \mathbb{Z}^+$.*

**Proof:** It can be shown that $S_{2n}(x)$ is a regular L-polynomial of L-degree $2n$ for each $n \in \mathbb{Z}^+$. Further, $S_{4m}(x)$ and $S_{4m+3}(x)$ are of L-degree $4m$ and $4m+3$, respectively, where each of these L-polynomials consists entirely of even powers of $x$ for all $m \in \mathbb{Z}_0^+$. So there exists a sequence, $\{R_n(x)\}_{n=0}^{\infty}$, of L-polynomials where

$$R_{2m}(x^2) = S_{4m}(x) \quad \text{and} \quad R_{2m+1}(x^2) = S_{4m+3}(x) \tag{40}$$

for all $m \in \mathbb{Z}_0^+$. By equations (37) and (40),

$$R_{2m}(x^2) = \frac{xS_{4m+1}(x)}{b_{2m}} \quad \text{and} \quad R_{2m+1}(x^2) = \frac{b_{2m+1}S_{4m+2}(x)}{x} \tag{41}$$

for all $m \in \mathbb{Z}_0^+$. From (38),

$$S_{4m+2}(x) = \left(x + \frac{c_{2m+1}}{x}\right) S_{4m}(x) - d_{2m+1}S_{4m-2}(x)$$

and

$$S_{4m+4}(x) = \left(x + \frac{c_{2m+2}}{x}\right) S_{4m+2}(x) - d_{2m+2}S_{4m}(x)$$

for all $m \in \mathbb{Z}_0^+$. Using (40) and (41) we have

$$\frac{xR_{2m+1}(x^2)}{b_{2m+1}} = \left(x + \frac{c_{2m+1}}{x}\right) R_{2m}(x^2) - d_{2m+1}\frac{xR_{2m-1}(x^2)}{b_{2m-1}}$$

and

$$R_{2m+2}(x^2) = \left(x + \frac{c_{2m+2}}{x}\right) \frac{xR_{2m+1}(x^2)}{b_{2m+1}} - d_{2m+2}R_{2m}(x^2)$$

or all $m \in \mathbb{Z}_0^+$, where $b_{2m-1}$ is an arbitrary constant. This in turn implies that

$$R_{2m+1}(x^2) = \left(\frac{b_{2m+1}c_{2m+1}}{x^2} + b_{2m+1}\right) R_{2m}(x^2) - \frac{b_{2m+1}d_{2m+1}}{b_{2m-1}}R_{2m-1}(x^2)$$

and

$$R_{2m+2}(x^2) = \left(\frac{x^2}{b_{2m+1}} + \frac{c_{2m+2}}{b_{2m+1}}\right) R_{2m+1}(x^2) - d_{2m+2}R_{2m}(x^2)$$

for all $m \in \mathbb{Z}_0^+$. By letting

$$\alpha_{2m+1} = \frac{1}{b_{2m+1}c_{2m+1}}, \quad \alpha_{2m+2} = b_{2m+1}, \tag{42}$$

$$\beta_{2m+1} = b_{2m+1}, \quad \beta_{2m+2} = \frac{c_{2m+2}}{b_{2m+1}}, \tag{43}$$

$$\lambda_{2m+1} = -\frac{b_{2m+1}d_{2m+1}}{b_{2m-1}} \quad \text{and} \quad \lambda_{2m+2} = -d_{2m+2} \tag{44}$$

for all $m \in \mathbb{Z}_0^+$ and by replacing $x^2$ with $x$,

$$R_{2m+1}(x) = \left(\frac{x^{-1}}{\alpha_{2m+1}} + \beta_{2m+1}\right) R_{2m}(x^2) + \lambda_{2m+1}R_{2m-1}(x)$$

and

$$R_{2m+2}(x^2) = \left(\frac{x}{\alpha_{2m+2}} + \beta_{2m+2}\right) R_{2m+1}(x^2) + \lambda_{2m+2} R_{2m}(x^2).$$

Now suppose $\{S_n(x)\}_{n=0}^{\infty}$ is a monic OLPS corresponding to a symmetric SMF $\mathcal{M}$ and let $\mathcal{L}$ be the SMF satisfying (17) for all $n \in \mathbb{Z}$. By (40) and (41), with

$$a_{2m} = \frac{1}{b_{2m}} \quad \text{and} \quad a_{2m+1} = b_{2m+1}, \quad m \in \mathbb{Z}_0^+,$$

(19) and (20) are satisfied for all $m \in \mathbb{Z}_0^+$. By Theorem 2.2 (c), $\{R_n(x)\}_{n=0}^{\infty}$ is a regular OLPS corresponding to $\mathcal{L}$ and it follows by Theorem 1.1 that $\beta_{m+1} = \alpha_{m+2}$ for all $m \in \mathbb{Z}_0^+$ and $\lambda_n \neq 0$ for all integers $n \geq 2$. By (44), $d_{n+1} \neq 0$ for all $n \in \mathbb{Z}_0^+$ and by Theorem 2.6, (39) holds for all $n \in \mathbb{Z}^+$.

Conversely, suppose (39) holds and $d_{n+1} \neq 0$. By (44), $\lambda_{n+1} \neq 0$ for all $n \in \mathbb{Z}_0^+$. By (39) and (43),

$$\beta_{2m+2} = \frac{c_{2m+2}}{b_{2m+1}} = \left(\prod_{k=1}^{2m+1} c_k\right) c_{2m+2} = \frac{1}{b_{2m+2}} = \frac{b_{2m+3}}{b_{2m+2}b_{2m+3}}$$

$$= \frac{\prod_{k=1}^{2m+2} c_k}{b_{2m+3} \prod_{k=1}^{2m+3} c_k} = \frac{1}{b_{2m+3}c_{2m+3}} = \alpha_{2m+3}$$

for all $m \in \mathbb{Z}_0^+$. This together with (42) and (43) shows that

$$\beta_n = \alpha_{n+1}$$

for all $n \in \mathbb{Z}_0^+$. Since $\{b_n\}_{n=1}^{\infty}$ and $\{c_n\}_{n=1}^{\infty}$ are sequences of nonzero numbers so are $\{\alpha_n\}_{n=1}^{\infty}$ and $\{\beta_n\}_{n=1}^{\infty}$. By (39),

$$\alpha_1 = \frac{1}{b_1 c_1} = 1.$$

So by Theorem 1.4, $\{R_n(x)\}_{n=0}^{\infty}$ is an OLPS corresponding to a SMF $\mathcal{L}$. Define $\mathcal{M}$ as in (17) and (18). Then Theorem 2.2 (c) concludes the proof. ∎

Suppose $\{S_n(x)\}_{n=0}^{\infty}$ is an OLPS corresponding to a symmetric SMF $\mathcal{M}$ which is positive-definite on $(-\infty, 0) \cup (0, \infty)$. From our previous work we know that $\mathcal{L}$, where $\mathcal{L}(x^n) = \mathcal{M}(x^{2n})$ for all $n \in \mathbb{Z}$, is positive-definite on $(0, \infty)$. Hence the OLPS, $\{R_n(x)\}_{n=0}^{\infty}$, corresponding to $\mathcal{L}$ is regular and for each $n \in \mathbb{Z}^+$, $R_n(x)$ contains $n$ zeros, $x_{n,1}, x_{n,2}, \ldots, x_{n,n}$ where

$$x_{n,1} < x_{n,2} < \ldots < x_{n,n}.$$

For each $n \in \mathbb{Z}^+$, the zeros of $R_n(x)$ separate the zeros of $R_{n+1}(x)$, i.e.,

$$x_{n+1,i} < x_{n,i} < x_{n+1,i+1}$$

for $i = 1, 2, \ldots, n$. By (19) and (20), the zeros of $S_{2n}(x)$ are $\pm\sqrt{x_{n,1}}$, $\pm\sqrt{x_{n,2}}$, ..., $\pm\sqrt{x_{n,n}}$ where the zeros of $S_{2n}(x)$ separate the zeros of $S_{2n+2}(x)$ or equivalently,

$$-\sqrt{x_{n+1,i+1}} < -\sqrt{x_{n,i}} < -\sqrt{x_{n+1,i}} < 0 < \sqrt{x_{n+1,i}} < \sqrt{x_{n,i}} < \sqrt{x_{n+1,i+1}}$$

for all $n \in \mathbb{Z}^+$. So the following theorem holds.

**THEOREM 2.10** *If $\mathcal{M}$ is a positive-definite, symmetric SMF with the corresponding monic OLPS $\{S_n(x)\}_{n=0}^{\infty}$ then, for each $n \in \mathbb{Z}^+$, $S_{2n}(x)$ has $2n$ distinct zeros of the form $\pm z_{n,i}$ for $i = 1, 2, \ldots, n$ such that*

$$0 < z_{n+1,k} < z_{n,k} < z_{n+1,k+1} \tag{45}$$

*for $k = 1, 2, \ldots, n$ and $n \in \mathbb{Z}^+$.*

# 3 CONTINUED FRACTIONS AND ASSOCIATED SYMMETRIC LAURENT POLYNOMIALS

The purpose of this section is to introduce *associated symmetric Laurent polynomials* in order to discuss the relationship between continued fractions and symmetric orthogonal Laurent polynomials.

Recall [1, 5, 10, 11] that a continued fraction

$$\frac{a_1}{b_1} + \frac{a_2}{b_2} + \frac{a_3}{b_3} + \cdots$$

has a n$^{\text{th}}$ approximant $C_n$ satisfying

$$C_n = \frac{A_n}{B_n}, \quad n \in \mathbb{Z}^+$$

where

$$A_n = b_n A_{n-1} + a_n A_{n-2}, \quad A_{-1} = 1, \quad A_0 = 0 \tag{46}$$

and

$$B_n = b_n B_{n-1} + a_n B_{n-2}, \quad B_{-1} = 0, \quad B_0 = 1 \tag{47}$$

for all $n \in \mathbb{Z}^+$. Furthermore,

$$A_n B_{n-1} - A_{n-1} B_n = (-1)^{n+1} a_1 a_2 \cdots a_n \tag{48}$$

for all $n \in \mathbb{Z}^+$.

Suppose $\mathcal{M}$ is a positive-definite, symmetric SMF with a corresponding monic OLPS $\{S_n(x)\}_{n=0}^{\infty}$. Then, by Theorem 2.6,

$$S_{-2}(x) \equiv 0, \quad S_0(x) \equiv 1, \tag{49}$$

$$S_{2n-1}(x) = \frac{b_{n-1}S_{2n-2}(x)}{x}, \tag{50}$$

$$S_{2n}(x) = (x + \frac{a_n}{x})S_{2n-2}(x) - d_n S_{2n-4}(x) \tag{51}$$

where

$$a_n < 0, \quad d_{n+1} > 0, \quad b_0 = 1 \text{ and } b_n = \frac{1}{\prod_{k=1}^{n} a_k} \tag{52}$$

for all $n \in \mathbb{Z}^+$.

For each $k \in \mathbb{Z}_0^+$, define $\{a_{n,k}\}_{n=1}^{\infty}$, $\{b_{n-1,k}\}_{n=0}^{\infty}$, and $\{d_{n,k}\}_{n=1}^{\infty}$ such that

$$a_{n,k} = a_{n+k}, \quad b_{n-1,k} = b_{n+k-1}\prod_{i=1}^{k} a_i \text{ and } d_{n,k} = d_{n+k} \tag{53}$$

for all $n \in \mathbb{Z}^+$ where it is understood that $\prod_{i=1}^{0} a_1 = 1$.

For each $k \in \mathbb{Z}_0^+$, define $\{S_{n,k}(x)\}_{n=0}^{\infty}$ such that

$$S_{-2,k}(x) \equiv 0, \quad S_{0,k}(x) \equiv 1, \tag{54}$$

$$S_{2n-1,k}(x) = \frac{b_{n-1,k}S_{2n-2,k}(x)}{x}, \tag{55}$$

$$S_{2n,k}(x) = (x + \frac{a_{n,k}}{x})S_{2n-2,k}(x) - d_{n,k}S_{2n-4,k}(x) \tag{56}$$

for all $n \in \mathbb{Z}^+$.

Note that $b_{0,k} = b_k \prod_{i=1}^{k} a_i = 1$ for all $k \in \mathbb{Z}_0^+$ and

$$b_{n,k} = b_{n+k}\prod_{i=1}^{k} a_i = \frac{\prod_{i=1}^{k} a_i}{\prod_{i=1}^{n+k} a_i} = \frac{1}{\prod_{i=k+1}^{n+k} a_i}$$
$$= \frac{1}{\prod_{i=1}^{n} a_{i,k}}$$

for all $n \in \mathbb{Z}^+$ and $k \in \mathbb{Z}_0^+$. We have just shown that $\{S_{n,k}(x)\}_{n=0}^{\infty}$ satisfies the hypothesis of Theorem 2.9 for all $k \in \mathbb{Z}_0^+$ and hence the following theorem holds:

**THEOREM 3.1** *Let $\mathcal{M}$ be a symmetric, positive-definite SMF with the corresponding monic OLPS $\{S_n(x)\}_{n=0}^{\infty}$ which satisfies (49) through (52) for all $n \in \mathbb{Z}^+$. For each $k \in \mathbb{Z}_0^+$, define $\{S_{n,k}(x)\}_{n=0}^{\infty}$ by (53) through (56). Then, for each $k \in \mathbb{Z}_0^+$, there exists a unique symmetric, positive-definite SMF $\mathcal{M}_k$ satisfying $\mathcal{M}_k(1) = d_{1,k}$ and whose corresponding monic OLPS is $\{S_{n,k}(x)\}_{n=0}^{\infty}$. Finally, $\mathcal{M}_0 = \mathcal{M}$ and $\{S_n(x)\}_{n=0}^{\infty} = \{S_{n,0}(x)\}_{n=0}^{\infty}$.*

By comparing equations (47) with (54) and (56), it is clear that $S_{2n,k}(x)$ represents the $n^{th}$ partial denominator for the continued fraction

$$\frac{-d_{1,k}}{x+\frac{a_{1,k}}{x}} - \frac{d_{2,k}}{x+\frac{a_{2,k}}{x}} - \frac{d_{3,k}}{x+\frac{a_{3,k}}{x}} - \cdots \tag{57}$$

for all $n \in \mathbb{Z}^+$ and $k \in \mathbb{Z}_0^+$.

To obtain the $n^{th}$ partial numerator for (57), define, for each $k \in \mathbb{Z}_0^+$,

$$A_{2n,k}(x) = (x + \frac{a_{n,k}}{x})A_{2n-2,k}(x) - d_{n,k}A_{2n-4,k}(x) \tag{58}$$

for all $n \in \mathbb{Z}^+$ with the initial conditions

$$A_{-2,k}(x) \equiv 1 \text{ and } A_{0,k}(x) \equiv 0. \tag{59}$$

Then $A_{2n,k}(x)$ is the $n^{th}$ partial numerator of (57) for all $n \in \mathbb{Z}^+$ and $k \in \mathbb{Z}_0^+$. The next theorem gives a relationship between a sequence of partial numerators and partial denominators.

**THEOREM 3.2** *For a given $k \in \mathbb{Z}_0^+$, suppose $\{S_{2n,k}(x)\}_{n=0}^{\infty}$ satisfies (54) and (56) for all $n \in \mathbb{Z}^+$ where the coefficients in the recurrence relations satisfy (52) and (53) for all $n \in \mathbb{Z}^+$. If $\{A_{2n,k}(x)\}_{n=0}^{\infty}$ is defined as in (58) and (59) for all $n \in \mathbb{Z}^+$, then*

$$A_{2n+2,k}(x) = -d_{1,k}S_{2n,k+1}(x) \tag{60}$$

*for all $n \in \mathbb{Z}_0^+$ and the $n^{th}$ approximant of (57), $C_{n,k}(x)$, satisfies*

$$C_{n,k}(x) = \frac{-d_{1,k}S_{2n-2,k+1}(x)}{S_{2n,k}(x)} \tag{61}$$

*for all $n \in \mathbb{Z}^+$.*

**Proof:** Let $k \in \mathbb{Z}_0^+$ be given. Induction will be used to verify that (60) holds for all $n \in \mathbb{Z}_0^+$. By (58) and (59)

$$A_{2,k}(x) = -d_{1,k}$$

and by (54)

$$-d_{1,k}S_{0,k+1}(x) = -d_{1,k}.$$

Therefore (60) holds for $n = 0$. Now suppose (60) holds for all $n$ such that $0 \leq n \leq j$ where $j \geq 0$. By (58),

$$A_{2j+4,k}(x) = (x + \frac{a_{j+2,k}}{x})A_{2j+2,k}(x) - d_{j+2,k}A_{2j,k}(x)$$

$$= (x + \frac{a_{j+k,2}}{x})[-d_{1,k}S_{2j,k+1}(x)] + d_{j+2,k}d_{1,k}S_{2j-2,k+1}(x)$$

$$= -d_{1,k}\left[\left[x + \frac{a_{j+2,k}}{x}\right]S_{2j,k+1}(x) - d_{j+2,k}S_{2j-2,k+1}(x)\right]$$

$$= -d_{1,k}\left[(x + \frac{a_{j+1,k+1}}{x})(S_{2j,k+1}(x)) - d_{j+1,k+1}S_{2j-2,k+1}(x)\right] = -d_{1,k}S_{2j+2,k+1}(x).$$

Therefore (60) is valid for all $n \in \mathbb{Z}_0^+$. As stated earlier, $A_{2n,k}(x)$ is the $n^{th}$ partial numerator of (57) and $S_{2n,k}(x)$ is the $n^{th}$ partial denominator of (57) for all $n \in \mathbb{Z}^+$. Thus

$$\begin{aligned} C_{n,k}(x) &= \frac{A_{2n,k}(x)}{S_{2n,k}(x)} \\ &= \frac{-d_{1,k}S_{2n-2,k+1}(x)}{S_{2n,k}(x)} \end{aligned}$$

for all $n \in \mathbb{Z}^+$. ■

Next, a separation of zeros property for $S_{2n+2,k}(x)$ and $S_{2n,k+1}(x)$ is established.

**THEOREM 3.3** *Let $\mathcal{M}$ be a symmetric, positive-definite SMF with the corresponding monic OLPS $\{S_n(x)\}_{n=0}^{\infty}$ which satisfies (49) through (52) for all $n \in \mathbb{Z}^+$. For each $k \in \mathbb{Z}_0^+$, define $\{S_{n,k}(x)\}_{n=0}^{\infty}$ by (53) through (56) for all $n \in \mathbb{Z}^+$. Then the zeros of $S_{2n,k+1}(x)$ separate the zeros of $S_{2n+2,k}(x)$ for all $n \in \mathbb{Z}^+$ and $k \in \mathbb{Z}_0^+$. I.e., if $\{x_{n,j}^{(k+1)}\}_{j=1}^{n}$ and $\{x_{n+1,j}^{(k)}\}_{j=1}^{n+1}$ represent the positive zeros of $S_{2n,k+1}(x)$ and $S_{2n+2,k}(x)$, respectively, then*

$$x_{n+1,j}^{(k)} < x_{n,j}^{(k+1)} < x_{n+1,j+1}^{(k)} \tag{62}$$

*for $j = 1, 2, \ldots, n$ where $n \in \mathbb{Z}^+$ and $k \in \mathbb{Z}_0^+$.*

**Proof:** Let $n \in \mathbb{Z}^+$ and $k \in \mathbb{Z}_0^+$ be given. By equation (48)

$$A_{2n+2,k}(x)S_{2n,k}(x) - A_{2n,k}(x)S_{2n+2,k}(x) = (-1)^{n+2}\prod_{i=1}^{n+1}(-d_i)$$

and by Theorem 3.2,

$$-d_{1,k}S_{2n,k+1}(x)S_{2n,k}(x) + d_{1,k}S_{2n-2,k+1}(x)S_{2n+2,k}(x) = -\prod_{i=1}^{n+1} d_i$$

Therefore

$$S_{2n,k+1}(x)S_{2n,k}(x) - S_{2n-2,k+1}(x)S_{2n+2,k}(x) = \frac{\prod_{i=1}^{n+1}(d_i)}{d_{1,k}} \tag{63}$$

for all $n \in \mathbb{Z}^+$ and $k \in \mathbb{Z}_0^+$. By Theorem 2.10, all the zeros of $S_{2n,k}(x)$ and $S_{2n+2,k}(x)$ are simple and of the form $\{\pm x_{n,j}^{(k)}\}_{j=1}^{n}$ and $\{\pm x_{n+1,j}^{(k)}\}_{j=1}^{n+1}$, respectively, such that

$$0 < x_{n+1,j}^{(k)} < x_{n,j}^{(k)} < x_{n+1,j+1}^{(k)}, \quad j = 1, 2, \ldots, n.$$

Therefore $S_{2n,k}(x_{n+1,j}^{(k)})$ and $S_{2n,k}(x_{n+1,j+1}^{(k)})$ have opposite signs for $j = 1, 2, \ldots, n-1$.

By (63),

$$S_{2n,k+1}(x_{n+1,j}^{(k)})S_{2n,k}(x_{n+1,j}^{(k)}) > 0$$

and

$$S_{2n,k+1}(x_{n+1,j+1}^{(k)})S_{2n,k}(x_{n+1,j+1}^{(k)}) > 0$$

for $j = 1, 2, \ldots, n-1$ implying $S_{2n,k+1}(x_{n+1,j}^{(k)})$ and $S_{2n,k+1}(x_{n+1,j+1}^{(k)})$ have opposite signs for $j = 1, 2, \ldots, n-1$. So there is at least one zero of $S_{2n,k+1}(x)$ located in $(x_{n+1,j}^{(k)}, x_{n+1,j+1}^{(k)})$ for $j = 1, 2, \ldots, n$ and since $S_{2n,k+1}(x)$ has exactly n positive zeros, $S_{2n,k+1}(x)$ has exactly one zero, namely $x_{n,j}^{(k+1)}$ satisfying (62). ∎

Note that by equations (55) and (61),

$$C_{n,k}(x) = \frac{-d_{1,k}S_{2n-2,k+1}(x)}{S_{2n,k}(x)} = \frac{-d_{1,k}xS_{2n-1,k}(x)b_{n,k}}{b_{n-1,k}xS_{2n+1,k}(x)}$$

or

$$C_{n,k}(x) = \frac{-b_{n,k}d_{1,k}S_{2n-1,k}(x)}{b_{n-1,k}S_{2n+1,k}(x)}, \quad n \in \mathbb{Z}^+.$$

Since the zeros of $S_{2n+2,k}(x)$ and $S_{2n,k+1}(x)$ interlace and $C_{n+1,k}(x)$ is defined in terms of $S_{2n+2,k}(x)$ and $S_{2n,k+1}(x)$, we call $\{S_{n,k+1}(x)\}_{n=0}^{\infty}$ the **associated symmetric Laurent polynomials** corresponding to $\{S_{n+1,k}(x)\}_{n=0}^{\infty}$ for all $k \in \mathbb{Z}_0^+$ (adopting terminology from [1]).

# 4 INTEGRAL REPRESENTATIONS FOR POSITIVE-DEFINITE SYMMETRIC SMF's

We conclude by deriving integral representations for positive-definite symmetric SMF's. The associated symmetric Laurent polynomials discussed in the previous section will play an important role in this investigation.

Suppose $\mathcal{M}$ is positive-definite and the terminology introduced in the last section continues to hold here. Let $k \in \mathbb{Z}_0^+$ and an integer $n > 1$ both be given. By Theorem 3.2,

$$C_{n,k}(x) = \frac{-d_{1,k}S_{2n-2,k+1}(x)}{S_{2n,k}(x)}.$$

Since $S_{2n-2,k+1}(x)$ and $S_{2n,k}(x)$ are regular Laurent polynomials of L-degrees $2n-2$ and $2n$, respectively, there exist monic polynomials $P_{2n-2,k+1}(x)$ and $P_{2n,k}(x)$ of exact degrees $2n-2$ and $2n$, respectively, such that

$$S_{2n-2,k+1}(x) = \frac{P_{2n-2,k+1}(x)}{x^{n-1}} \quad \text{and} \quad S_{2n,k}(x) = \frac{P_{2n,k}(x)}{x^n}.$$

By Theorem 2.10, the zeros of $S_{2n-2,k+1}(x)$ and $S_{2n,k}(x)$ are all nonzero. Therefore the zeros of $S_{2n-2,k+1}(x)$ and $P_{2n-2,k+1}(x)$ are identical and the same can be said for $S_{2n,k}(x)$ and $P_{2n,k}(x)$. By Theorem 3.3, the zeros of $P_{2n-2,k+1}(x)$ are $\pm x_{n-1,i}^{(k+1)}$, $i = 1, 2, \ldots, n-1$ and the zeros of $P_{2n,k}(x)$ are $\pm x_{n,i}^{(k)}$, $i = 1, 2, \ldots, n$ where the following interlacing property holds:

$$0 < x_{n,i}^{(k)} < x_{n-1,i}^{(k+1)} < x_{n,i+1}^{(k)}, \quad i = 1, 2, \ldots, n-1. \tag{64}$$

Since $P_{2n,k}(x)$ and $P_{2n-2,k+1}(x)$ are monic,

$$\lim_{x\to\infty} P_{2n,k}(x) = \infty \quad \text{and} \quad \lim_{x\to\infty} P_{2n-2,k+1}(x) = \infty$$

and combining this with (64) yields

$$\frac{P_{2n-2,k+1}(x_{n,j}^{(k)})}{P'_{2n,k}(x_{n,j}^{(k)})} > 0 \tag{65}$$

for $j = 1, 2, \ldots, n$. By Theorem 2.1,

$$\frac{P_{2n,k}(x)}{x^n} = \frac{(-1)^n P_{2n,k}(-x)}{(-x)^n} = \frac{P_{2n,k}(-x)}{x^n}$$

implying

$$P_{2n,k}(x) = P_{2n,k}(-x).$$

Similarly,

$$P_{2n-2,k+1}(x) = P_{2n-2,k+1}(-x). \tag{66}$$

Therefore,

$$P'_{2n-2,k+1}(x) = -P'_{2n-2,k+1}(-x)$$

and

$$P'_{2n,k}(x) = P'_{2n,k}(-x) \tag{67}$$

and hence

$$\frac{P_{2n-2,k+1}(-x_{n,j}^{(k)})}{P'_{2n,k}(-x_{n,j}^{(k)})} < 0$$

for $j = 1, 2, \ldots, n$. Now

$$C_{n,k}(x) = \frac{-d_{1,k}S_{2n-2,k+1}(x)}{S_{2n,k}(x)} = \frac{-d_{1,k}x^n P_{2n-2,k+1}(x)}{x^{n-1}P_{2n,k}(x)}$$

or

$$C_{n,k}(x) = \frac{-d_{1,k}xP_{2n-2,k+1}(x)}{P_{2n,k}(x)}. \tag{68}$$

Note that the zeros of $xP_{2n-2,k+1}(x)$ separate the zeros of $P_{2n,k}(x)$, for all $n \in \mathbb{Z}^+$, since zero is a root of $xP_{2n-2,k+1}(x)$, $xP_{2n-2,k+1}(x)$ is of exact degree $2n-1$, and $P_{2n,k}(x)$ is of exact degree $2n$. So there exists numbers $E_{n,-n}^{(k)}, \ldots, E_{n,-1}^{(k)}, E_{n,1}^{(k)}, \ldots, E_{n,n}^{(k)}$ such that

$$\frac{d_{1,k}xP_{2n-2,k+1}(x)}{P_{2n,k}(x)} = \sum_{j=-n}^{-1} \frac{E_{n,j}^{(k)}}{(x + x_{n,-j}^{(k)})} + \sum_{j=1}^{n} \frac{E_{n,j}^{(k)}}{(x - x_{n,j}^{(k)})}. \tag{69}$$

Multiplying both sides of this equation by $P_{2n,k}(x)$ yields

$$d_{1,k}xP_{2n-2,k+1}(x) = \sum_{j=-n}^{-1} \frac{E_{n,j}^{(k)}P_{2n,k}(x)}{(x + x_{n,-j}^{(k)})} + \sum_{j=1}^{n} \frac{E_{n,j}^{(k)}P_{2n,k}(x)}{(x - x_{n,j}^{(k)})}$$

and therefore

$$\lim_{x \to -x_{n,j}} (d_{1,k} x P_{2n-2,k+1}(x)) = E_{n,-j}^{(k)} P'_{2n,k}(-x_{n,j})$$

and

$$\lim_{x \to x_{n,j}} (d_{1,k} x P_{2n-2,k+1}(x)) = E_{n,j}^{(k)} P'_{2n,k}(x_{n,j})$$

for $j = 1, 2, \ldots, n$. Thus,

$$E_{n,-j}^{(k)} = \frac{-d_{1,k} x_{n,j} P_{2n-2,k+1}(-x_{n,j})}{P'_{2n,k}(-x_{n,j})} \tag{70}$$

and

$$E_{n,j}^{(k)} = \frac{d_{1,k} x_{n,j} P_{2n-2,k+1}(x_{n,j})}{P'_{2n,k}(x_{n,j})} \tag{71}$$

for $j = 1, 2, \ldots, n$. Combining equations (67), (66), (70) and (71), produces

$$E_{n,-j}^{(k)} = E_{n,j}^{(k)}, \quad j = 1, 2, \ldots n. \tag{72}$$

So by equations (68), (69) and (72),

$$C_{n,k}(x) = \frac{-d_{1,k} x P_{2n-2,k+1}(x)}{P_{2n,k}(x)} = \sum_{j=1}^{n} \left( \frac{-E_{n,j}^{(k)}}{(x + x_{n,j}^{(k)})} + \frac{-E_{n,j}^{(k)}}{(x - x_{n,j}^{(k)})} \right) \tag{73}$$

where $E_{n,j}^{(k)} > 0$ for $j = \pm 1, \pm 2, \ldots, \pm n$. The fact that $E_{n,j}^{(k)} > 0$ follows from equations (71) and (65). Now

$$\sum_{j=-n}^{-1} E_{n,j}^{(k)} + \sum_{j=1}^{n} E_{n,j}^{(k)} = \sum_{j=-n}^{-1} \left( \lim_{x\to\infty} \frac{x E_{n,j}^{(k)}}{x + x_{n,-j}^{(k)}} \right) + \sum_{j=1}^{n} \left( \lim_{x\to\infty} \frac{x E_{n,j}^{(k)}}{x - x_{n,j}^{(k)}} \right) =$$

$$= \lim_{x\to\infty} x \left( \sum_{j=-n}^{-1} \frac{E_{n,j}^{(k)}}{x + x_{n,-j}^{(k)}} + \sum_{j=1}^{n} \frac{E_{n,j}^{(k)}}{x - x_{n,j}^{(k)}} \right) =$$

$$\lim_{x\to\infty} \frac{x^2 d_{1,k} P_{2n-2,k+1}(x)}{P_{2n,k}(x)} = d_{1,k} \lim_{x\to\infty} \frac{x^2 P_{2n-2,k+1}(x)}{P_{2n,k}(x)} = d_{1,k}.$$

The last equality follows from the fact that $x^2 P_{2n-2,k+1}(x)$ and $P_{2n,k}(x)$ are both monic polynomials of exact degree $2n$. Thus

$$2 \sum_{j=1}^{n} E_{n,j}^{(k)} = \sum_{j=-n}^{-1} E_{n,j}^{(k)} + \sum_{j=1}^{n} E_{n,j}^{(k)} = d_{1,k}.$$

Define $E_{n,0}^{(k)} = 0$, $x_{n,0} = 0$, and $x_{n,-j} = -x_{n,j}$ for $j = 1, 2, \ldots, n$. Also let

$$\Psi_{n,k}(x) = \begin{cases} 0 & \text{if } x < x_{n,-n} \\ \displaystyle\sum_{j=-n}^{p} E_{n,j}^{(k)} & \text{if } x_{n,p} \le x < x_{n,p+1},\ p = -n, -n+1, \ldots, n-1 \\ d_{1,k} & \text{if } x \ge x_{n,n}. \end{cases} \tag{74}$$

Then by (73),

$$-C_{n,k}(x) = \int_{-\infty}^{\infty} \frac{d\Psi_{n,k}(t)}{x-t}$$

or

$$C_{n,k}(x) = \int_{-\infty}^{\infty} \frac{d\Psi_{n,k}(t)}{t-x}$$

for $n \in \mathbb{Z}^+$ and $k \in \mathbb{Z}_0^+$.

For each $k \in \mathbb{Z}_0^+$, define

$$l_{2n,j}^{(k)}(x) = \frac{S_{2n,k}(x)}{(x-x_{n,j})S'_{2n,k}(x_{n,j})}$$

for $j = \pm 1, \pm 2, \ldots, \pm n$. Note that $l_{2n,j}^{(k)}(x)$ is of L-degree $2n-1$ and

$$l_{2n,j}^{(k)}(x_{n,i}) = \delta_{i,j}$$

for $i, j = \pm 1, \pm 2, \ldots, \pm n$. For a given set of numbers $\{y_{-n}, \ldots, y_{-1}, y_1, \ldots, y_n\}$, let

$$L(x) = \sum_{\substack{j=-n \\ j \neq 0}}^{n} y_j l_{2n,j}^{(k)}(x).$$

Then $L(x_{n,i}) = y_i$ for $i = \pm 1, \pm 2, \ldots, \pm n$ and $L(x)$ is of L-degree at most $2n-1$. Let $R(x)$ be a given Laurent polynomial of L-degree at most $4n-1$. Define

$$L_{2n,k}(x) = \sum_{\substack{j=-n \\ j \neq 0}}^{n} R(x_{n,j}) l_{2n,j}^{(k)}(x)$$

and

$$Q(x) = L_{2n,k}(x) - R(x).$$

Then $Q(x)$ is of L-degree $4n-1$ or less. Since

$$Q(x_{n,i}) = L_{2n,k}(x_{n,i}) - R(x_{n,i}) = R(x_{n,i}) - R(x_{n,i}) = 0, \quad i = \pm 1, \pm 2, \ldots, \pm n,$$

$$Q(x) = S(x) S_{2n}(x)$$

where $S(x)$ is some Laurent polynomial of L-degree $2n-1$ or less.

Therefore $\mathcal{M}_k[Q(x)] = 0$ or

$$\mathcal{M}_k[R(x)] = \mathcal{M}_k[L_{2n,k}(x)] = \sum_{\substack{j=-n \\ j \neq 0}}^{n} R(x_{n,j}) \mathcal{M}_k[l_{2n,j}^{(k)}(x)]. \tag{75}$$

Define

$$F_{n,j}^{(k)} = \mathcal{M}_k[l_{2n,j}^{(k)}(x)] = \frac{1}{S'_{2n,k}(x_{n,j})} \mathcal{M}_k\left(\frac{S_{2n,k}(x)}{x-x_{n,j}}\right)$$

for $j = \pm 1, \pm 2, \ldots, \pm n$. Now it is claimed that

$$E_{n,j}^{(k)} = F_{n,j}^{(k)}, \quad j = \pm 1, \pm 2, \ldots, \pm n.$$

Recall that $P_{2n,k}(x) = x^n S_{2n,k}(x)$ and it follows that

$$P'_{2n,k}(x) = nx^{n-1}S_{2n,k}(x) + x^n S'_{2n,k}(x).$$

Therefore,

$$P'_{2n,k}(x_{n,j}) = x^n_{n,j} S'_{2n,k}(x_{n,j})$$

and so

$$E^{(k)}_{n,j} = \frac{d_{1,k}x_{n,j}P_{2n-2,k+1}(x_{n,j})}{P'_{2n,k}(x_{n,j})} = \frac{d_{1,k}x_{n,j}\left[x^{n-1}_{n,j}S_{2n-2,k+1}(x_{n,j})\right]}{x^n_{n,j}S'_{2n,k}(x_{n,j})}$$

$$= \frac{d_{1,k}S_{2n-2,k+1}(x_{n,j})}{S'_{2n,k}(x_{n,j})}$$

or

$$E^{(k)}_{n,j} = \frac{d_{1,k}S_{2n-2,k+1}(x_{n,j})}{S'_{2n,k}(x_{n,j})}.$$

Therefore, it will be shown that

$$\mathcal{M}_k\left(\frac{S_{2n,k}(x)}{x - x_{n,j}}\right) = d_{1,k}S_{2n-2,k+1}(x_{n,j}).$$

More generally, it will be verified that

$$\mathcal{M}_k\left(\frac{S_{2n,k}(x) - S_{2n,k}(t)}{x-t}\right) = d_{1,k}S_{2n-2,k+1}(t) \tag{76}$$

using induction. The method of proof is modeled after the proof found in [6, p.113]. It should be noted here that the domain of $\mathcal{M}_k$ is the set of all Laurent polynomials. Therefore, it should first be verified that

$$\frac{S_{2n,k}(x) - S_{2n,k}(t)}{x-t},$$

in equation (76) is a Laurent polynomial in the variable $x$ if $t$ is some fixed constant. There exist constants $b_0, b_1, \ldots, b_n$ such that

$$S_{2n,k}(x) = b_0x^{-n} + b_1x^{-n+2} + \ldots + b_nx^n.$$

Therefore,

$$\frac{S_{2n,k}(x) - S_{2n,k}(t)}{x-t} = \frac{b_0(x^{-n} - t^{-n}) + b_1(x^{-n+2} - t^{-n+2}) + \ldots + b_n(x^n - t^n)}{x-t}.$$

If $n$ is a positive integer, then

$$x^n - y^n = (x-y)(x^{n-1} + x^{n-2}y + \cdots + xy^{n-2} + y^{n-1})$$

so that $x^j - t^j$ is equal to a factor of $x - t$ times a polynomial of degree $j - 1$ for $j = 1, 2, \ldots, n$. Now

$$x^{-j} - t^{-j} = \frac{t^j - x^j}{x^j t^j}$$

and therefore

$$\frac{x^{-j}-t^{-j}}{x-t}$$

is a Laurent polynomial in $x$. Therefore

$$\frac{S_{2n,k}(x)-S_{2n,k}(t)}{x-t},$$

is indeed a Laurent polynomial in the variable $x$. Now return to the problem of verifying (76). It is clear that (76) holds for $n=0$ by (54). For $n=1$,

$$\mathcal{M}_k\left[\frac{S_{2,k}(x)-S_{2,k}(t)}{x-t}\right]=\mathcal{M}_k\left(\frac{(x+\frac{a_{1,k}}{x})-(t+\frac{a_{1,k}}{t})}{x-t}\right)$$

$$=\mathcal{M}_k\left[\frac{(x-t)+\frac{(t-x)a_{1,k}}{tx}}{x-t}\right]$$

$$=\mathcal{M}_k\left[1-\frac{a_{1,k}}{tx}\right]=\mathcal{M}_k(1)-\frac{a_{1,k}}{t}\mathcal{M}_k\left[\frac{1}{x}\right]=\mathcal{M}_k(1)=d_{1,k}S_{0,k+1}(t).$$

Thus (76) holds for $n=1$. Now assume that (76) holds for $n=0,1,\ldots,j$ $(j\geq 1)$ and show that

$$\mathcal{M}_k\left[\frac{S_{2j+2,k}(x)-S_{2j+2,k}(t)}{x-t}\right]=d_{1,k}S_{2j,k+1}(t).$$

Using (56),

$$\mathcal{M}_k\left[\frac{S_{2j+2,k}(x)-S_{2j+2,k}(t)}{x-t}\right]$$

$$=\mathcal{M}_k\left[\frac{(x+\frac{a_{j+1,k}}{x})S_{2j,k}(x)-d_{j+1,k}S_{2j-2,k}(x)-(t+\frac{a_{j+1,k}}{t})S_{2j,k}(t)}{x-t}\right.$$

$$\left.+\frac{d_{j+1,k}S_{2j-2,k}(t)}{x-t}\right]$$

$$=\mathcal{M}_k\left[\frac{d_{j+1,k}[S_{2j-2,k}(t)-S_{2j-2,k}(x)]+(x+\frac{a_{j+1,k}}{x})S_{2j,k}(x)}{x-t}\right.$$

$$\left.+\frac{-(t+\frac{a_{j+1,k}}{t})S_{2j,k}(x)+(t+\frac{a_{j+1,k}}{t})[S_{2j,k}(x)-S_{2j,k}(t)]}{x-t}\right]$$

$$=\mathcal{M}_k\left[\frac{d_{j+1,k}[S_{2j-2,k}(t)-S_{2j-2,k}(x)]+(x-t)S_{2j,k}(x)}{x-t}\right.$$

$$\left.+\frac{(\frac{a_{j+1,k}}{x}-\frac{a_{j+1,k}}{t})S_{2j,k}(x)+\left[t+\frac{a_{j+1,k}}{t}\right](S_{2j,k}(x)-S_{2j,k}(t))}{x-t}\right]$$

$$= -d_{j+1,k}d_{1,k}S_{2j-4,k+1}(t) + \mathcal{M}_k[S_{2j,k}(x)] + \frac{-a_{j+1,k}}{t}\mathcal{M}_k\left(\frac{1}{x}S_{2j,k}(x)\right) +$$

$$\left(t + \frac{a_{j+1,k}}{t}\right)\mathcal{M}_k\left[\frac{S_{2j,k}(x) - S_{2j,k}(t)}{x-t}\right]$$

$$= -d_{j+1,k}d_{1,k}S_{2j-4,k+1}(t) + \left(t + \frac{a_{j+1,k}}{t}\right)d_{1,k}S_{2j-2,k+1}(t)$$

$$= d_{1,k}\left[\left(t + \frac{a_{j+1,k}}{t}\right)S_{2j-2,k+1}(t) - d_{j+1,k}S_{2j-4,k+1}(t)\right]$$

$$= d_{1,k}\left[\left(t + \frac{a_{j,k+1}}{t}\right)S_{2j-2,k+1}(t) - d_{j,k+1}S_{2j-4,k+1}(t)\right] = d_{1,k}S_{2j,k+1}(t).$$

Therefore this inductive argument shows that (76) holds for all $n \in \mathbb{Z}^+$ and $k \in \mathbb{Z}_0^+$. Letting $t = x_{n,j}$ in (76) yields

$$\mathcal{M}_k\left(\frac{S_{2n,k}(x)}{x - x_{n,j}}\right) = d_{1,k}S_{2n-2,k+1}(x_{n,j})$$

and therefore $E_{n,j}^{(k)} = F_{n,j}^{(k)}$ for $j = \pm 1, \pm 2, \ldots, \pm n$. Thus, (75) implies

$$\mathcal{M}_k[R(x)] = \int_{-\infty}^{\infty} R(x)d\Psi_{n,k}(x)$$

for every L-polynomial, $R(x)$, of L-degree $4n-1$ or less. The following theorem summarizes this work.

**THEOREM 4.1** *Let $k \in \mathbb{Z}_0^+$ be given. Suppose $\mathcal{M}_k$ is a symmetric, positive-definite SMF whose monic OLPS $\{S_{n,k}(x)\}_{n=0}^{\infty}$ satisfies (54) through (56). Then, for each $n \in \mathbb{Z}^+$, there exist positive numbers $E_{n,1}^{(k)}, \ldots, E_{n,n}^{(k)}$ such that*

$$2\sum_{j=1}^{n} E_{n,j}^{(k)} = d_{1,k}$$

*and*

$$E_{n,j}^{(k)} = \frac{1}{S'_{2n,k}(x_{n,j})}\mathcal{M}_k\left(\frac{S_{2n,k}(x)}{x - x_{n,j}}\right) = \frac{d_{1,k}S_{2n-2,k+1}(x_{n,j})}{S'_{2n,k}(x_{n,j})}$$

*for $j = 1, 2, \ldots, n$.*

*If, for each $n \in \mathbb{Z}^+$, $\Psi_{n,k}(x)$ is defined as in (74), then*

$$C_{n,k}(x) = \int_{-\infty}^{\infty} \frac{d\Psi_{n,k}(t)}{t-x} = \sum_{j=1}^{n}\left(\frac{-E_{n,j}^{(k)}}{(x + x_{n,j}^{(k)})} + \frac{-E_{n,j}^{(k)}}{(x - x_{n,j}^{(k)})}\right),$$

*where $C_{n,k}(x)$ is the $n^{th}$ approximant of (57) and*

$$\mathcal{M}_k[R(x)] = \int_{-\infty}^{\infty} R(x)d\Psi_{n,k}(x)$$

*for every L-polynomial, $R(x)$, of L-degree $4n-1$ or less.*

To proceed, the following theorems, known as *Grommer's Selection Theorem* (see [8, pp. 514,515]) and *Helly's Selection Theorem* (see [1, pp. 53,54]), respectively, will be used.

**THEOREM 4.2** *(Grommer's Selection Theorem) Let $\{\psi_n(t)\}$ be a sequence of real-valued nondecreasing functions defined on $-\infty < t < \infty$, such that $c \leq \psi_n(t) \leq C$ for all $-\infty < t < \infty$ where $n \in \mathbb{Z}^+$. Then there exists a real-valued nondecreasing function $\psi(t)$ defined on $-\infty < t < \infty$ such that $c \leq \psi(t) \leq C$ for all $-\infty < t < \infty$, and there exists a subsequence, $\{n_k\}$, of positive integers such that $\lim_{k\to\infty} \psi_{n_k}(t) = \psi(t)$ for $-\infty < t < \infty$. Moreover, if $g(t)$ is a continuous complex-valued function of the real variable $t$ such that $\lim_{t\to\pm\infty} g(t) = 0$, then*

$$\lim_{k\to\infty} \int_{-\infty}^{\infty} g(t)d\psi_{n_k}(t) = \int_{-\infty}^{\infty} g(t)d\psi(t).$$

**THEOREM 4.3** *(Helly's Selection Theorem) If $\{\phi_n(x)\}_{n=1}^{\infty}$ is a sequence of uniformly bounded, nondecreasing functions on a compact interval $[a,b]$ which converges on $[a,b]$ to some limit function $\phi(x)$, then for every real function $f(x)$ which is continuous on $[a,b]$,*

$$\lim_{n\to\infty} \int_a^b f(x)d\phi_n(x) = \int_a^b f(x)d\phi(x).$$

According to Grommer's Selection Theorem, since $\{\Psi_{n,k}(x)\}_{n=0}^{\infty}$ is a sequence of real-valued nondecreasing functions defined on $-\infty < t < \infty$ for all $n \in \mathbb{Z}^+$, such that
$0 \leq \Psi_{n,k}(x) \leq d_{1,k}$ for all $-\infty < t < \infty$ and $n \in \mathbb{Z}^+$, there exists a subsequence $\{\Psi_{n_j,k}(x)\}_{j=0}^{\infty}$ and a real-valued non-decreasing function $\Psi_k(t)$ defined on $-\infty < t < \infty$ such that $0 \leq \Psi_k(t) \leq d_{1,k}$ for all $-\infty < t < \infty$ and

$$\lim_{j\to\infty} \Psi_{n_j,k}(t) = \Psi_k(t)$$

for $-\infty < t < \infty$.

Let $x \notin \mathbb{R}$ be given. Define $g(t) = \dfrac{1}{t-x}$. Then $g(t)$ is a continuous, complex-valued function of the real variable $t$ and

$$\lim_{t\to\pm\infty} g(t) = 0.$$

So again by Grommer's Selection Theorem,

$$\lim_{j\to\infty} C_{n_j,k}(x) = \lim_{j\to\infty} \int_{-\infty}^{\infty} \frac{d\Psi_{n_j,k}(t)}{t-x} = \int_{-\infty}^{\infty} \frac{d\Psi_k(t)}{t-x}.$$

By (75) and Theorem 4.1,

$$\mathcal{M}_k[x^{2p}] = \int_{-\infty}^{\infty} x^{2p} d\Psi_{n,k}(x) = \sum_{j=-n}^{n} E_{n,j}^{(k)} x_{n,j}^{2p} = \sum_{j=1}^{n} 2E_{n,j}^{(k)} x_{n,j}^{2p} =$$

$$2\int_0^\infty x^{2p}\,d\Psi_{n,k}(x)$$

for $p=-n,-n+1,\ldots,n-1$. It is claimed that

$$\mathcal{M}_k[x^{2p}]=2\int_0^\infty x^{2p}\,d\Psi_k(x)$$

for all $p\in\mathbb{Z}$. Assume $p\in\mathbb{Z}_0^+$ and let $\alpha>0$ be given. Then, for a sufficiently large $n_j$,

$$\left|2\int_0^\alpha x^{2p}\,d\Psi_k(x)-\mathcal{M}_k[x^{2p}]\right|=\left|2\int_0^\alpha x^{2p}\,d\Psi_k(x)-2\int_0^\infty x^{2p}\,d\Psi_{n_j,k}(x)\right|$$

$$\le 2\left|\int_0^\alpha x^{2p}\,d\Psi_k(x)-\int_0^\alpha x^{2p}\,d\Psi_{n_j,k}(x)\right|+2\left|\int_\alpha^\infty \frac{x^{2p+2}}{x^2}\,d\Psi_{n_j,k}(x)\right|$$

$$\le 2\left|\int_0^\alpha x^{2p}\,d\Psi_k(x)-\int_0^\alpha x^{2p}\,d\Psi_{n_j,k}(x)\right|+\frac{2}{\alpha^2}\mathcal{M}_k[x^{2p+2}].$$

Let $j\to\infty$ and apply Theorem 4.3 to see that

$$\left|2\int_0^\alpha x^{2p}\,d\Psi_k(x)-\mathcal{M}_k[x^{2p}]\right|\le\frac{2}{\alpha^2}\mathcal{M}_k[x^{2p+2}].$$

Now let $\alpha\to\infty$ to see that

$$\mathcal{M}_k[x^{2p}]=2\int_0^\infty x^{2p}\,d\Psi_k(x),\quad p\in\mathbb{Z}_0^+.$$

Suppose $p\in\mathbb{Z}^-$. Let $\alpha$ and $\beta$ be given such that $0<\alpha<1<\beta<\infty$. Then for sufficiently large $n_j$,

$$\left|2\int_\alpha^\beta x^{2p}\,d\Psi_k(x)-\mathcal{M}_k[x^{2p}]\right|=\left|2\int_\alpha^\beta x^{2p}\,d\Psi_k(x)-2\int_0^\infty x^{2p}\,d\Psi_{n_j,k}(x)\right|$$

$$\le 2\left|\int_\alpha^\beta x^{2p}\,d\Psi_k(x)-\int_\alpha^\beta x^{2p}\,d\Psi_{n_j,k}(x)\right|+2\int_0^\alpha x^{2p}\,d\Psi_{n_j,k}(x)+2\int_\beta^\infty x^{2p}\,d\Psi_{n_j,k}(x).$$

Since

$$\int_0^\alpha x^{2p}\,d\Psi_{n_j,k}(x)=\int_0^\alpha x^2x^{2p-2}\,d\Psi_{n_j,k}(x)\le$$

$$\alpha^2\int_0^\alpha x^{2p-2}\,d\Psi_{n_j,k}(x)\le\alpha^2\mathcal{M}_k[x^{2p-2}]$$

and

$$\int_\beta^\infty x^{2p}\,d\Psi_{n_j,k}(x)\le\beta^{2p}d_{1,k},$$

$$\left|2\int_\alpha^\beta x^{2p}\,d\Psi_k(x)-\mathcal{M}_k[x^{2p}]\right|\le$$

$$\left|2\int_{\alpha}^{\beta} x^{2p}d\Psi_k(x) - 2\int_{\alpha}^{\beta} x^{2p}d\Psi_{n_j,k}(x)\right| + 2\alpha^2\mathcal{M}_k[x^{2p-2}] + 2\beta^{2p}d_{1,k}.$$

Let $j \to \infty$ first, then let $\alpha \to 0$ and $\beta \to \infty$ to see that

$$\mathcal{M}_k[x^{2p}] = 2\int_0^{\infty} x^{2p}d\Psi_k(x), \quad p \in \mathbb{Z}^-.$$

Finally, it is claimed that $\Psi_k(x)$ has infinitely many points of increase on $(0,\infty)$. In order to reach a contradiction, suppose $\Psi_k(x)$ has exactly $N$ points of increase in $(0,\infty)$, say $x_1, x_2, \ldots, x_N$, where $N \in \mathbb{Z}^+$. Let $P(x) = \prod_{i=1}^{N}(x^2 - x_i^2)^2$ so that $P(x) \geq 0$ and $P(x) \not\equiv 0$ on $(-\infty,\infty)$. Then $P(x)$ vanishes at every point of increase of $\Psi(x)$ and it contains only even powers of $x$. Therefore

$$\mathcal{M}[P(x)] = 2\int_0^{\infty} P(x)d\Psi_k(x) = 0$$

contradicting the positive-definiteness of $\mathcal{M}_k$. This concludes the proof of the final theorem in this paper.

**THEOREM 4.4** *Let $k \in \mathbb{Z}_0^+$ be given. Suppose $M_k$ is a symmetric, positive-definite SMF whose monic OLPS $\{S_{n,k}(x)\}_{n=0}^{\infty}$ satisfies (54) through (56) and, for each $n \in \mathbb{Z}^+$, suppose $C_{n,k}(x)$ is the $n^{th}$ approximant of (57). Then there exists a subsequence, $\{C_{n_j,k}(x)\}_{j=1}^{\infty}$, of $\{C_{n,k}(x)\}_{n=1}^{\infty}$ satisfying*

$$\lim_{j\to\infty} C_{n_j,k}(x) = \int_{-\infty}^{\infty} \frac{d\Psi_k(t)}{t-x}$$

*for all $x \notin (-\infty,\infty)$ where $\Psi_k(t)$ is a nondecreasing, real-valued function with infinitely many points of increase on $(0,\infty)$. Furthermore,*

$$\mathcal{M}_k[x^{2n}] = 2\int_0^{\infty} x^{2n}d\Psi_k(x)$$

*for all $n \in \mathbb{Z}$.*

# References

[1] Theodore S. Chihara. An Introduction to Orthogonal Polynomials, New York/London/Paris: Gordon and Breach, Science Publishers, 1978.

[2] Lyle Cochran and S. Clement Cooper. Orthogonal Laurent Polynomials on the Real Line, Continued Fractions and Orthogonal Functions: Theory and Applications, Lecture Notes in Pure and Applied Mathematics, Vol. 154, pp. 47-100, Marcel Dekker, 1994.

[3] S. Clement Cooper and Philip E. Gustafson. The Strong Chebyshev Distribution and Orthogonal Laurent Polynomials, to appear in J. Approximation Theory.

[4] E. Hendriksen and H. van Rossum. Orthogonal Laurent Polynomials, Proceedings of the Koninklijke Nederlandse Akademie von Wetenschappen, Proceedings A 89(1), 1986, 17-36.

[5] William B. Jones and W.J. Thron. Continued Fractions: Analytic Theory and Applications, Encyclopedia of Mathematics and Its Applications; Vol.11, Reading, MA: Addison-Wesley Publ. Co., 1980, (distributed now by Cambridge Univ. Press, NY).

[6] William B. Jones, Olav Njåstad and W.J. Thron. Two-point Padé expansions for a family of analytic functions, JCAM 9, 1983, 105-123.

[7] William B. Jones and W.J. Thron. Orthogonal Laurent Polynomials and the Strong Hamburger Moment Problem, J. Math Anal. and Appl. Vol. 98,1984, 528-554.

[8] William B. Jones, W.J. Thron and H. Waadeland. A Strong Stieltjes Moment Problem, Trans. of the AMS Vol. 261, 1980, 503-528.

[9] Olav Njåstad and W. J. Thron. The Theory of Sequences of Orthogonal L-polynomials. Det Kongelige Norske Videnskaber Selskab, Skrifter 1, 1983, pp. 54-91.

[10] O. Perron, Die Lehre von of Kettenbrüchen II, New York: Van Nostrand, 1948.

[11] H. S. Wall. Analytic Theory of Continued Fractions, Toronto/New York/London: D. Van Nostrand Company, Inc., 1948.

# Interpolating Laurent Polynomials

S. CLEMENT COOPER Department of Pure and Applied Mathematics, Washington State University, Pullman, WA, 99164-3113, USA

PHILIP E. GUSTAFSON[1] Division of Mathematics and Computer Science, Emporia State University, Emporia, KS, 66801, USA

## 1. INTRODUCTION

The study of strong moment problems began in 1980 with an examination of the strong Stieltjes moment problem [8]. In [7], strong moment functionals and orthogonal Laurent polynomials were introduced in connection with the strong Hamburger moment problem. Since then, the general theory surrounding these problems has developed rapidly. Several survey articles [3, 5, 6, 9] link many of the advances in the quickly growing field.

Important analyses in the classical moment problem setting involved interpolating polynomials constructed from the zeros of quasi-orthogonal polynomials, [1, 2, 4, 10, 11]. Here we develop interpolating Laurent polynomials. These were originally developed with a specific application in mind, but were eclipsed by a more efficient approach. However, the authors found them interesting and offer them in the spirit of a "splinter" to the proceedings in the hope that others will also find them interesting and perhaps recognize a use for them in their work

We start with some basic terminology and notation. A Laurent polynomial,

[1]Research supported in part by a grant from the Washington State University Graduate School.

or L-polynomial, is a rational function of a nonzero, real variable $x$ with the form $R(x) = \sum_{i=m}^{n} r_i x^i$, where $m, n \in \mathbb{Z}$ with $m \leq n$ and $r_i$ complex for $i = m, \ldots, n$. $R(x)$ is said to be real if $r_i \in \mathbb{R}$ for $i = m, \ldots n$. We use $\mathcal{R}$ to denote the vector space of all Laurent polynomials and $\mathcal{R}_{m,n}$ the set of all Laurent polynomials of the form $R(x) = \sum_{i=m}^{n} r_i x^i$. Two classes of L-polynomials that are particularly important in the study of orthogonal Laurent polynomials are

$$\mathcal{R}_{2m} = \{R \in \mathcal{R}_{-m,m} : \text{ the coefficient of } x^m \text{ is nonzero}\}$$

and

$$\mathcal{R}_{2m+1} = \left\{R \in \mathcal{R}_{-(m+1),m} : \text{ the coefficient of } x^{-(m+1)} \text{ is nonzero}\right\}$$

for all integers $m \geq 0$. For every L-polynomial, $R(x)$, there exists a unique $n$ such that $R(x) \in \mathcal{R}_n$. This number $n$ is called the L-degree of $R(x)$ and is denoted by L-deg $(R(x)) = n$, [3].

## 2. INTERPOLATING L-POLYNOMIALS

The existance and construction of interpolating L-polynomials is given in the following theorem and its proof.

**Theorem 1** *Let $t_1, t_2, \ldots, t_n$ denote arbitrary distinct nonzero real or complex numbers, $\nu_1, \nu_2, \ldots, \nu_n$ arbitrary nonnegative integers, $y_k^{(r_k)}, k = 1, 2, \ldots, n,\ r_k = 0, 1, \ldots, \nu_k$, arbitrary real or complex numbers and let $N = \sum_{k=1}^{n}(\nu_k + 1)$. Then there exists a unique L-polynomial, $F(x)$, of L-degree at most $N - 1$ such that*

$$F(t_k) = y_k^{(0)},\ F'(t_k) = y_k^{(1)},\ \ldots,\ F^{(\nu_k)}(t_k) = y_k^{(\nu_k)},\ k = 1, 2, \ldots, n. \quad (1)$$

**Proof:** We consider (1) as a system of $N$ linear equations with respect to the $N$ unknown coefficients of $F(x)$. This system is uniquely determined if and only if the corresponding homogeneous system of equations possesses only the trivial solution. The homogeneous system requires that $t_k$ be a zero of $F(x)$ with multiplicity of at least $\nu_k + 1$. Thus, taking into account multiplicities, $F(x)$ has at least $\sum_{k=1}^{n} (\nu_k + 1) = N$ zeros. Since the L-degree of $F(x)$ is at most $N - 1$, the L-polynomial must vanish identically. Thus, the homogeneous system possesses only the trivial solution implying that the system is uniquely determined. ■

For the remainder of the paper we will assume that $[a, b]$ is a bounded interval in $\mathbb{R}\backslash\{0\}$ and that $M = \max\{|a|, |b|\}$. Let $t_1, t_2, \ldots, t_n$ denote real numbers satisfying $t_1 < t_2 < \cdots < t_n$, $n-1$ of which are in $[a, b]$ and either $\tau = t_1 < -M$ or $\tau = t_n > M$. Let $j$ be fixed such that $\tau = t_{nj}$ ($j$ is either 1 or $n$). By Theorem (1), there exist unique L-polynomials $\phi_n(x, \tau)$ and $\Phi_n(x, \tau)$ of L-degree at most $2n - 2$ such that

$$\phi_n(t_i, \tau) = \begin{cases} 1 & \text{for } i < j \\ 0 & \text{for } i \geq j \end{cases}, \quad \phi_n'(t_i, \tau) = 0 \text{ for } i \neq j \tag{2}$$

$$\Phi_n(t_i, \tau) = \begin{cases} 1 & \text{for } i \leq j \\ 0 & \text{for } i > j \end{cases}, \quad \Phi_n'(t_i, \tau) = 0 \text{ for } i \neq j \tag{3}$$

The graphs of $\phi_n(x, \tau)$ and $\Phi_n(x, \tau)$ have interesting shapes, as we show next. We consider each shape separately.

Case 1. $\tau = t_1 < -M$. Conditions (2) become

$$\begin{aligned} \phi_n(t_i, \tau) &= 0, \ i = 1, \ldots, n \\ \phi_n'(t_i, \tau) &= 0, \ i = 2, \ldots, n \end{aligned}$$

Therefore $t_i$, $i = 2, \ldots, n$, are zeros of $\phi_n(x, \tau)$ of multiplicity at least two. Also, $\tau = t_1$ is a zero of $\phi_n(x, \tau)$ of multiplicity at least one. Altogether then, $\phi_n(x, \tau)$ has at least $2(n-1) + 1 = 2n - 1$ zeros, counting multiplicities. Since L-deg

$\phi_n(x,\tau) \leq 2n-2$, we conclude that

$$\phi_n(x,\tau) \equiv 0. \tag{4}$$

The analysis of $\Phi_n(x,\tau)$ is less straighforward. Conditions (3) become

$$\begin{aligned} \Phi_n(t_1,\tau) &= \Phi_n(\tau,\tau) = 1 \\ \Phi_n(t_i,\tau) &= \Phi_n'(t_i,\tau) = 0,\ i = 2,\ldots,n. \end{aligned}$$

Since L-deg $\Phi_n(x,\tau) \leq 2n-2$, it has at most $2n-2$ zeros. We see that $t_i$, $i = 2,\ldots,n$, are double zeros of $\Phi_n(x,\tau)$, so we conclude that these points are its $2n-2$ zeros. Therefore

$$\Phi_n(x,\tau) = \frac{c_n(x-t_2)^2\cdots(x-t_n)^2}{x^{n-1}},$$

with $c_n = \tau^{n-1}\left[(\tau-t_2)\cdots(\tau-t_n)\right]^{-2}$. Since $\tau < 0$, $c_n < 0$ if $n$ is even and $c_n > 0$ if $n$ is odd. We will focus on $n$ odd, and thus $\Phi_n(x,\tau) \geq 0$ for all real nonzero $x$.

We now show that $\Phi_n(x,\tau) \geq 1$ on $(-\infty,\tau]$. Let $x \in (-\infty,\tau]$ and

$$\frac{(x-t_k)^2}{(\tau-t_k)^2} = \min_{2\leq i\leq n} \frac{(x-t_i)^2}{(\tau-t_i)^2}.$$

Then

$$\Phi_n(x,\tau) \geq \left[\frac{\tau(x-t_k)^2}{x(\tau-t_k)^2}\right]^{n-1}.$$

Now $-\tau(x-t_k)^2 \geq -x(\tau-t_k)^2$ is equivalent to $(\tau-x)(x\tau-t_k^2) \geq 0$, and since $x \leq \tau < -M$, we have $\tau - x \geq 0$ and $x\tau - t_k^2 > 0$. Hence, for n odd,

$$\Phi_n(x,\tau) \geq 1 \text{ on } (-\infty,\tau] \text{ for } \tau < -M. \tag{5}$$

Also, when $n$ is odd, $\Phi_n(x,\tau)$ is nonnegative for all real nonzero $x$, so that

$$\Phi_n(x,\tau) \geq 0 \text{ on } (\tau,0)\cup(0,\infty) \text{ for } \tau < -M. \tag{6}$$

Case 2. $\tau = t_n > M$. Conditions (3) become

$$\begin{aligned} \Phi_n(t_i, \tau) &= 1, \quad i = 1, \ldots, n \\ \Phi'_n(t_i, \tau) &= 0, \quad 1 = 1, \ldots, n-1. \end{aligned}$$

Defining $\Phi^*_n(x,\tau) = 1 - \Phi_n(x,\tau)$, it follows from the above conditions that $\Phi^*_n(x,\tau) \equiv 0$, and hence

$$\Phi_n(x,\tau) \equiv 1. \tag{7}$$

By an analysis similar to that used in the previous case, it can be shown that for $n$ odd, $\phi^*_n(x,\tau) = 1 - \phi_n(x,\tau) \geq 0$ on $(-\infty,0) \cup (0,\infty)$ and $\phi^*_n(x,\tau) \geq 1$ on $[\tau,\infty)$. Therefore,

$$\phi_n(x,\tau) \leq 1 \quad \text{on } (-\infty,0) \cup (0,\tau), \tag{8}$$

$$\phi_n(x,\tau) \leq 0 \quad \text{on } [\tau,\infty). \tag{9}$$

Summarizing results (4) - (9), we have, for $n$ odd and $\tau < -M$,

$$\begin{aligned} \phi_n(x,\tau) &\equiv 0 \\ \Phi_n(x,\tau) &\geq 1 \text{ on } (-\infty,\tau] \\ \Phi_n(x,\tau) &\geq 0 \text{ on } (\tau,0) \cup (0,\infty) \end{aligned}$$

while for $\tau > M$,

$$\begin{aligned} \phi_n(x,\tau) &\leq 1 \text{ on } (-\infty,0) \cup (0,\tau) \\ \phi_n(x,\tau) &\leq 0 \text{ on } [\tau,\infty) \\ \Phi_n(x,\tau) &\equiv 1. \end{aligned}$$

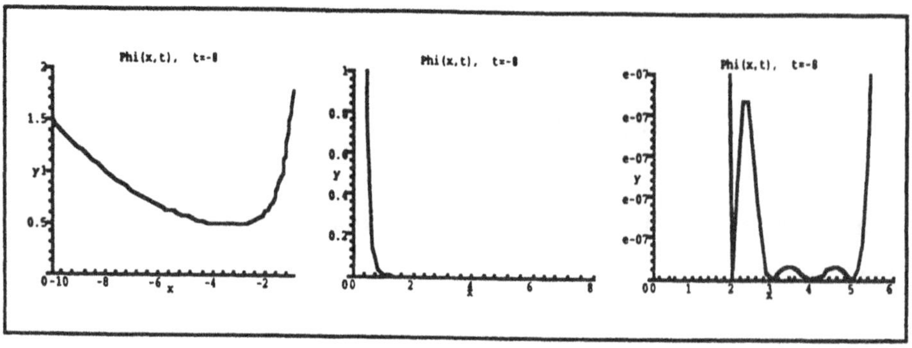


Figure 1: Graph of $\Phi_n(x, \tau)$ for $\tau = -8$.

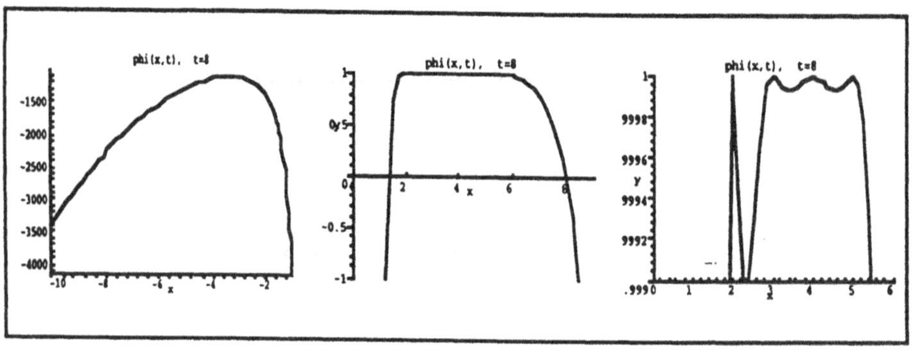


Figure 2: Graph of $\phi_n(x, \tau)$ for $\tau = 8$.

**Example.**

Let $n = 5$, $[a, b] = [1, 6]$ and $\{t_i\} = \{2, 3, 4, 5\}$, $i \neq j$. If we then take $\tau = -8$ and $\tau = 8$, we obtain the graphs in Figure (1) and Figure (2), respectively.

# References

[1] N. I. Akhiezer. The Classical Moment Problem. Oliver and Boyd, London, 1965.

[2] T.S. Chihara. An Introduction to Orthogonal Polynomials. Gordon and Breach, New York, 1978.

[3] L. Cochran, S. Clement Cooper. Orthogonal Laurent polynomials on the real line. Continued Fractions and Orthogonal Functions: Theory and Applications, Proceedings Loen, Norway, 1992, S. Clement Cooper, W.J. Thron, (eds.), Lecture Notes in Pure and Applied Mathematics, Marcel Dekker, 47-100, 1993.

[4] G. Freud. Orthogonal Polynomials. Pergamon Press, New York, 1971.

[5] W. B. Jones, W.J. Thron. Survey of continued fraction methods of solving moment problems and related topics, W.B. Jones, W.J. Thron and H. Waadeland (eds.), Analytic Theory of Continued Fractions, Proceedings Loen, Norway 1981, Lecture Notes in Mathematics, Springer - Verlag, 932: 4-37, 1982.

[6] W. B. Jones, W.J. Thron. Orthogonal Laurent polynomials and Gaussian quadrature, Karl E. Gustafson and William P. Reinhardt (eds.), Quantum Mechanics in Mathematics, Chemistry, and Physics, Plenum, New York, 449-445, 1981.

[7] W. B. Jones, O. Njåstad, W.J. Thron. Orthogonal Laurent polynomials and the strong Hamburger moment problem, J. Math. Anal. and Appl. 98: 528-554, 1984.

[8] W. B. Jones, W.J. Thron, H. Waadeland. A strong Stieltjes moment problem, Trans. of the AMS, 261: 503-528, 1980.

[9] O. Njåstad, W.J. Thron. The theory of sequences of orthogonal L-polynomials. Det Kongelige Norske Videnskabers Selskab, 1 54-91, 1983.

[10] J.A. Shohat, J.D. Tamarkin. The Problem of Moments, AMS Mathematical Surveys, No. I, Providence, R.I., 1963.

[11] Gabor Szegö. Orthogonal Polynomials, AMS Colloquium Publications, Vol. 23, Providence, RI, 1975.

# Computation of the Binet and Gamma Functions by Stieltjes Continued Fractions

CATHLEEN M. CRAVIOTTO* Dept. of Mathematical Sciences, University of Northern Colorado, Greeley, CO 80639 U.S.A.

WILLIAM B. JONES* Dept. of Mathematics, University of Colorado, Boulder, CO 80309–0395 U.S.A.

NANCY J. WYSHINSKI Dept. of Mathematics, Trinity College, Hartford, CT 06106 USA

## 1 INTRODUCTION

One of the most important functions in mathematics and mathematical physics is the gamma function $\Gamma(z)$. Among several equivalent definitions of $\Gamma(z)$, we state the following

$$\Gamma(z) := \sqrt{2\pi}\, z^{z-(1/2)} e^{-z} e^{J(z)}, \tag{1.1}$$

which is given in terms of the Binet function $J(z)$ defined by

$$\frac{J(\sqrt{z})}{\sqrt{z}} := \int_0^\infty \frac{v(t)}{z+t}\, dt =: G^J(z), \quad \text{for} \quad z \in S_\pi, \tag{1.2a}$$

where

$$v(t) := \frac{1}{2\pi} \frac{1}{\sqrt{t}} \operatorname{Log} \frac{1}{1-e^{-2\pi\sqrt{t}}}, \quad 0 < t < \infty, \tag{1.2b}$$

and

$$S_\theta := [z \in \mathbb{C} : |\arg z| < \theta], \quad 0 \le \theta \le \pi. \tag{1.2c}$$

Here $z^{z-(1/2)}$ and $\sqrt{z}$ denote principal branches. The function $G^J(z)$ defined in (1.2) is analytic in $S_\pi$ and is represented at each $z \in S_\pi$ by a convergent modified Stieltjes continued fraction (MSF) of the form

$$\frac{a_1^J}{z} \,+\, \frac{a_2^J}{1} \,+\, \frac{a_3^J}{z} \,+\, \frac{a_4^J}{1} \,+\, \cdots, \quad \text{where} \quad a_n^J > 0 \text{ for } n \in \mathbb{N}. \tag{1.3}$$

*Research supported in part by the U.S. National Science Foundation under Grant Number DMS–9302584.

Stieltjes [13, §6.4] computed a few of the coefficients $a_k$ of the MSF (1.3) and remarked that these calculations were very difficult. (More precisely, he computed coefficients $b_k$ which satisfy the relations $b_1 := 1/12$ and $a_k^J = 1/(b_k b_{k+1})$ for $k \in \mathbb{N}$.) Char [1] published $a_k^J$, for $1 \leq k \leq 41$, correct to 40 decimal places and Ruehr [18] has computed $a_k^J$, for $1 \leq k \leq 300$, using multiple precision arithmetic with 290 decimal digits. Cizek and Vrscay [2] investigated the behavior of the $a_k^J$ as $k \to \infty$ and they formulated the following conjecture:

$$\lim_{n\to\infty} \frac{a_n^J}{n^2} = \frac{1}{16}. \tag{1.4}$$

This conjecture (1.4) has recently been proved by Jones and Van Assche in an article [10] that is contained in the present volume. They also obtained the following asymptotic error estimates: *For each $\theta$ with $0 \leq \theta < \pi$, there exist constants $A$ and $B$ (independent of $|z|$ and $n$) such that for $n = 1, 2, 3, \ldots,$ $0 \leq |z| < \infty$ and $|\arg z| = \theta$,*

$$\left| \frac{J(\sqrt{z})}{\sqrt{z}} - g_n(z) \right| < \frac{A}{n^{B\sqrt{|z|}}}, \tag{1.5a}$$

*or, equivalently,*

$$|J(z) - zg_n(z^2)| < \frac{A|z|}{n^{B|z|}}, \tag{1.5b}$$

*where $g_n(z)$ denotes the $n$th approximant of the MSF (1.3).* By using (1.5), they showed that $g_n(z)$ is a better asymptotic approximation of $G^J(z)$ as $|z| \to \infty$ than is a partial sum of the asymptotic power series expansion of $G^J(z)$ as $|z| \to \infty$.

The primary purpose of the present paper is to investigate the number of significant decimal digits that can be obtained from an approximation of $G^J(z)$ that is derived from an $n$th approximant $g_n(z)$ of the MSF (1.3). More precisely, for a fixed $z \in S_\pi$ we let

$$u := \operatorname{Re} G^J(z), \;\; u_n := \operatorname{Re} g_n(z), \;\; v := \operatorname{Im} G^J(z), \;\; v_n := \operatorname{Im} g_n(z). \tag{1.6}$$

We let $\hat{u}_n$ and $\hat{v}_n$ denote approximate values of $u_n$ and $v_n$, respectively, where $\hat{u}_n$ and $\hat{v}_n$ are computed by machine floating decimal arithmetic, using a machine that carries $\nu$ decimal digits in the mantissa of the floating decimal approximation of a number. The machine unit $\omega$ is therefore given by

$$\omega := \left(\frac{1}{2}\right) 10^{1-\nu}. \tag{1.7}$$

The values $\hat{u}_n$ and $\hat{v}_n$ are computed by the backward recurrence algorithm (see Section 2). We let

$$SD(\hat{u}_n, u) \quad \text{and} \quad SD(\hat{v}_n, v)$$

denote the number of significant digits in the approximation of $u$ and $v$ by $\hat{u}_n$ and $\hat{v}_n$, respectively. A method for determining $SD(\hat{u}_n, u)$ and $SD(\hat{v}_n, v)$ is described in Section 3. That method makes use of sharp error bounds $B_T^{(n)}(\omega)$ and $B_r^{(n)}(\omega)$ of the truncation error $|G^J(z) - g_n(z)|$ and of the roundoff error $|g_n(z) - \hat{g}_n(z)|$, respectively, where

$$\hat{g}_n(z) := \hat{u}_n + i\hat{v}_n, \quad g_n(z) = u_n + iv_n \quad \text{and} \quad G^J(z) = u + iv.$$

The error bounds $B_T^{(n)}(\omega)$ and $B_R^{(n)}(\omega)$, satisfying

$$|G^J(z) - g_n(z)| \leq B_T^{(n)}(\omega) \quad \text{and} \quad |g_n(z) - \hat{g}_n(z)| \leq B_R^{(n)}(\omega)$$

were derived by Henrici and Pfluger [7], Jones and Thron [8], and [4].For completeness, Stieltjes transforms, Stieltjes continued fractions and related error bounds $B^{(n)}(\omega) = B_T^{(n)}(\omega) + B_R^{(n)}(\omega)$ are descibed in Section 2. Numerical examples are given in Section 4 to illustrate the convergence behavior of the MSF approximants $g_n(z)$ and to serve as a guide in applying the results given in previous sections to other special functions.

## 2 STIELTJES TRANSFORMS AND STIELTJES CONTINUED FRACTIONS

For later use we summarize here some known properties of Stieltjes transforms and related continued fractions. Proofs of many of these properties can be found in the books: [6], [9], [11], [12] and [15]. References to more recent properties are cited when needed. This material is treated with more generality than is needed for the Binet function, since we plan future applications to other special functions of these types.

A function $\psi(t)$ is called a moment distribution function (MDF) on $\mathbb{R}_0^+ := [0, \infty)$ if $\psi$ is real-valued, bounded and non-decreasing, with infinitely many points of increase on $\mathbb{R}_0^+$ and if all of the improper Riemann–Stieltjes integrals

$$c_n := \int_0^\infty t^n d\psi(t), \qquad n = 0, 1, 2, 3, \ldots \tag{2.1}$$

are convergent. For $n \in \mathbb{N}$, $c_n$ is called the $n$th moment of $\psi$. For a given MDF $\psi$, we consider functions defined by the following Stieltjes transforms:

$$G(z) := \int_0^\infty \frac{d\psi(t)}{z+t}, \quad z \in S_\pi, \tag{2.2a}$$

$$H(z) := \int_0^\infty \frac{z d\psi(t)}{z+t}, \quad z \in S_\pi, \tag{2.2b}$$

$$F(z) := \int_0^\infty \frac{z d\psi(t)}{1+zt}, \quad z \in S_\pi, \tag{2.2c}$$

$$P(z) := \int_0^\infty \frac{z d\psi(t)}{1+z^2 t}, \quad z \in S_{\frac{\pi}{2}}. \tag{2.2d}$$

It is readily seen that $G, H$ and $F$ are analytic functions for $z \in S_\pi$ and $P(z)$ is analytic for $z \in S_{\frac{\pi}{2}}$. Moreover,

$$G(z) = \frac{H(z)}{z} = F\left(\frac{1}{z}\right) = \frac{P(\sqrt{z})}{\sqrt{z}}, \quad \text{for} \quad z \in S_\pi. \tag{2.3}$$

Our reason for considering these four related functions is that at times it is convenient to work with their associated continued fraction expansions which have the following forms, respectively:

$$\widehat{G}(z) := \frac{a_1}{z} \, \frac{a_z}{+ \, 1} \, \frac{a_3}{+ \, z} \, \frac{a_4}{+ \, 1 \, +} \cdots, \quad a_n > 0 \text{ for } n \in \mathbb{N} \tag{2.4a}$$

$$\widehat{H}(z) := \frac{a_1}{1} \, \frac{a_2}{+ \, z} \, \frac{a_3}{+ \, 1} \, \frac{a_4}{+ \, z \, +} \cdots, \quad a_n > 0 \text{ for } n \in \mathbb{N}, \tag{2.4b}$$

$$\widehat{F}(z) := \frac{a_1 z}{1} \, \frac{a_2 z}{+ \, 1} \, \frac{a_3 z}{+ \, 1} \, \frac{a_4 z}{+ \, 1 \, +} \cdots, \quad a_n > 0 \text{ for } n \in \mathbb{N}, \tag{2.4c}$$

and

$$\widehat{P}(z) := \frac{a_1}{z} \, \frac{a_2}{+ \, z} \, \frac{a_3}{+ \, z \, +} \cdots, \quad a_n > 0 \quad \text{for} \quad n \in \mathbb{N}, \tag{2.4d}$$

We denote the $n$th approximants of these continued fractions, respectively, by the symbols

$$g_n(z), \quad h_n(z), \qquad f_n(z) \qquad \text{and} \quad p_n(z), \quad n \in \mathbb{N}. \tag{2.5}$$

It is then readily shown that

$$g_n(z) = \frac{h_n(z)}{z} = f_n\left(\frac{1}{z}\right) = \frac{p_n(\sqrt{z})}{\sqrt{z}}, \quad \text{for} \quad n \in \mathbb{N}. \tag{2.6}$$

A continued fraction of the form $\widehat{F}(z)$ in (2.4c) is called a Stieltjes continued fraction (or S-fraction). The other continued fractions $\widehat{G}, \widehat{H}$ and $\widehat{P}$ in (2.4) are called modified S-fractions (MSF's).

From (2.1) and (2.2a) one can show that $G(z)$ in (2.2a) admits an asymptotic power series expansion $L(z)$ defined by

$$G(z) \approx L(z) := \sum_{k=1}^{\infty} (-1)^k \frac{c_k}{z^{k+1}}, \quad z \to \infty, \ z \in \widehat{S}_\theta \tag{2.7a}$$

for $0 \le \theta < \pi$. Similarly one obtains the asymptotic expansions (for $0 \le \theta < \pi$)

$$H(z) \approx zL(z) := \sum_{k=0}^{\infty}(-1)^k \frac{c_k}{z^k}, \quad z \to \infty, \ z \in \widehat{S}_\theta, \tag{2.7b}$$

$$F(z) \approx L\left(\frac{1}{z}\right) := \sum_{k=0}^{\infty}(-1)^k c_k z^{k+1}, \quad z \to 0, \ z \in \widehat{S}_\theta, \tag{2.7c}$$

$$P(z) \approx zL(z^2) := \sum_{k=0}^{\infty}(-1)^k \frac{c_k}{z^{2k}}, \quad z \to \infty, \ z \in \widehat{S}_{\frac{\theta}{2}}. \tag{2.7d}$$

The associations of the continued fractions in (2.4) with the functions in (2.2) are obtained by choosing the $a_m$ in (2.4) so that the $n$th approximants (2.5) have Laurent (or Maclaurin) series with coefficients that agree with those of the asymptotic series expansion for as many terms as possible. In fact, the $a_m$ are uniquely determined so that one obtains the following correspondences; for $n = 1, 2, 3, \ldots$,

$$g_n(z) = \sum_{k=0}^{n-1}(-1)^k \frac{c_k}{z^{k+1}} + \frac{c_n^{(n)}}{z^{n+1}} + \frac{c_{n+1}^{(n)}}{z^{n+2}} + \cdots, \tag{2.8a}$$

$$h_n(z) = \sum_{k=0}^{n-1}(-1)^k \frac{c_k}{z^k} + \frac{c_n^{(n)}}{z^n} + \frac{c_{n+1}^{(n)}}{z^{n+1}} + \cdots, \tag{2.8b}$$

$$f_n(z) = \sum_{k=0}^{n-1}(-1)^k c_k z^{k+1} + c_n^{(n)} z^{n+1} + c_{n+1}^{(n)} z^{n+2} + \cdots, \tag{2.8c}$$

$$p_n(z) = \sum_{k=0}^{n-1}(-1)^k \frac{c_k}{z^{2k}} + \frac{c_n^{(n)}}{z^{2n}} + \frac{c_{n+1}^{(n)}}{z^{2n+1}} + \cdots, \tag{2.8d}$$

Each $a_m$ can be expressed in terms of Hankel determinants associated with the moment sequence $\{c_k\}_{k=0}^{\infty}$. These determinant formulas lead to the, so-called, quotient-difference ($qd-$) algorithm given as follows: We begin by setting

$$e_0^{(m)} := 0. \quad \text{for } m = 1, 2, 3, \ldots \quad \text{and } q_1^{(m)} := \frac{c_{m+1}}{c_m}, \quad m = 0, 1, 2, \ldots, \tag{2.9a}$$

and then compute successively, for $k = 1, 2, 3, \ldots$ and $m = 0, 1, 2, \ldots$,

$$e_k^{(m)} = q_k^{(m+1)} - q_k^{(m)} + e_{k-1}^{(m+1)} \tag{2.9b}$$

and

$$q_{k+1}^{(m)} = \frac{e_k^{(m+1)}}{e_k^{(m)}} q_k^{(m+1)}. \tag{2.9c}$$

Finally, one obtains the desired coefficients

$$a_1 = c_0, \quad a_{2k} = -q_k^{(0)}, \quad a_{2k+1} = -e_k^{(0)}, \quad k = 1, 2, 3, \ldots. \tag{2.10}$$

Although the $qd$-algorithm is not stable, by using high-degree precision in the machine's arithmetic operations, one can compute the $a_m$ with desired precision (assuming that the $c_k$ are known with high precision).

For any given MDF $\psi$ on $\mathbb{R}_0^+$, the corresponding continued fractions (2.4) are uniquely determined, but they may be divergent. A necessary and sufficient condition for the convergence of (2.4) is that

$$\sum_{m=2}^{\infty} \frac{a_2 a_4 \cdots a_{2m}}{a_1 a_3 \cdots a_{2m-1}} = \infty \quad \text{or} \quad \sum_{m=1}^{\infty} \frac{a_1 a_3 \cdots a_{2m-1}}{a_2 a_4 \cdots a_{2m}} = \infty. \tag{2.11}$$

When (2.11) holds the continued fractions in (2.4) converge to the corresponding functions in (2.2) throughout their domains of analyticity. A sufficient condition for the convergence of the continued fractions (2.4) is that Carleman's criterion

$$\sum_{k=1}^{\infty} \frac{1}{|c_k|^{\frac{1}{2k}}} = \infty \tag{2.12}$$

holds.

Conversely, if (2.4c) is a convergent S-fraction (i.e. (2.11) holds), then there exists a unique MDF $\psi$ on $\mathbb{R}_0^+$ such that the S-fraction converges to the Stieltjes transform (2.2c).

When the continued fractions (2.4) are convergent, it is important for computation to have a sharp bound on the trunction error $|G(z) - g_n(z)|$ that results when the value $G(z)$ is replaced by the $n$th approximant $g_n(z)$. Sharp *a posteriori* truncation error bounds were given by Henrici and Pfluger [6] as follows: For $z \in S_\pi$ we define

$$K(z) := \begin{cases} 1, \text{ if } 0 \le |\arg z| \le \pi/2 \\ |\csc(\arg z)|, \text{ if } \frac{\pi}{2} < |\arg z| < \pi. \end{cases} \tag{2.13}$$

Then, for $n = 2, 3, 4, \ldots,$

$$|G(z) - g_n(z)| \le K(z)|g_n(z) - g_{n-1}(z)|, \quad z \in S_\pi, \tag{2.14a}$$

$$|H(z) - h_n(z)| \le K(z)|h_n(z) - h_{n-1}(z)|, \quad z \in S_\pi, \tag{2.14b}$$

$$|F(z) - f_n(z)| \le K(z)|f_n(z) - f_{n-1}(z)|, \quad z \in S_\pi, \tag{2.14c}$$

and

$$|P(z) - p_n(z)| \le K(z^2)|p_n(z) - p_{n-1}(z)|, \quad z \in S_{\frac{\pi}{2}}. \tag{2.14d}$$

Somewhat sharper (but more complicated) truncation error bounds can be found in [3] and [4].

To make a rigorous error analysis for the computation of a function, say $F(z)$ in (2.2c), by the continued fraction $\widehat{F}(z)$ in (2.4c), one also needs bounds for the roundoff error produced when the true value of an approximant $f_n(z)$ is replaced by the value $\hat{f}_n(z)$ obtained by a machine using floating point arithmetic with mantissas having a fixed number $\nu$ of decimal digits. We describe for later use some known results for S-fractions (2.2c) when $\hat{f}_n(z)$ is obtained by using the backward recurrence algorithm (BR-algorithm). Connections are then made for use of the MSF's in (2.4).

The BR-algorithm for computing the $n$th approximant

$$f_n := \frac{a_1}{b_1} \underset{+}{} \frac{a_2}{b_2} \underset{+}{} \cdots \underset{+}{} \frac{a_n}{b_n}, \tag{2.15a}$$

of a continued fraction

$$F = \underset{j=1}{\overset{\infty}{K}} \left(\frac{a_j}{b_j}\right) = \frac{a_1}{b_1} \underset{+}{} \frac{a_2}{b_2} \underset{+}{} \frac{a_3}{b_3} \underset{+}{} \cdots, \quad 0 \neq a_j \in \mathbb{C},\ b_j \in \mathbb{C}, \tag{2.15b}$$

is the following: Compute successively

$$D_{n+1}^{(n)} := 0, \quad D_k^{(n)} := \frac{a_k}{b_k + D_{k+1}^{(n)}}, \quad k = n,\ n-1, \ldots, 1. \tag{2.16a}$$

Then set

$$f_n := D_1^{(n)}. \tag{2.16b}$$

Results on backward error analysis for roundoff error in the BR-algorithm are now described. For $n \in \mathbb{N}$ and $k = 1, \ldots, n$, we let $\hat{a}_k, \hat{b}_k$ and $\widehat{D}_k^{(n)}$ denote the floating decimal approximations, respectively, of $a_k, b_k$ and $D_k^{(n)}$ resulting from machine computations. We assume that the machine unit $\omega$ is given by (1.7). We also assume that, if $x, y \in \mathbb{R}$ and $f\ell(x \text{ op } y)$ denotes the machine's computed value of $x$ op $y$, where "op" stands for any of the 4 arithmetic operations $+, -, \times$ and $\div$, then

$$|f\ell(x \text{ op } y) - (x \text{ op } y)| \le |x \text{ op } y| \cdot \omega. \tag{2.17}$$

That is, the relative error of $f\ell(x \text{ op } y)$ is not greater than the machine number $\omega := (\frac{1}{2})\cdot 10^{1-\nu}$. (See, e.g. [5, §2.3.3].) Let $\alpha_k, \beta_k$ and $\epsilon_k^{(n)}$ denote the relative errors, respectively, in $\hat{a}_k, \hat{b}_k$ and $\widehat{D}_k^{(n)}$, so that, for $k = 1, 2, \ldots, n$,

$$\hat{a}_k = a_k(1+\alpha_k), \quad \hat{b}_k = b_k(1+\beta_k), \quad \widehat{D}_k^{(n)} = D_k^{(n)}(1+\varepsilon_k^{(n)}). \tag{2.18}$$

Let $\hat{f}_n = \widehat{D}_1^{(n)}$ denote the floating decimal approximation of $f_n$ computed by the BR-algorithm. Then

$$\hat{f}_n = f_n(1+\varepsilon_1^{(n)}) \quad \text{and hence} \quad \hat{f}_n - f_n = f_n \varepsilon_1^{(n)}. \tag{2.19}$$

We define $\gamma_k^{(n)}$ and $d_k^{(n)}$, for $k = 1, \ldots, n$, by

$$\widehat{D}_k^{(n)} = \frac{\hat{a}_k}{\hat{b}_k + \widehat{D}_{k+1}^{(n)}}(1+\gamma_k^{(n)}) \quad \text{and} \quad d_k^{(n)} := \frac{D_{k+1}^{(n)}}{b_k + D_{k+1}^{(n)}}, \tag{2.20}$$

so that $\gamma_k^{(n)}$ is the relative error resulting from the computation of $\widehat{D}_k^{(n)}$ from $\hat{a}_k, \hat{b}_k$ and $\widehat{D}_{k+1}^{(n)}$. For later use we state the following basic theorem on roundoff given by [8, Theorem 3.1].

THEOREM 2.1. *Let $F$ be a continued fraction (2.15b) and let $n$ be a given positive integer. Let $f_n$ denote the $n$th approximant of $F$ and let $\hat{f}_n$ denote the floating decimal approximation of $f_n$, where $\hat{f}_n$ is computed by a machine with machine number $\omega$, using the BR-algorithm. Let $\alpha^{(n)}, \beta^{(n)}, \gamma^{(n)}$ and $\eta_n$ be real numbers satisfying, for $k = 1, \ldots, n$,*

$$|\alpha_k| \le \omega\alpha^{(n)}, \quad |\beta_k| \le \omega\beta^{(n)}, \quad |\gamma_k^{(n)}| \le \omega\gamma^{(n)}, \quad |d_k^{(n)}| \le \eta_n, \tag{2.21a}$$

$$\alpha^{(n)}, \beta^{(n)} \in [0] \cup [x \in \mathbb{R} : x \ge 1], \qquad \gamma^{(n)} \ge 1, \ \eta_n > 0, \tag{2.21b}$$

*and*

$$\alpha^{(n)} + \beta^{(n)} + \gamma^{(n)} \ge 2. \tag{2.21c}$$

*Let $\varepsilon_k^{(n)}$, $k = 1, 2, \ldots, n$, be defined by (2.18) so that $\varepsilon_1^{(n)}$ is the relative error in the approximation of $f_n$ by $\hat{f}_n$ (see (2.19)). Then, for $k = 1, 2, \ldots, n$,*

$$|\varepsilon_k^{(n)}| \le E_k^{(n)} := \omega(1 + \alpha^{(n)} + \beta^{(n)}(1+\eta_n) + \gamma^{(n)}) \sum_{j=0}^{n-k} \eta_n^j, \tag{2.22a}$$

*provided the following conditions holds:*

$$0 \le \omega < [4(\alpha^{(n)} + \beta^{(n)} + \gamma^{(n)}]^{-2} \tag{2.22b}$$

*and*

$$0 \leq \omega < \frac{1}{2}\left[1 + \beta^{(n)}\eta_n + \eta_n(1 + \alpha^{(n)} + \beta^{(n)}(1 + \eta_n) + \gamma^{(n)})\sum_{j=0}^{n-2}\eta_n^j\right]^{-1}. \tag{2.22c}$$

*Moreover, if* $0 < E_1^{(n)} < 1$, *then*

$$|f_n - \hat{f}_n| \leq \frac{|\hat{f}_n|E_1^{(n)}}{1 - E_1^{(n)}} =: B_R^{(n)}(\omega). \tag{2.23}$$

### Remark

Inequality (2.23) can be obtained from (2.19) and (2.22a) since

$$|f_n| = \left|\frac{\hat{f}_n}{1 + \varepsilon_1^{(n)}}\right| \leq \frac{|\hat{f}_n|}{1 - |\varepsilon_1^{(n)}|} \leq \frac{|\hat{f}_n|}{1 - E_1^{(n)}}$$

and

$$|f_n - \hat{f}_n| = |f_n\varepsilon_1^{(n)}| \leq |f_n|E_1^{(n)}.$$

Theorem 2.1 was applied to non-negative T-fractions in [4, Theorem 4.2]. We state here the special case for S-fractions. Let $\widehat{F}(z)$ be a given S-fraction (2.4c), let $n$ be a positive integer, let

$$z \in S_\pi, \quad \theta := \arg z, \quad r := |z| \tag{2.24}$$

and

$$f_n(z) = \frac{a_1 z}{1} + \frac{a_2 z}{1} + \cdots + \frac{a_n z}{1}, \tag{2.25}$$

Let $\hat{f}_n(z)$ denote the floating decimal approximation of $f_n(z)$ obtained by applying the BR-algorithm with a machine whose machine number is $\omega = (\frac{1}{2}) \cdot 10^{1-\nu}$. For this case we may assume that

$$\alpha^{(n)} = \gamma^{(n)} = 1 \quad \text{and} \quad \beta^{(n)} = 0 \tag{2.26}$$

so that conditions (2.21) reduce (for $k = 1, \ldots, n$) to

$$|\alpha_k| \leq \omega, \quad \beta_k = 0, \quad |\gamma_k^{(n)}| \leq 1 \quad \text{and} \quad |d_k^{(n)}| \leq \eta_n. \tag{2.27}$$

We then obtain (from [4, Theorem 4.2])

THEOREM 2.2. *Let $f_n(z)$ be the nth approximant (2.25) of a convergent S-fraction (2.4c) and let $\hat{f}_n(z)$ denote the computed approximation of $f_n(z)$ as described above. Then: For $k = 1, 2, \ldots, n$,*
(A)

$$|\varepsilon_k^{(n)}| \leq E_k^{(n)} := 3\omega \sum_{j=0}^{n-k} \eta_n^j, \tag{2.28a}$$

*provided*

$$0 \leq \omega < \min\left[\frac{1}{64}, \frac{1}{2}\left(1 + 3\eta_n \sum_{j=0}^{n-2} \eta_n^j\right)^{-1}\right]. \tag{2.28b}$$

*(B)*

$$|f_n(z) - \hat{f}_n(z)| \leq \frac{|\hat{f}_n| E_1^{(n)}}{1 - E_1^{(n)}} =: B_R^{(n)}(\omega) \tag{2.29a}$$

*provided*

$$0 < E_1^{(n)} < 1. \tag{2.29b}$$

*(C) Let $a^{(n)} := \max[a_1, a_2, \ldots, a_n]$. Then for the upper bounds $\eta_n$ in (2.27) and (2.28) we can use the following expressions.*

*($C_1$) For $0 \leq |\theta| \leq \pi/2$, we can let*

$$\eta_n = \frac{a^{(n)} r}{1 + 2r\cos\theta + r^2} \tag{2.30a}$$

*or*

$$\eta_n = \frac{a^{(n)} r}{[1 + 2r(1 + a^{(n)})\cos\theta + r^2(1 + a^{(n)})^2]^{1/2}} < 1. \tag{2.30b}$$

*($C_2$) For $\frac{\pi}{2} < |\theta| < \pi$, we can let*

$$\eta_n = a^{(n)} r \csc^2\theta, \quad \text{if } \frac{3\pi}{2} - |\arg(-1 - z)| \leq |\theta|; \tag{2.31a}$$

$$\eta_n = \frac{a^{(n)} r}{1 + 2r\cos\theta + r^2}, \quad \text{if } |\theta| \leq \frac{3\pi}{2} - |\arg(-1 - z)|; \tag{2.31b}$$

$$\eta_n = \frac{a^{(n)} \csc^2\theta}{r}, \quad \text{if } 1 + r\cos\theta > 0; \tag{2.31c}$$

$$\eta_n = \frac{a^{(n)} r \csc|\theta|}{[1 + 2(1 + a^{(n)}) r\cos\theta + (a + a^{(n)})^2 r^2]^{1/2}}, \quad \text{if } (2 + a^{(n)}) r + \cos\theta > 0. \tag{2.31d}$$

From (2.14c) and (2.23) we obtain

$$\begin{aligned}|F(z) - \hat{f}_n(z)| &\leq |F(z) - f_n(z)| + |f_n(z) - \hat{f}_n(z)| \\ &\leq K_n(z)|f_n(z) - f_{n-1}(z)| + \frac{|\hat{f}_n(z)|E_1^{(n)}}{1 - E_1^{(n)}} \\ &\leq K_n(z)[|f_n(z) - \hat{f}_n(z)| + |\hat{f}_n(z) - \hat{f}_{n-1}(z)| \\ &\quad + |\hat{f}_{n-1}(z) - f_{n-1}(z)|] + B_R^{(n)}(\omega).\end{aligned}$$

Applying (2.23) to this yields

$$|F(z) - \hat{f}_n(z)| \leq \widehat{B}_T^{(n)} + B_{TR}^{(n)}(\omega), \tag{2.32a}$$

where

$$\widehat{B}_T^{(n)} := K(z)|\hat{f}_n(z) - \hat{f}_{n-1}(z)|, \quad \text{(see (2.13) for } K(z)) \tag{2.32b}$$

and

$$\begin{aligned}B_{TR}^{(n)}(\omega) &:= B_R^{(n)}(\omega)[1 + K(z)] + B_R^{(n)}(\omega) \\ &= 3\omega\left[\frac{|\hat{f}_n(z)|(1 + K(z))}{1 - E_1^{(n)}(\omega)}\sum_{j=0}^{n-1}\eta_n^j + \frac{|\hat{f}_{n-1}(z)|}{1 - E_1^{(n-1)}(\omega)}\sum_{j=0}^{n-2}\eta_{n-1}^j\right],\end{aligned} \tag{2.32c}$$

where (from (2.28a))

$$E_1^{(m)}(\omega) := 3\omega\sum_{j=0}^{m-1}\eta_m^j, \quad (\eta_m \text{ determined by Theorem 2.2}). \tag{2.32d}$$

We summarize the above in the following

THEOREM 2.3. *Let $z \in S_\pi$ and $n \in \mathbb{N}$ be given and let $f_n(z)$ denote the nth approximant of an S-fraction (2.4c) which converges to a limit $F(z)$. Let $\hat{f}_n(z)$ denote the approximation to $f_n(z)$ computed by a machine with machine unit $\omega = (\frac{1}{2}) \cdot 10^{1-\nu}$, applying the BR-algorithm. Let $\omega$ satisfy*

$$0 \leq \omega < \min\left[\frac{1}{64}, \frac{1}{2}\left(1 + 3\eta_m\sum_{j=0}^{m-2}\eta_m^j\right)^{-1}\right], \quad m = n, n-1, \tag{2.33}$$

*where the $\eta_m$ are determined by Theorem 2.2, and let $E_1^{(m)}(\omega)$ satisfy*

$$0 < E_1^{(m)}(\omega) < 1. \quad \text{for} \quad m = n, n-1. \tag{2.34}$$

Then

$$|F(z) - \hat{f}_n(z)| \le \widehat{B}_T^{(n)} + B_{TR}^{(n)}(\omega) =: B^{(n)}(\omega) \tag{2.35}$$

*where* $E_1^{(m)}(\omega)$, $\widehat{B}_T^{(n)}$ *and* $B_{TR}^{(n)}(\omega)$ *are defined by (2.32).*

**Remarks on Theorem 2.3**

(a) In (2.35), $\widehat{B}_T^{(n)}$ is the computed (approximate) value of the upper bound of the truncation error (see (2.14c)). The symbol $B_{TR}^{(n)}(\omega)$ denotes the upper bound of the error due to roundoff in the computation of $\hat{f}_n(z)$ and $\hat{f}_{n-1}(z)$ as approximations of $f_n(z)$ and $f_{n-1}(z)$, respectively. In practice one normally expects that, as $n$ increases, $\widehat{B}_T^{(n)}$ decreases and $B_{TR}^{(n)}(\omega)$ increases. The sum $B^{(n)}(\omega)$ of these two bounds (see (2.35)) would therefore normally decrease for the initial values of $n = 1, 2, 3, \ldots$, but for some values, say $n = n_0$, $B^{(n)}(\omega)$ increases for $n = n_0, n_0 + 1, \ldots$. This occurrence would suggest that approximants $\hat{f}_n(z)$ should not be computed for $n > n_0$, since such approximations of $f_n(x)$ are significantly affected by roundoff error.

(b) If there exists an $n_0 \in \mathbb{N}$ such that

$$\widehat{B}_T^{(n)} > B_{TR}^{(n)}(\omega) \quad \text{for} \quad n \le n_0$$

and

$$\widehat{B}_T^{(n)} \le B_{TR}^{(n)}(\omega) \quad \text{for} \quad n > n_0.$$

then the roundoff error has probably become dominant and so one should terminate the computations and use $\hat{f}_n(z)$ for only values of $n \le n_0$.

(c) We note that the bounds $\eta_m$, determined by Theorem 2.2 and used in (2.32), are independent of $\omega$. Therefore if one wishes to compute reliable values of $\hat{f}_n(z)$ for $n > n_0$, one should decrease the machine unit $\omega = (\frac{1}{2}) \cdot 10^{1-\nu}$. This could be accomplished, for example, by replacing $\nu$ by $2\nu$ (i.e., using double precision arithmetic). The expressions given in Theorem 2.3 and (2.32) shed light upon the manner in which the parameters $n, z$ and $\omega$ affect the error $|F(z) - \hat{f}_n(z)|$ that results when the value $F(z)$ is replaced by a computed approximation $\hat{f}_n(z)$ of the $n$th approximant $f_n(z)$.

(d) Finally we note that the term $K(z)$ in (2.34) is independent of $n$ and $\omega$ and $K(z)$ tends to infinity as $|\arg z| \to \pi$. The term $K(z)$ can therefore have a major effect on $B^{(n)}(\omega)$.

Before going to Section 3 we show how the three modified S-fractions $\widehat{G}(z)$, $\widehat{H}(z)$ and $\widehat{P}(z)$ in (2.4) can be transformed to S-fractions by means of equivalence transformations (see, e.g., [9, §2.3]). We define continued fractions

$$\widetilde{G}(z) := \widehat{F}(z^{-1}) = \frac{a_1 z^{-1}}{1} + \frac{a_2 z^{-1}}{1} + \frac{a_3 z^{-1}}{1} + \cdots, \tag{2.36a}$$

$$\widetilde{H}(z) := z\widehat{F}(z^{-1}) = z\left[\frac{a_1 z^{-1}}{1} + \frac{a_2 z^{-1}}{1} + \frac{a_3 z^{-1}}{a} + \cdots\right], \tag{2.36b}$$

$$\widetilde{P}(z) := z\widehat{F}(z^{-2}) = z\left[\frac{a_1 z^{-2}}{1} + \frac{a_2 z^{-2}}{1} + \frac{a_3 z^{-3}}{1} + \cdots\right]. \tag{2.36c}$$

Then $\widetilde{G}(z), \widetilde{H}(z)$ and $\widetilde{P}(z)$ are equivalent to $\widehat{G}(z), \widehat{H}(z)$ and $\widehat{P}(z)$, respectively. It follows that, if $\tilde{g}_n(z), \tilde{h}_n(z)$ and $\tilde{p}_n(z)$ are the corresponding $n$th approximants, respectively, then

$$g_n(z) = \tilde{g}_n(z) = f_n(z^{-1}), \quad \text{for } n \in \mathbb{N}, \tag{2.37a}$$

$$h_n(z) = \tilde{h}_n(z) = z f_n(z^{-1}), \quad \text{for } n \in \mathbb{N}, \tag{2.37b}$$

$$p_n(z) = \tilde{p}_n(z) = z f_n(z^{-2}), \quad \text{for } n \in \mathbb{N}. \tag{2.37c}$$

With the results in (2.36) and (2.37), Theorem 2.3 can be applied to the modified S-fractions in (2.4 a,b and d).

The next section is used to describe a procedure for computing the numbers of significant digits in the approximations

$$\hat{u}_n := \operatorname{Re} \hat{f}_n(z) \quad \text{and} \quad \hat{v}_n := \operatorname{Im} \hat{f}_n(z) \tag{2.38a}$$

of the numbers

$$u := \operatorname{Re} F(z) \quad \text{and} \quad v := \operatorname{Im} F(z). \tag{2.38b}$$

The bounds $B^{(n)}(\omega)$ of Theorem 2.3 play an essential role in determining the number of significant digits.

## 3 SIGNIFICANT DIGITS IN COMPUTATION WITH FLOATING DECIMAL ARITHMETIC

A non-zero real number $u$ can be expressed as an infinite series

$$u = \sigma(u) \sum_{j=-\infty}^{\mu(u)} D_j(u) \cdot 10^j, \tag{3.1a}$$

where

$$\sigma(u) := \begin{cases} 1, & \text{if } u > 0, \\ -1, & \text{if } u < 0, \end{cases} \tag{3.1b}$$

and

$$D_j(u) \in \begin{cases} [0, 1, 2, \ldots, 9], & \text{for } -\infty < j < \mu(u) \\ [1, 2, \ldots, 9], & \text{for } j = \mu(u). \end{cases} \tag{3.1c}$$

$\sigma(u)$ is called the sign of $u$, $D_j(u)$ is called the decimal digit of $u$ in the $j$th decimal place, and $\mu(u)$ denotes the largest decimal place where $u$ has a

$\sigma(u)$ is called the sign of $u$, $D_j(u)$ is called the decimal digit of $u$ in the $j$th decimal place, and $\mu(u)$ denotes the largest decimal place where $u$ has a non-zero decimal digit. To avoid ambiguities (such as $1.000\cdots = 0.999\ldots$) we choose the representation $(1.000\ldots)$ with infinitely many zero decimal digits. The floating decimal representation of $u$ has the form

$$u = \sigma(u)M(u)\cdot 10^{\mu(u)+1} \tag{3.2a}$$

where (if we let $\mu := \mu(u)$)

$$\begin{aligned} M(u) &= 0.D_\mu(u)D_{\mu+1}(u)D_{\mu+2}(u)\ldots \\ &= \sum_{j=-\infty}^{\mu} D_j(u)\cdot 10^{j-\mu-1}. \end{aligned} \tag{3.2b}$$

We note that $M(u)$, called the mantissa of $u$, satisfies

$$0.1 \le M(u) < 1.0. \tag{3.2c}$$

Thus, for example, $M(1.000\ldots) = 0.1$.

We consider a sequence $\{u_n\}_{n=1}^{\infty}$ of real numbers converging to $u$. For each $n \in \mathbb{Z}_0^+ := [0,1,2,3,\ldots]$, we call $u_n$ the $n$th *approximant* of $u$. It can be seen that, for each pair of real numbers $(u_n, u)$, where $u_n \neq u$, there exist numbers

$$\Delta(u_n, u) \quad \text{and} \quad t(u_n, u)$$

that are uniquely defined by the following conditions:

$$|u_n - u| = \Delta(u_n, u)\cdot 10^{-t(u_n,u)}, \tag{3.3a}$$

where

$$0.05 < \Delta(u_n, u) \le 0.5 \quad \text{and} \quad t(u_n, u) \in \mathbb{Z}. \tag{3.3b}$$

Thus $\Delta$ and $t$ are functions defined on the set

$$S := [(a,b) : a \in \mathbb{R}, b \in \mathbb{R}, a \neq b]. \tag{3.4}$$

For each $n \in \mathbb{Z}_0^+$ with $u_n \neq u$, we call $t(u_n, u)$ the *number of correct decimal digits in the $n$th approximant $u_n$*. The decimal digits

$$D_j(u_n) \quad \text{such that} \quad -t(u_n, u) \le j \le \mu(u_n) \tag{3.5}$$

are called the *significant digits of the $n$th approximant $u_n$*. Therefore

$$SD(u_n, u) := t(u_n, u) + \mu(u_n) + 1 \tag{3.6}$$

is the *number of significant (decimal) digits of* $u_n$ [5, §2.1.4]. The following three theorems are useful to determine (or approximate) $SD(u_n, u)$.

THEOREM 3.1. *Let* $u$ *and* $u_n$ *satisfy*

$$0 \neq u \in \mathbb{R}, \quad 0 \neq u_n \in \mathbb{R}, \quad u_n \neq u. \tag{3.7}$$

*Let*

$$\widetilde{SD}(u_n, u) := \log_{10} \left| \frac{u_n}{2(u_n - u)} \right|. \tag{3.8}$$

*Then*

$$-1 \leq \widetilde{SD}(u_n, u) - SD(u_n, u) < 1. \tag{3.9}$$

PROOF. It follows from (3.2), (3.6) and (3.8), that

$$\begin{aligned}
&\widetilde{SD}(u_n, u) - SD(u_n, u) \\
&= \log_{10} M(u_n) + \mu(u_n) + 1 + \log_{10}\left(\frac{1}{2}\right) - \log_{10} \Delta(u_n, u) + t(u_n, u) \\
&\quad - t(u_n, u) - \mu(u_n) - 1 \\
&= \log_{10}\left(\frac{M(u_n)}{2\Delta(u_n, u)}\right)
\end{aligned}$$

From (3.2c) and (3.3b) we obtain

$$10^{-1} = 0.1 \leq \frac{M(u_n)}{2\Delta(u_n, u)} < 10$$

and hence

$$-1 \leq \log_{10} \frac{M(u_n)}{2\Delta(u_n, u)} < 1. \quad \square$$

One can see from Theorem 1 that $\widetilde{SD}(u_n, u)$ is a good estimate of $SD(u_n, u)$. However, $\widetilde{SD}(u_n, u)$ is not computable, since it is defined explicitly in terms of $u$, which is an unknown (infinite decimal (3.1)). If the $n$th approximant $u_n$ is computable, then our next result (Theorem 2) yields a computable estimate $SDLB(u_n, u, b)$ defined in terms of an upper bound $B$ of $|u_n - u|$. We use Theorem 3.1 in our proof of Theorem 3.2. (SDLB stands for *significant digits lower bound.*)

THEOREM 3.2. *Let* $u, u_n$ *and* $B$ *satisfy*

$$0 \neq u \in \mathbb{R}, \quad 0 \neq u_n \in \mathbb{R}, \quad u_n \neq u, \quad B > 0 \tag{3.10a}$$

*and*

$$0 < |u_n - u| \leq B < |u_n|. \tag{3.10b}$$

*Let*

$$SDLB(u_n, u, B) := \log_{10}\left(\frac{|u_n| - B}{20B}\right). \tag{3.11}$$

*Then*

$$SDLB(u_n, u, B) \leq SD(u_n, u) \tag{3.12}$$

*and*

$$\sigma(u_n) = \sigma(u). \tag{3.13}$$

PROOF. It is readily seen from (3.10b) that $u_n$ and $u$ have the same sign, which is equivalent to (3.13). We now prove (3.12). From (3.10) we have

$$0 < \frac{|u_n| - B}{2B} < \frac{|u_n|}{2|u_n - u|},$$

and hence by (3.8)

$$\log_{10}\left(\frac{|u_n| - B}{2B}\right) < \log_{10}\left(\frac{|u_n|}{2|u_n - u|}\right) =: \widetilde{SD}(u_n, u).$$

It follows from this and (3.9) that

$$\log_{10}\left(\frac{|u_n| - B}{2B}\right) - 1 < \widetilde{SD}(u_n, u) - 1 < SD(u_n, u).$$

From this and (3.11) we obtain (3.12). □

Our next result (Theorem 3) describes a procedure which (subject to sufficient conditions that are given) leads to the determination of $SD(u_n, u)$. The main idea involved in this result is that we can find an $m$th approximant $u_m$ that is near enough to $u$ so that it satisfies

$$t(u_n, u) = t(u_n, u_m).$$

If $u_n$ and $u_m$ are computable, we are able to determine $t(u_n, u_m)$ and hence $t(u_n, u)$ and $SD(u_n, u)$.

THEOREM 3.3. *Let $u$ and $u_n$ satisfy*

$$0 \neq u \in \mathbb{R}, \quad 0 \neq u_n \in \mathbb{R} \quad \text{and} \quad u \neq u_n. \tag{3.14}$$

*Let $u_m$ and $B^{(m)}$ satisfy*

$$0 \neq u_m \in \mathbb{R}, \quad u_n \neq u_m \neq u, \quad B^{(m)} > 0, \tag{3.15a}$$

$$0 \neq |u_m \dot{-} u| \leq B^{(m)}, \tag{3.15b}$$

$$0.05 < \Delta(u_n, u_m) < 0.5, \tag{3.15c}$$

*and*

$$0 < B^{(m)} \cdot 10^{t(u_n, u_m)} \leq M(u_n, u_m, \varepsilon), \tag{3.15d}$$

*where*

$$M(u_n, u_m, \varepsilon) := \min[\Delta(u_n, u_m) - 0.05 - \epsilon,\ 0.5 - \Delta(u_n, u_m)] \tag{3.15e}$$

*and*

$$0 < \varepsilon < \Delta(u_n, u_m) - 0.05. \tag{3.15f}$$

*Then*

$$t(u_n, u) = t(u_n, u_m) \tag{3.16a}$$

*and*

$$SD(u_n, u) = t(u_n, u_m) + \mu(u_n) + 1. \tag{3.16b}$$

PROOF. We recall that the functions $\Delta$ and $t$ are defined on the set $S$ in (3.4) by (3.3). It is convenient to define

$$\delta := \delta(u_n, u_m, u) := |u_m - u| \cdot 10^{t(u_n, u_m)}. \tag{3.17}$$

From (3.15 d and e) and (3.17) it follows that

$$\delta \leq B^{(m)} \cdot 10^{t(u_n, u_m)} \leq M(u_n, u_m, \varepsilon). \tag{3.18}$$

By using the triangle inequalities together with (3.17) and the definitions of $\Delta$ and $t$, we obtain

$$|u - u_n| \leq |u - u_m| + |u_m - u_n| = [\delta + \Delta(u_n, u_m)] \cdot 10^{-t(u_n, u_m)} \tag{3.19a}$$

and

$$|u - u_n| \geq |u_n - u_m| - |u_m - u| = [\Delta(u_n, u_m) - \delta] \cdot 10^{-t(u_n, u_m)} \tag{3.19b}$$

Now (3.19a), (3.18) and (3.15e) yield

$$\begin{aligned} |u - u_n| \cdot 10^{t(u_n, u_m)} &\leq \delta + \Delta(u_n, u_m) \\ &\leq M(u_n, u_m, \varepsilon) + \Delta(u_n, u_m) \\ &\leq [0.5 - \Delta(u_n, u_m)] + \Delta(u_n, u_m) \\ &= 0.5. \end{aligned}$$

Similarly (3.19b), (3.18) and (3.15e) yield

$$\begin{aligned}|u-u_n|\cdot 10^{t(u_n,u_m)} &\geq \Delta(u_n,u_m)-\delta\\ &\geq \Delta(u_n,u_m)-M(u_n,u_m,\varepsilon)\\ &\geq \Delta(u_n,u_m)-[\Delta(u_n,u_m)-0.05-\varepsilon]\\ &= 0.05+\varepsilon > 0.05.\end{aligned}$$

Thus

$$0.05 < |u-u_n|\cdot 10^{t(u_n,u_m)} \leq 0.5. \qquad (3.20)$$

By definition of the function $t$ we obtain

$$0.05 < |u-u_n|\cdot 10^{t(u_n,u)} \leq 0.5. \qquad (3.21)$$

The inequalities (3.20) and (3.21) imply $t(u_n,u_m)=t(u_n,u)$ and hence also (3.16b). □

### 3.1 Significant Digits with Roundoff

Let $\{f_n(z)\}_{n=1}^{\infty}$ denote the sequence of approximants of a convergent S-fraction (2.4c) and let $F(z):=\lim_{n\to\infty} f_n(z)\in\mathbb{C}$ for $z\in S_\pi$. For each $n\in\mathbb{N}$, let $\hat{f}_n(z)$ denote the approximation of $f_n(z)$ resulting from computation using the BR-algorithm and machine unit $\omega=(\frac{1}{2})\cdot 10^{1-\nu}$. Let

$$u_n=\operatorname{Re} f_n(z),\quad \hat{u}_n=\operatorname{Re}\hat{f}_n(z),\quad v_n=\operatorname{Im} f_n(z),\quad \hat{v}_n=\operatorname{Im} f_n(z) \qquad (3.22a)$$

and

$$u=\operatorname{Re} F(z),\qquad v=\operatorname{Im} F(z) \qquad (3.22b)$$

Let

$$|F(z)-\hat{f}_n(z)|\leq \widehat{B}_T^{(n)}(\omega)+B_{TR}^{(n)}(\omega)=:B^{(n)}(\omega),\quad n\in\mathbb{N} \qquad (3.23)$$

where $\widehat{B}_T^{(n)}(\omega)$ and $B_{TR}^{(n)}(\omega)$ are defined as in (2.32). It follows that

$$|u_n-\hat{u}_n|\leq B^{(n)}(\omega)\quad\text{and}\quad |v_n-\hat{v}_n|\leq B^{(n)}(\omega) \qquad (3.24)$$

for all $n\in\mathbb{N}$ and $z\in S_\pi$ such that $\widehat{B}_T^{(n)}(\omega)\geq B_{TR}^{(n)}(\omega)$. Note that $\hat{u}_n$ has a floating decimal representation of the form

$$\hat{u}_n=\sigma(\hat{u}_n)M_\omega(u)\cdot 10^{\mu(\hat{u}_n)+1}, \qquad (3.25a)$$

where (we let $\mu:=\mu(\hat{u}_n)$)

$$M_\omega(\hat{u}_n):=0.D_\mu(\hat{u}_n)D_{\mu+1}(\hat{u}_n)\cdots D_{\mu+\nu-1}(\hat{u}_n). \qquad (3.25b)$$

We now restate Theorems 3.2 ad 3.3 in terms of $\hat{u}_n$, $\widehat{B}_T^{(n)}(\omega)$, $B_{TR}^{(n)}(\omega)$, $B^{(n)}(\omega)$, $E_1^{(n)}$ and $\omega$.

Similar results can be stated for $\hat{v}_n$.

THEOREM 3.4. *Let* $u, \hat{u}_n, \widehat{B}_T^{(n)}(\omega), B_{TR}^{(n)}(\omega)$ *and* $B^{(n)}(\omega) := \widehat{B}_T^{(n)}(\omega) + B_{TR}^{(n)}(\omega)$ *satisfy*

$$0 \neq u \in \mathbb{R}, \quad 0 \neq \hat{u}_n \in \mathbb{R}, \quad \hat{u}_n \neq u, \quad \widehat{B}_T^{(n)}(\omega) > B_{TR}^{(n)}(\omega)$$

*and*

$$0 < |u - \hat{u}_n| \leq B^{(n)}(\omega) < |\hat{u}_n| \tag{3.26}$$

*and let* $E_1^{(n)}$ *and* $\omega$ *satisfy (2.28b) and* $0 < E_1^{(n)} < 1$. *Let*

$$SDLB(\hat{u}_n, u, B^{(n)}(\omega)) := \log_{10}\left(\frac{|\hat{u}_n| - B^{(n)}(\omega)}{20B^{(n)}(\omega)}\right) \tag{3.27}$$

*Then*

$$SDLB(\hat{u}_n, u, B^{(n)}(\omega)) \leq SD(\hat{u}_n, u) \tag{3.28}$$

*and*

$$\sigma(\hat{u}_n) = \sigma(u). \tag{3.29}$$

The following result is used to determine $SD(\hat{u}_n, u)$.

THEOREM 3.5. *Let* $u$ *and* $\hat{u}_n$ *satisfy*

$$0 \neq u \in \mathbb{R}, \quad 0 \neq \hat{u}_n \in \mathbb{R}, \quad \text{and} \quad \hat{u}_n \neq u. \tag{3.30}$$

*Let*

$$0 \neq \hat{u}_m \in \mathbb{R}, \quad \hat{u}_n \neq \hat{u}_m \neq u, \quad \widehat{B}_T^{(m)}(\omega) > B_{TR}^{(m)}(\omega), \tag{3.31a}$$

$$0 < |\hat{u}_m - u| \leq B^{(m)}(\omega) := \widehat{B}_T^{(m)}(\omega) + B_{TR}^{(m)}(\omega), \tag{3.31b}$$

$$0.05 < \Delta(\hat{u}_n, \hat{u}_m) < 0.5, \tag{3.31c}$$

*and*

$$0 < B^{(m)}(\omega) \cdot 10^{t(\hat{u}_n, \hat{u}_m)} \leq M(\hat{u}_n, \hat{u}_m, \varepsilon), \tag{3.31d}$$

*where*

$$M(\hat{u}_n, \hat{u}_m, \varepsilon) := \min[\Delta(\hat{u}_n, \hat{u}_m) - 0.05 - \varepsilon,\ 0.5 - \Delta(\hat{u}_n, \hat{u}_m)] \tag{3.31e}$$

*and*

$$0 < \varepsilon < \Delta(\hat{u}_n, \hat{u}_m) - 0.05. \tag{3.31f}$$

*Let* $E_1^{(m)}, E_1^{(m-1)}$ *(in (2.28a)) and* $\omega$ *satisfy* $0 < E_1^{(m)} < 1$, $0 < E_1^{(m-1)} < 1$, *and (2.28b). Then*

$$t(\hat{u}_n, u) = t(\hat{u}_n, \hat{u}_m) \tag{3.32a}$$

*and*

$$SD(\hat{u}_n, u) = t(\hat{u}_n, \hat{u}_m) + \mu(u_n) + 1. \tag{3.32b}$$

## 4. NUMERICAL ILLUSTRATIONS

Two examples are given in this section to illustrate the application of methods described in this paper (particularly Theorem 3.5) applied to the MSF expansion, for $|\arg z| < \pi/2$

$$J(z) = z\left[\frac{a_1^J z^{-2}}{1} + \frac{a_2^J z^{-2}}{1} + \frac{a_3^J z^{-2}}{1} + \cdots\right] = z\widehat{F}^J(z^{-2}) \tag{4.1}$$

(see (1.3), (2.4) and (2.36)). For computation we set

$$w := z^{-2} \quad \text{and} \quad z = w^{-1/2}, \quad \text{where} \quad 0 \le |\arg w| < \pi$$

$$f_n^J(w) := \frac{a_1^J w}{1} + \frac{a_2^J w}{1} + \cdots + \frac{a_n^J w}{1}. \tag{4.2}$$

Then in (2.25), (2.29a) and (2.32) we replace $z$ by $w$ to obtain, for $0 \le |\arg w| < \pi$, $0 \le |\arg z| < \pi/2$,

$$|J(z) - z\hat{f}_n^J(w)| \le B^{(n)}(\omega) := |z|[\widehat{B}_T^{(n)} + B_{TR}^{(n)}(\omega)], \tag{4.3a}$$

where

$$\widehat{B}_T^{(n)} := K(w)|\hat{f}_n^J(w) - \hat{f}_{n-1}(z)|, \tag{4.3b}$$

$$B_{TR}^{(n)}(\omega) := B_R^{(n)}[1 + K(\omega)] + B_R^{(n)}(\omega), \tag{4.3c}$$

$$B_R^{(n)}(\omega) := \frac{\hat{f}_n^J(w) E_1^{(n)}}{1 - E_1^{(n)}}, \quad E_1^{(n)} := 3\omega \sum_{j=0}^{n-1} \eta_n^j. \tag{4.3d}$$

Here $\hat{f}_n^J(w)$ denotes the computed value of $f_n^J(w)$ in (4.2) and $\omega := (.5)10^{1-\nu}$ is the machine unit. For the captions of the tables we employ the notation

$$\hat{f}_n(z) := z\hat{f}_n(w), \qquad w := z^{-2}.$$

Results given in Tables 1–4 are for $z = 7/2$ and in Tables 5–10 for $z = 1 + 5i$. For each of these values of $z$, for two values of the machine unit $\omega := (.5)10^{1-\nu}$ and for $n = 2, 3, 4, \ldots, 10, 15, 20, 25, 30, 35, 40,$ we give the following calculations: $\hat{f}_n(z) =: \hat{u}_n + i\hat{v}_n$, $|\hat{u}_n - u|$, $|\hat{v}_n - v|$ (where $u + iv = f(z)$), $SDLB(\hat{u}_n, u, B^{(n)})$, $SDLB(\hat{v}_n, v, B^{(n)})$, $SD(\hat{u}_n, u)$, $SD(\hat{v}_n, v)$, $\widehat{B}_T^{(n)}$ and $B_{TR}^{(n)}(\omega)$.

In Table 4 it can be seen that $\widehat{B}_T^{(n)} < B_{TR}^{(n)}(\omega)$ for $n \geq 8$ when $\omega = (.5)10^{-10}$ and for $n \geq 30$ when $\omega = (.5)10^{-20}$. These results are reflected in Table 3, where we see that $SD(u_n, u) = 10$ for $n \geq 8$ and $\omega = (.5)10^{-10}$; with $\omega = (.5)10^{-20}$ we have $\widehat{B}_T^{(n)} < B_{TR}^{(n)}(\omega)$ for $n = 35$ and 40 and $SD(\hat{u}_n, u) = 18$ for $n = 35$ and $SD(\hat{u}_n, u) = 19$ for $n = 40$. The method has broken down at $n = 40$ due to roundoff error and one has $SD(\hat{u}_{40}, u) = 18$. The results $|\hat{u}_n - u| = 0$ for $n \geq 7$ are a consequence of the machine unit $\omega = (.5)10^{-10}$.

Similar phenomena can be seen in Tables 5–10 with $z = 1 + 5i$. In Table 10 with $\omega = (.5)10^{-10}$ we obtain $\widehat{B}_T^{(n)} < B_{TR}^{(n)}(\omega)$ for $n \geq 6$. Consequently in Tables 8 and 9 we have $SD(\hat{u}_n, u) = SD(\hat{v}_n, v) = 10$ for $n \geq 6$. Consequently in Table 5 $|\hat{u}_n - u| = |\hat{v}_n - v| = 0$ for $n \geq 5$ (where $\omega = (.5)10^{-10}$).

| $n$ | $\widehat{u}_n$ | $\|\widehat{u}_n - u\|$ |
|---|---|---|
| 2 | .02374 49118 04 | .13(-05) |
| 3 | .02374 62126 35 | .49(-07) |
| 4 | .02374 61601 51 | .35(-08) |
| 5 | .02374 61610 05 | .35(-09) |
| 6 | .02374 61636 08 | .48(-10) |
| 7 | .02374 61636 64 | 0. |
| 8 | .02374 61636 56 | 0. |
| 9 | .02374 61636 57 | 0. |
| 10 | .02374 61636 56 | 0. |
| 15 | .02374 61636 56 | 0. |
| 20 | .02374 61636 56 | 0. |
| 25 | .02374 61636 56 | 0. |
| 30 | .02374 61636 56 | 0. |
| 35 | .02374 61636 56 | 0. |
| 40 | .02374 61636 56 | 0. |

Table 1. Computed approximants and actual errors in the approximation of $J(z)$ at $z = 7/2$ by the computed $n$th approximant $\hat{f}_n(z) = \hat{u}_n$ of the MSF (1.3) with $\omega = .5(10)^{-10}$. Note that $u = .02374\ 61636\ 56297\ 49597\ 13302\ 79090\ 08587\ 1\ldots$

| $n$ | $\widehat{u}_n$ | $\|\widehat{u}_n - u\|$ |
|---|---|---|
| 2 | .02374 49118 04613 29715 06 | .13(-05) |
| 3 | .02374 62126 351021 52399 04 | .49(-07) |
| 4 | .02374 61601 51103 55920 85 | .35(-08) |
| 5 | .02374 61610 04665 16792 04 | .35(-09) |
| 6 | .02374 61636 07894 59614 33 | .48(-10) |
| 7 | .02374 61636 64436 86685 30 | .81(-11) |
| 8 | .02374 61636 54597 00967 82 | .17(-11) |
| 9 | .02374 61636 56698 20081 24 | .40(-12) |
| 10 | .02374 61636 56187 39667 32 | .11(-12) |
| 15 | .02374 61636 56298 09466 01 | .60(-15) |
| 20 | .02374 61636 56297 48265 24 | .13(-16) |
| 25 | .02374 61636 56297 49661 12 | .64(-18) |
| 30 | .02374 61636 56297 49590 51 | .66(-19) |
| 35 | .02374 61636 56297 49596 68 | .50(-20) |
| 40 | .02374 61636 56297 49595 91 | .12(-19) |

Table 2. Computed approximants and actual errors in the approximation of $J(z)$ at $z = 7/2$ by the computed $n$th approximant $\hat{f}_n(z) = \hat{u}_n$ of the MSF (1.3) with $\omega = .5(10)^{-20}$. Note that $u = .02374\ 61636\ 56297\ 49597\ 13302\ 79090\ 08587\ 1\ldots$

| $n$ | $SDLB(\widehat{u}_n, u, B^{(n)})$ | $SD(\widehat{u}_n, u)$ | $SDLB(\widehat{u}_n, u, B^{(n)})$ | $SD(\widehat{u}_n, u)$ |
|---|---|---|---|---|
| | $\omega = .5(10)^{-10}$ | $\omega = .5(10)^{-10}$ | $\omega = .5(10)^{-20}$ | $\omega = .5(10)^{-20}$ |
| 2 | 1.26 | 4 | 1.26 | 4 |
| 3 | 2.96 | 5 | 2.96 | 5 |
| 4 | 4.35 | 7 | 4.35 | 7 |
| 5 | 5.49 | 8 | 5.49 | 8 |
| 6 | 6.45 | 8 | 6.48 | 8 |
| 7 | 7.16 | 9 | 7.32 | 9 |
| 8 | 7.45 | 10 | 8.08 | 10 |
| 9 | 7.47 | 10 | 8.75 | 10 |
| 10 | 7.42 | 10 | 9.37 | 11 |
| 15 | 7.15 | 10 | 11.8 | 13 |
| 20 | 6.96 | 10 | 13.5 | 15 |
| 25 | 6.82 | 10 | 14.8 | 16 |
| 30 | 6.72 | 10 | 15.9 | 17 |
| 35 | 6.63 | 10 | 16.4 | 18 |
| 40 | 6.56 | 10 | 16.5 | 19 |

Table 3. Significant digits $SD(\hat{u}_n, u)$ and lower bounds on the number of significant digits $SDLB(\hat{u}_n, u, B^{(n)}(\omega))$ in computed approximants $\hat{f}_n(z) = \hat{u}_n$, with $z = 7/2$, of the MSF (1.3).

| $n$ | $\widehat{B}_T^{(n)}(\omega)$ | $B_{TR}^{(n)}(\omega)$ | $\widehat{B}_T^{(n)}(\omega)$ | $B_{TR}^{(n)}(\omega)$ |
|---|---|---|---|---|
| | $\omega = .5(10)^{-10}$ | $\omega = .5(10)^{-10}$ | $\omega = .5(10)^{-20}$ | $\omega = .5(10)^{-20}$ |
| 2 | .65(-04) | .11(-10) | .65(-04) | .11(-20) |
| 3 | .13(-05) | .12(-10) | .13(-05) | .12(-20) |
| 4 | .52(-07) | .14(-10) | .52(-07) | .14(-20) |
| 5 | .39(-08) | .18(-10) | .39(-08) | .18(-20) |
| 6 | .40(-09) | .21(-10) | .40(-09) | .21(-20) |
| 7 | .57(-10) | .26(-10) | .57(-10) | .26(-20) |
| 8 | 0. | .32(-10) | .98(-11) | .32(-20) |
| 9 | 0. | .38(-10) | .21(-11) | .38(-20) |
| 10 | 0. | .45(-10) | .51(-12) | .45(-20) |
| 15 | 0. | .84(-10) | .21(-14) | .84(-20) |
| 20 | 0. | .13(-09) | .39(-16) | .13(-19) |
| 25 | 0. | .18(-09) | .18(-17) | .18(-19) |
| 30 | 0. | .23(-09) | .14(-18) | .23(-19) |
| 35 | 0. | .28(-09) | .17(-19) | .28(-19) |
| 40 | 0. | .33(-09) | .26(-20) | .33(-19) |

Table 4. Computed truncation and roundoff error bounds in the approximation of $J(z)$ at $z = 7/2$ by the computed $n$-th approximant $\hat{f}_n(z) = \hat{u}_n$ of the MSF (1.3) for two different values of $\omega$.

| $n$ | $\widehat{u}_n$ | $\lvert\widehat{u}_n - u\rvert$ | $\widehat{v}_n$ | $\lvert\widehat{v}_n - v\rvert$ |
|---|---|---|---|---|
| 2 | .00321 68458 93 | .17(-06) | -.01604 30406 57 | .11(-06) |
| 3 | .00321 70180 69 | .43(-08) | -.01604 31532 29 | .68(-09) |
| 4 | .003217 02228 25 | .17(-09) | -.01604 31539 63 | .51(-10) |
| 5 | .00321 70224 53 | 0. | -.01604 31539 20 | 0. |
| 6 | .00321 70224 62 | 0. | -.01604 31539 13 | 0. |
| 7 | .00321 70224 62 | 0. | -.01604 31539 11 | 0. |
| 8 | .00321 70224 62 | 0. | -.01604 31539 11 | 0. |
| 9 | .00321 70224 62 | 0. | -.01604 31539 11 | 0. |
| 10 | .00321 70224 62 | 0. | -.01604 31539 11 | 0. |
| 15 | .00321 70224 62 | 0. | -.01604 31539 11 | 0. |
| 20 | .00321 70224 62 | 0. | -.01604 31539 11 | 0. |
| 25 | .00321 70224 62 | 0. | -.01604 31539 11 | 0. |
| 30 | .00321 70224 62 | 0. | -.01604 31539 11 | 0. |
| 35 | .00321 70224 62 | 0. | -.01604 31539 11 | 0. |
| 40 | .00321 70224 62 | 0. | -.01604 31539 11 | 0. |

Table 5. Computed approximants and actual errors in the approximation of $J(z)$ at $z = 1 + 5i$ by the computed $n$th approximant $\hat{f}_n(z) = \hat{u}_n + i\hat{v}_n$ of the MSF (1.3) with $\omega = .5(10)^{-10}$. Note here that $u = .00321\ 70224\ 62287\ 23958\ 77925\ 70481\ 40945\ 72...$ and $v = -.01604\ 31539\ 11288\ 56630\ 78850\ 92504\ 43439\ 21....$

| $n$ | $\widehat{u}_n$ | $\lvert\widehat{u}_n - u\rvert$ |
|---|---|---|
| 2 | .00321 68458 92899 21428 230 | .17(-06) |
| 3 | .00321 70180 69433 64744 696 | .43(-08) |
| 4 | .00321 70222 82480 26957 183 | .17(-09) |
| 5 | .00321 70224 53110 89884 096 | .91(-11) |
| 6 | .00321 70224 61859 47414 610 | .42(-12) |
| 7 | .00321 70224 62329 51309 052 | .42(-13) |
| 8 | .00321 70224 62318 19063 935 | .31(-13) |
| 9 | .00321 70224 62298 90636 905 | .12(-13) |
| 10 | .00321 70224 62290 02259 213 | .28(-14) |
| 15 | .00321 70224 62287 23282 405 | .68(-17) |
| 20 | .00321 70224 62287 28178 284 | .42(-16) |
| 25 | .00321 70224 62287 20707 300 | .33(-16) |
| 30 | .00321 70224 62287 22089 006 | .19(-16) |
| 35 | .00321 70224 62287 22982 635 | .98(-17) |
| 40 | .00321 70224 62287 23507 497 | .45(-17) |

Table 6. Computed approximants and actual errors in the approximation of $Re(J(z))$ at $z = 1 + 5i$ by the computed $n$th approximant $\hat{f}_n(z) = \hat{u}_n + i\hat{v}_n$ of the MSF (1.3) with $\omega = .5(10)^{-20}$. Note here that $u = .00321\ 70224\ 62287\ 23958\ 77925\ 70481\ 40945\ 72....$

| $n$ | $\widehat{v}_n$ | $\lvert\widehat{v}_n - v\rvert$ |
|---|---|---|
| 2 | -.01604 30406 56648 45022 992 | .11(-06) |
| 3 | -.01604 31532 28881 59959 888 | .68(-09) |
| 4 | -.01604 31539 62578 62509 412 | .51(-10) |
| 5 | -.01604 31539 19632 17633 593 | .83(-11) |
| 6 | -.01604 31539 12512 52184 243 | .12(-11) |
| 7 | -.01604 31539 11477 08307 443 | .19(-12) |
| 8 | -.01604 31539 11315 55181 949 | .28(-13) |
| 9 | -.01604 31539 11288 29128 029 | .28(-15) |
| 10 | -.01604 31539 11285 17256 588 | .34(-14) |
| 15 | -.01604 31539 11288 16257 956 | .40(-15) |
| 20 | -.01604 31539 11288 66436 144 | .98(-16) |
| 25 | -.01604 31539 11288 59033 217 | .24(-16) |
| 30 | -.01604 31539 11288 56865 017 | .23(-17) |
| 35 | -.01604 31539 11288 56829 300 | .20(-17) |
| 40 | -.01604 31539 11288 56993 580 | .36(-17) |

Table 7. Computed approximants and actual errors in the approximation of $Im(J(z))$ at $z = 1 + 5i$ by the computed $n$th approximant $\hat{f}_n(z) = \hat{u}_n + i\hat{v}_n$ of the MSF (1.3) with $\omega = .5(10)^{-20}$. Note here that $v = -.01604\ 31539\ 11288\ 56630\ 78850\ 92504\ 43439\ 21\ldots$.

| $n$ | $SDLB(\widehat{u}_n, u, B^{(n)})$ | $SD(\widehat{u}_n, u)$ | $SDLB(\widehat{u}_n, u, B^{(n)})$ | $SD(\widehat{u}_n, u)$ |
|---|---|---|---|---|
| | $\omega = .5(10)^{-10}$ | $\omega = .5(10)^{-10}$ | $\omega = .5(10)^{-20}$ | $\omega = .5(10)^{-20}$ |
| 2 | 0.88 | 4 | 0.88 | 4 |
| 3 | 2.89 | 6 | 2.89 | 6 |
| 4 | 4.57 | 7 | 4.58 | 7 |
| 5 | 5.93 | 8 | 5.96 | 8 |
| 6 | 6.77 | 10 | 7.15 | 10 |
| 7 | 6.88 | 10 | 8.15 | 11 |
| 8 | 6.81 | 10 | 9.00 | 11 |
| 9 | 6.74 | 10 | 9.68 | 11 |
| 10 | 6.67 | 10 | 9.37 | 12 |
| 15 | 6.40 | 10 | 11.3 | 13 |
| 20 | 6.22 | 10 | 11.9 | 15 |
| 25 | 6.09 | 10 | 12.3 | 13 |
| 30 | 5.99 | 10 | 12.6 | 14 |
| 35 | 5.90 | 10 | 12.9 | 14 |
| 40 | 5.83 | 10 | 13.1 | 14 |

Table 8. Significant digits $SD(\hat{u}_n, u)$ and lower bounds on the number of significant digits $SDLB(\hat{u}_n, u, B^{(n)}(\omega))$ in computed approximants $\hat{f}_n(z) = \hat{u}_n + i\hat{v}_n$, with $z = 1 + 5i$, of the MSF (1.3).

| $n$ | $SDLB(\widehat{v}_n, v, B^{(n)})$ | $SD(\widehat{v}_n, v)$ | $SDLB(\widehat{v}_n, v, B^{(n)})$ | $SD(\widehat{v}_n, v)$ |
|---|---|---|---|---|
| | $\omega = .5(10)^{-10}$ | $\omega = .5(10)^{-10}$ | $\omega = .5(10)^{-20}$ | $\omega = .5(10)^{-20}$ |
| | $.5(10)^{-10}$ | $.5(10)^{-10}$ | $.5(10)^{-20}$ | $.5(10)^{-20}$ |
| 2 | 1.58 | 5 | 1.58 | 5 |
| 3 | 3.59 | 7 | 3.59 | 7 |
| 4 | 5.27 | 9 | 5.27 | 9 |
| 5 | 6.63 | 9 | 6.66 | 9 |
| 6 | 7.47 | 10 | 7.85 | 10 |
| 7 | 7.57 | 10 | 8.85 | 11 |
| 8 | 7.51 | 10 | 9.69 | 12 |
| 9 | 7.43 | 10 | 10.4 | 13 |
| 10 | 7.36 | 10 | 10.9 | 13 |
| 15 | 7.10 | 10 | 12.0 | 14 |
| 20 | 6.92 | 10 | 12.6 | 14 |
| 25 | 6.79 | 10 | 13.0 | 15 |
| 30 | 6.68 | 10 | 13.3 | 15 |
| 35 | 6.60 | 10 | 13.6 | 16 |
| 40 | 6.53 | 10 | 13.8 | 16 |

Table 9. Significant digits $SD(\hat{v}_n, v)$ and lower bounds on the number of significant digits $SDLB(\hat{v}_n, v, B^{(n)}(\omega))$ in computed approximants $\hat{f}_n(z) = \hat{u}_n + i\hat{v}_n$, with $z = 1 + 5i$, of the MSF (1.3).

| $n$ | $\widehat{B}_T^{(n)}(\omega)$ | $B_{TR}^{(n)}(\omega)$ | $\widehat{B}_T^{(n)}(\omega)$ | $B_{TR}^{(n)}(\omega)$ |
|---|---|---|---|---|
| | $\omega = .5(10)^{-10}$ | $\omega = .5(10)^{-10}$ | $\omega = .5(10)^{-20}$ | $\omega = .5(10)^{-20}$ |
| 2 | .21(-04) | 0. | .21(-04) | .75(-21) |
| 3 | .21(-06) | 0. | .21(-06) | .86(-21) |
| 4 | .43(-08) | .10(-10) | .43(-08) | .10(-21) |
| 5 | .18(-09) | .13(-10) | .18(-09) | .13(-20) |
| 6 | .11(-10) | .16(-10) | .11(-10) | .16(-20) |
| 7 | 0. | .20(-10) | .11(-11) | .20(-20) |
| 8 | 0. | .25(-10) | .16(-12) | .25(-20) |
| 9 | 0. | .30(-10) | .33(-13) | .30(-20) |
| 10 | 0. | .35(-10) | .94(-14) | .35(-20) |
| 15 | 0. | .64(-10) | .77(-15) | .64(-20) |
| 20 | 0. | .97(-10) | .21(-15) | .97(-20) |
| 25 | 0. | .13(-09) | .81(-16) | .13(-19) |
| 30 | 0. | .17(-09) | .38(-16) | .17(-19) |
| 35 | 0. | .20(-09) | .20(-16) | .20(-19) |
| 40 | 0. | .24(-09) | .12(-16) | .24(-19) |

Table 10. Computed truncation and roundoff error bounds in the approximation of $J(z)$ at $z = 1 + 5i$ by the computed $n$-th approximant $\hat{f}_n(z) = \hat{u}_n + i\hat{v}_n$ of the MSF (1.3) for two different values of $\omega$.

## REFERENCES


1. Bruce W. Char. On Stieltjes' continued fraction for the gamma function, *Math. of Comp.* **34**, No. 150 (April 1980), 547-551.
2. J. Cizek and E. R. Vrscay. Asymptotic estimation of the coefficients of the continued fraction representing the Binet function, *C.R. Math. Rep. Acad. Sci. Canada* Vol. IV, No. 4 (August 1982), 201-206.
3. Cathleen Craviotto, William B. Jones, W. J. Thron. A survey of truncation error analysis for Padé and continued fraction approximants, *Acta Applicandae Mathematicae* **33**, Nos. 2 and 3 (December 1993), 211-272.
4. Cathleen M. Craviotto, William B. Jones, W. J. Thron and Nancy J. Wyshinski. Computation of special functions by Padé approximants with orthogonal polynomial denominators, *Numerical Algorithms* **11** (1966), 117-141.
5. G. Dahlquist and A. Björck. *Numerical Methods*, Prentice-Hall, Inc., Englewood Cliffs, (1974).
6. P. Henrici. *Applied and Computational Complex Analysis*, Vol. 2, Special Functions, Integral Transforms, Asymptotics and Continued Fractions, John Wiley and Sons, New York (1977).
7. P. Henrici and Pia Pfluger. Truncation error estimates for Stieltjes fractions, *Numer. Math.* **9** (1966), 120-138.
8. William B. Jones and W. J. Thron. Numerical stability in evaluating continued fractions, *Math of Comp.* **28**, No. 127 (July 1974), 795-810.
9. William B. Jones and W. J. Thron. *Continued Fractions: Analytic Theory and Applications*, Encyclopedia of Mathematics and its Applications **11**, Addison-Wesley Publishing Company, Reading, Mass. (1980), distributed now by Cambridge University Press, New York.
10. William B. Jones and Walter Van Assche. Asymptotic behavior of the continued fraction coefficients of a class of Stieltjes transforms including the Binet function, *Orthogonal Functions, Moment Theory and Continued Fractions: Theory and Applications*, (Eds., W.B. Jones and A. Sri Ranga), Lecture Notes in Pure and Applied Math., Marcel Dekker, Inc., New York (1998). To appear.
11. Lisa Lorentzen and Haakon Waadeland, *Continued Fractions with Applications*, Studies in Computational Math, Vol. 3, North-Holland, New York (1992)
12. O. Perron. *Die Lehre von den Kettenbrüchen* Band II, Teubner, Stuttgart (1957).
13. Otto G. Ruehr. Personal communication.
14. T. J. Stieltjes. *Recherches sur les fraction continue*, Ann. Fac. Sci. Toulouse **8** (1894), J, 1-122; 9 (1895), A, 1-47; (G. Van Dijk, ed.) Oeuvres Complètes-Collected Papers, Vol. II, Springer-Verlag, Berlin, 1993, pp.

406–570, English translation on pp. 609–745.
15. H. S. Wall. *Analytic Theory of Continued Fractions*, D. Van Nostrand Co., Inc., New York (1948).

# Formulas for the Moments of Some Strong Moment Distributions

BRIAN A. HAGLER* Department of Mathematics, University of Colorado, Boulder, Colorado

## 1 INTRODUCTION

In this paper we are concerned with the development of explicit expressions relating the moments of moment distribution functions and their strong moment distribution function extensions. To further explain our intentions, we recall (see [1]) several definitions and theorems.

If $f: D \to \mathbf{R}$, where $D$ is a subset of the set of real numbers $\mathbf{R}$, then the set

$$\sigma\ (f) := [x \in D : \textit{There is an } \epsilon > 0 \textit{ such that } (x-\epsilon,\ x+\epsilon) \subseteq D \textit{ and } f(x+\delta) - f(x-\delta) > 0 \textit{ for all } \delta > 0 \textit{ such that } \delta < \epsilon]$$

is called the *spectrum* of $f$. If $\psi: \mathbf{R} \to \mathbf{R}$ is a bounded, non-decreasing function with an infinite spectrum $\sigma(\psi)$ such that the *moments* $\mu_n(\psi)$ defined by the Riemann-Stieltjes integrals

$$\mu_n(\psi) := \int_{-\infty}^{\infty} x^n \, d\psi(x)$$

exist for all $n \in \mathbf{Z}_0^+ := [0, 1, 2, ...]$, then $\psi$ is called a *moment distribution function (MDF)*. If $\phi: (\mathbf{R}^- \cup \mathbf{R}^+) \to \mathbf{R}$ is a bounded function, non-decreasing on the negative reals $\mathbf{R}^-$ and the positive reals $\mathbf{R}^+$ separately, with infinite spectrum $\sigma(\phi)$ such that the *moments*

$$\mu_n(\phi) := \int_{-\infty}^{0} x^n \, d\phi(x) + \int_{0}^{\infty} x^n \, d\phi(x)$$

exist for all $n \in \mathbf{Z} := [0, \pm 1, \pm 2, ...]$, then $\phi$ is called a *strong moment distribution function (SMDF)*.

We denote by $\mathcal{P}$ the space of all polynomials in $x$ with real coefficients, and we denote by $\mathcal{R}$ the space of all *Laurent polynomials (L-polynomials)*, $\sum_{k=m}^{n} r_k x^k$ for

* Research supported in part by the U.S. National Science Foundation under Grants Nos. DMS-93002584 and DMS-9701028.

$m, n \in \mathbf{Z}$ with $m \le n$, having real coefficients $r_k$ for $m \le k \le n$. For $R \in \mathcal{R}$ we sometimes write $C_k(R)$ for the coefficient of $x^k$ in $R$. We will find it useful to define the following subsets of $\mathcal{R}$:

$$\mathcal{R}_{m,n} := \left[\sum_{k=m}^{n} r_k x^k \ : \ r_k \in \mathbf{R} \ , \ m \le k \le n\right] \text{ for } m, n \in \mathbf{Z} \text{ with } m \le n;$$

$$\mathcal{R}_{2n} := [R \in \mathcal{R}_{-n,n} : \ C_n(R) \neq 0] \text{ for } n \in \mathbf{Z}_0^+;$$

$$\mathcal{R}_{2n+1} := [R \in \mathcal{R}_{-n-1,n} : \ C_{-n-1}(R) \neq 0] \text{ for } n \in \mathbf{Z}_0^+.$$

THEOREM 1.1 For any non-zero $R \in \mathcal{R}$, there is a unique $d \in \mathbf{Z}_0^+$, called the *L-degree* of $R$, such that $R \in \mathcal{R}_d$.
Proof: See [2], p. 48, Theorem 1.1.

Notice that for a polynomial in $\mathcal{P}$, its L-degree is twice its polynomial degree. If $R \in \mathcal{R}_{2n}$, then $C_n(R)$ and$C_{-n}(R)$ are called the *leading coefficient* and *trailing coefficient* of R, respectively, and if $R \in \mathcal{R}_{2n+1}$, then the *leading coefficient* and *trailing coefficient* of $R$ are $C_{-n-1}(R)$ and $C_n(R)$, respectively. An L-polynomial is called *monic* if its leading coefficient is 1, *regular* if its trailing coefficient is non-zero, and *singular* if its trailing coefficient is 0.

For any MDF $\psi$ and polynomials $P, Q \in \mathcal{P}$, we define

$$(P,Q)_\psi := \int_{-\infty}^{\infty} P(x)Q(x)\, d\psi(x),$$

and, for any SMDF $\phi$ and L-polynomials $R, S \in \mathcal{R}$, we set

$$(R,S)_\phi := \int_{-\infty}^{0} R(x)S(x)\, d\phi(x) + \int_{0}^{\infty} R(x)S(x)\, d\phi(x).$$

THEOREM 1.2 If $\psi$ is an MDF, then $(\cdot,\cdot)_\psi$ is an inner-product on $\mathcal{P}$.
Proof: See [3], p.13 and p.16.

THEOREM 1.3 If $\phi$ is an SMDF, then $(\cdot,\cdot)_\phi$ is an inner-product on $\mathcal{R}$.
Proof: See [2], p.58 and p.62, Theorem 3.6.

A sequence $\{P_n(x)\}_{n=0}^{\infty}$ in $\mathcal{P}$ such that, for $\psi$, an MDF, and all $m, n \in \mathbf{Z}_0^+$,

$$P_n \textit{ has polynomial degree } n \textit{ and } (P_m, P_n)_\psi = 0 \textit{ if } m \neq n,$$

is called an *orthogonal polynomial sequence (OPS) with respect to* $\psi$. If in addition each of the polynomials $P_n$ is monic, $\{P_n(x)\}_{n=0}^{\infty}$ is called *monic*. It is not difficult to show that a monic OPS with respect to a given MDF $\psi$ exists and is unique (see [3], Theorem 3.3, p.14, and Corollary, p.9). Analogously in the Laurent case, a sequence $\{R_n(x)\}_{n=0}^{\infty}$ in $\mathcal{R}$ is called an *orthogonal Laurent polynomial sequence*

*(OLPS) with respect to* $\phi$, an SMDF, if

$$R_n \textit{ has L-degree } n \textit{ and } (R_m, R_n)_\phi = 0 \textit{ if } m \neq n.$$

$\{R_n(x)\}_{n=0}^{\infty}$ is called *monic* if each of the L-polynomials $R_n$ is monic. Again, it is not hard to show that a unique monic OLPS with respect to a given SMDF $\phi$ exists (see [2], Theorem 3.2, p.59, and Corollary 1.6, p.53).

The following theorem, which appears in [1], explains what we mean by *a SMDF extension* $\widetilde{\psi}$ *of an MDF* $\psi$.

THEOREM 1.4 (Transformation Theorem) Let $\psi$ be a moment distribution function, let $\sigma(\psi)$ denote the spectrum of $\psi$, and let $\{P_n(x)\}_{n=0}^{\infty}$ denote the monic orthogonal polynomial sequence with respect to $\psi$. Let $\lambda, \gamma \in \mathbf{R}^+$, and set $v(x) := \frac{1}{\lambda}(x - \frac{\gamma}{x})$ and $v_{\mp}^{-1}(y) := \frac{\lambda}{2}(y \pm \sqrt{y^2 + \frac{4\gamma}{\lambda^2}})$. Then:

(A) $\widetilde{\psi}(x) := \begin{cases} \int_{-\infty}^{x} \frac{1}{v'(t)}\, d(\psi \circ v)(t), & x \in \mathbf{R}^- \\ \int_{0}^{x} \frac{1}{v'(t)}\, d(\psi \circ v)(t), & x \in \mathbf{R}^+ \end{cases}$ is a strong moment distribution function.

(B) $\sigma(\widetilde{\psi}) = v_{-}^{-1}(\sigma(\psi)) \cup v_{+}^{-1}(\sigma(\psi))$ is the spectrum of $\widetilde{\psi}$.

(C) $\{\widetilde{P}_m(x)\}_{m=0}^{\infty}$, where $\widetilde{P}_{2n}(x) := \lambda^n P_n(v(x))$ and $\widetilde{P}_{2n+1}(x) := \left(-\frac{\lambda}{\gamma}\right)^n \frac{1}{x} P_n(v(x))$ for $n = 0, 1, 2, \ldots$, is the monic orthogonal Laurent polynomial sequence with respect to $\widetilde{\psi}$.

Proof: See [1].

Many examples of MDF's and SMDF's are given by weight functions (for a brief survey, see [1], [3], [4], [5] and [6]). If $\psi$ is an MDF which is differentiable, then $w(x) = \frac{d\psi}{dx}$ is called the weight function for $\psi$. Similarly, if $\phi$ is an SMDF which is differentiable, then $\omega(x) = \frac{d\phi}{dx}$ is called the weight function for $\phi$. If $\psi$ is an MDF, and if $\psi$ is differentiable with $\frac{d\psi}{dx} = w(x)$, then $\widetilde{\psi}$ is differentiable with $\frac{d\widetilde{\psi}}{dx} = w(v(x))$ (see [1]).

Our purpose in this paper is to forge explicit formulas between the moments $\mu_n(\psi)$ and the moments $\mu_n(\widetilde{\psi})$ (Section 2) and to illustrate the application of these formulas (Section 3).

## 2 THEOREMS

Throughout this section, we adopt the assumptions and notation of the Transformation Theorem. Within this structure, it is shown in [1] that

$$\mu_n(\widetilde{\psi}) = (-1)^n \gamma^{n+1} \mu_{-n-2}(\widetilde{\psi}), \textit{ for } n = 0, \pm 1, \pm 2, \ldots, \tag{2.1}$$

and, hence,

$$\mu_{-1}(\widetilde{\psi}) = 0. \tag{2.2}$$

Thus, $\{\mu_n(\widetilde{\psi})\}_{n=0}^{\infty}$ determines the entire bisequence $\{\mu_n(\widetilde{\psi})\}_{n=-\infty}^{\infty}$, and we will concentrate our efforts here on formulas giving $\{\mu_n(\widetilde{\psi})\}_{n=0}^{\infty}$ in terms of $\{\mu_n(\psi)\}_{n=0}^{\infty}$, and vice versa. Our first result, Theorem 2.1, however, gives $\mu_n(\psi)$ as a linear combination of $\mu_n(\widetilde{\psi})$, $\mu_{n-2}(\widetilde{\psi})$, $\mu_{n-4}(\widetilde{\psi})$, ..., and $\mu_{-n}(\widetilde{\psi})$. Athough interesting on its own, we anticipate combining Theorem 2.1 with Equation (2.1) to write $\mu_n(\psi)$ in terms of non-negative moments of $\widetilde{\psi}$.

THEOREM 2.1 $\mu_n(\psi) = \lambda^{-n-1}\sum_{k=0}^{n}\binom{n}{k}(-\gamma)^{n-k}\mu_{2k-n}(\widetilde{\psi})$, for each non-negative integer $n$.

Proof: It is shown in [1] that $v'(\frac{-\gamma}{x}) = \frac{x^2}{\gamma}v'(x)$ (Theorem 2.2.1 (E)) and that $\int R(x)\,d\widetilde{\psi}(x) = \int R(x)\frac{1}{v'(x)}\,d(\psi\circ v)(x)$ for any L-polynomial $R(x)$ (Theorem 2.2.7). Hence, using $v(\frac{-\gamma}{x}) = v(x)$ and the substitution $x \to \frac{-\gamma}{x}$, we find that

$$\begin{aligned}
(v(x)^n,1)_{\widetilde{\psi}} &:= \int_{-\infty}^{0} v(x)^n\,d\widetilde{\psi}(x) + \int_{0}^{\infty} v(x)^n\,d\widetilde{\psi}(x)\\
&= \int_{-\infty}^{0} v(x)^n\frac{1}{v'(x)}\,d(\psi\circ v)(x) + \int_{0}^{\infty} v(x)^n\frac{1}{v'(x)}\,d(\psi\circ v)(x)\\
&= \gamma\int_{0}^{\infty}\frac{v(x)^n}{x^2v'(x)}\,d\psi(v(x)) + \gamma\int_{-\infty}^{0}\frac{v(x)^n}{x^2v'(x)}\,d\psi(v(x)) \qquad (2.3)\\
&= \gamma\int_{0}^{\infty}\frac{v(x)^n}{v_+^{-1}(v(x))^2v'(v_+^{-1}(v(x)))}\,d\psi(v(x)) +\\
&\qquad \gamma\int_{-\infty}^{0}\frac{v(x)^n}{v_-^{-1}(v(x))^2v'(v_-^{-1}(v(x)))}\,d\psi(v(x)).
\end{aligned}$$

Then, the substitution $v(x) \to x$ leads to

$$(v(x)^n,1)_{\widetilde{\psi}} = \gamma\int_{-\infty}^{\infty} x^n\frac{v_-^{-1}(x)^2v'(v_-^{-1}(x)) + v_+^{-1}(x)^2v'(v_+^{-1}(x))}{v_-^{-1}(x)^2v_+^{-1}(x)^2v'(v_-^{-1}(x))v'(v_+^{-1}(x))}\,d\psi(x)\,. \qquad (2.4)$$

But a direct computation shows

$$\frac{v_-^{-1}(x)^2v'(v_-^{-1}(x)) + v_+^{-1}(x)^2v'(v_+^{-1}(x))}{v_-^{-1}(x)^2v_+^{-1}(x)^2v'(v_-^{-1}(x))v'(v_+^{-1}(x))} = \frac{\lambda}{\gamma}. \qquad (2.5)$$

It thus follows that

$$(v(x)^n,1)_{\widetilde{\psi}} = \lambda(x^n,1)_{\psi}. \qquad (2.6)$$

Using $v(x) = \frac{1}{\lambda}(x - \frac{\gamma}{x})$, the Binomial Theorem, and linearity of the integral, we find

$$(v(x)^n,1)_{\widetilde{\psi}} = \frac{1}{\lambda^n}\sum_{k=0}^{n}\binom{n}{k}(-\gamma)^{n-k}(x^{2k-n},1)_{\widetilde{\psi}}. \qquad (2.7)$$

Equations (2.6) and (2.7) yield

$$\mu_n(\psi) = \frac{1}{\lambda^{n+1}}\sum_{k=0}^{n}\binom{n}{k}(-\gamma)^{n-k}\mu_{2k-n}(\widetilde{\psi}),$$

which concludes the proof.

In the following, let $[r]$ denote the greatest integer less than $r$, and let $\binom{n}{-1}=0$.

THEOREM 2.2 $\mu_n(\psi) = \lambda^{-n-1}\sum_{k=0}^{[n/2]}\left(\binom{n}{k}-\binom{n}{k-1}\right)(-\gamma)^k\mu_{n-2k}(\widetilde{\psi})$, for each non-negative integer $n$.
Proof: This result follows by an obvious use of Equation (2.1) in Theorem 2.1.

We now endeavor to invert the formula of Theorem 2.2 in order to find an expression for the moments of $\widetilde{\psi}$ in terms of those of $\psi$.

THEOREM 2.3 $\mu_n(\widetilde{\psi}) = \sum_{k=0}^{[n/2]}\binom{n-k}{k}\gamma^k\lambda^{n-2k+1}\mu_{n-2k}(\psi)$, for each non-negative integer $n$.
Proof: First, note that Theorem 2.2 implies

$$\mu_n(\widetilde{\psi}) = \lambda^{n+1}\mu_n(\psi) + \sum_{k=1}^{[\frac{n}{2}]}\left(\binom{n}{k}-\binom{n}{k-1}\right)(-1)^{k+1}\gamma^k\mu_{n-2k}(\widetilde{\psi}). \tag{2.8}$$

We proceed by induction on $n$. By (2.8), $\mu_0(\widetilde{\psi}) = \lambda\mu_0(\psi)$, which equals

$$\sum_{k=0}^{[\frac{n}{2}]}\binom{n-k}{k}\gamma^k\lambda^{n-2k+1}\mu_{n-2k}(\psi)$$

with $n=0$. Let $n\geq 1$, and assume

$$\mu_m(\widetilde{\psi}) = \sum_{k=0}^{[\frac{m}{2}]}\binom{m-k}{k}\gamma^k\lambda^{m-2k+1}\mu_{m-2k}(\psi),\ for\ \ 0\leq m\leq n-1. \tag{2.9}$$

Use of (2.9) in (2.8) yields

$$\mu_n(\widetilde{\psi}) = \lambda^{n+1}\mu_n(\psi) -$$
$$\sum_{k=1}^{[\frac{n}{2}]}\left(\binom{n}{k}-\binom{n}{k-1}\right)(-\gamma)^k\sum_{j=0}^{[\frac{n}{2}]-k}\binom{n-2k-j}{j}\gamma^j\lambda^{n-2k-2j+1}\mu_{n-2k-2j}(\psi).$$

Re-indexing over $i=j+k$, we find that

$$\mu_n(\widetilde{\psi}) = \lambda^{n+1}\mu_n(\psi) + \sum_{i=1}^{[\frac{n}{2}]}B_{n,i}\gamma^i\lambda^{n-2i+1}\mu_{n-2i}(\psi), \tag{2.10}$$

where

$$B_{n,i} := \sum_{l=1}^{i}(-1)^{l+1}\left(\binom{n}{l}-\binom{n}{l-1}\right)\binom{n-i-l}{i-l}. \tag{2.11}$$

Hence, to complete the proof it suffices to show that $\binom{n-k}{k}$ equals $B_{n,k}$ for each $k$ such that $1 \le k \le [n/2]$. But this follows by the formula

$$\binom{s}{r} = \sum_{l=1}^{r}(-1)^{l+1}\left(\binom{s+r}{l} - \binom{s+r}{l-1}\right)\binom{s-l}{r-l}, \text{ when } 1 \le r \le s, \tag{2.12}$$

which can be shown by induction on $s$ using the fact that $\binom{q+1}{p} = \binom{q}{p} + \binom{q}{p-1}$ for $1 \le p \le q$. The proof is complete.

## 3 FORMULAS FOR THE MOMENTS OF SOME SMDF'S

We conclude by exemplifying the use of Theorem 2.3 for finding explicit expressions for the moments of strong moment distribution functions, both in the direct and indirect range of the Transformation Theorem.

### 3.1 Legendre Distributions

The classical Legendre distribution $\psi_P$ and its extension $\widetilde{\psi}_p$ of parameters $a$ and $b$, $0 < a < b < \infty$ ($\gamma = ab$, $\lambda = b - a$), are given by

$$\frac{d\psi_P}{dx} = \begin{cases} 1 & , \text{if } x \in (-1,1) \\ 0 & , \text{otherwise} \end{cases}$$

*and*

$$\frac{d\widetilde{\psi}_P}{dx} = \begin{cases} 1 & , \text{if } x \in (-b,-a) \cup (a,b) \\ 0 & , \text{if } x \in (-\infty,-b] \cup [-a,0) \cup (0,a] \cup [b,\infty) \end{cases}.$$

The moments can be computed directly from the definitions:

$$\mu_{2l+1}(\psi_P) = 0 \text{ and } \mu_{2l}(\psi_P) = \frac{2}{2l+1}, \quad l = 0,1,2,\ldots, \tag{3.1}$$

*and*

$$\mu_{2l+1}(\widetilde{\psi}_P) = 0 \text{ and } \mu_{2l}(\widetilde{\psi}_P) = 2\frac{b^{2l+1} - a^{2l+1}}{2l+1}, \quad l = 0,\pm1,\pm2,\ldots. \tag{3.2}$$

Theorem 2.3 and Equation (3.1) combine to yield another expression for the non-negative even moments of $\widetilde{\psi}_P$, but we omit it here in view of (3.2).

### 3.2 Tchebycheff Distributions

Applying the Transformation Theorem to the classical Tchebycheff distribution $\psi_T$, which is given by

$$\frac{d\psi_T}{dx} = \begin{cases} \frac{1}{\sqrt{1-x^2}}, & \text{if } x \in (-1,1) \\ 0, & \text{otherwise} \end{cases},$$

yields a SMDF $\widetilde{\psi}_T$ of parameters $a$ and $b$, $0 < a < b < \infty$ ($\gamma = ab$, $\lambda = b - a$),

$$\frac{d\widetilde{\psi}_T}{dx} = \begin{cases} \frac{(b-a)\ |x|}{\sqrt{b^2-x^2}\sqrt{x^2-a^2}}, & \text{if } x \in (-b,-a) \cup (a,b) \\ 0, & \text{if } x \in (-\infty,-b] \cup [-a,0) \cup (0,a] \cup [b,\infty) \end{cases}.$$

The moments of $\psi_T$ can be calculated as

$$\mu_{2l+1}(\psi_T) = 0 \ and \ \mu_{2l}(\psi_T) = \frac{\pi}{4^l}\binom{2l}{l}, \ l = 0,1,2,\ldots. \tag{3.3}$$

Then (2.1), (3.3) and Theorem 2.3 combine to give

$$\mu_{2l+1}(\widetilde{\psi}_T) = 0, \ l = 0, \pm1, \pm2, \ldots, \tag{3.4}$$

$$\mu_{-2l-2}(\widetilde{\psi}_T) = (ab)^{-2l-1}\mu_{2l}(\widetilde{\psi}_T), \ l = 0,1,2,\ldots, \tag{3.5}$$

*and*

$$\mu_{2l}(\widetilde{\psi}_T) = \frac{\pi}{4^l}\sum_{k=0}^{l}\binom{2l-k}{k}\binom{2l-2k}{l-k}(4ab)^k(b-a)^{2l-2k+1}, \ l = 0,1,2,\ldots. \tag{3.6}$$

### 3.3 Generalized Hermite Distributions

The generalized Hermite distribution $\psi_H^{(\alpha)}$ of parameter $\alpha > -1/2$ is given by

$$\frac{d\psi_H^{(\alpha)}}{dx} = |x|^{2\alpha}\ e^{-x^2}, \ x \in \mathbf{R}.$$

It is not very difficult to show that the moments of $\psi_H^{(\alpha)}$ satisfy

$$\mu_{2l+1}(\psi_H^{(\alpha)}) = 0 \ and \ \mu_{2l}(\psi_H^{(\alpha)}) = \Gamma\left(\alpha + l + \frac{1}{2}\right), \ l = 0,1,2,\ldots, \tag{3.7}$$

where $\Gamma$ is the gamma function. $\psi_H^{(\alpha)}$ has an extension, $\widetilde{\psi}_H^{(\alpha)}$ of parameters $\gamma$ and $\lambda$ ($0 < \gamma, \lambda < \infty$), defined by

$$\frac{d\widetilde{\psi}_H^{(\alpha)}}{dx} = \frac{1}{\lambda^{2\alpha}}\left|x^2 - 2\gamma + \frac{\gamma^2}{x^2}\right|^{\alpha} exp\left(-\frac{1}{\lambda^2}\left(x^2 - 2\gamma + \frac{\gamma^2}{x^2}\right)\right), \ x \in \mathbf{R}^- \cup \mathbf{R}^+.$$

Again using (2.1) and applying Theorem 2.3, considering (3.7), we find

$$\mu_{2l+1}(\widetilde{\psi}_H^{(\alpha)}) = 0, \; l = 0, \pm 1, \pm 2, \ldots, \tag{3.8}$$

$$\mu_{-2l-2}(\widetilde{\psi}_H^{(\alpha)}) = \gamma^{-2l-1}\mu_{2l}(\widetilde{\psi}_H^{(\alpha)}), \; l = 0, 1, 2, \ldots, \tag{3.9}$$

and

$$\mu_{2l}(\widetilde{\psi}_H^{(\alpha)}) = \sum_{k=0}^{l} \binom{2l-k}{k} \gamma^k \lambda^{2l-2k+1} \Gamma(\alpha + l - k + 1/2), \; l = 0, 1, 2, \ldots. \tag{3.10}$$

### 3.4 Generalized Laguerre Distributions

Following the example of the connection between the classical Hermite and Laguerre distributions, we consider lastly a SMDF of parameters $\gamma$ and $\lambda$ ($0 < \gamma, \lambda < \infty$), which we denote by $\phi_A^{(\alpha)}(x)$ for $\alpha > -1/2$, given by

$$\frac{d\phi_A^{(\alpha)}}{dx} := \begin{cases} \frac{1}{\lambda^{2\alpha}} x^{-\frac{1}{2}} \left| x - 2\gamma + \frac{\gamma^2}{x} \right|^{\alpha} exp\left( -\frac{1}{\lambda^2} \left( x - 2\gamma + \frac{\gamma^2}{x} \right) \right), & \text{if } x \in \mathbf{R}^+ \\ 0, & \text{if } x \in \mathbf{R}^- \end{cases}. \tag{3.11}$$

The moments of this generalized Laguerre type distribution satisfy

$$\mu_l(\phi_A^{(\alpha)}) = \mu_{2l}(\widetilde{\psi}_H^{(\alpha)}), \; l = 0, \pm 1, \pm 2, \ldots, \tag{3.12}$$

by the change of variable $x \to x^2$.

## REFERENCES


1. Brian A. Hagler, William B. Jones and W. J. Thron. Orthogonal Laurent Polynomials of Jacobi, Hermite and Laguerre Types. These Proceedings.
2. L. Cochran and S. Clement Cooper. Orthogonal Laurent Polynomials on the Real Line. Continued Fractions and Orthogonal Functions: Theory and Applications. Proceedings Loen, Norway, 1992 (S. Clement Cooper and W. J, Thron, eds.), Lecture Notes in Pure and Applied Mathematics: Marcel Dekker, 1993, pp 47-100.
3. T. S. Chihara. An Introduction to Orthogonal Polynomials: Gordon and Breach, 1978.
4. S. Clement Cooper, William B. Jones and W. J. Thron. Orthogonal Laurent Polynomials and Continued Fractions Associated with Log-Nornal Distributions. J CAM 32:39-46, 1990.
5. A. Sri Ranga and J. H. McCabe. On the Extensions of Some Classical Distributions. Proc. of the Edinburgh Math. Soc. 34:19-29, 1991.
6. Gabor Szegő. Orthogonal Polynomials. Providence, RI: AMS Colloquim Publications, Vol. 23, 4th Ed., 1975.

# Orthogonal Laurent Polynomials of Jacobi, Hermite, and Laguerre Types

BRIAN A. HAGLER* Department of Mathematics, University of Colorado, Boulder, Colorado

WILLIAM B. JONES† Department of Mathematics, University of Colorado, Boulder, Colorado

W. J. THRON‡ Department of Mathematics, University of Colorado, Boulder, Colorado

## 1 INTRODUCTION

We will outline in this paper a general procedure for constructing whole systems of orthogonal Laurent polynomials on the real line from systems of orthogonal polynomials. To further explain our intentions, we proceed with some basic definitions and results, all of which appear, or are modifications of those that appear, in the literature. In particular, [1,2,3,4,5,6,7,8,9,10,11,12] were consulted in our preparation.

If $f: D \to \mathbf{R}$, where $D$ is a subset of the set of real numbers $\mathbf{R}$, then the set

$$\sigma\ (f) := [x \in D : \textit{There is an } \epsilon > 0 \textit{ such that } (x-\epsilon,\ x+\epsilon) \subseteq D \textit{ and } f(x+\delta) - f(x-\delta) > 0 \textit{ for all } \delta > 0 \textit{ such that } \delta < \epsilon]$$

is called the *spectrum* of $f$. If $\psi: \mathbf{R} \to \mathbf{R}$ is a bounded, non-decreasing function with an infinite spectrum $\sigma(\psi)$ such that the *moments* $\mu_n(\psi)$ defined by the Riemann-Stieltjes integrals

$$\mu_n(\psi) := \int_{-\infty}^{\infty} x^n \, d\psi(x)$$

* Research supported in part by the U.S. National Science Foundation under Grants Nos. DMS-93002584 and DMS-9701028.

† Research supported in part by the U.S. National Science Foundation under Grants No. DMS-93002584.

‡ Research supported in part by the U.S. National Science Foundation under Grants No. DMS-93002584.

exist for all $n \in \mathbf{Z}_0^+ := [0, 1, 2, ...]$, then $\psi$ is called a *moment distribution function (MDF)*. If $\phi$: $(\mathbf{R}^- \cup \mathbf{R}^+) \to \mathbf{R}$ is a bounded function, non-decreasing on the negative reals $\mathbf{R}^-$ and the positive reals $\mathbf{R}^+$ separately, with infinite spectrum $\sigma(\phi)$ such that the *moments*

$$\mu_n(\phi) := \int_{-\infty}^{0} x^n \, d\phi(x) + \int_{0}^{\infty} x^n \, d\phi(x)$$

exist for all $n \in \mathbf{Z} := [0, \pm 1, \pm 2, ...]$, then $\phi$ is called a *strong moment distribution function (SMDF)*.

We denote by $\mathcal{P}$ the space of all polynomials in $x$ with real coefficients, and we denote by $\mathcal{R}$ the space of all *Laurent polynomials (L-polynomials)*, $\sum_{k=m}^{n} r_k x^k$ for $m, n \in \mathbf{Z}$ with $m \leq n$, having real coefficients $r_k$ for $m \leq k \leq n$. For $R \in \mathcal{R}$ we sometimes write $C_k(R)$ for the coefficient of $x^k$ in $R$. We will find it useful to define the following subsets of $\mathcal{R}$:

$$\mathcal{R}_{m,n} := \left[\sum_{k=m}^{n} r_k x^k \; : \; r_k \in \mathbf{R} \; , \; m \leq k \leq n\right] \text{ for } m, n \in \mathbf{Z} \text{ with } m \leq n;$$

$$\mathcal{R}_{2n} := [R \in \mathcal{R}_{-n,n} : \; C_n(R) \neq 0] \text{ for } n \in \mathbf{Z}_0^+;$$

$$\mathcal{R}_{2n+1} := [R \in \mathcal{R}_{-n-1,n} : \; C_{-n-1}(R) \neq 0] \text{ for } n \in \mathbf{Z}_0^+.$$

THEOREM 1.1 For any non-zero $R \in \mathcal{R}$, there is a unique $d \in \mathbf{Z}_0^+$, called the *L-degree* of $R$, such that $R \in \mathcal{R}_d$.
Proof: See [4], p. 48, Theorem 1.1.

Notice that for a polynomial in $\mathcal{P}$, its L-degree is twice its polynomial degree. If $R \in \mathcal{R}_{2n}$, then $C_n(R)$ and$C_{-n}(R)$ are called the *leading coefficient* and *trailing coefficient* of $R$, respectively, and if $R \in \mathcal{R}_{2n+1}$, then the *leading coefficient* and *trailing coefficient* of $R$ are $C_{-n-1}(R)$ and $C_n(R)$, respectively. An L-polynomial is called *monic* if its leading coefficient is 1, *regular* if its trailing coefficient is non-zero, and *singular* if its trailing coefficient is 0.

For any MDF $\psi$ and polynomials $P, Q \in \mathcal{P}$, we define

$$(P, Q)_\psi := \int_{-\infty}^{\infty} P(x)Q(x) \, d\psi(x),$$

and, for any SMDF $\phi$ and L-polynomials $R, S \in \mathcal{R}$, we set

$$(R, S)_\phi := \int_{-\infty}^{0} R(x)S(x) \, d\phi(x) + \int_{0}^{\infty} R(x)S(x) \, d\phi(x).$$

THEOREM 1.2 If $\psi$ is an MDF, then $(\cdot, \cdot)_\psi$ is an inner-product on $\mathcal{P}$.
Proof: See [3], p.13 and p.16.

THEOREM 1.3 If $\phi$ is an SMDF, then $(\cdot, \cdot)_\phi$ is an inner-product on $\mathcal{R}$.

Proof: See [4], p.58 and p.62, Theorem 3.6.

A sequence $\{P_n(x)\}_{n=0}^{\infty}$ in $\mathcal{P}$ such that, for $\psi$, an MDF, and all $m, n \in \mathbf{Z}_0^+$,

$$P_n \textit{ has polynomial degree } n \textit{ and } (P_m, P_n)_\psi = 0 \textit{ if } m \neq n,$$

is called an *orthogonal polynomial sequence (OPS) with respect to* $\psi$. If in addition each of the polynomials $P_n$ is monic, $\{P_n(x)\}_{n=0}^{\infty}$ is called *monic*. It is not difficult to show that a monic OPS with respect to a given MDF $\psi$ exists and is unique (see [4], Theorem 3.3, p.14, and Corollary, p.9). Analogously in the Laurent case, a sequence $\{R_n(x)\}_{n=0}^{\infty}$ in $\mathcal{R}$ is called an *orthogonal Laurent polynomial sequence (OLPS) with respect to* $\phi$, an SMDF, if

$$R_n \textit{ has L-degree } n \textit{ and } (R_m, R_n)_\phi = 0 \textit{ if } m \neq n.$$

$\{R_n(x)\}_{n=0}^{\infty}$ is called *monic* if each of the L-polynomials $R_n$ is monic. Again, it is not hard to show that a unique monic OLPS with respect to a given SMDF $\phi$ exists (see [4], Theorem 3.2, p.59, and Corollary 1.6, p.53).

Our main goal in the next section is to show that, given positive real parameters $\gamma$ and $\lambda$, the L-polynomial transformation $x \to 1/\lambda(x - \gamma/x)$ takes systems of OPS's into systems of OLPS's. This and other results of the preceding sections are preparations for our exposition of orthogonal L-polynomials of Jacobi, Hermite and Laguerre types, contained in the third, and final, part of this paper.

## 2 THE TRANSFORMATION

In this section, we present a rigorous exposition of a general, parameterized L-polynomial transformation of OPS's into OLPS's. Although our work here is of a general nature, we anticipate specific applications, presented in Section 3, to Jacobi, Hermite and Laguerre systems of OPS's.

### 2.1 Definitions

For the rest of Section 2, we assume that

$$\psi \textit{ is an MDF with monic OPS } \{P_n(x)\}_{n=0}^{\infty}$$

and that

$$\lambda \textit{ and } \gamma \textit{ are fixed positive real numbers.}$$

Our development will be facilitated by a series of definitions: We set

$$v(x) := \frac{1}{\lambda}\left(x - \frac{\gamma}{x}\right),$$

$$v_{\pm}^{-1}(y) := \frac{\lambda}{2}\left(y \pm \sqrt{y^2 + \frac{4\gamma}{\lambda^2}}\right),$$

$$\widetilde{\psi}(x) := \begin{cases} \int_{-\infty}^{x} \frac{1}{v'(t)}\, d(\psi \circ v)(t), & for \;\; x \in \mathbf{R}^{-} \\ \int_{0}^{x} \frac{1}{v'(t)}\, d(\psi \circ v)(t) & , for \;\; x \in \mathbf{R}^{+} \end{cases},$$

and, for $n = 0, 1, 2, \ldots$,

$$\widetilde{P}_{2n}(x) := \lambda^n P_n\left(v(x)\right) \quad and \quad \widetilde{P}_{2n+1}(x) := \left(-\frac{\lambda}{\gamma}\right)^n \frac{1}{x} P_n\left(v(x)\right).$$

## 2.2 Preliminary Theorems

We begin with an exploration of the essential component, $v$, and the related functions, $v_{-}^{-1}$, $v_{+}^{-1}$, and $v'$, of our transformation.

THEOREM 2.2.1

(A) $v|_{\mathbf{R}^-}$, $v|_{\mathbf{R}^+}$, $v_{-}^{-1}$, and $v_{+}^{-1}$ are differentiable, monotone increasing functions.

(B) $v|_{\mathbf{R}^+}$ is a diffeomorphism from $\mathbf{R}^{+}$ to $\mathbf{R}$. Its inverse is $v_{+}^{-1}$.

(C) $v|_{\mathbf{R}^-}$ is a diffeomorphism from $\mathbf{R}^{-}$ to $\mathbf{R}$. Its inverse is $v_{-}^{-1}$.

(D) For all $x \in \mathbf{R}^{-} \cup \mathbf{R}^{+}$, $v(-\frac{\gamma}{x}) = v(x)$.

(E) For all $t \in \mathbf{R}^{-} \cup \mathbf{R}^{+}$, $\frac{dv}{dx}|_{x=-\frac{\gamma}{t}} = \frac{t^2}{\gamma}\frac{dv}{dx}|_{x=t}$.

Proof: (A) Evidently, $v$ has domain $\mathbf{R}^{-} \cup \mathbf{R}^{+}$, and $\frac{dv}{dx} = \frac{1}{\lambda}(1 + \frac{\gamma}{x^2}) > \frac{1}{\lambda} > 0$; that is, $v$ is differentiable and monotone increasing on $\mathbf{R}^{-}$ and on $\mathbf{R}^{+}$ by elementary calculus techniques. Similarly, it is clear that the domain of $v_{\pm}^{-1}$ is $\mathbf{R}$, and, since $\left|y/\sqrt{y^2 + \frac{4\gamma}{\lambda^2}}\right| < 1$, $\frac{dv_{\pm}^{-1}}{dy} = \frac{\lambda}{2}\left(1 \pm y/\sqrt{y^2 + \frac{4\gamma}{\lambda^2}}\right) > 0$; that is, again by elementary calculus, $v_{\pm}^{-1}$ is differentiable and monotone increasing. See a graph of $v$, $v_{-}^{-1}$, and $v_{+}^{-1}$ in Figure 1.

(B) By inspection of the definitions, $v|_{\mathbf{R}^+}$ maps $\mathbf{R}^{+}$ onto $\mathbf{R}$, and $v_{+}^{-1}$ maps $\mathbf{R}$ onto $\mathbf{R}^{+}$. By Theorem 2.2.1 (A), $v|_{\mathbf{R}^+}$ and $v_{+}^{-1}$ are injective and differentiable. Hence, it suffices to show that $v_{+}^{-1}(v(x)) = x$, for any $x$ in $\mathbf{R}^{+}$. But, for $x$ in $\mathbf{R}^{+}$,

$$\begin{aligned} v_{+}^{-1}(v(x)) &= \frac{\lambda}{2}\left(\frac{1}{\lambda}(x - \frac{\gamma}{x}) + \sqrt{(\frac{1}{\lambda}(x - \frac{\gamma}{x}))^2 + \frac{4\gamma}{\lambda^2}}\right) \\ &= \frac{\lambda}{2}\left(\frac{1}{\lambda}(x - \frac{\gamma}{x}) + \sqrt{\frac{1}{\lambda^2}(x^2 + 2\gamma + \frac{\gamma^2}{x^2})}\right) \\ &= \frac{\lambda}{2}\left(\frac{1}{\lambda}(x - \frac{\gamma}{x}) + \sqrt{(\frac{1}{\lambda}(x + \frac{\gamma}{x}))^2}\right) \\ &= \frac{\lambda}{2}\left(\frac{1}{\lambda}(x - \frac{\gamma}{x}) + \frac{1}{\lambda}(x + \frac{\gamma}{x})\right) \\ &= x. \end{aligned}$$

(C) A proof can be given in exactly the same way as that for Theorem 2.2.1 (B).

(D) $v(x) := \frac{1}{\lambda}(x - \frac{\gamma}{x}) = \frac{1}{\lambda}(\frac{-\gamma}{x} - \gamma/(\frac{-\gamma}{x})) = v(-\frac{\gamma}{x})$, for any non-zero $x$ in $\mathbf{R}$.

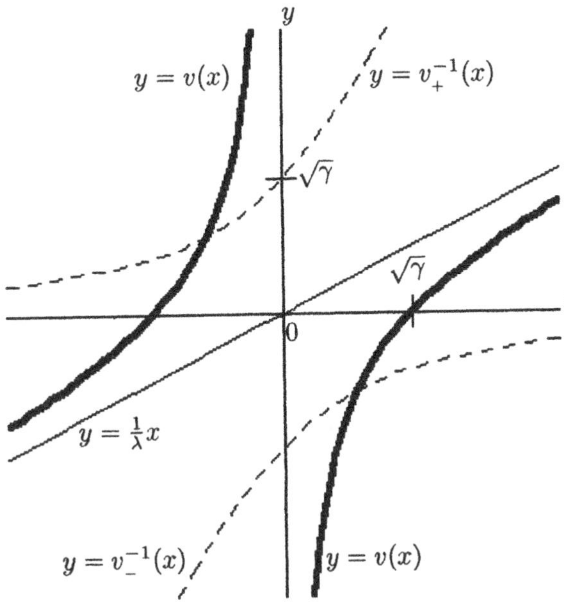


**Figure 1** Graph of $v$, $v_-^{-1}$, and $v_+^{-1}$.

(E) $\frac{dv}{dx}|_{x=-\frac{\gamma}{t}} = \frac{1}{\lambda}(1+\frac{\gamma}{x^2})|_{x=-\frac{\gamma}{t}} = \frac{1}{\lambda}(1+\frac{t^2}{\gamma}) = \frac{t^2}{\gamma}\frac{1}{\lambda}(1+\frac{\gamma}{t^2}) = \frac{t^2}{\gamma}\frac{dv}{dx}|_{x=t}$, for any non-zero $t$ in $\mathbf{R}$.

THEOREM 2.2.2

(A) $\widetilde{\psi}$ is a bounded function on $\mathbf{R}^- \cup \mathbf{R}^+$.

(B) $\widetilde{\psi}$ is non-decreasing on $\mathbf{R}^-$ and $\mathbf{R}^+$ separately.

Proof: (A) Inspection of the definitions shows that $0 < \frac{1}{v'(t)} < \lambda$, for all non-zero $t$ in $\mathbf{R}$. Hence, by comparison, for $x$ in $\mathbf{R}^-$,

$$0 \leq \int_{-\infty}^{v(x)} \frac{1}{v'(v_-^{-1}(y))}\, d\psi(y) \leq \int_{-\infty}^{\infty} \lambda\, d\psi(y) = \lambda\mu_0(\psi).$$

But, for $x$ in $\mathbf{R}^-$, with $y = v(t)$,

$$\widetilde{\psi}(x) := \int_{-\infty}^{x} \frac{1}{v'(t)}\, d(\psi \circ v)(t) = \int_{-\infty}^{v(x)} \frac{1}{v'(v_-^{-1}(y))}\, d\psi(y).$$

Hence, $0 \leq \widetilde{\psi}(x) \leq \lambda\mu_0(\psi)$, for $x$ in $\mathbf{R}^-$; that is, $\widetilde{\psi}$ is a bounded map from $\mathbf{R}^-$ to $\mathbf{R}$. A similar argument shows that $\widetilde{\psi}$ is a bounded map from $\mathbf{R}^+$ to $\mathbf{R}$.

(B) To verify that $\widetilde{\psi}$ is non-decreasing on $\mathbf{R}^-$, suppose $-\infty < x \leq y < 0$. For all $t \leq y$, $\frac{1}{v'(t)} \geq \frac{1}{v'(y)}$. Hence, $\widetilde{\psi}(y) - \widetilde{\psi}(x) = \int_x^y \frac{1}{v'(t)}\, d(\psi \circ v)(t) \geq \frac{1}{v'(y)} \int_x^y d(\psi \circ v)(t) = \frac{\psi(v(y)) - \psi(v(x))}{v'(y)}$. But, $v'(y) = \frac{1}{\lambda}(1 + \frac{\gamma}{y^2}) > 0$, and $\psi(v(y)) - \psi(v(x)) \geq 0$ by the monotonicity of $\psi$ and $v$. Thus, $\widetilde{\psi}(y) - \widetilde{\psi}(x) \geq 0$; that is, $\widetilde{\psi}$ is non-decreasing on

$\mathbf{R}^-$. Likewise, it can be shown that $\widetilde{\psi}$ is non-decreasing on $\mathbf{R}^+$.

THEOREM 2.2.3 $\sigma(\widetilde{\psi}) = v_-^{-1}(\sigma(\psi)) \cup v_+^{-1}(\sigma(\psi))$.

Proof: Suppose $x \in v_-^{-1}(\sigma(\psi))$. Then $x \in \mathbf{R}^-$, and $v(x) \in v(v_-^{-1}(\sigma(\psi))) = \sigma(\psi)$. Therefore, there is an $\epsilon > 0$ such that $(x-\epsilon, x+\epsilon) \subset \mathbf{R}^-$, and $\psi(v(x)+\delta) - \psi(v(x)-\delta) > 0$ for all $\delta > 0$. Let $\delta_1$ satisfy $0 < \delta_1 < \epsilon$. We have $v(x-\delta_1) < v(x) < v(x+\delta_1)$ by Theorem 2.2.1 (A). Hence, there exists a $\delta_2 > 0$ such that $v(x-\delta_1) < v(x) - \delta_2 < v(x) + \delta_2 < v(x+\delta_1)$, and the estimates

$$\begin{aligned}
\widetilde{\psi}(x+\delta_1) - \widetilde{\psi}(x-\delta_1) &= \int_{v(x-\delta_1)}^{v(x+\delta_1)} \frac{1}{v'(v_-^{-1}(y))}\, d\psi(y) \\
&\geq \frac{1}{v'(x+\delta_1)} \int_{v(x-\delta_1)}^{v(x+\delta_1)} d\psi(y) \\
&= \frac{1}{v'(x+\delta_1)} (\psi(v(x+\delta_1)) - \psi(v(x-\delta_1))) \\
&\geq \frac{1}{v'(x+\delta_1)} (\psi(v(x)+\delta_2) - \psi(v(x)-\delta_2))
\end{aligned}$$

hold. But $\frac{1}{v'(x+\delta_1)} > 0$, and, since $v(x)$ is in $\sigma(\psi)$, $\psi(v(x)+\delta_2) - \psi(v(x)-\delta_2) > 0$. Thus, $\widetilde{\psi}(x+\delta_1) - \widetilde{\psi}(x-\delta_1) > 0$. It follows that $v_-^{-1}(\sigma(\psi)) \subseteq \sigma(\widetilde{\psi})$. A similar argument shows that $v_+^{-1}(\sigma(\psi)) \subseteq \sigma(\widetilde{\psi})$. Thus, $\left(v_-^{-1}(\sigma(\psi)) \cup v_+^{-1}(\sigma(\psi))\right) \subseteq \sigma(\widetilde{\psi})$.

Next, suppose $x \in (\sigma(\widetilde{\psi}) \cap \mathbf{R}^-)$. By Theorem 2.2.1 (C), there is a unique $y$ in $\mathbf{R}$ such that $v_-^{-1}(y) = x$. But estimates similar to those above show that $y \in \sigma(\psi)$. Thus, $x \in v_-^{-1}(\sigma(\psi))$ if $x \in (\sigma(\widetilde{\psi}) \cap \mathbf{R}^-)$, and hence $(\sigma(\widetilde{\psi}) \cap \mathbf{R}^-) \subseteq v_-^{-1}(\sigma(\psi))$. In an analogous manner, it can be shown that $(\sigma(\widetilde{\psi}) \cap \mathbf{R}^+) \subseteq v_+^{-1}(\sigma(\psi))$. Thus, it follows that $\sigma(\widetilde{\psi}) = \left((\sigma(\widetilde{\psi}) \cap \mathbf{R}^-) \cup (\sigma(\widetilde{\psi}) \cap \mathbf{R}^+)\right) \subseteq \left(v_-^{-1}(\sigma(\psi)) \cup v_+^{-1}(\sigma(\psi))\right)$.

THEOREM 2.2.4 Let $n$ be any integer. Then:

(A) $\int_{-\infty}^{0} x^n \frac{1}{v'(x)}\, d(\psi \circ v)(x)$ and $\int_{0}^{\infty} x^n \frac{1}{v'(x)}\, d(\psi \circ v)(x)$ exist.

(B) $\int_{-\infty}^{0} x^n \frac{1}{v'(x)}\, d(\psi \circ v)(x) = (-1)^n \gamma^{n+1} \int_{0}^{\infty} x^{-n-2} \frac{1}{v'(x)}\, d(\psi \circ v)(x)$.

(C) $\int_{-\infty}^{0} x^n\, d\widetilde{\psi}(x) = \int_{-\infty}^{0} x^n \frac{1}{v'(x)}\, d(\psi \circ v)(x)$ and $\int_{0}^{\infty} x^n\, d\widetilde{\psi}(x) = \int_{0}^{\infty} x^n \frac{1}{v'(x)}\, d(\psi \circ v)(x)$.

(D) $\int_{-\infty}^{0} x^n\, d\widetilde{\psi}(x)$ and $\int_{0}^{\infty} x^n\, d\widetilde{\psi}(x)$ exist.

(E) $\int_{-\infty}^{0} x^n\, d\widetilde{\psi}(x) = (-1)^n \gamma^{n+1} \int_{0}^{\infty} x^{-n-2}\, d\widetilde{\psi}(x)$.

Proof: (A) Since $0 < \frac{1}{v'(x)} < \lambda$, and, for all $x$ in $\mathbf{R}^-$,

$$\begin{aligned} 0 &\le |x| \\ &= \left|v_-^{-1}(v(x))\right| \\ &= \left|\frac{\lambda}{2}\left(v(x) - \sqrt{v^2(x) + \frac{4\gamma}{\lambda^2}}\right)\right| \\ &\le \frac{\lambda}{2}\left(|v(x)| + |v(x)| + \frac{2\sqrt{\gamma}}{\lambda}\right) \\ &\le \lambda|v(x)| + \sqrt{\gamma}, \end{aligned}$$

we have $0 \le |x|^n \frac{1}{v'(x)} \le \lambda(\lambda|v(x)| + \sqrt{\gamma})^n$ for all $x$ in $\mathbf{R}^-$ and $n$ in $\mathbf{Z}_0^+$. Since the moments $\mu_n(\psi)$ are finite for $n$ in $\mathbf{Z}_0^+$, we can deduce that the integrals $\int_{-\infty}^{\infty} |t|^n \, d\psi(t)$ are finite by comparing $|t|^n$ to $t^N+1$, for $N$ an even integer greater than $n$. Hence, for $n$ in $\mathbf{Z}_0^+$, the integrals $\int_{-\infty}^{\infty} \lambda(\lambda|t| + \sqrt{\gamma})^n \, d\psi(t) = \int_{-\infty}^{0} \lambda(\lambda|v(x)| + \sqrt{\gamma})^n \, d(\psi \circ v)(x)$ exist, and the integrals $\int_{-\infty}^{0} x^n \frac{1}{v'(x)} \, d(\psi \circ v)(x)$ exist, by comparison. A similar argument shows that the integrals $\int_0^{\infty} x^n \frac{1}{v'(x)} \, d(\psi \circ v)(x)$, for $n$ in $\mathbf{Z}_0^+$, exist.

The substitution $x \to -\frac{\gamma}{x}$ in $\int_{-\infty}^{0} x^n \frac{1}{v'(x)} \, d(\psi \circ v)(x)$ yields, by Theorem 2.2.1, parts (D) and (E),

$$\int_{-\infty}^{0} x^n \frac{1}{v'(x)} \, d(\psi \circ v)(x) = (-1)^n \gamma^{n+1} \int_0^{\infty} x^{-n-2} \frac{1}{v'(x)} \, d(\psi \circ v)(x).$$

Hence, the integrals $\int_{-\infty}^{0} x^n \frac{1}{v'(x)} \, d(\psi \circ v)(x)$ and $\int_0^{\infty} x^n \frac{1}{v'(x)} \, d(\psi \circ v)(x)$ exist for all integers $n$, with the possible exception of the case $n = -1$. But, a comparison of $|x|^{-1}$ to $x^{-2} + 1$ now shows that the integrals exist also for $n = -1$.

(B) By Theorem 2.2.4 (A), the integral $\int_{-\infty}^{0} x^n \frac{1}{v'(x)} \, d(\psi \circ v)(x)$ exists for any integer $n$. As in the proof of Theorem 2.2.4 (A), the subtitution $x \to -\frac{\gamma}{x}$ now yields

$$\int_{-\infty}^{0} x^n \frac{1}{v'(x)} \, d(\psi \circ v)(x) = (-1)^n \gamma^{n+1} \int_0^{\infty} x^{-n-2} \frac{1}{v'(x)} \, d(\psi \circ v)(x),$$

for any integer $n$.

(C, D) Let $n$ be an integer, and suppose $\infty < a < b < 0$. Then, since the integrands are continuous and the integrators are non-decreasing and bounded on the closed interval $[a, b]$, the integrals

$$\int_a^b x^n \, d\widetilde{\psi}(x) \quad \textit{and} \quad \int_a^b x^n \frac{1}{v'(x)} \, d(\psi \circ v)(x) \quad \textit{exist for all} \quad n \in \mathbf{Z}.$$

Next, set $x_{m,k} := k\frac{b-a}{m} + a$ for $m \ge 1$ and $k = 0, 1, 2, ..., m$. By the Mean Value

Theorem, there is a $c_{m,k}$ in the closed interval $[x_{m,k-1}, x_{m,k}]$ such that

$$\int_{x_{m,k-1}}^{x_{m,k}} \frac{1}{v'(x)}\, d(\psi \circ v)(x) = \frac{1}{v'(c_{m,k})}(\psi(v(x_{m,k})) - \psi(v(x_{m,k-1}))) ,$$

for each $k = 1, 2, ..., m$. Since the integrals exist, we can choose to take

$$\int_a^b x^n\, d\widetilde{\psi}(x) = \lim_{m\to\infty} \sum_{k=1}^{m} c_{m,k}^n (\widetilde{\psi}(x_{m,k}) - \widetilde{\psi}(x_{m,k-1}))$$

and

$$\int_a^b x^n \frac{1}{v'(x)}\, d(\psi \circ v)(x) = \lim_{m\to\infty} \sum_{k=1}^{m} c_{m,k}^n \frac{1}{v'(c_{m,k})}(\psi(v(x_{m,k})) - \psi(v(x_{m,k-1}))) .$$

But, the definition of $\widetilde{\psi}$ and additivity of the integral imply

$$\widetilde{\psi}(x_{m,k}) - \widetilde{\psi}(x_{m,k-1}) = \int_{x_{m,k-1}}^{x_{m,k}} \frac{1}{v'(x)}\, d(\psi \circ v)(x).$$

It follows that

$$\begin{aligned} \int_a^b x^n\, d\widetilde{\psi}(x) &= \lim_{m\to\infty} \sum_{k=1}^{m} c_{m,k}^n \frac{1}{v'(c_{m,k})}(\psi(v(x_{m,k})) - \psi(v(x_{m,k-1}))) \\ &= \int_a^b x^n \frac{1}{v'(x)}\, d(\psi \circ v)(x). \end{aligned}$$

Since $\int_{-\infty}^0 x^n \frac{1}{v'(x)}\, d(\psi \circ v)(x)$ exists by Theorem 2.2.4 (A), we then have $\int_{-\infty}^0 x^n\, d\widetilde{\psi}(x)$ exists and equals $\int_{-\infty}^0 x^n \frac{1}{v'(x)}\, d(\psi \circ v)(x)$.

Likewise, it follows that $\int_0^\infty x^n\, d\widetilde{\psi}(x)$ exists and is equal to $\int_0^\infty x^n \frac{1}{v'(x)}\, d(\psi \circ v)(x)$.

(E) The result follows by substition of the integrals in Theorem 2.2.4 (C) in the equation of Theorem 2.2.4 (B).

THEOREM 2.2.5 Let $n$ be any integer. Then:

(A) $\mu_n(\widetilde{\psi})$ exists.

(B) $\mu_n(\widetilde{\psi}) = (-1)^n \gamma^{n+1} \mu_{-n-2}(\widetilde{\psi})$.

(C) $\mu_{-1}(\widetilde{\psi}) = 0$.

Proof: (A) Since $\mu_n(\widetilde{\psi}) := \int_{-\infty}^0 x^n\, d\widetilde{\psi}(x) + \int_0^\infty x^n\, d\widetilde{\psi}(x)$ by definition, Theorem 2.2.4 (D) implies $\mu_n(\widetilde{\psi})$ exists for any integer $n$.

(B) The result follows immediately by applying Theorem 2.2.4 (E) to the definition $\mu_n(\widetilde{\psi}) := \int_{-\infty}^0 x^n\, d\widetilde{\psi}(x) + \int_0^\infty x^n\, d\widetilde{\psi}(x)$.

(C) By Theorem 2.2.5 (A), $\mu_{-1}(\widetilde{\psi})$ exists, and, by Theorem 2.2.5 (B) with $n = -1$, $\mu_{-1}(\widetilde{\psi}) = -\mu_{-1}(\widetilde{\psi})$. Hence, $\mu_{-1}(\widetilde{\psi}) = 0$.

THEOREM 2.2.6 $\widetilde{\psi}$ is a SMDF.
Proof: By Theorem 2.2.2, $\widetilde{\psi}$ is a bounded function, non-decreasing on $\mathbf{R}^-$ and $\mathbf{R}^+$ separately. Theorem 2.2.1 (C) implies $v_-^{-1}$ and $v_+^{-1}$ are one-to-one, and Theorem 2.2.3 says $\sigma(\widetilde{\psi}) = v_-^{-1}(\sigma(\psi)) \cup v_+^{-1}(\sigma(\psi))$. Hence, $\sigma(\widetilde{\psi})$ is infinite since $\sigma(\psi)$ is infinite, $\widetilde{\psi}$ being a MDF. Lastly, the moments $\mu_n(\widetilde{\psi})$, for each integer $n$, exists by Theorem 2.2.5 (A).

THEOREM 2.2.7 Let $R$ and $S$ be Laurent polynomials. Then the inner-product $(R,S)_{\widetilde{\psi}} = \int_{-\infty}^{0} R(x)S(x)\frac{1}{v'(x)}\,d(\psi\circ v)(x) + \int_0^{\infty} R(x)S(x)\frac{1}{v'(x)}\,d(\psi\circ v)(x)$.
Proof: By Theorem 2.2.6, $\widetilde{\psi}$ is a SMDF. Hence, by defintion,

$$(R,S)_{\widetilde{\psi}} = \int_{-\infty}^{0} R(x)S(x)\frac{1}{v'(x)}\,d\widetilde{\psi}(x) + \int_0^{\infty} R(x)S(x)\frac{1}{v'(x)}\,d\widetilde{\psi}(x).$$

Thus, the result follows from Theorem 2.2.4 (C) and linearity of the integral.

THEOREM 2.2.8 Let $j$ and $k$ be non-negative integers. Then:
(A) $(\widetilde{P}_{2j},\widetilde{P}_{2k})_{\widetilde{\psi}} = \lambda^{j+k+1}(P_j,P_k)_{\psi}$.
(B) $(\widetilde{P}_{2j+1},\widetilde{P}_{2k+1})_{\widetilde{\psi}} = \left(\frac{\lambda}{\gamma}\right)^{j+k+1}(P_j,P_k)_{\psi}$.
(C) $(\widetilde{P}_{2j+1},\widetilde{P}_{2k})_{\widetilde{\psi}} = 0$.
Proof: (A) The definition of $(\widetilde{P}_{2j},\widetilde{P}_{2k})_{\widetilde{\psi}}$, Theorem 2.2.7, the subtitution $x = -\frac{\gamma}{t}$, Theorem 2.2.1 (E), linearity of the integral, the definition of $\widetilde{P}_{2j}(x)$ and $\widetilde{P}_{2k}(x)$, the substitution $y = v(x)$, and the definition of $(P_j,P_k)_{\psi}$ justify

$$\begin{aligned}
(\widetilde{P}_{2j},\widetilde{P}_{2k})_{\widetilde{\psi}} &:= \int_{-\infty}^{0} \widetilde{P}_{2j}(x)\widetilde{P}_{2k}(x)\,d\widetilde{\psi}(x) + \int_0^{\infty} \widetilde{P}_{2j}(x)\widetilde{P}_{2k}(x)\,d\widetilde{\psi}(x) \\
&= \int_{-\infty}^{0} \widetilde{P}_{2j}(x)\widetilde{P}_{2k}(x)\frac{1}{v'(x)}\,d(\psi\circ v)(x) + \\
&\qquad \int_0^{\infty} \widetilde{P}_{2j}(x)\widetilde{P}_{2k}(x)\frac{1}{v'(x)}\,d(\psi\circ v)(x) \\
&= \int_0^{\infty} \widetilde{P}_{2j}(t)\widetilde{P}_{2k}(t)\frac{\gamma}{t^2}\frac{1}{v'(t)}\,d(\psi\circ v)(t) + \\
&\qquad \int_0^{\infty} \widetilde{P}_{2j}(x)\widetilde{P}_{2k}(x)\frac{1}{v'(x)}\,d(\psi\circ v)(x) \\
&= \lambda\int_0^{\infty} \widetilde{P}_{2j}(x)\widetilde{P}_{2k}(x)\frac{1}{\lambda}(1+\frac{\gamma}{x^2})\frac{1}{v'(x)}\,d(\psi\circ v)(x) \\
&= \lambda\int_0^{\infty} \widetilde{P}_{2j}(x)\widetilde{P}_{2k}(x)\,d(\psi\circ v)(x) \\
&= \lambda^{j+k+1}\int_0^{\infty} P_j(v(x))P_k(v(x))\,d(\psi\circ v)(x) \\
&= \lambda^{j+k+1}\int_{-\infty}^{\infty} P_j(y)P_k(y)\,d\psi(y) \\
&= \lambda^{j+k+1}(P_j,P_k)_{\psi}.
\end{aligned}$$

(B) By similar means as used in the proof of Theorem 2.2.8 (A),

$$(\widetilde{P}_{2j+1}, \widetilde{P}_{2k+1})_{\widetilde{\psi}} = (-1)^{j+k} \left(\frac{\lambda}{\gamma}\right)^{j+k+1} (P_j, P_k)_\psi.$$

If $j \neq k$, then $(P_j, P_k)_\psi = 0$ by orthogonality. If $j = k$, then $(-1)^{j+k} = 1$. In either case,

$$(-1)^{j+k} \left(\frac{\lambda}{\gamma}\right)^{j+k+1} (P_j, P_k)_\psi = \left(\frac{\lambda}{\gamma}\right)^{j+k+1} (P_j, P_k)_\psi.$$

The result therefore follows.

(C) By arguments similar to those used in the proofs of the previous two parts of Theorem 2.2.8,

$$(\widetilde{P}_{2j+1}, \widetilde{P}_{2k})_{\widetilde{\psi}} = -(\widetilde{P}_{2j+1}, \widetilde{P}_{2k})_{\widetilde{\psi}}.$$

Then, since $(\widetilde{P}_{2j+1}, \widetilde{P}_{2k})_{\widetilde{\psi}}$ is finite by Theorem 2.2.6, we must have $(\widetilde{P}_{2j+1}, \widetilde{P}_{2k})_{\widetilde{\psi}} = 0$.

THEOREM 2.2.9 $\{\widetilde{P}_n(x)\}_{n=0}^{\infty}$ is the monic OLPS with respect to $\widetilde{\psi}$.
Proof: Inspection of the definition of $\widetilde{P}_n(x)$ shows that it is a monic L-polynomial of L-degree $n$, and Theorem 2.2.8 implies orthogonality of $\{\widetilde{P}_n(x)\}_{n=0}^{\infty}$ with respect to $\widetilde{\psi}$.

## 2.3 The Transformation Theorem

For ease of reference and discussion we collect several of the results obtained in the previous section into the following theorem.

THEOREM 2.3.1 (The Transformation Theorem) Let $\psi$ be a moment distribution function, let $\sigma(\psi)$ denote the spectrum of $\psi$, and let $\{P_n(x)\}_{n=0}^{\infty}$ denote the monic orthogonal polynomial sequence with respect to $\psi$. Let $\lambda, \gamma \in \mathbf{R}^+$, and set $v(x) := \frac{1}{\lambda}(x - \frac{\gamma}{x})$ and $v_{\pm}^{-1}(y) := \frac{\lambda}{2}(y \pm \sqrt{y^2 + \frac{4\gamma}{\lambda^2}})$. Then:

(A) $\widetilde{\psi}(x) := \begin{cases} \int_{-\infty}^{x} \frac{1}{v'(t)}\, d(\psi \circ v)(t), x \in \mathbf{R}^- \\ \int_{0}^{x} \frac{1}{v'(t)}\, d(\psi \circ v)(t), \quad x \in \mathbf{R}^+ \end{cases}$ is a strong moment distribution function.

(B) $\sigma(\widetilde{\psi}) = v_-^{-1}(\sigma(\psi)) \cup v_+^{-1}(\sigma(\psi))$ is the spectrum of $\widetilde{\psi}$.

(C) $\{\widetilde{P}_m(x)\}_{m=0}^{\infty}$, where $\widetilde{P}_{2n}(x) := \lambda^n P_n(v(x))$ and $\widetilde{P}_{2n+1}(x) := \left(-\frac{\lambda}{\gamma}\right)^n \frac{1}{x} P_n(v(x))$ for $n = 0, 1, 2, \ldots$, is the monic orthogonal Laurent polynomial sequence with respect to $\widetilde{\psi}$.
Proof: See the proofs of Theorem 2.2.3, Theorem 2.2.6 and Theorem 2.2.9.

We call $v$ the *doubling transformation* because it is a monotone increasing function of both $\mathbf{R}^-$ and $\mathbf{R}^+$ onto $\mathbf{R}$. In effect, $(f \circ v)|_{\mathbf{R}^-}$ and $(f \circ v)|_{\mathbf{R}^+}$ are copies of $f : \mathbf{R} \to \mathbf{R}$ living on the negative reals and the positive reals, respectively. In this sense, $f \circ v$ is a doubling of $f$. See Figure 2 for an example.

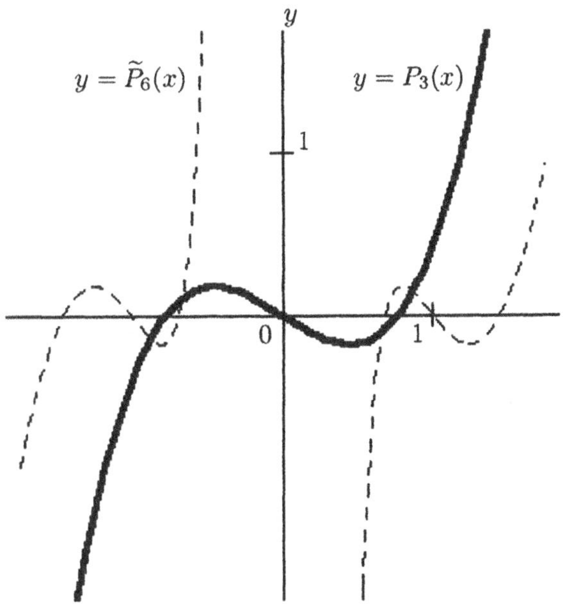


**Figure 2** Graph of the monic Legendre polynomial $P_3(x) = x^3 - \frac{3}{5}x$ and $\widetilde{P}_6(x) = \lambda^3 P_3(v(x))$ with $\lambda = \gamma = 1$.

Of course, $v$ is not the only monotone increasing function of both $\mathbf{R}^-$ and $\mathbf{R}^+$ onto $\mathbf{R}$; that is, $v$ is not the only doubling transformation. However, $v(x) = \frac{1}{\lambda}(x - \frac{\gamma}{x})$ is a Laurent polynomial. This feature, along with those given in Theorem 2.2.1, make $v$ especially useful for the purpose of transforming systems of OPS's into systems of OLPS's. Inspection of the Transformation Theorem shows that the L-polynomial $\widetilde{P}_{2n}$ is a doubling of the polynomial $P_n$, and, in a slightly looser sense, $\widetilde{\psi}$ with spectrum $\sigma(\widetilde{\psi}) = v_-^{-1}(\sigma(\psi)) \cup v_+^{-1}(\sigma(\psi))$ is a doubling of $\psi$ with spectrum $\sigma(\psi)$.

The doubling of the spectrum in particular can be used to discuss to what extent $\widetilde{\psi}$ is an extension of $\psi$. For example, if $\sigma(\psi)$ is a symmetric set about the origin, it can be seen by Theorem 2.3.1 (B) and the definitions of $v_-^{-1}$ and $v_+^{-1}$ that $\sigma(\widetilde{\psi})$ is symmetric about the origin. When $\sigma(\psi)$ is a symmetric interval about the origin, $\sigma(\widetilde{\psi})$ is the union of two disjoint intervals forming a set symmetric about the origin. In particular, if $\sigma(\psi) = \mathbf{R}$, then $\sigma(\widetilde{\psi}) = \mathbf{R}^- \cup \mathbf{R}^+$. If $\sigma(\psi) \subseteq \mathbf{R}_0^+ = [0, \infty)$, we would like an extension of $\psi$ to a SMDF to have its spectrum contained in $\mathbf{R}^+$. However, a direct application of the Transformation Theorem to a MDF $\psi$ having spectrum $\sigma(\psi) \subseteq \mathbf{R}_0^+$ yields the SMDF $\widetilde{\psi}$ with its doubled spectrum $\sigma(\widetilde{\psi}) = v_-^{-1}(\sigma(\psi)) \cup v_+^{-1}(\sigma(\psi))$ half contained in $\mathbf{R}^-$. Similarly, if $\sigma(\psi) \subseteq \mathbf{R}_0^- = (-\infty, 0]$, then $\sigma(\widetilde{\psi})$ is half contained in $\mathbf{R}^+$.

In a further effort to discover to what extent the transformed objects given by the Transformation Theorem are extensions of the corresponding original objects, it is worth examining the limiting case of $\lambda = 1$ and $\gamma = 0$. In this situation, it can be seen by inspection of the definitions that $v(x) = x$ and $v_\pm^{-1}(x) = xI_{\mathbf{R}^\pm}(x)$, where $I_A(x)$ is the indicator function for a set $A$. Hence, in this limiting case, we see that $\widetilde{P}_{2n}(x) = P_n(x)$, $v_{\pm}^{-1}(\sigma(\psi)) = \sigma(\psi) \cap \mathbf{R}_0^{\pm}$ and $\sigma(\widetilde{\psi}) = \sigma(\psi)$.

## 3 OLPS'S OF JACOBI, HERMITE AND LAGUERRE TYPES

For a detailed treatment of the consequences of the Tranformation Theorem, see [13]. For an example of the expositions presented there, see the study of moments in [14]. Here, we will content ourselves with the application of the Transformation Theorem to several systems with MDF's given by weight functions. If $\psi$ is an MDF which is differentiable, then $w(x) = \frac{d\psi}{dx}$ is called the weight function for $\psi$. Similarly, if $\phi$ is an SMDF which is differentiable, then $\omega(x) = \frac{d\phi}{dx}$ is called the weight function for $\phi$. If $\psi$ is an MDF, and if $\psi$ is differentiable with $\frac{d\psi}{dx} = w(x)$, then $\widetilde{\psi}$ is differentiable with $\frac{d\widetilde{\psi}}{dx} = w(v(x))$ (see [3]).

### 3.1 The Jacobi Class

#### The General Class

The monic *Jacobi polynomials* of parameters $\alpha > -1$ and $\beta > -1$ are denoted by $\widehat{P}_n^{(\alpha,\beta)}(x)$ and can be defined by the explicit formula

$$\widehat{P}_n^{(\alpha,\beta)}(x) = \binom{2n+\alpha+\beta}{n}^{-1} \sum_{k=0}^{n} \binom{n+\alpha}{n-k}\binom{n+\beta}{k}(x-1)^k(x+1)^{n-k},$$
$$n = 0, 1, 2, \ldots, \qquad (3.1)$$

([3], equations (2.6) and (2.7), p. 144). The Jacobi MDF we denote by $\psi_P^{(\alpha,\beta)}$. It is given by

$$\frac{d\psi_P^{(\alpha,\beta)}}{dx} := \begin{cases} (1-x)^\alpha(1+x)^\beta, & \text{if } x \in (-1,1) \\ 0, & \text{otherwise} \end{cases} \qquad (3.2)$$

To be explicit, the orthogonality relation is

$$\begin{aligned}(\widehat{P}_m^{(\alpha,\beta)}, \widehat{P}_n^{(\alpha,\beta)})_{\psi_P^{(\alpha,\beta)}} &= \int_{-1}^{1} \widehat{P}_m^{(\alpha,\beta)}(x)\widehat{P}_n^{(\alpha,\beta)}(x)(1-x)^\alpha(1+x)^\beta\,dx \\ &= 2^{2n+\alpha+\beta+1}\binom{2n+\alpha+\beta}{n}^{-1} B(n+\alpha+1, n+\beta+1)\,\delta_{mn}\end{aligned} \qquad (3.3)$$

(see the discussion in [3] beginning at the bottom of page 146 and ending on the next page) where $B$ denotes the beta function, which can be given in terms of the gamma function as

$$B(x,y) = \frac{\Gamma(x)\Gamma(y)}{\Gamma(x+y)},$$

and where $\delta_{mn}$ is the Kronecker delta,

$$\delta_{mn} = \begin{cases} 0, & \text{if } m \neq n \\ 1, & \text{if } m = n \end{cases}.$$

To facilitate the reporting of the results of applying the Transformation Theorem and its consequences to the Jacobi class of systems of orthogonal polynomials, we find it useful to make the following definitions:

$$a := v_+^{-1}(-1) = \frac{\lambda}{2}\left(-1+\sqrt{1+\frac{4\gamma}{\lambda^2}}\right)$$

*and*

$$b := v_+^{-1}(1) = \frac{\lambda}{2}\left(1+\sqrt{1+\frac{4\gamma}{\lambda^2}}\right),$$

where we continue to maintain the assumption made at the beginning of Section 2 that $\lambda$ and $\gamma$ are fixed positive real numbers. Thus, we have assumed that $a$ and $b$ are the unique fixed positive real numbers having the following properties:

$$-a = -v_+^{-1}(-1) = \frac{\lambda}{2}\left(1-\sqrt{1+\frac{4\gamma}{\lambda^2}}\right) = v_-^{-1}(1),$$

$$-b = -v_+^{-1}(1) = \frac{\lambda}{2}\left(-1-\sqrt{1+\frac{4\gamma}{\lambda^2}}\right) = v_-^{-1}(-1),$$

$$\lambda = b - a$$

*and*

$$\gamma = ab.$$

Using the Transformation Theorem and equation (3.2), we find that $\widetilde{\psi}_P^{(\alpha,\beta)}(x)$, given by

$$\frac{d\widetilde{\psi}_P^{(\alpha,\beta)}}{dx} = \begin{cases} \frac{|x+a|^\alpha |b-x|^\alpha |x-a|^\beta |b+x|^\beta}{(b-a)^{\alpha+\beta}|x|^{\alpha+\beta}}, & \text{if } x \in (-b,-a)\cup(a,b) \\ 0, & \text{if } x \in (-\infty,-b]\cup[-a,0)\cup(0,a]\cup[b,\infty) \end{cases}, \tag{3.4}$$

is a SMDF for each choice of parameters. The monic OLPS with respect to $\widetilde{\psi}_P^{(\alpha,\beta)}$ we denote by $\{\widetilde{P}_m^{(\alpha,\beta)}(x)\}_{m=0}^{\infty}$. By Theorem 2.3.1 (C), formula (3.1) and algebra, we find the explicit formulas

$$\widetilde{P}_{2n}^{(\alpha,\beta)}(x) = \frac{1}{\binom{2n+\alpha+\beta}{n}}\frac{1}{x^n}\sum_{k=0}^{n}\binom{n+\alpha}{n-k}\binom{n+\beta}{k}(x+a)^k(x-b)^k(x-a)^{n-k}(x+b)^{n-k},$$
$$n = 0,1,2,\ldots, \tag{3.5}$$

*and*

$$\widetilde{P}_{2n+1}^{(\alpha,\beta)}(x) =$$

$$\frac{(-1/ab)^n}{\binom{2n+\alpha+\beta}{n}} \frac{1}{x^{n+1}} \sum_{k=0}^{n} \binom{n+\alpha}{n-k} \binom{n+\beta}{k} (x+a)^k (x-b)^k (x-a)^{n-k} (x+b)^{n-k},$$

$$n = 0, 1, 2, \ldots. \tag{3.6}$$

Considering Theorem 2.2.8 and equation (3.3), the orthogonality relation is

$$(\widetilde{P}_m^{(\alpha,\beta)}, \widetilde{P}_n^{(\alpha,\beta)})_{\widetilde{\psi}_P^{(\alpha,\beta)}} = k_m \, \delta_{mn}, \tag{3.7}$$

where

$$k_m =$$

$$\begin{cases} (b-a)^{2j+1} 2^{2j+\alpha+\beta+1} \binom{2j+\alpha+\beta}{j}^{-1} B(j+\alpha+1, j+\beta+1), & \text{if } m = 2j \\ (\frac{b-a}{ab})^{2j+1} 2^{2j+\alpha+\beta+1} \binom{2j+\alpha+\beta}{j}^{-1} B(j+\alpha+1, j+\beta+1), & \text{if } m = 2j+1 \end{cases} \tag{3.8}$$

### The Tchebycheff Polynomials of the First Kind

These polynomials can be defined by

$$\widehat{T}_n(x) = \widehat{P}_n^{(-1/2,-1/2)}(x), \; n = 0, 1, 2, \ldots, \tag{3.9}$$

hence the Tchebycheff polynomials of the first kind are an example from the class of Jacobi polynomials with $\alpha = \beta = -1/2$.

$\{\widehat{T}_n(x)\}_{n=0}^{\infty}$ is the monic OPS with respect to the MDF $\psi_T$ given by

$$\frac{d\psi_T}{dx} = \begin{cases} \frac{1}{\sqrt{1-x^2}}, & \text{if } x \in (-1,1) \\ 0, & \text{otherwise} \end{cases} ; \tag{3.10}$$

To be specific, the orthogonality relation is

$$\begin{aligned} (\widehat{T}_m, \widehat{T}_n)_{\psi_T} &= \int_{-1}^{1} \widehat{T}_m(x) \widehat{T}_n(x) \frac{1}{\sqrt{1-x^2}} \, dx \\ &= k_n \, \delta_{mn}, \end{aligned} \tag{3.11}$$

where

$$k_n = \begin{cases} \pi, & \text{if } n = 0 \\ 2^{1-2n} \, \pi, & \text{if } n \geq 1 \end{cases} \tag{3.12}$$

Applying the Transformation Theorem with $\lambda = b - a$ and $\gamma = ab$ to the monic Tchebycheff polynomials of the first kind results in a monic orthogonal Laurent polynomial sequence $\{\widetilde{T}_m(x)\}_{m=0}^{\infty}$ which can be defined by

$$\widetilde{T}_m(x) = \widetilde{P}_m^{(-1/2,-1/2)}(x), \; m = 0, 1, 2, \ldots . \tag{3.13}$$

These L-polynomials, examined by Cooper and Gustafson, [5], are orthogonal with respect to the SMDF, which we denote by $\widetilde{\psi}_T$, given by

$$\frac{d\widetilde{\psi}_T}{dx} = \begin{cases} \frac{(b-a)\,|x|}{\sqrt{b^2-x^2}\sqrt{x^2-a^2}}, & \text{if } x \in (-b,-a) \cup (a,b) \\ 0, & \text{if } x \in (-\infty,-b] \cup [-a,0) \cup (0,a] \cup [b,\infty) \end{cases} . \tag{3.14}$$

Considering (3.11), (3.12) and Theorem 2.2.8, the orthogonality relation is

$$\begin{aligned} (\widetilde{T}_m, \widetilde{T}_n)_{\widetilde{\psi}_T} &= \int_{-b}^{-a} \widetilde{T}_m(x)\widetilde{T}_n(x) \frac{(b-a)\,|x|}{\sqrt{b^2-x^2}\sqrt{x^2-a^2}}\,dx \\ &\quad + \int_a^b \widetilde{T}_m(x)\widetilde{T}_n(x) \frac{(b-a)\,|x|}{\sqrt{b^2-x^2}\sqrt{x^2-a^2}}\,dx \\ &= k_m\,\delta_{mn} \end{aligned} \tag{3.15}$$

where

$$k_m = \begin{cases} (b-a)\pi, & \text{if } m = 0 \\ (b-a)^{2j+1}\,2^{1-2j}\,\pi, & \text{if } m = 2j \geq 2\,. \\ (\frac{b-a}{ab})^{2j+1}\,2^{1-2j}\,\pi, & \text{if } m = 2j+1 \end{cases} \tag{3.16}$$

### The Legendre Polynomials

The monic *Legendre polynomials* we denote by $\widehat{P}_n(x)$, $n = 0, 1, 2, \ldots$. These polynomials can be defined by

$$\widehat{P}_n(x) = \widehat{P}_n^{(0,0)}(x), \; n = 0, 1, 2, \ldots, \tag{3.17}$$

hence they are an instance of the Jacobi classes of polynomials. The monic Legendre polynomials thus form an OPS with respect to an MDF, which we denote by $\psi_P$, given by

$$\frac{d\psi_P}{dx} = \begin{cases} 1 & , \text{if } x \in (-1,1) \\ 0 & , \text{otherwise} \end{cases} . \tag{3.18}$$

The orthogonality relation is

$$(\widehat{P}_m, \widehat{P}_n)_{\psi_P} = \int_{-1}^{1} \widehat{P}_m(x)\widehat{P}_n(x)\,dx = \frac{2^{2n+1}(n!)^4}{(2n)!\,(2n+1)!}\delta_{mn}\,. \tag{3.19}$$

Applying the Transformation Theorem, with $\lambda = b - a$ and $\gamma = ab$, to the system of monic Legendre polynomials results in a system of orthogonal Laurent polynomials. We denote the L-polynomials by $\widetilde{P}_m(x)$.

$$\widetilde{P}_m(x) = \widetilde{P}_m^{(0,0)}(x), \; m = 0, 1, 2, \ldots . \tag{3.20}$$

We denote the resulting SMDF by $\widetilde{\psi}_P$, where

$$\frac{d\widetilde{\psi}_P}{dx} = \begin{cases} 1 & , \text{if } x \in (-b,-a) \cup (a,b) \\ 0 & , \text{if } x \in (-\infty,-b] \cup [-a,0) \cup (0,a] \cup [b,\infty) \end{cases}. \tag{3.21}$$

By Theorem 2.2.8 and equation (3.19), we see that the orthogonality relation is

$$(\widetilde{P}_m, \widetilde{P}_n)_{\widetilde{\psi}_P} = \int_{-b}^{-a} \widetilde{P}_m(x)\widetilde{P}_n(x)\,dx + \int_a^b \widetilde{P}_m(x)\widetilde{P}_n(x)\,dx = k_m\,\delta_{mn} \tag{3.22}$$

where

$$k_m = \begin{cases} (b-a)^{2j+1} \frac{2^{2j+1}(j!)^4}{(2j)!\,(2j+1)!}, & \text{if } m = 2j \\ \left(\frac{b-a}{ab}\right)^{2j+1} \frac{2^{2j+1}(j!)^4}{(2j)!\,(2j+1)!}, & \text{if } m = 2j+1 \end{cases}. \tag{3.23}$$

## 3.2 The Generalized Hermite Class

### The General Class

We denote the monic *generalized Hermite polynomials* of parameter $\alpha > -1/2$ by $\widehat{H}_n^{(\alpha)}(x)$, $n = 0, 1, 2, \ldots$.

$$\widehat{H}_{2k}^{(\alpha)}(x) = (-1)^k\,k! \sum_{j=0}^{k} \binom{k+\alpha-1/2}{k-j} \frac{(-1)^j}{j!}\,x^{2j},\ k = 0,1,2,\ldots, \tag{3.24}$$

and

$$\widehat{H}_{2k+1}^{(\alpha)}(x) = (-1)^k\,k! \sum_{j=0}^{k} \binom{k+\alpha+1/2}{k-j} \frac{(-1)^j}{j!}\,x^{2j+1},\ k = 0,1,2,\ldots, \tag{3.25}$$

(see [3], (2.43), p. 156, and (2.11), p. 145). $\{\widehat{H}_n^{(\alpha)}(x)\}_{n=0}^{\infty}$ is the monic OPS with respect to the MDF $\psi_H^{(\alpha)}$ which is given by

$$\frac{d\psi_H^{(\alpha)}}{dx} = |x|^{2\alpha}\,e^{-x^2},\ x \in \mathbf{R}. \tag{3.26}$$

The orthogonality relation is

$$\begin{aligned}(\widehat{H}_m^{(\alpha)}, \widehat{H}_n^{(\alpha)})_{\psi_H^{(\alpha)}} &= \int_{-\infty}^{\infty} \widehat{H}_m^{(\alpha)}(x)\widehat{H}_n^{(\alpha)}(x)|x|^{2\alpha}e^{-x^2}\,dx \\ &= \left[\frac{n}{2}\right]!\,\Gamma\left(\left[\frac{n+1}{2}\right] + \alpha + \frac{1}{2}\right)\delta_{mn}\end{aligned} \tag{3.27}$$

([3], eq. (2.45), p. 157), where $[z]$ denotes the integer part of $z$.

For each choice of $\alpha > -1/2$, the Transformation Theorem applied to the system of monic generalized Hermite polynomials of parameter $\alpha$ results in a system of monic orthogonal Laurent polynomials, for each choice of parameters $\lambda > 0$ and $\gamma > 0$. The L-polynomials, which we denote by $\widetilde{H}_m^{(\alpha)}(x)$, have the explicit expressions

$$\widetilde{H}_{4k}^{(\alpha)}(x) = (-1)^k\ k!\ \lambda^{2k} \frac{1}{x^{2k}} \sum_{j=0}^{k} \binom{k+\alpha-1/2}{k-j} \frac{(-1)^j}{j!\ \lambda^{2j}}\ (x^2-\gamma)^{2j}\ x^{2(k-j)}, \quad k = 0,1,2,\ldots, \tag{3.28}$$

$$\widetilde{H}_{4k+1}^{(\alpha)}(x) = (-1)^k\ k! \left(\frac{\lambda}{\gamma}\right)^{2k} \frac{1}{x^{2k+1}} \sum_{j=0}^{k} \binom{k+\alpha-1/2}{k-j} \frac{(-1)^j}{j!\ \lambda^{2j}}\ (x^2-\gamma)^{2j}\ x^{2(k-j)}, \quad k = 0,1,2,\ldots, \tag{3.29}$$

$$\widetilde{H}_{4k+2}^{(\alpha)}(x) = (-1)^k\ k!\ \lambda^{2k+1} \frac{1}{x^{2k+1}} \sum_{j=0}^{k} \binom{k+\alpha+1/2}{k-j} \frac{(-1)^j}{j!\ \lambda^{2j+1}}\ (x^2-\gamma)^{2j+1}\ x^{2(k-j)}, \quad k = 0,1,2,\ldots, \tag{3.30}$$

*and*

$$\widetilde{H}_{4k+3}^{(\alpha)}(x) = (-1)^k\ k! \left(\frac{\lambda}{\gamma}\right)^{2k+1} \frac{1}{x^{2(k+1)}} \sum_{j=0}^{k} \binom{k+\alpha+1/2}{k-j} \frac{(-1)^{j+1}}{j!\ \lambda^{2j+1}}\ (x^2-\gamma)^{2j+1}\ x^{2(k-j)}, \quad k = 0,1,2,\ldots, \tag{3.31}$$

given by equations (3.24) and (3.25) and Theorem 2.3.1 (C). According to the Transformation Theorem and equation (3.26), $\{\widetilde{H}_m^{(\alpha)}(x)\}_{m=0}^{\infty}$ is the monic OLPS with respect to a SMDF, which we denote by $\widetilde{\psi}_H^{(\alpha)}$, defined by

$$\frac{d\widetilde{\psi}_H^{(\alpha)}}{dx} = \frac{1}{\lambda^{2\alpha}} \left|x^2 - 2\gamma + \frac{\gamma^2}{x^2}\right|^{\alpha} exp\left(-\frac{1}{\lambda^2}\left(x^2 - 2\gamma + \frac{\gamma^2}{x^2}\right)\right),\ x \in \mathbf{R}^{-} \cup \mathbf{R}^{+}. \tag{3.32}$$

By (3.26), (3.27) and Theorem 2.2.8, the orthogonality relation is

$$(\widetilde{H}_m^{(\alpha)}, \widetilde{H}_n^{(\alpha)})_{\widetilde{\psi}_H^{(\alpha)}} = k_m\ \delta_{mn}\ , \tag{3.33}$$

where

$$k_m = \begin{cases} \lambda^{2j+1}\left(\left[\frac{j}{2}\right]!\right)\ \Gamma\left(\left[\frac{j+1}{2}\right]+\alpha+\frac{1}{2}\right), & \text{if } m = 2j \\ \left(\frac{\lambda}{\gamma}\right)^{2j+1}\left(\left[\frac{j}{2}\right]!\right)\ \Gamma\left(\left[\frac{j+1}{2}\right]+\alpha+\frac{1}{2}\right), & \text{if } m = 2j+1 \end{cases}. \tag{3.34}$$

### The Hermite Polynomials

The monic *Hermite polynomials*, which we denote by $\widehat{H}_n(x)$, are the monic generalized Hermite polynomials of parameter $\alpha = 0$:

$$\widehat{H}_n(x) = \widehat{H}_n^{(0)}(x),\ n = 0,1,2,\ldots. \tag{3.35}$$

$\{\widehat{H}_n(x)\}_{n=0}^{\infty}$ is the monic OPS with respect to the MDF $\psi_H$ which is given by

$$\frac{d\psi_H}{dx} = e^{-x^2}, \quad x \in \mathbf{R}. \tag{3.36}$$

The orthogonality relation is

$$(\widehat{H}_m, \widehat{H}_n)_{\psi_H} = \int_{-\infty}^{\infty} \widehat{H}_m(x)\widehat{H}_n(x)\, e^{-x^2}\, dx = \frac{n!}{2^n}\sqrt{\pi}\, \delta_{mn}\, , \tag{3.37}$$

We apply the Transformation Theorem to obtain a system of orthogonal Laurent polynomials. We denote the L-polynomials by $\widetilde{H}_m(x)$, and, considering (3.35),

$$\widetilde{H}_m(x) = \widetilde{H}_m^{(0)}(x), \; m = 0, 1, 2, \ldots. \tag{3.38}$$

$\{\widetilde{H}_m(x)\}_{m=0}^{\infty}$ is the monic OLPS with respect to the SMDF $\widetilde{\psi}_H = \widetilde{\psi}_H^{(0)}$, given by

$$\frac{d\widetilde{\psi}_H}{dx} = e^{-\frac{1}{\lambda^2}\left(x^2 - 2\gamma + \frac{\gamma^2}{x^2}\right)}, \; x \in \mathbf{R}^- \cup \mathbf{R}^+, \tag{3.39}$$

according to (3.32) with $\alpha = 0$. The orthogonality relation is

$$\begin{aligned}(\widetilde{H}_m, \widetilde{H}_n)_{\widetilde{\psi}_H} &= \int_{-\infty}^{0} \widetilde{H}_m(x)\widetilde{H}_n(x)\, e^{-\frac{1}{\lambda^2}\left(x^2 - 2\gamma + \frac{\gamma^2}{x^2}\right)}\, dx \\ &\quad + \int_{0}^{\infty} \widetilde{H}_m(x)\widetilde{H}_n(x)\, e^{-\frac{1}{\lambda^2}\left(x^2 - 2\gamma + \frac{\gamma^2}{x^2}\right)}\, dx \\ &= k_m\, \delta_{mn}\, ,\end{aligned} \tag{3.40}$$

where

$$k_m = \begin{cases} \lambda^{2j+1} \frac{j!}{2^j} \sqrt{\pi}, & \text{if } m = 2j \\ \left(\frac{\lambda}{\gamma}\right)^{2j+1} \frac{j!}{2^j} \sqrt{\pi}. & \text{if } m = 2j+1 \end{cases}. \tag{3.41}$$

## 3.3 The Generalized Laguerre Class

The systems of orthogonal polynomials associated with the names of Jacobi, Hermite, and Laguerre are collectively called the *classical orthogonal polynomials*. In this section we finish our treatment of applying the Tranformation Theorem and some of its consequences to the systems of classical polynomials by considering those polynomials associated with Laguerre.

### A Direct Application of the Transformation Theorem

The monic *Sonine-Laguerre* or *generalized Laguerre polynomials* of parameter $\alpha > -1$ are denoted by $\widehat{L}_n^{(\alpha)}(x)$. These polynomials can be defined by the explicit expressions

$$\widehat{L}_n^{(\alpha)}(x) = (-1)^n n! \sum_{k=0}^{n} \binom{n+\alpha}{n-k} \frac{(-x)^k}{k!}, \; n = 0, 1, 2, \ldots \tag{3.42}$$

([3], (2.11) and (2.12), p. 154), and they are orthogonal with respect to the MDF $\psi_L^{(\alpha)}$, given by

$$\frac{d\psi_L^{(\alpha)}}{dx} = \begin{cases} x^\alpha e^{-x}, & \text{if } x \in \mathbf{R}^+ \\ 0, & \text{otherwise} \end{cases}. \tag{3.43}$$

The orthogonlity relation is

$$\left(\widehat{L}_m^{(\alpha)}, \widehat{L}_n^{(\alpha)}\right)_{\psi_L^{(\alpha)}} = \int_0^\infty \widehat{L}_m^{(\alpha)}(x) \widehat{L}_n^{(\alpha)}(x) \; x^\alpha e^{-x} \, dx = n! \, \Gamma\left(n + \alpha + 1\right) \delta_{mn} \tag{3.44}$$

([3], (2.18), p. 148, and (2.12), p. 145).

As mentioned at the end of Section 2, applying the Transformation Theorem to an MDF results in an SMDF, for each choice of $\lambda > 0$ and $\gamma > 0$, with part of the spectrum in the negative reals and part in the positive reals. This raises serious doubts as to the appropriateness of the resulting SMDF as an analogue to the corresponding MDF whose spectrum is contained, say, in the non-negative reals. One may consider such MDF's as examples of the limitations of this transformation. The generalized Laguerre class here provides a collection of such MDF's, since we see that the resulting SMDF, which we denote dy $\widetilde{\psi}_L^{(\alpha)}$, is given by

$$\frac{d\widetilde{\psi}_L^{(\alpha)}}{dx} = \begin{cases} \left(\frac{1}{\lambda}(x - \frac{\gamma}{x})\right)^\alpha e^{-\frac{1}{\lambda}(x - \frac{\gamma}{x})}, & \text{if } x \in (-\sqrt{\gamma}, 0) \cup (\sqrt{\gamma}, \infty) \\ 0, & \text{if } x \in (-\infty, -\sqrt{\gamma}\,] \cup (0, \sqrt{\gamma}\,] \end{cases}, \tag{3.45}$$

according to (3.43) and the Transformation Theorem. It is evident from the definitions that the spectrum $\sigma(\widetilde{\psi}_L^{(\alpha)})$ is $[-\sqrt{\gamma}, 0) \cup [\sqrt{\gamma}, \infty)$, while $\sigma(\psi_L^{(\alpha)}) = [0, \infty)$. We will return to the problem of Laguerre-like strong moment distributions at the end of this section.

Regardless of our views about the question of appropriateness in this case, we can proceed to transfer the formulas for the monic generalized Laguerre polynomials of parameter $\alpha > -1$ over to the corresponding OLPS given by the Transformation Theorem. We denote the L-polynomials by $\widetilde{L}_m^{(\alpha)}(x)$, and we find the explicit expressions

$$\widetilde{L}_{2n}^{(\alpha)}(x) = (-1)^n n! \, \lambda^n \frac{1}{x^n} \sum_{k=0}^{n} \binom{n+\alpha}{n-k} \frac{(-1)^k}{k! \lambda^k} (x^2 - \gamma)^k x^{n-k}, \; n = 0, 1, 2, \ldots, \tag{3.46}$$

*and*

$$\widetilde{L}_{2n+1}^{(\alpha)}(x) = n! \, \left(\frac{\lambda}{\gamma}\right)^n \frac{1}{x^{n+1}} \sum_{k=0}^{n} \binom{n+\alpha}{n-k} \frac{(-1)^k}{k! \lambda^k} (x^2 - \gamma)^k x^{n-k}, \; n = 0, 1, 2, \ldots, \tag{3.47}$$

by (3.42) and Theorem 2.3.1 (C). Considering (3.45), the inner products (3.44) and Theorem 2.2.8, the orthogonality relation is

$$\begin{aligned}(\widetilde{L}_m^{(\alpha)}, \widetilde{L}_n^{(\alpha)})_{\widetilde{\psi}_L^{(\alpha)}} &= \int_{-\sqrt{\gamma}}^{0} \widetilde{L}_m^{(\alpha)}(x)\widetilde{L}_n^{(\alpha)}(x) \left(\frac{1}{\lambda}(x - \frac{\gamma}{x})\right)^{\alpha} e^{-\frac{1}{\lambda}(x-\frac{\gamma}{x})}\, dx \\ &\quad + \int_{\sqrt{\gamma}}^{\infty} \widetilde{L}_m^{(\alpha)}(x)\widetilde{L}_n^{(\alpha)}(x) \left(\frac{1}{\lambda}(x - \frac{\gamma}{x})\right)^{\alpha} e^{-\frac{1}{\lambda}(x-\frac{\gamma}{x})}\, dx \\ &= k_m\, \delta_{mn}\ ,\end{aligned} \tag{3.48}$$

where

$$k_m = \begin{cases} \lambda^{2j+1}\ (\alpha+1)_j\ \Gamma(\alpha+1), & \text{if } m = 2j \\ (\frac{\lambda}{\gamma})^{2j+1}\ (\alpha+1)_j\ \Gamma(\alpha+1), & \text{if } m = 2j+1 \end{cases}. \tag{3.49}$$

## A Laguerre Class Related to the Generalized Hermite Class

Let $\alpha > -1/2$ and define $\{A_m^{(\alpha)}(x)\}_{m=0}^{\infty}$ by

$$A_{2n}^{(\alpha)}(x^2) := \widetilde{H}_{4n}^{(\alpha)}(x),\ n = 0, 1, 2, \ldots, \tag{3.50}$$

*and*

$$A_{2n+1}^{(\alpha)}(x^2) := \widetilde{H}_{4n+3}^{(\alpha)}(x),\ n = 0, 1, 2, \ldots. \tag{3.51}$$

By (3.28),

$$A_{2n}^{(\alpha)}(x) = (-1)^n\ n!\ \lambda^{2n} \frac{1}{x^n} \sum_{j=0}^{n} \binom{n+\alpha-1/2}{n-j} \frac{(-1)^j}{j!\ \lambda^{2j}}\ (x-\gamma)^{2j}\ x^{n-j}, \quad n = 0, 1, 2, \ldots, \tag{3.52}$$

and by (3.31),

$$A_{2n+1}^{(\alpha)}(x) = (-1)^{n+1}\ n! \left(\frac{\lambda}{\gamma}\right)^{2n+1} \frac{1}{x^{n+1}} \sum_{j=0}^{n} \binom{n+\alpha+1/2}{n-j} \frac{(-1)^j}{j!\ \lambda^{2j+1}}\ (x-\gamma)^{2j+1}\ x^{n-j}, \quad n = 0, 1, 2, \ldots. \tag{3.53}$$

Inspection of (3.52) and (3.53) shows that $A_m^{(\alpha)}(x)$ is a monic L-polynomial of L-degree $m$, for each $m = 0, 1, 2, \ldots$.

Next, define $\phi_A^{(\alpha)}(x)$ by

$$\frac{d\phi_A^{(\alpha)}}{dx} := \begin{cases} \frac{1}{\lambda^{2\alpha}}\ x^{-\frac{1}{2}} \left|x - 2\gamma + \frac{\gamma^2}{x}\right|^{\alpha}\ exp\left(-\frac{1}{\lambda^2}\left(x - 2\gamma + \frac{\gamma^2}{x}\right)\right), & \text{if } x \in \mathbf{R}^+ \\ 0, & \text{if } x \in \mathbf{R}^- \end{cases}. \tag{3.54}$$

Considering (3.54), $\phi_A^{(\alpha)}$ is a SMDF with spectrum $\sigma(\phi_A^{(\alpha)}) = \mathbf{R}^+$, whose moments are

$$\mu_l(\phi_A^{(\alpha)}) = \mu_{2l}(\widetilde{\psi}_H^{(\alpha)}),\ l = 0, \pm 1, \pm 2, \ldots, \tag{3.55}$$

by the change of variables $x^2 \to x$ in

$$\mu_{2l}(\widetilde{\psi}_H^{(\alpha)}) = 2\int_0^\infty x^{2l}\frac{1}{\lambda^{2\alpha}}\left|x^2 - 2\gamma + \frac{\gamma^2}{x^2}\right|^\alpha e^{-\frac{1}{\lambda^2}\left(x^2-2\gamma+\frac{\gamma^2}{x^2}\right)}\,dx\ .$$

If $k, l, m$ and $n$ are non-negative integers such that $A_m^{(\alpha)}(x^2) = \widetilde{H}_k^{(\alpha)}(x)$ and $A_n^{(\alpha)}(x^2) = \widetilde{H}_l^{(\alpha)}(x)$, then, by linearity of the integral and (3.55),

$$(A_m^{(\alpha)}, A_n^{(\alpha)})_{\phi_A^{(\alpha)}} = (\widetilde{H}_k^{(\alpha)}, \widetilde{H}_l^{(\alpha)})_{\widetilde{\psi}_H^{(\alpha)}}\ . \tag{3.56}$$

Hence, $\{A_m^{(\alpha)}(x)\}_{m=0}^\infty$ is the monic OLPS with respect to $\phi_A^{(\alpha)}$, the orthogonality relation being

$$(A_m^{(\alpha)}, A_n^{(\alpha)})_{\widetilde{\psi}_A^{(\alpha)}} = k_m\ \delta_{mn}\ , \tag{3.57}$$

where

$$k_m = \begin{cases} \lambda^{4j+1}(j!)\ \Gamma\left(\left[\frac{2j+1}{2}\right] + \alpha + \frac{1}{2}\right), & \text{if } m = 2j \\ \left(\frac{\lambda}{\gamma}\right)^{4j+3}\left(\left[\frac{2j+1}{2}\right]!\right)\ \Gamma\left(j + \alpha + \frac{3}{2}\right), & \text{if } m = 2j+1 \end{cases}. \tag{3.58}$$

## REFERENCES


1. P. Barrucand and D. Dickinson. On Cubic Transformations of Orthogonal Polynomials. Proc. Amer. Math. Soc. 17:810-814, 1966.
2. P. Barrucand and P. Moussa. Orthogonal Properties of Iterated Polynomial Mappings. Comm. Math. Phys. 88:503-529, 1983.
3. T. S. Chihara. An Introduction to Orthogonal Polynomials: Gordon and Breach, 1978.
4. L. Cochran and S. Clement Cooper. Orthogonal Laurent Polynomials on the Real Line. Continued Fractions and Orthogonal Functions: Theory and Applications. Proceedings Loen, Norway, 1992 (S. Clement Cooper and W. J, Thron, eds.), Lecture Notes in Pure and Applied Mathematics: Marcel Dekker, 1993, pp 47-100.
5. S. Clement Cooper and Philip E. Gustafson. The Strong Tchebycheff Distribution and Orthogonal Laurent Polynomials. Journal of Approximation Theory, to appear.
6. S. Clement Cooper, William B. Jones and W. J. Thron. Orthogonal Laurent Polynomials and Continued Fractions Associated with Log-Nornal Distributions. J CAM 32:39-46, 1990.
7. J. S. Geronimo and W. Van Assche. Orthogonal Polynomials on Several Intervals Via a Polynomial Mapping. Trans. Amer. Math. Soc. 308:559-581, 1988.
8. E. Hendriksen and H. van Rossum. Orthogonal Laurent Polynomials. Proceedings of the Koninklijke Nedelandse Akademie van Wetenschappen 89(1):17-36, March 24, 1986.

9. William B. Jones, W. J. Thron and H. Waadeland. A Strong Stieltjes Moment Problem. Trans. Amer. Math. Soc. 261:503-528, 1980.
10. O. Njåstad and W. J. Thron. The Theory of Sequences of Orthogonal L-polynomials. Det Kongelige Norske Videnskabers Selskab 1:54-91, 1983.
11. A. Sri Ranga and J. H. McCabe. On the Extensions of Some Classical Distributions. Proc. of the Edinburgh Math. Soc. 34:19-29, 1991.
12. Gabor Szegő. Orthogonal Polynomials. Providence, RI: AMS Colloquim Publications, Vol. 23, 4th Ed., 1975.
13. Brian A. Hagler. A Transformation of Orthogonal Polynomial Sequences into Orthogonal Laurent Polynomial Sequences. Ph.D. thesis, University of Colorado, 1997.
14. Brian A. Hagler. Formulas for the Moments of Some Strong Moment Distribution Functions. These Proceedings.

# Regular Strong Hamburger Moment Problem

WILLIAM B. JONES† and GUOXIANG SHEN* Department of Mathematics, University of Colorado, Boulder, CO 80309–0395, U.S.A.

## 1. INTRODUCTION

Moment theory, orthogonal functions, linear functionals and integral transforms are found in many parts of twentieth century mathematics and its applications in mathematical physics, chemistry, statistics and engineering. The analytic theory of continued fractions plays a central role in both the origin and the development of these closely related topics. Moment theory has been developed by using tools from different mathematical areas including: (a) continued fractions, (b) orthogonal polynomials (and Laurent polynomials) and (c) functional analysis. The present paper is an expository survey of the theory of regular, strong Hamburger moment problems developed by means of continued fractions called APT-fractions (alternating, positive-term continued fractions). Although connections with orthogonal Laurent polynomials are pointed out, no essential use is made of orthogonal functions or of quadrature formulas.

The principal results on regular strong moment problems are given in Section 5. These include: (a) conditions to ensure the existence of a solution (Theorem 5.1), (b) conditions to ensure the uniqueness of the solution (Theorem 5.2) and (c) a method for characterizing all solutions to a given moment problem when more solutions than one exist (Theorem 5.3). An example giving explicit formulas used to characterize all solutions (case (c) above) is described in Section 6. The example treats the moment bisequence generated by the log-normal distribution function. This example is believed

†Research supported in part by the U.S. National Science Foundation under grant DMS–9302584.

*Research supported in part by the U.S. National Science Foundation under grant DMS–9701028.

to be the first non-trivial, indeterminate strong Hamburger moment problem for which one can give such explicit formulas to characterize all solutions. Sections 2, 3 and 4 are used to develop properties of APT-fractions used in Sections 5 and 6. Some basic definitions and remarks on strong moment problems are contained in the remainder of this introduction. Before going to those topics, we note that another survey article on strong moment theory is contained in this volume. It is entitled: *Remarks on Canonical Solutions of Strong Moment Problems*, by Olav Njåstad [19].

Let $S$ be a non-empty subset of $\mathbb{R}$ and let $\psi : S \to \mathbb{R}$. The *spectrum* $\sigma(\psi)$ of the function $\psi$ is defined by

$$\begin{aligned}\sigma(\psi) :=[&x \in S : \text{there exists an } \epsilon > 0 \text{ such that } (x-\varepsilon, x+\varepsilon) \subseteq S \\ &\text{and } \psi(x-\delta) < \psi(x+\delta) \text{ if } 0 < \delta < \varepsilon].\end{aligned} \tag{1.1}$$

Let $\mathbb{R}^* := \mathbb{R}^- \cup \mathbb{R}^+$, where $\mathbb{R}^- := [x \in \mathbb{R} : x < 0]$ and $\mathbb{R}^+ =: [x \in \mathbb{R} : x > 0]$. A function $\psi : \mathbb{R}^* \to \mathbb{R}$ is called a ***strong moment distribution function* (SMDF)** if $\psi$ is bounded and non-decreasing both on $\mathbb{R}^-$ and on $\mathbb{R}^+$, if the spectrum $\sigma(\psi)$ is an infinite set and if the Riemann–Stieltjes integrals

$$\mu_n(\psi) := \int_{-\infty}^{\infty} t^n d\psi(t), \quad \text{for } n \in \mathbb{Z} := [0, \pm 1, \pm 2, \dots] \tag{1.2}$$

are all convergent. Each integral in (1.2) is called the *nth moment with respect to* $\psi$ and is denoted by $\mu_n(\psi)$. We use the symbol $\Psi$ to denote the family of all SMDF's. The ***strong Hamburger moment problem* (SHMP)** *for a bisequence* $\{\mu_n\}_{n=-\infty}^{\infty}$ *of real numbers* is to find a $\psi \in \Psi$ such that

$$\mu_n = \mu_n(\psi) := \int_{-\infty}^{\infty} t^n d\psi(t), \quad \text{for all } n \in \mathbb{Z}. \tag{1.3}$$

If a function $\psi \in \Psi$ satisfies (1.3), then $\psi$ is called a *solution to the SHMP for* $\{\mu_n\}_{-\infty}^{\infty}$. Two solutions of a SHMP are not considered to be distinct if their difference is a constant at all points where that difference is continuous. A SHMP is called *determinate* if there is exactly one (distinct) solution; it is called *indeterminate* if there exist more solutions than one. A SHMP is called a ***strong Stieltjes moment problem* (SSMP)** if one imposes the additional condition that, for any solution $\psi$ one has $\sigma(\psi) \subseteq \mathbb{R}^+ = (0, \infty)$. A SHMP for a bisequence $\{\mu_n\}_{-\infty}^{\infty}$ is called *regular* if there exists an APT-fraction corresponding to pair $(L_0, L_\infty)$ of formal power series

$$L_0(z) := -\sum_{j=1}^{\infty} (-1)^j \mu_{-j} z^j \text{ and } L_\infty(z) := \sum_{j=0}^{\infty} (-1)^j \frac{\mu_j}{z^j}. \tag{1.4}$$

## 2 APT-FRACTIONS

### Continued Fractions

Let $\{a_n(z)\}$ and $\{b_n(z)\}$ be sequences of functions from a set $S \subseteq \mathbb{C}$ into $\mathbb{C}$, such that $a_n(z) \neq 0$ for $z \in S$. These functions generate a continued fraction (CF)

$$\langle\langle\{a_n(z)\}, \{b_n(z)\}\rangle, \{S_n(z,0)\}\rangle \tag{2.1}$$

by means of the linear fractional transformations ($\ell$.f.t.'s)

$$s_n(z,w) := \frac{a_n(z)}{b_n(z)+w}, \quad \text{for } n \in \mathbb{N}, \tag{2.2a}$$

and

$$S_1(z,w) := s_1(z,w), \ S_n(z,w) := S_{n-1}(z, s_n(z,w)), \ n \geq 2. \tag{2.2b}$$

For each $n \in \mathbb{N}$, $f_n(z) := S_n(z,0)$ is called the $n$th approximant and $a_n(z)$ and $b_n(z)$ are called the $n$th *elements* of (2.1). For convenience we denote the CF (2.1) by

$$\underset{j=1}{\overset{\infty}{K}} \left(\frac{a_j(z)}{b_j(z)}\right) \text{ or } \frac{a_1(z)}{b_1(z)} + \frac{a_2(z)}{b_2(z)} + \frac{a_3(z)}{b_3(z)} + \cdots, \tag{2.3}$$

where $K$ in the first symbol stands for the German word "Kettenbrüch"; in the second symbol the fraction lines are to be extended above the expressions to the right of them. In keeping with (2.3) we also write

$$S_n(z,w) = \frac{a_1(z)}{b_1(z)} + \frac{a_2(z)}{b_2(z)} + \cdots + \frac{a_n(z)}{b_n(z)+w} \tag{2.4}$$

and

$$f_n(z) := S_n(z,0) =: \underset{j=1}{\overset{n}{K}} \left(\frac{a_j(z)}{b_j(z)}\right) = \frac{a_1(z)}{b_1(z)} + \frac{a_2(z)}{b_2(z)} + \cdots + \frac{a_n(z)}{b_n(z)}. \tag{2.5}$$

The $n$th numerator $A_n(z)$ and $n$th denominator $B_n(z)$ are defined by the *second order linear difference equations (3-term recurrence relations)*

$$A_{-1}(z) \equiv 1, \ B_{-1}(z) \equiv 0, \ A_0(z) \equiv 0, B_0(z) \equiv 1, \tag{2.6a}$$

$$A_n(z) := b_n(z)A_{n-1}(z) + a_n(z)A_{n-2}(z), \quad n \in \mathbb{N} \tag{2.6b}$$

$$B_n(z) := b_n(z)B_{n-1}(z) + a_n(z)B_{n-2}(z), \quad n \in \mathbb{N} \tag{2.6c}$$

These functions satisfy the *determinant equations*

$$A_n(z)B_{n-1}(z) - A_{n-1}(z)B_n(z) = (-1)^{n-1}\prod_{j=1}^{n} a_j(z), \quad n \in \mathbb{N}, \tag{2.7}$$

as well as

$$S_n(z,w) = \frac{A_n(z) + wA_{n-1}(z)}{B_n(z) + wB_{n-1}(z)}, \quad n \in \mathbb{N} \tag{2.8a}$$

and

$$\frac{A_n(z)}{B_n(z)} = f_n(z) := S_n(z,0), \quad n \in \mathbb{N} \tag{2.8b}$$

Two CF's $\underset{j=1}{\overset{\infty}{K}}(a_j/b_j)$ and $\underset{j=1}{\overset{\infty}{K}}(a_j^*/b_j^*)$ with $n$th approximants $f_n$ and $f_n^*$, respectively, are said to be *equivalent* if

$$f_n = f_n^*, \quad \text{for} \quad n \in \mathbb{N}. \tag{2.9}$$

It is known (see, e.g., [14, Theorem 2.6]) that the two CF's are equivalent if and only if there exists a sequence $\{r_n\}_{n=1}^{\infty}$ of non-zero complex numbers (called the *multiplying factors*)such that $r_0 := 1$ and

$$a_n^* = r_n r_{n-1} a_n \quad \text{and} \quad b_n^* = r_n b_n, \quad \text{for } n \in \mathbb{N}. \tag{2.10}$$

When (2.10) holds, one can show that

$$A_n^* = A_n \prod_{j=1}^{n} r_j \quad \text{and} \quad B_n^* = B_n \prod_{j=1}^{n} r_j, \quad n \in \mathbb{N},$$

where $A_n, B_n$ (and $A_n^*, B_n^*$) denote the $n$th numerator and denominator, respectively, of the CF $\underset{j=1}{\overset{\infty}{K}}(a_j/b_j)$ (and $\underset{j=1}{\overset{\infty}{K}}(a_j^*/b_j^*)$).

**General T-Fractions**

A special class of CF's, called APT-fractions, is of fundamental importance in this paper. Since an APT-fraction is a special type of *general T-fraction* **(GT-fraction)**

$$\underset{j=1}{\overset{\infty}{K}}\left(\frac{F_j z}{1+G_j z}\right), \quad \text{where } 0 \neq F_j \in \mathbb{C} \text{ and } 0 \neq G_j \in \mathbb{C}, \ j \in \mathbb{N}, \tag{2.11}$$

we summarize here some properties of GT-fractions (2.11) that are subsequently used. From the determinant formulas (2.7) one can derive the following basic result (see, e.g., [14, §7.3]):

THEOREM 2.1. (A) *Let a GT-fraction* (2.11) *be given. Then there exists a unique bisequence* $\{\mu_n\}_{n=-\infty}^{\infty}$ *of complex numbers with the property that the GT-fraction* (2.11) *(with nth approximant $f_n(z)$) corresponds to the pair* $(L_0, L_\infty)$ *of formal power series* **(f.p.s.)**

$$L_0(z) := -\sum_{k=1}^{\infty}(-1)^k \mu_{-k} z^k \quad \text{and} \quad L_\infty(z) := \sum_{k=0}^{\infty}(-1)^k \frac{\mu_k}{z^k} \tag{2.12}$$

*in the sense that, for each $n \in \mathbb{N}$, $f_n(z)$ is a rational function, analytic at $z = 0$ and at $z = \infty$, represented in neighborhoods of these two points by convergent series of the forms*

$$f_n(z) = -\sum_{k=1}^{n}(-1)^k \mu_{-k} z^k + \mu_{-n-1}^{(n)} z^{n+1} + \mu_{-n-2}^{(n)} z^{n+2} + \cdots \tag{2.13a}$$

*and*

$$f_n(z) = \sum_{k=0}^{n-1}(-1)^k \frac{\mu_k}{z^k} + \frac{\mu_n^{(n)}}{z^n} + \frac{\mu_{n+1}^{(n)}}{z^{n+1}} + \cdots . \tag{2.13b}$$

*In addition, the Hankel determinants $H_0^{(m)} := 1$ and*

$$H_k^{(m)} := \det(\mu_{m+i+j})_{i,j=0}^{k-1}, \; k \in \mathbb{N}, \; m \in \mathbb{Z}, \tag{2.14}$$

*associated with $\{\mu_n\}_{-\infty}^{\infty}$ satisfy the conditions*

$$H_{2n}^{(-2n)} \neq 0, \; H_{2n+1}^{(-2n)} \neq 0, \; H_{2n}^{(-2n+1)} \neq 0, \; H_{2n+1}^{-2n-1} \neq 0, \; n \in \mathbb{Z}_0^+ := [0, 1, 2, \ldots], \tag{2.15}$$

*and the coefficients of the GT-fraction* (2.11) *are given by*

$$F_1 = H_1^{(-1)} \quad \text{and} \quad F_n = \frac{H_{n-2}^{(-n+3)} H_n^{(-n)}}{H_{n-1}^{(-n+2)} H_{n-1}^{(-n+1)}}, \quad n \geq 2, \tag{2.16a}$$

$$G_1 = \frac{H_1^{(-1)}}{H_1^{(0)}} \quad \text{and} \quad G_n = \frac{H_{n-1}^{(-n+2)} H_n^{(-n)}}{H_n^{(-n+1)} H_{n-1}^{(-n+1)}}, \quad n \geq 2. \tag{2.16b}$$

*(B) Conversely, let $\{\mu_n\}_{n=-\infty}^{\infty}$ satisfy* (2.15) *and let the GT-fraction* (2.11) *be determined by* (2.16). *Then the GT-fraction* (2.11) *corresponds to the pair* $(L_0, L_\infty)$ *of f.p.s.* (2.12) *in the sense of* (2.13).

A bisequence $\{\mu_n\}_{n=-\infty}^{\infty}$ is called *regular* if it satisfies (2.15).

### Equivalent Forms of GT-fractions

It is useful to introduce here two additional CF's that are equivalent to a given GT-fraction (2.11). A third equivalent form is described in Section 3.

THEOREM 2.2. *Let* (2.11) *be a given GT-fraction and let its nth numerator and denominator be denoted by $A_n(z)$ and $B_n(z)$, respectively. Let $\{e_n\}$ and $\{d_n\}$ be defined by*

$$e_1 := \frac{1}{F_1}, \; e_{2n-1} := \frac{\prod_{k=1}^{n-1} F_{2k}}{\prod_{k=1}^{n} F_{2k-1}}, \; e_{2n} := \prod_{k=1}^{n} \frac{F_{2k-1}}{F_{2k}}, \quad n \in \mathbb{N} \tag{2.17a}$$

*and*

$$d_n = e_n G_n, \quad \textit{for} \quad n \in \mathbb{N}. \tag{2.17b}$$

*Then: (A) The CF*

$$\operatorname*{K}_{j=1}^{\infty} \left( \frac{z}{e_j + d_j z} \right) \tag{2.18}$$

*is equivalent to the GT-fraction* (2.11). *(B)*

$$F_1 = \frac{1}{e_1}, \; F_n = \frac{1}{e_{n-1}e_n}, \; \textit{for } n \geq 2 \textit{ and } G_n = \frac{d_n}{e_n}, \quad n \in \mathbb{N}. \tag{2.19}$$

*(C) If $C_n(z)$ and $D_n(z)$ denote the nth numerator and denominator, respectively, of the modified GT-fraction* (2.18), *then for all $n \in \mathbb{N}$,*

$$C_n(z) = \left( \prod_{k=1}^{n} e_k \right) A_n(z) \quad \text{and} \quad D_n(z) = \left( \prod_{k=1}^{n} e_k \right) B_n(z). \tag{2.20}$$

**Remark.** We note that the factors in (2.20) can be expressed by

$$\prod_{k=1}^{2n-1} e_k = \frac{1}{\prod_{j=1}^{n} F_{2n-1}} \quad \text{and} \quad \prod_{k=1}^{2n} e_k = \frac{1}{\prod_{j=1}^{n} F_{2j}}, \quad n \in \mathbb{N}. \tag{2.21}$$

Another CF that is equivalent to a given GT-fraction is described in the following:

THEOREM 2.3. *Let* (2.11) *be a given GT-fraction and let $A_n(z)$ and $B_n(z)$ denote its nth numerator and denominator, respectively. Let $\{\beta_n\}$ and $\{\lambda_n\}$ be defined by*

$$\lambda_{2k-1} := F_{2k-1} \quad \textit{and} \quad \lambda_{2k} := \frac{F_{2k}}{G_{2k-1}G_{2k}}, \quad \textit{for} \quad k \in \mathbb{N}, \tag{2.22a}$$

*and*

$$\beta_0 := 1,\ \beta_{2k-1} := \prod_{j=1}^{2k-1} G_j,\ \beta_{2k} := \left[\prod_{j=1}^{2k} G_j\right]^{-1},\ \textit{for } k \in \mathbb{N}. \tag{2.22b}$$

*Then: (A) The CF*

$$\frac{\lambda_1}{\frac{z^{-1}}{\beta_0} + \beta_1} + \frac{\lambda_2}{\frac{z}{\beta_1} + \beta_2} + \frac{\lambda_3}{\frac{z^{-1}}{\beta_2} + \beta_3} + \frac{\lambda_4}{\frac{z}{\beta_3} + \beta_4} + \cdots \tag{2.22c}$$

*is equivalent to the GT-fraction* (2.11) *(B)*

$$F_{2k-1} = \lambda_{2k-1},\ F_{2k} = \frac{\lambda_{2k}\beta_{2k-2}}{\beta_{2k}},\ \textit{for } k \in \mathbb{N}, \tag{2.23a}$$

*and*

$$G_{2k-1} = \beta_{2k-2}\beta_{2k-1},\ G_{2k} = \frac{1}{\beta_{2k-1}\beta_{2k}},\ \textit{for } k \in \mathbb{N}. \tag{2.23b}$$

*(C) If $P_n(z)$ and $Q_n(z)$ are the nth numerator and denominator, respectively, of the CF* (2.22), *then*

$$P_{2n-1}(z) = \frac{A_{2n-1}(z)}{z^n} = \frac{C_{2n-1}(z)}{z^n \prod_{j=1}^{2n-1} e_j}, \quad \text{for} \quad n \in \mathbb{N}, \tag{2,24a}$$

$$Q_{2n-1}(z) = \frac{B_{2n-1}(z)}{z^n} = \frac{D_{2n-1}(z)}{z^n \prod_{j=1}^{2n-1} e_j}, \quad \text{for} \quad n \in \mathbb{N}, \tag{2.24b}$$

$$P_{2n}(z) = \frac{A_{2n}(z)}{z^n \prod_{j=1}^{2n} G_j} = \frac{C_{2n}(z)}{z^n \prod_{j=1}^{2n} d_j}, \quad \text{for} \quad n \in \mathbb{N}, \tag{2.24c}$$

$$Q_{2n}(z) = \frac{B_{2n}(z)}{z^n \prod_{j=1}^{2n} G_j} = \frac{D_{2n}(z)}{z^n \prod_{j=1}^{2n} d_j}, \quad \text{for} \quad n \in \mathbb{N}. \tag{2.24d}$$

The CF (2.22) is also called a modified GT-fraction. The next result pertains to the functions $A_n, B_n, C_n, D_n, P_n$ and $Q_n$.

THEOREM 2.4. *Let* (2.11) *be a given GT-fraction and let $A_n, B_n, C_n, D_n, P_n$ and $Q_n$ denote the functions appearing in Theorems 2.2 and 2.3. Then there exist complex numbers $A_{n,j}, B_{n,j}, C_{n,j}, D_{n,j}, P_{n,j}$ and $Q_{n,j}$ such that, for $n \in \mathbb{N}$:*
*(A)*

$$A_n(z) = \sum_{j=1}^{n} A_{n,j} z^j, \quad B_n(z) = \sum_{j=1}^{n} B_{n,j} z^j, \tag{2.25a}$$

*where*

$$A_{n,1} = F_1, \; A_{n,n} = F_1 \prod_{k=2}^{n} G_k, \; B_{n,0} = 1, \; B_{n,n} = \prod_{k=1}^{n} G_k. \tag{2.25b}$$

*(B)*

$$C_n(z) = \sum_{j=1}^{n} C_{n,j} z^j, \quad D_n(z) = \sum_{j=0}^{n} D_{n,j} z^j, \tag{2.26a}$$

*where*

$$C_{n,1} = \prod_{k=2}^{n} e_k, \; C_{n,n} = \prod_{k=2}^{n} d_k, \; D_{n,0} = \prod_{k=1}^{n} e_k, \; D_{n,n} = \prod_{k=1}^{n} d_k. \tag{2.26b}$$

*(C) For $n \geq 0$,*

$$P_{2n+1}(z) = \sum_{j=-n}^{n} P_{2n+1,j} z^j, \qquad Q_{2n+1}(z) = \sum_{j=-n-1}^{n} Q_{2n+1,j} z^j, \tag{2.27a}$$

*where*

$$P_{2n+1,-n} = \lambda_1, \; P_{2n+1,n} = \frac{\lambda_1 \beta_{2n+1}}{\beta_1}, \; Q_{2n+1,-n-1} = 1, \; Q_{2n+1,n} = \beta_{2n+1}, \tag{2.27b}$$

*and, for $n \geq 1$,*

$$P_{2n}(z) = \sum_{j=-n+1}^{n} P_{2n,j} z^j, \qquad Q_{2n}(z) = \sum_{j=-n}^{n} Q_{2n,j} z^j, \tag{2.27c}$$

*where*

$$P_{2n,-n+1} = \lambda_1 \beta_{2n}, \quad P_{2n,n} = \frac{\lambda_1}{\beta_1}, \quad Q_{2n,-n} = \beta_{2n}, \quad Q_{2n,n} = 1. \tag{2.27d}$$

*(D) Let $\{\mu_n\}_{n=-\infty}^{\infty}$ be the unique bisequence of complex numbers determined by a given GT-fraction (2.11) (see Theorem 2.1). Then, for all $m \in \mathbb{Z}_0^+ :=$ $[0, 1, 2, \ldots]$,*

$$Q_{2m}(z) = \frac{1}{H_{2m}^{(-2m)}} \begin{vmatrix} \mu_{-2m} & \mu_{-2m+1} & \cdots & \mu_{-1} & z^{-m} \\ \mu_{-2m+1} & \mu_{-2m+2} & \cdots & \mu_0 & z^{-m+1} \\ \vdots & \vdots & & \vdots & \vdots \\ \mu_{-1} & \mu_0 & \cdots & \mu_{2m-2} & z^{m-1} \\ \mu_0 & \mu_1 & \cdots & \mu_{2m-1} & z^m \end{vmatrix}, \tag{2.28a}$$

*and*

$$Q_{2m+1}(z) = \frac{-1}{H_{2m+1}^{(-2m)}} \begin{vmatrix} \mu_{-2m-1} & \mu_{-2m} & \cdots & \mu_{-1} & z^{-m-1} \\ \mu_{-2m} & \mu_{-2m+1} & \cdots & \mu_0 & z^{-m} \\ \vdots & \vdots & & \vdots & \vdots \\ \mu_{-1} & \mu_0 & \cdots & \mu_{2m-1} & z^{m-1} \\ \mu_0 & \mu_1 & \cdots & \mu_{2m} & z^m \end{vmatrix}. \quad (2.28b)$$

*(E) (Three-term recurrence relations)*

$$A_n(z) := (1 + G_n z)A_{n-1}(z) + F_n z A_{n-2}(z), \quad n \in \mathbb{N}, \quad (2.29a)$$

$$B_n(z) := (1 + G_n z)B_{n-1}(z) + F_n z B_{n-2}(z), \quad n \in \mathbb{N}, \quad (2.29b)$$

$$A_{-1}(z) := 1,\ B_{-1}(z) := 0,\ A_0(z) := 0,\ B_0(z) := 1. \quad (2.29c)$$

$$C_n(z) := (e_n + d_n z)C_{n-1}(z) + zC_{n-2}(z),\ n \in \mathbb{N}, \quad (2.30a)$$

$$D_n(z) := (e_n + d_n z)D_{n-1}(z) + zD_{n-2}(z),\ n \in \mathbb{N}, \quad (2.30b)$$

$$C_{-1}(z) := 1,\ D_{-1}(z) := 0,\ C_0(z) := 0,\ D_0(z) := 1. \quad (2.30c)$$

$$P_{2n+1}(z) := \left(\frac{1}{\beta_{2n} z} + \beta_{2n+1}\right) P_{2n}(z) + \lambda_{2n+1} P_{2n-1}(z),\ n \in \mathbb{Z}_0^+, \quad (2.31a)$$

$$Q_{2n+1}(z) := \left(\frac{1}{\beta_{2n} z} + \beta_{2n+1}\right) Q_{2n}(z) + \lambda_{2n+1} Q_{2n-1}(z),\ n \in \mathbb{Z}_0^+, \quad (2.31b)$$

$$P_{2n}(z) := \left(\frac{z}{\beta_{2n-1}} + \beta_{2n}\right) P_{2n-1}(z) + \lambda_{2n} P_{2n-2}(z),\ n \in \mathbb{N}, \quad (2.31c)$$

$$Q_{2n}(z) := \left(\frac{z}{\beta_{2n-1}} + \beta_{2n}\right) Q_{2n-1}(z) + \lambda_{2n} Q_{2m-2}(z),\ n \in \mathbb{N}, \quad (2.31d)$$

$$P_{-1}(z) := 1,\ Q_{-1}(z) := 0,\ P_0(z) := 0, Q_0(z) := 1. \quad (2.31e)$$

**Remarks on Proof.** The recurrence relations (E) follow from (2.6), (2.11), (2.18) and (2.22). (A), (B) and (C) are readily derived from (E). The determinant formulas (2.28) can be deduced from the correspondence property (2.13), since

$$f_n(z) = \frac{A_n(z)}{B_n(z)} = \frac{C_n(z)}{D_n(z)} = \frac{P_n(z)}{Q_n(z)}, \quad \text{for} \quad n \in \mathbb{N}, \quad (2.32)$$

(see, e.g. [13, Theorems 2.8 and 4.1]).

**APT-Fractions**

A GT-fraction (2.11) is called an *alternativing-positive-term continued fraction* **(APT-fraction)** if the associated Hankel determinants $H_k^{(m)}$ satisfy the conditions

$$H_{2n}^{(-2n)} > 0,\ H_{2n+1}^{(-2n)} > 0,\ H_{2n}^{(-2n+1)} \neq 0,\ H_{2n+1}^{(-2n-1)} \neq 0,\ \text{for } n \in \mathbb{N}. \quad (2.33)$$

Here the $H_k^{(m)}$ are defined by (2.14) (see Theorem 2.1) in terms of the unique bisequence $\{\mu_n\}_{n=-\infty}^{\infty}$. We call (2.33) *the APT-condition.*

THEOREM 2.5. (A) *A continued fraction*

$$\underset{j=1}{\overset{\infty}{K}} \left( \frac{F_j z}{1 + G_j z} \right) \quad (2.34a)$$

*is an APT-fraction if and only if*

$$F_j \in \mathbb{R} - [0],\ G_j \in \mathbb{R} - [0],\ F_{2j-1}F_{2j} > 0,\ F_{2j-1}G_{2j-1} > 0 \ \textit{for } j \in \mathbb{N}. \quad (2.34b)$$

*(B) A continued fraction*

$$\underset{j=1}{\overset{\infty}{K}} \left( \frac{z}{e_j + d_j z} \right) \quad (2.35a)$$

*is equivalent to an APT-fraction* (2.34) *if and only if*

$$e_j \in \mathbb{R} - [0],\ d_j \in \mathbb{R} - [0],\ e_{2j} > 0,\ d_{2j-1} > 0,\ \textit{for } j \in \mathbb{N}. \quad (2.35b)$$

*The relationships between the coefficients $F_j, G_j, e_j$ and $d_j$ are given by* (2.17) *and* (2.19). *(C) A continued fraction*

$$\frac{\lambda_1}{\frac{z^{-1}}{\beta_0} + \beta_1 +} \frac{\lambda_2}{\frac{z}{\beta_1} + \beta_2 +} \frac{\lambda_3}{\frac{z^{-1}}{\beta_2} + \beta_3 +} \frac{\lambda_4}{\frac{z}{\beta_3} + \beta_4 + \cdots} \quad (2.36a)$$

*is equivalent to an APT-fraction* (2.34) *if and only if*

$$\beta_1 := 1,\ \beta_j \in \mathbb{R} - [0],\ \lambda_j \in \mathbb{R} - [0],\ \frac{\lambda_j \beta_{j-1}}{\beta_j} > 0,\ \textit{for } j \in \mathbb{N}. \quad (2.36b)$$

*The relationships between the coefficients $F_j, G_j, \lambda_j$ and $\beta_j$ are given by* (2.22) *and* (2.23) [7].

Continued fractions of the forms (2.35) and (2.36) are called *modified APT-fractions* or, simply, APT-fractions if the meaning is clear from the context in which they are used.

## Nested Disks

We define notation for the following subsets of $\mathbb{C}$:

$$U := [w \in \mathbb{C} : \operatorname{Im} w > 0], \quad W := [w \in \mathbb{C} : \operatorname{Im} w < 0]. \tag{2.37a}$$

and for each $z \in \mathbb{C} - \mathbb{R}$,

$$V_0(z) := \begin{cases} [w \in \mathbb{C} : -\pi + \arg z < \arg w < \arg z], & \text{if } 0 < \arg z < \pi, \\ [w \in \mathbb{C} : \arg z < \arg w < \arg z + \pi], & \text{if } -\pi < \arg z < 0, \end{cases} \tag{2.37b}$$

and

$$V_1(z) := \begin{cases} U, & \text{if } z \in U \\ W, & \text{if } z \in W. \end{cases} \tag{2.37c}$$

It is readily shown that

$$\frac{V_0(z)}{z} = \begin{cases} W, & \text{if } z \in U \\ U, & \text{if } z \in W. \end{cases} \tag{2.38}$$

Let (2.35) be a given APT-fraction and (see (2.2) and (2.8)) let

$$s_n(z, w) := \frac{z}{e_n + d_n z + w}, \quad \text{for all} \quad n \in \mathbb{N}, \tag{2.39a}$$

and

$$S_n(z, w) = \frac{C_n(z) + wC_{n-1}(z)}{D_n(z) + wD_{n-1}(z)}, \quad n \in \mathbb{N}, \tag{2.39b}$$

where (see Theorem 2.2) $C_n(z)$ and $D_n(z)$ denote the $n$th numerator and denominator, respectively, of the APT-fraction. For each $z \in \mathbb{C} - \mathbb{R}$ let $\{K_n(z)\}_{n=1}^{\infty}$ be defined by

$$K_{2n-1}(z) := S_{2n-1}(z, V_1(z)), \; K_{2n}(z) := S_{2n}(z, V_0(z)), \; n \in \mathbb{N}, \tag{2.40}$$

and let

$\widehat{K}_n(z) := \partial K_n(z) \cup K_n(t)$, where $\partial K_n(z)$ denotes the boundary of $K_n(z)$.

Using mapping properties of linear fractional transformations, one can prove

THEOREM 2.6. [11] *Let an APT-fraction* (2.35) *be given and let* $z \in \mathbb{C} - \mathbb{R}$. *Then for all* $n \in \mathbb{N}$:

$$e_{2n-1} + d_{2n-1} z \in V_1(z), \quad e_{2n} + d_{2n} z \in V_0(z); \tag{2.41a}$$

$$s_{2n+1}(z, V_1(z)) \subset V_0(z), \quad s_{2n}(z, V_0(z)) \subset V_1(z); \tag{2.41b}$$

$$S_{2n-1}(z,\tau) \in V_0(z), \quad S_{2n}(z,\tau z) \in V_0(z), \quad \text{for} \quad \tau \in \mathbb{R}; \tag{2.41c}$$

$$S_n(z,0) = \frac{C_n(z)}{D_n(z)} \in V_0(z); \tag{2.41d}$$

$$\{\widehat{K}_n(z)\}_{n=1}^{\infty} \text{ is a nested sequence of closed circular disks;} \tag{2.41e}$$

$$S_{2n-1}(z,\tau) \in \widehat{K}_{2n-1}(z), \quad S_{2n}(z,\tau z) \in \widehat{K}_{2n}(z), \text{ for } \tau \in \mathbb{R}. \tag{2.41f}$$

It follows from (2.41d) that all of the zeros of $C_n(z)$ and $D_n(z)$ are real. Motivated by (2.41c), we define, for $\tau \in \mathbb{R}$, $z \in \mathbb{C}$

$$C_{2m}(z,\tau) := C_{2m}(z) + \tau z C_{2m-1}(z), \quad m \in \mathbb{N}, \tag{2.42a}$$

$$C_{2m+1}(z,\tau) := C_{2m+1}(z) + \tau C_{2m}(z), \quad m \in \mathbb{Z}_0^+, \tag{2.42b}$$

$$D_{2m}(z,\tau) := D_{2m}(z) + \tau z D_{2m-1}(z), \quad m \in \mathbb{N}, \tag{2.42c}$$

$$D_{2m+1}(z,\tau) := D_{2m+1}(z) + \tau D_{2m}(z), \quad m \in \mathbb{Z}_0^+, \tag{2.42d}$$

where

$$C_0(z) \equiv 0 \quad \text{and} \quad D_0(z) \equiv 1. \tag{2.42e}$$

For each $n \in \mathbb{N}$ and $\tau \in \mathbb{R}$, the quotient

$$\frac{C_n(z,\tau)}{D_n(z,\tau)} \tag{2.43}$$

is called an $n$th *generalized approximant of the APT-fraction* (2.35). It is readily seen that

$$\frac{C_n(z,\tau)}{D_n(z,\tau)} = \begin{cases} S_{2m-1}(z,\tau), & \text{if} \quad n = 2m-1 \\ S_{2m}(z,\tau z), & \text{if} \quad n = 2m, \end{cases} \tag{2.44}$$

and hence, by (2.41f)

$$\frac{C_n(z,\tau)}{D_n(z,\tau)} \in \widehat{K}_n(z), \quad \text{for} \quad z \in \mathbb{C} - \mathbb{R}, \ \tau \in \mathbb{R}, \ n \in \mathbb{N}. \tag{2.45}$$

Important properties of these approximants are summarized in the following

THEOREM 2.7. [11] *Let an APT-fraction* (2.35) *be given and let* $\{\tau_n\}_{n=1}^{\infty}$ *be an arbitrary sequence of real numbers. For each* $n \in \mathbb{N}$, *let* $r(n,\tau_n)$ *denote the number of zeros of* $D_n(z,\tau_n)$. *We denote these zeros by* $-t_1^{(n)}(\tau_n), -t_2^{(n)}(\tau_n), \ldots, -t_{r(n,\tau_n)}^{(n)}(\tau_n)$. *Then for each* $n \in \mathbb{N}$:

*(A) All of the zeros of $D_n(z,\tau_n)$ are real and simple and*

$$r(n,\tau_n) = \begin{cases} n-1, & \textit{if} \quad n=2k \textit{ and } \tau_n = -d_{2k}, \\ n-1, & \textit{if} \quad n=2k+1 \textit{ and } \tau_n = -e_{2k+1}, \\ n, & \textit{otherwise.} \end{cases} \tag{2.46}$$

*(B) There exist $r(n,\tau_n)$ positive real numbers $k_j^{(n)}(\tau_n)$ such that*

$$\frac{C_n(z,\tau_n)}{D_n(z,\tau_n)} = \sum_{j=1}^{r(n,\tau_n)} \frac{z k_j^{(n)}(\tau_n)}{z + t_j^{(n)}(\tau_n)} \tag{2.47a}$$

*and*

$$\sum_{j=1}^{r(n,\tau_n)} k_j^{(n)}(\tau_n) = \frac{1}{d_1} = \mu_0 > 0 \tag{2.47b}$$

*(C) There exists a subsequence $\{\tau_{n_k}\}_{k=1}^{\infty}$ of $\{\tau_n\}_{n=1}^{\infty}$ such that the sequence*

$$\left\{ \frac{C_{n_k}(z,\tau_{n_k})}{D_{n_k}(z,\tau_{n_k})} \right\}_{k=1}^{\infty} \tag{2.48}$$

*converges to a function (of $z$) analytic on $U$ (and on $W$), locally uniformly on $U$ (and on $W$).*

It follows from Theorem 2.7 (C) that the family of all generalized approximants of an APT-fraction (2.35) is a normal family on $U$ (and on $W$). It follows from Theorem 2.3 that the same statement applies to APT-fractions that are equivalent to (2.35) (such as (2.34) and (2.36)).

## 3 NORMAL APT-FRACTIONS

### Connections with Orthogonal L-Polynomials

For later use we describe connections between APT-fractions (2.36) and orthogonal Laurent polynomials. Recall that the $n$th numerator and denominator of (2.36) are denoted by $P_n(z)$ and $Q_n(z)$, respectively.

Let $\{\mu_n\}_{n=-\infty}^{\infty}$ denote the unique bisequence of real numbers determined (see Theorem 2.1) by an APT-fraction (2.36). Let

$$\Lambda^{\mathbb{R}} := \left[ \sum_{j=p}^{q} r_j z^j : p, q \in \mathbb{Z},\ p \leq q \text{ and } r_j \in \mathbb{R} \text{ for } p \leq j \leq q \right], \tag{3.1}$$

so that $\Lambda^{\mathbb{R}}$ consists of the space of all *Laurent polynomials* (*L-polynomials*) with real coefficients. Let $\mathcal{L}$ denote the linear functional defined on $\Lambda^{\mathbb{R}}$ as follows:

$$R(z) := \sum_{j=p}^{q} r_j z^j \in \Lambda^{\mathbb{R}} \Rightarrow \mathcal{L}[R(z)] := \sum_{j=p}^{q} r_j \mu_j. \tag{3.2}$$

We say that, for each $n \in \mathbb{Z}$, $\mu_n =: \mathcal{L}[z^n]$ is the $n$th moment with respect to $\mathcal{L}$ and that $\mathcal{L}$ is the *strong moment functional* (**SMF**) determined by $\{\mu_n\}_{n=-\infty}^{\infty}$. A SMF $\mathcal{L}$ is said to be *positive definite* (or **PDSMF**) if

$$[0 \not\equiv R(z) \in \Lambda^{\mathbb{R}} \text{ and } R(x) \geq 0 \text{ for } x \in \mathbb{R} - [0]] \Rightarrow \mathcal{L}[R(z)] > 0.$$

It is well known [12] that a SMF $\mathcal{L}$ and the bisequence $\{\mu_n\}_{-\infty}^{\infty}$ are positive definite if and only if the Hankel determinants $H_k^{(m)}$ associated with $\{\mu_n\}_{-\infty}^{\infty}$ satisfy

$$H_{2m}^{(-2n)} > 0 \quad \text{and} \quad H_{2n+1}^{(-2n)} > 0, \quad \text{for} \quad n \in \mathbb{Z}_0^+. \tag{3.3}$$

(See (2.14) for the definition of $H_k^{(m)}$.) Since the bisequence $\{\mu_n\}_{-\infty}^{\infty}$ is generated by an APT-fraction, it follows from the definition (see (2.33)) that the $H_k^{(m)}$ satisfy (3.3) and hence that $\mathcal{L}$ is a PDSMF on $\Lambda^{\mathbb{R}}$. Therefore an inner product $\langle\cdot,\cdot\rangle_{\mathcal{L}}$ on $\Lambda^{\mathbb{R}}$ is defined by

$$\langle F, G\rangle_{\mathcal{L}} := \mathcal{L}[F(z)G(z)], \quad \text{for} \quad F, G \in \Lambda^{\mathbb{R}}. \tag{3.4}$$

From (2.28) and (3.4) it can be readily shown that, for $n \in \mathbb{N}$,

$$\langle Q_{2n}(z), z^m\rangle_{\mathcal{L}} = 0, \quad \text{for} \quad m = -n, -n+1, \dots, n-1, \tag{3.5a}$$

$$\|Q_{2n}(z)\|_{\mathcal{L}} := \sqrt{\langle Q_{2n}(z), Q_{2n}(z)\rangle_{\mathcal{L}}} = \sqrt{\langle Q_{2n}(z), z^n\rangle_{\mathcal{L}}} > 0, \tag{3.5b}$$

and, for $n \in \mathbb{Z}_0^+$,

$$\langle Q_{2n+1}(z), z^m\rangle_{\mathcal{L}} = 0, \quad \text{for} \quad m = -n, -n+1, \dots, n, \tag{3.5c}$$

$$\|Q_{2n+1}(z)\|_{\mathcal{L}} := \sqrt{\langle Q_{2n+1}(z), Q_{2n+1}(z)\rangle_{\mathcal{L}}} = \sqrt{\langle Q_{2n+1}(z), z^{-n-1}\rangle_{\mathcal{L}}} > 0. \tag{3.5d}$$

Thus $\{Q_n(z)\}_{n=1}^{\infty}$ is a monic orthogonal L-polynomial sequence (**OLPS**) with respect to $\mathcal{L}$. The sequence $\{Q_n(z)\}_0^{\infty}$ is called *monic*, since the leading coefficient of $Q_n(z)$ is 1 for all $n \geq 0$ (see (2.27b) and (2.27d)). The $Q_n(z)$ are given explicitly by the determinants (2.28); they could also be generated by applying Gram–Schmidt to the sequence $\{1, z^{-1}, z, z^{-2}, z^2, \dots\}$; a third way to generate the sequence $\{Q_n(z)\}_{n=0}^{\infty}$ is by the three-term recurrence relations (2.31).

The coefficients $\beta_n$ and $\lambda_n$ can be expressed explicitly in terms of the Hankel determinant $H_k^{(m)}$ associated with $\{\mu_n\}_{n=-\infty}^{\infty}$ by use of the formulas (2.16) and (2.22). The recurrence relations (2.31) can be used to obtain the expressions

$$\|Q_{2n-1}(z)\|_{\mathcal{L}}^2 = \frac{1}{\beta_{2n}} \prod_{j=1}^{2n} \lambda_j \quad \text{and} \quad \|Q_{2n}(z)\|_{\mathcal{L}}^2 = \frac{1}{\beta_{2n+1}} \prod_{v=1}^{2n+1} \lambda_j, n \in \mathbb{N}, \tag{3.6}$$

(see, e.g., [7, (5.10) and (5.11)] and [20]). Therefore the sequence $\{q_n(z)\}_{n=0}^{\infty}$ defined by

$$q_n(z) := \frac{Q_n(z)}{\|Q_n(z)\|_{\mathcal{L}}} = \kappa_n Q_n(z), \text{ for } n \in \mathbb{Z}_0^+, \text{ where} \tag{3.7a}$$

$$\kappa_n := \frac{1}{\|Q_n(z)\|_{\mathcal{L}}} = \left[\frac{\beta_{n+1}}{\prod_{j=1}^{n+1} \lambda_j}\right]^{1/2}, \; n \in \mathbb{Z}_0^+, \tag{3.7b}$$

is an *orthonormal L-polynomial sequence* **(ONLPS)** with respect to $\mathcal{L}$; that is, for $n, m \in \mathbb{Z}_0^+$,

$$\langle q_n(z), q_m(z) \rangle_{\mathcal{L}} = \delta_{n,m} := \begin{cases} 0 \text{ if } n \neq m \\ 1 \text{ if } n = m. \end{cases} \tag{3.7c}$$

We note that $\kappa_0 := 1/\sqrt{\mu_0}$. Similarly we define

$$p_n(z) := \kappa_n P_n(z) \text{ and } p_n^*(z) := p_n(z)/z, \text{ for } n \in \mathbb{Z}_0^+. \tag{3.7d}$$

The sequences $\{P_n(z)\}_0^{\infty}$, $\{Q_n(z)\}_0^{\infty}$, $\{C_n(z)\}_0^{\infty}$, $\{D_n(z)\}_0^{\infty}$ are related by (2.24); this together with (3.7) yields

$$p_{2n-1}(z) = \sqrt{e_{2n}} \frac{C_{2n-1}(z)}{z^n}, \quad p_{2n}(z) = \sqrt{d_{2n+1}} \frac{C_{2n}(z)}{z^n}, \; n \in \mathbb{N}, \tag{3.8a}$$

$$q_{2n-1}(z) = \sqrt{e_{2n}} \frac{D_{2n-1}(z)}{z^n}, \quad q_{2n}(z) = \sqrt{d_{2n+1}} \frac{D_{2n}(z)}{z^n}, \; n \in \mathbb{N}. \tag{3.8b}$$

By analogy with (2.42) we define

$$p_{2n}^*(z, \tau) := p_{2n}^*(z) - \tau z p_{2n-1}^*(z), \quad n \in \mathbb{N}, \tag{3.9a}$$

$$p_{2n+1}^*(z, \tau) := p_{2n+1}^*(z) - \tau z^{-1} p_{2n}^*(z), \quad n \in \mathbb{Z}_0^+, \tag{3.9b}$$

$$q_{2n}(z, \tau) := q_{2n}(z) - \tau z q_{2n-1}(z), \quad n \in \mathbb{N}, \tag{3.9c}$$

$$q_{2n+1}(z,\tau) := q_{2n+1}(z) - \tau z^{-1} q_{2n}(z), \quad n \in \mathbb{Z}_0^+. \tag{3.9d}$$

where

$$p_0^*(z) \equiv p_0(z) \equiv 0, \quad q_0(z) \equiv \kappa_0 = 1/\sqrt{\mu_0}. \tag{3.9e}$$

and

$$p_n(z,\tau) := z p_n^*(z,\tau) \quad \text{for} \quad n \in \mathbb{Z}_0^+. \tag{3.9f}$$

## Normal APT-Fractions

We introduce another equivalent form of APT-fractions. Let (2.36) be a given APT-fraction and let $\{\alpha_n^*\}, \{\beta_n^*\}$ and $\{\lambda_n^*\}$ be defined by

$$\alpha_1^* := \kappa_1, \ \beta_1^* := \beta_1 \kappa_1, \ \lambda_1^* := \lambda_1 \kappa_1, \tag{3.10a}$$

$$\alpha_2^* := \frac{\kappa_2}{\beta_1 \kappa_1}, \ \beta_2^* := \frac{\beta_2 \kappa_2}{\kappa_1}, \ \lambda_2^* := \lambda_2 \kappa_2, \tag{3.10b}$$

$$\alpha_n^* := \frac{\kappa_n}{\beta_{n-1} \kappa_{n-1}}, \ \beta_n^* := \frac{\beta_n \kappa_n}{\kappa_{n-1}}, \ \lambda_n^* := \frac{\lambda_n \kappa_n}{\kappa_{n-1}}, \ n \geq 3, \tag{3.10c}$$

where $\kappa_n$ is defined by (3.7) for $n \in \mathbb{Z}_0^+$. Then

$$\mathop{K}_{n=1}^{\infty} \left( \frac{\lambda_n^*}{\alpha_n^* z^{(-1)^n} + \beta_n^*} \right) \tag{3.10d}$$

is called a *normal APT-fraction.*

THEOREM 3.1. *Let (3.10) be a given normalized APT-fraction (defined in terms of an APT-fraction (2.36)). Let $\{p_n(z)\}$ and $\{q_n(z)\}$ be defined by (3.7). Then:*

*(A) The CF (3.10) is equivalent to the APT-fraction (2.36).*

*(B) For $n \in \mathbb{N}$, $p_n(z)$ and $q_n(z)$ are the nth numerator and denominator, respectively, of (3.10).*

*(C) If $p_{-1}(z) \equiv 1$, $q_{-1}(z) \equiv 0$, $p_0(z) \equiv 0$, $q_0(z) \equiv 1$, then for $n \in \mathbb{N}$,*

$$p_{2n+1}(z) = \left( \frac{\alpha_{2n+1}^*}{z} + \beta_{2n+1}^* \right) p_{2n}(z) + \lambda_{2n+1}^* p_{2n-1}(z), \ n \geq 0, \tag{3.11a}$$

$$q_{2n+1}(z) = \left( \frac{\alpha_{2n+1}^*}{z} + \beta_{2n+1}^* \right) q_{2n}(z) + \lambda_{2n+1}^* q_{2n-1}(z), \ n \geq 0, \tag{3.11b}$$

$$p_{2n}(z) = (\alpha_{2n}^* z + \beta_{2n}^*) p_{2n-1}(z) + \lambda_{2n}^* p_{2n-1}(z), \ n \geq 1, \tag{3.11c}$$

$$q_{2n}(z) = (\alpha_{2n}^* z + \beta_{2n}^*) q_{2n-1}(z) + \lambda_{2n}^* q_{2n-1}(z), \ n \geq 1. \tag{3.11d}$$

*(D) For* $n \in \mathbb{Z}_0^+$,

$$p_{2n+1}(z) = \sum_{j=-n}^{n} p_{2n+1,j} z^j, \quad q_{2n+1}(z) = \sum_{j=-n-1}^{n} q_{2n+1,j} z^j, \tag{3.12a}$$

*where*

$$p_{2n+1,-n} = \lambda_1 \kappa_{2n+1}, \quad p_{2n+1,n} = \frac{\lambda_1}{\beta_1} \beta_{2n+1} \kappa_{2n+1}, \tag{3.12b}$$

$$q_{2n+1,-n-1} = \kappa_{2n+1}, \quad q_{2n+1,n} = \beta_{2n+1} \kappa_{2n+1}, \tag{3.12c}$$

*and, for* $n \in \mathbb{N}$,

$$p_{2n}(z) = \sum_{j=-n+1}^{n} p_{2n,j} z^j, \quad q_{2n}(z) = \sum_{j=-n}^{n} q_{2n,j} z^j, \tag{3.12d}$$

*where*

$$p_{2n,-n+1} = \lambda_1 \beta_{2n} \kappa_{2n}, \quad p_{2n,n} = \frac{\lambda_1}{\beta_1} \kappa_{2n}, \tag{3.12e}$$

$$q_{2n,-n} = \beta_{2n} \kappa_{2n}, \quad q_{2n,n} = \kappa_{2n}. \tag{3.12f}$$

*(E) [22] For* $n \in \mathbb{N}$,

$$p_{2n}(z) q_{2n-1}(z) - p_{2n-1}(z) q_{2n}(z) = \frac{-q_{2n,-n}}{q_{2n-1,-n}}, \tag{3.13a}$$

*and, for* $n \in \mathbb{Z}_0^+$,

$$p_{2n+1}(z) q_{2n}(z) - p_{2n}(z) q_{2n+1}(z) = \frac{q_{2n+1,n}}{q_{2n,n}}. \tag{3.13b}$$

PROOF. (A) follows from (2.10), where the multiplying factors given by

$$r_0 := 1, \; r_1 := \kappa_1, \; r_n = \frac{\kappa_n}{\kappa_{n-1}}, \; n = 2, 3, 4, \ldots \tag{3,14}$$

are applied to the APT-fraction (3.10). Hence $\lambda_n^* = r_n r_{n-1} \lambda_n$, $\alpha_n^* = r_n \alpha_n$, $\beta_n^* = r_n \beta_n$, $n \in \mathbb{N}$. (B) is a consequence of (3.7) and (3.10). (C) follows from (2.6) and (3.10). (D) can be derived from the recurrence relations (3.11) and (3.10). To prove (E) we apply the determinant formula for continued fractions (2.7) to obtain

$$p_{2n}(z) q_{2n-1}(z) - p_{2n-1}(z) q_{2n}(z) = -\prod_{j=1}^{2n} \lambda_j^*, \quad n \in \mathbb{N}, \tag{3.15a}$$

$$p_{2n+1}(z)q_{2n}(z) - p_{2n}(z)q_{2n+1}(z) = \prod_{j=1}^{2n+1} \lambda_j^*, \quad n \in \mathbb{Z}_0^+. \tag{3.15b}$$

For $n \in \mathbb{N}$, by (3.7b) and (3.15),

$$\prod_{j=1}^{2n} \lambda_j^* = \left(\prod_{j=1}^{2n} \lambda_j\right) \kappa_{2n-1}\kappa_{2n} = \kappa_{2n}\sqrt{\beta_{2n} \prod_{j=1}^{2n} \lambda_j}$$

and

$$\frac{q_{2n,-n}}{q_{2n-1,-n}} = \frac{\beta_{2n}\kappa_{2n}}{\kappa_{2n-1}} = \kappa_{2n}\sqrt{\beta_{2n} \prod_{j=1}^{2n} \lambda_j},$$

which proves (3.13a). Similarly, for $n \in \mathbb{N}$, by (3.7b) and (3.15)

$$\prod_{j=1}^{2n+1} \lambda_j^* = \left(\prod_{j=1}^{2n} \lambda_j\right) \kappa_{2n}\kappa_{2n+1} = \kappa_{2n+1}\sqrt{\beta_{2n+1} \prod_{j=1}^{2n+1} \lambda_j}$$

and

$$\frac{q_{2n+1,n}}{q_{2n,n}} = \frac{\beta_{2n+1}\kappa_{2n+1}}{\kappa_{2n}} = \kappa_{2n+1}\sqrt{\beta_{2n+1} \prod_{j=1}^{2n+1} \lambda_j},$$

which yields (3.13b). □

For each $n \in \mathbb{N}$, $\tau \in \widehat{\mathbb{R}} := \mathbb{R} \cup [\infty]$ and $z \in \mathbb{C} - \mathbb{R}$,

$$\mathcal{R}_n(z,\tau) := \frac{p_n(z,\tau)}{q_n(z,\tau)} = \frac{zp_n^*(z,\tau)}{q_n(z,\tau)} =: zR_n^*(z,\tau), \tau \in \mathbb{R}, \tag{3.16a}$$

and

$$\mathcal{R}_n(z,\infty) := \frac{p_{2n-1}(z)}{q_{2n-1}(z)} \tag{3.16b}$$

is called a *generalized nth approximant of the normal APT-fraction* (3.10) and $R_n^*(z,\tau)$ the *modified generalized nth approximant of* (3.10). From (2.44), (3.7) and (3.8) we obtain

$$\frac{p_{2n}(z,\tau)}{q_{2n}(z,\tau)} = \frac{C_{2n}(z) - \rho_{2n}\tau z C_{2n-1}(z)}{D_{2n}(z) - \rho_{2n}\tau z D_{2n-1}(z)} = S_{2n}(z, -\rho_{2n}\tau z), \tag{3.17a}$$

and

$$\frac{p_{2n+1}(z,\tau)}{q_{2n+1}(z,\tau)} = \frac{C_{2n+1}(z) - \rho_{2n+1}\tau C_{2n}(z)}{D_{2n+1}(z) - \rho_{2n+1}\tau D_{2n}(z)} = S_{2n+1}(z, -\rho_{2n+1}\tau), \tag{3.17b}$$

where

$$\rho_{2n} := \sqrt{\frac{e_{2n}}{d_{2n+1}}} \text{ for } n \in \mathbb{N} \quad \text{and} \quad \rho_{2n+1} := \sqrt{\frac{d_{2n+1}}{e_{2n+1}}} \text{ for } n \geq 0. \tag{3.17c}$$

Hence from (2.44) and (3.11) we have

$$R_n(z,\tau) := \frac{p_n(z,\tau)}{q_n(z,\tau)} = \frac{C_n(z,-\rho_n\tau)}{D_n(z,-\rho_n\tau)}, \text{ for } n \in \mathbb{N} \tag{3.18}$$

and therefore by (2.45)

$$R_n(z,\tau) := \frac{p_n(z,\tau)}{q_n(z,\tau)} \in \widehat{K}_n(z), \text{ for } z \in \mathbb{C} - \mathbb{R},\ \tau \in \widehat{\mathbb{R}},\ n \in \mathbb{N}. \tag{3.19}$$

From (3.12) and Theorem 2.7 we arrive at

THEOREM 3.2. *Let (3.10) be a given normal APT-fraction and let* $\{\tau_n\}_{n=1}^{\infty}$ *be an arbitrary sequence in* $\widehat{\mathbb{R}} := \mathbb{R} \cup [\infty]$. *For* $n \in \mathbb{N}$, *let* $\nu(n,\tau_n)$ *denote the number of zeros of* $q_n(z,\tau_n)$ *(see (3.9)). Let* $-\zeta_1^{(n)}(\tau_n), \ldots, -\zeta_{\nu(n,\tau_n)}^{(n)}(\tau_n)$ *denote these zeros. Then:*

*(A) For* $n \in \mathbb{N}$, *all zeros of* $q_n(z,\tau_n)$ *are real and simple and can be arranged so that*

$$-\zeta_1^{(n)}(\tau_n) < -\zeta_2^{(n)}(\tau_n) < \cdots < -\zeta_{\nu(n,\tau_n)}^{(n)}(\tau_n). \tag{3.20}$$

*Moreover,*

$$\nu(n,\tau_n) = \begin{cases} n-1, & \text{if } n = 2k \text{ and } \tau_n = q_{2k,k}/q_{2k-1,k-1}, \\ n-1, & \text{if } n = 2k+1 \text{ and } \tau_n = q_{2k+1,-(k+1)}/q_{2k,-k}, \\ n, & \text{otherwise.} \end{cases} \tag{3.21}$$

*(B) For* $n \in \mathbb{N}$, *there exist positive numbers* $\lambda_1^{(n)}(\tau_n), \ldots, \lambda_{\nu(n,\tau_n)}^{(n)}(\tau_n)$ *such that*

$$R_n(z,\tau_n) := \frac{p_n(z,\tau_n)}{q_n(z,\tau_n)} = \sum_{j=1}^{\nu(n,\tau_n)} \frac{z\lambda_j^{(n)}(\tau_n)}{z+\zeta_j^{(n)}(\tau_n)}, \tag{3.22a}$$

*and*

$$\sum_{j=1}^{\nu(n,\tau_n)} \lambda_j^{(n)}(\tau_n) = \mu_0 > 0. \tag{3.22b}$$

*(C) There exists a subsequence $\{\tau_{n_k}\}_{k=1}^{\infty}$ of $\{\tau_n\}_{n=1}^{\infty}$ such that $\{R_{n_k}(z,\tau_{n_k})\}_{k=1}^{\infty}$ converges locally uniformly on $U$ (and on $W$) to a function (of $z$) analytic on $U$ (and on $W$).*

*(D) For $m \in \mathbb{Z}_0^+$, $z \in \mathbb{C} - \mathbb{R}$ and $\tau \in \widehat{\mathbb{R}} := \mathbb{R} \cup [\infty]$,*

$$R_{2m}(z,\tau) = -\sum_{k=1}^{2m}(-1)^k \mu_{-k} z^k + O(z^{2m+1}) \text{ at } 0, \tag{3.23a}$$

$$R_{2m}(z,\tau) = \sum_{k=0}^{2m-2}(-1)^k \frac{\mu_k}{z^k} + O\left(\frac{1}{z^{2m-1}}\right) \text{ at } \infty, \tag{3.23b}$$

$$R_{2m+1}(z,\tau) = -\sum_{k=1}^{2m}(-1)^k \mu_{-k} z^k + O(z^{2m+1}) \text{ at } 0, \tag{3.23c}$$

$$R_{2m+1}(z,\tau) = \sum_{k=0}^{2m}(-1)^k \frac{\mu_k}{z^k} + O\left(\frac{1}{z^{2m+1}}\right) \text{ at } \infty. \tag{3.23d}$$

**Remarks on Proof.** (A): See [11, Theorem 2.2]. (B): From (A) and Theorem 3.1 (D) we obtain

$$\sum_{j=1}^{\nu(n,\tau_n)} \lambda_j^{(n)}(\tau_n) = \lim_{z\to\infty} \sum_{j=1}^{\nu(n,\tau_n)} \frac{\lambda_j^{(n)}(\tau_n)}{1+\frac{\lambda_j^{(n)}(\tau_n)}{z}} = \lim_{z\to\infty} \frac{p_n(z,\tau_n)}{q_n(z,\tau_n)}, \tag{3.24a}$$

and from (3.7), (3.9) and (3.12) one has

$$\lim_{z\to\infty} \frac{p_{2m}(z,\tau_{2m})}{q_{2m}(z,\tau_{2m})} = \frac{p_{2m,m} - \tau_{2m} p_{2m-1,m-1}}{q_{2m,m} - \tau_{2m} q_{2m-1,m-1}} = \frac{\lambda_1}{\beta_1} = \frac{F_1}{G_1} = \mu_0. \tag{3.24b}$$

(C): Let $K$ be an arbitrary subset of $U$ (or of $W$). Let $\delta$ and $B$ denote positive numbers satisfying

$$|z| \le B \text{ and } |z + \zeta_j^{(n)}(\tau_n)| \ge \delta, \text{ for } 1 \le j \le \nu(n,\tau_n), n \in \mathbb{N}, z \in K. \tag{3.25}$$

Then, for all $n \in \mathbb{N}$, $\tau_n \in \widehat{\mathbb{R}} := \mathbb{R} \cup [\infty]$ and $z \in K$,

$$\left|\frac{p_n(z,\tau_n)}{q_n(z,\tau_n)}\right| = \left|\sum_{j=1}^{\nu(n,\tau_n)} \frac{z\lambda_j^{(n)}(\tau_n)}{z+\zeta_j^{(n)}(\tau_n)}\right| \le \frac{B}{\delta} \sum_{j=1}^{\nu(n,\tau_n)} \lambda_j^{(n)}(\tau_n) = \frac{B\mu_0}{\delta}. \tag{3.26}$$

Therefore $\{R_n(z,\tau_n)\}_{n=1}^{\infty}$ is a sequence of rational functions (of $z$), analytic on $U$ (or on $W$) and uniformly bounded on every compact subset of $U$ (or of

$W$) (see, e.g., [1, p. 224]) and hence that there exists a subsequence $\{n_k\}_{k=1}^{\infty}$ of $\{n\}_{n=1}^{\infty}$ such that $\{R_{n_k}(z, \tau_{n_k})\}$ is locally uniformly convergent on $U$ (or on $W$). (D): Apply (2.13) in Theorem 2.1 together with (3.7), (3.5) and $f_n(z) = \frac{A_n(z)}{B_n(z)} = \frac{p_n(z)}{q_n(z)}$. □

# 4 COMPLETE CONVERGENCE

### Christoffel–Darboux Formulas

An APT-fraction (3.10) is called *completely convergent for a fixed* $z \in \mathbb{C}$ if there exists an $M(z) \in \mathbb{C}$ such that, for every sequence of real numbers $\{\tau_n\}_{n=1}^{\infty}$,

$$\lim_{n\to\infty} R_n^*(z, \tau_n) = \lim_{n\to\infty} \frac{p_n^*(z, \tau_n)}{q_n(z, \tau_n)} = M(z). \tag{4.1}$$

In (4.1) the limit $M(z)$ is independent of $\{\tau_n\}$. It can be seen that if a normal APT-fraction (3.10) is completely convergent for a fixed $z \in \mathbb{C}$, then all APT-fractions equivalent to (3.10) are also completely convergent at $z$. This section is used to describe important consequences of complete convergence that are subsequently used for strong moment theory. By using the recurrence formulas (3.11) and (3.7d) one can derive the following *Christoffel–Darboux formulas*.

THEOREM 4.1. (Christoffel–Darboux formulas [22]) *Let* $p_n^*(z) := p_n(z)/z$, *where* $p_n(z)$ *and* $q_n(z)$ *denote the nth numerator and denominator, respectively, of a normal APT-fraction (3.10). Then, for all* $z, \zeta \in \mathbb{C}$, *the following identities hold*:

$$z q_{2n-1}(z) q_{2n}(\zeta) - \zeta q_{2n}(z) q_{2n-1}(\zeta) = \frac{q_{2n,-n}}{q_{2n-1,-n}} (z - \zeta) \sum_{j=0}^{2n-1} q_j(z) q_j(\zeta),$$
$$\text{for } n \in \mathbb{N}; \tag{4.2a}$$

$$z q_{2n+1}(z) q_{2n}(\zeta) - \zeta q_{2n}(z) q_{2n+1}(z) = \frac{q_{2n+1,n}}{q_{2n,n}} (z - \zeta) \sum_{j=0}^{2n} q_j(z) q_j(\zeta),$$
$$\text{for } n \in \mathbb{Z}_0^+; \tag{4.2b}$$

$$z p_{2n-1}^*(z) q_{2n}(\zeta) - \zeta p_{2n}^*(z) q_{2n-1}(\zeta) = \frac{q_{2n,-n}}{q_{2n-1,-n}} \left[ 1 + (z - \zeta) \sum_{j=1}^{2n-1} p_j^*(z) q_j(\zeta) \right],$$
$$\text{for } n \in \mathbb{N}; \tag{4.2c}$$

$$zp^*_{2n+1}(z)q_{2n}(\zeta) - \zeta p^*_{2n}(z)q_{2n+1}(\zeta) = \frac{q_{2n+1,n}}{q_{2n,n}} \left[ 1 + (z-\zeta) \sum_{j=1}^{2n} p^*_j(z) q_j(\zeta) \right],$$
$$\text{for } n \in \mathbb{Z}_0^+; \qquad (2.4d)$$

$$zp^*_{2n-1}(z)p^*_{2n}(\zeta) - \zeta p^*_{2n}(z)p^*_{2n-1}(\zeta) = \frac{q_{2n,-n}}{q_{2n-1,-n}} (z-\zeta) \sum_{j=1}^{2n-1} p^*_j(z) p^*_j(\zeta),$$
$$\text{for } n \in \mathbb{N}; \qquad (4.2e)$$

$$zp^*_{2n+1}(z)p^*_{2n}(\zeta) - \zeta p^*_{2n}(z)p^*_{2n+1}(\zeta) = \frac{q_{2n+1,n}}{q_{2n,n}} (z-\zeta) \sum_{j=1}^{2n} p^*_j(z) p^*_j(\zeta),$$
$$\text{for } n \in \mathbb{Z}_0^+. \qquad (4.2f)$$

Our next result is a useful generalization of Theorem 4.1. We define

$$T_n(z,a,b) := ap^*_n(z) + bq_n(z), \text{ for } a,b \in \mathbb{C}, 0 \neq z \in \mathbb{C}, \qquad (4.3)$$

where $p^*_n(z)$ and $q_n(z)$ are as in Theorem 4.1. Then with the use of Theorems 3.1 (E) and 4.1, one can obtain the following:

THEOREM 4.2. (General Christoffel–Darboux Formulas [22]) *Let $a, b, \alpha, \beta \in \mathbb{C}$, $0 \neq z \in \mathbb{C}$ and $0 \neq \zeta \in \mathbb{C}$. For $n \in \mathbb{N}$, let $p_n(z)$ and $q_n(z)$ denote the nth numerator and denominator, respectively, of a normal APT-fraction (3.10) and let $p^*_n(z) := p_n(z)/z$. Let $T_n(z,a,b)$ be defined by (4.3). Then:*

$$zT_{2n-1}(z,a,b)T_{2n}(\zeta,\alpha,\beta) - \zeta T_{2n}(z,a,b)T_{2n-1}(\zeta,\alpha,\beta) =$$
$$\frac{q_{2n,-n}}{q_{2n-1,-n}} \left[ (\alpha\beta - ab) + (z-\zeta) \sum_{j=0}^{2n-1} T_j(z,a,b) T_j(\zeta,\alpha,\beta) \right], \qquad (4.4a)$$
*for $n \in \mathbb{N}$*

*and*

$$zT_{2n+1}(z,a,b)T_{2n}(\zeta,\alpha,\beta) - \zeta T_{2n}(z,a,b)T_{2n+1}(\zeta,\alpha,\beta) =$$
$$\frac{q_{2n+1,n}}{q_{2n,n}} \left[ (\alpha\beta - ab) + (z-\zeta) \sum_{j=0}^{2n} T_j(z,a,b) T_j(\zeta,\alpha,\beta) \right], \qquad (4.4b)$$
for $n \in \mathbb{Z}_0^+$.

Setting $a = \alpha = 1$, $b = w$ and $\zeta = \bar{z}$ in Theorem 4.2 yields:

COROLLARY 4.3. ([22]) *Let $T_n(z,a,b)$ be defined as in Theorem 4.2. Let $0 \neq w \in \mathbb{C}$ and $0 \neq z \in \mathbb{C} - \mathbb{R}$. Then:*

$$\begin{aligned} &zT_{2n-1}(z,1,w)\overline{T_{2n}(z,1,w)} - \bar{z}\overline{T_{2n-1}(z,1,w)}T_{2n}(z,1,w) \\ &= (z-\bar{z})\frac{q_{2n,-n}}{q_{2n-1,-n}}\left[-\frac{w-\bar{w}}{z-\bar{z}} + \sum_{j=0}^{2n-1} |T_j(z,1,w)|^2\right], \text{ for } n \in \mathbb{N} \end{aligned} \tag{4.5a}$$

*and*

$$\begin{aligned} &zT_{2n+1}(z,1,w)\overline{T_{2n}(z,1,w)} - \bar{z}\overline{T_{2n+1}(z,1,w)}T_{2n}(z,1,w) \\ &= (z-\bar{z})\frac{q_{2n+1,n}}{q_{2n,n}}\left[-\frac{w-\bar{w}}{z-\bar{z}} + \sum_{j=0}^{2n} |T_j(z,1,w)|^2\right], \text{ for } n \in \mathbb{Z}_0^+. \end{aligned} \tag{4.5b}$$

**Limit Point Case and Limit Circle Case**

For a given normal APT-fraction (3.10), for each $z \in \mathbb{C} - \mathbb{R}$ and for each $n \in \mathbb{N}$, we define

$$\Gamma_n(z) := \left[w = R_n^*(z,\tau) := \frac{p_n^*(z.\tau)}{q_n(z,\tau)} : \tau \in \widehat{\mathbb{R}} := \mathbb{R} \cup [\infty]\right], \tag{4.6a}$$

and

$$\Delta_n(z) := \Gamma_n(z) \cup \operatorname{Int} \Gamma_n(z). \tag{4.6b}$$

From mapping properties of linear fractional transformations one can see that $\Delta_n(z)$ is a closed circular disk with boundary $\Gamma_n(z)$

THEOREM 4.4. *Let (3.10) be a given normal APT-fraction and let $z \in \mathbb{C} - \mathbb{R}$. Then:*
*(A) for $n \in \mathbb{N}$,*

$$\Delta_n(z) = \left[w \in \mathbb{C} : \sum_{j=1}^{n-1} |T_j(z,1,w)|^2 \leq \frac{w-\bar{w}}{z-\bar{z}}\right]. \tag{4.7}$$

*(B) $\{\Delta_n(z)\}_{n=1}^{\infty}$ is a nested sequence of closed circular disks.*
*(C) If $\rho_n(z)$ denotes the radius of $\Delta_n(z)$, then for $n \in \mathbb{N}$,*

$$\rho_n(z) = \frac{1}{|z-\bar{z}| \sum_{j=0}^{n-1} |q_j(z)|^2}. \tag{4.8}$$

PROOF OUTLINE. (A): From (4.6a), (4.3) and (3.11) and $p_n^*(z) := p_n(z)/z$ we obtain

$$\Gamma_{2m}(z) = \left[w \in \mathbb{C} : -\mathrm{Im}\left(\frac{T_{2m}(z,1,w)}{zT_{2m-1}(z,1,w)}\right) = 0\right], \quad m \in \mathbb{N}. \tag{4.9}$$

It follows that the left side of (4.5a) vanishes if and only if $w \in \Gamma_{2m}(z)$. This combined with (4.5a) yields

$$\Gamma_{2m}(z) = \left[w \in \mathbb{C} : \sum_{j=0}^{2m-1} |T_j(z,1.w)| = \frac{w - \bar{w}}{z - \bar{z}}\right].$$

A standard mapping argument can be used to conclude that

$$\mathrm{Int}\, \Delta_{2m}(z) = \left[w \in \mathbb{C} : \sum_{j=0}^{2m-1} |T_j(z.1,w)|^2 < \frac{w - \bar{w}}{z - \bar{z}}\right].$$

A similar argument applies for $n = 2m + 1$.

(B): Recall from (2.41e) in Theorem 2.6 that, for $z \in \mathbb{C} - \mathbb{R}$, $\{\widehat{K}_n(z)\}_1^\infty$ is a nested sequence of closed circular disks, where $\widehat{K}_n(z) := \partial K_n(z) \cup K_n(z)$ and $K_n(z)$ is defined by (2.40). It follows from (2.39), (2.40), (3.11), (3.18) and $p_n(z) = zp_n^*(z)$ that

$$\partial K_n(z) = z\Gamma_n(z) \text{ and } \widehat{K}_n(z) = z\Delta_n(z), \quad n \in \mathbb{N}. \tag{4.10}$$

It follows that $\{\Delta_n(z)\}_1^\infty$ is a nested sequence of closed circular disks.

(C) can be derived from considerations of linear fractional transformations and applications of the Christoffel–Darboux formulas (4.2). The argument employed in [11, Theorem 4.3] makes use of an integral expressions for the circumference of a circle. □

Theorem 4.4 (C) implies that, for a given normal APT-fraction (3.10), the sequence of radii $\{\rho_n(z)\}_{n=1}^\infty$ is nonincreasing and bounded below by zero. Therefore, for $z \in \mathbb{C} - \mathbb{R}$,

$$\rho(z) := \lim_{n\to\infty} \rho_n(z) \geq 0. \tag{4.11}$$

We define

$$\Delta(z) := \bigcap_{n=1}^{\infty} \Delta_n(z), \quad z \in \mathbb{C} - \mathbb{R}, \tag{4.12}$$

and we say that the *limit circle case occurs* at $z \in \mathbb{C} - \mathbb{R}$ if

$$\rho(z) > 0 \quad \text{and} \quad \Delta(z) \quad \text{is a (proper) disk} \tag{4.13a}$$

and the *limit point case occurs at* $z \in \mathbb{C} - \mathbb{R}$ if

$$\rho(z) = 0 \quad \text{and} \quad \Delta(z) \quad \text{is a point in } \mathbb{C}. \tag{4.13b}$$

An expression for $\rho_n(z)$ can be derived that is similar to (4.8), but with $q_j(z)$ replaced by $p_j^*(z)$. From this and (4.8) one obtains the following:

THEOREM 4.5. *Let (3.10) be a given normal APT-fraction and let $z \in \mathbb{C} - \mathbb{R}$. Then:*
*(A) (Limit Circle Case) If $\rho(z) > 0$, then*

$$\sum_{j=1}^{\infty} |p_j^*(z)|^2 < \infty \text{ and } \sum_{j=1}^{\infty} |q_j(z)|^2 < \infty. \tag{4.14a}$$

*(B) (Limit Point Case) If $\rho(z) = 0$, then*

$$\sum_{j=1}^{\infty} |p_j^*(z)|^2 = \infty \text{ and } \sum_{j=1}^{\infty} |q_j(z)|^2 = \infty. \tag{4.14b}$$

*(C) The APT-fraction converges completely at $z \in \mathbb{C} - \mathbb{R}$ if and only if the limit point case occurs at $z$.*

*Remark.* Part (C) of Theorem 4.5 follows from (A) and (B), since $\Gamma_n(z) = \partial\Delta_n(z)$ consists of the generalized approximants $R_n^*(z, \tau)$ of the APT-fraction (3.10).

Using an idea that originated in work of H. Weyl [24] (see [11, Theorem 4.5] and references cited therein), one can prove:

THEOREM 4.6. (Theorem of Invariability [11, Theorem 4.5]) *Let (3.10) be a given normal APT-fraction. Then:*

*(A) The limit point case occurs at all $z \in \mathbb{C} - \mathbb{R}$ if and only if it occurs at one $z \in \mathbb{C} - \mathbb{R}$.*

*(B) If the limit point case occurs at all $z \in \mathbb{C} - \mathbb{R}$, then the infinite series*

$$\sum_{j=1}^{\infty} |p_j(z)|^2 \quad \text{and} \quad \sum_{j=1}^{\infty} |q_j(z)|^2$$

*converge locally uniformly on $\mathbb{C} - [0]$.*

*(C) If an APT-fraction converges completely for one value of $z \in \mathbb{C} - \mathbb{R}$, then it converges completely for all $z \in \mathbb{C} - \mathbb{R}$.*

## 5 STRONG MOMENT THEORY

### Main Results

We recall that a bisequence of real numbers $\{\mu_n\}_{n=-\infty}^{\infty}$ is called *regular and positive definite* if it satisfies the APT-conditions

$$H_{2n}^{(-2m)} > 0,\ H_{2n+1}^{(-2n)} > 0,\ H_{2n}^{(-2n+1)} \neq 0,\ H_{2n+1}^{(-2n-1} \neq 0,\ \text{for } n \in \mathbb{N} \quad (5.1)$$

(see (2.14) and (2.33)). Moreover, the APT-conditions (5.1) are necessary and sufficient for the existence of an APT-fraction that corresponds to the pair $(L_0, L_\infty)$ of formal power series

$$L_0(z) := -\sum_{k=1}^{\infty}(-1)^k \mu_{-k} z^k \text{ and } L_\infty(z) := \sum_{k=0}^{\infty}(-1)^k \frac{\mu_k}{z^k} \quad (5.2)$$

(see Theorem 2.1). Equivalent forms of APT-fractions are given by (2.34), (2.35), (2.36) and (3.10). This section is used to describe strong moment theory that can be obtained for regular bisequences using APT-fractions. We begin by stating the main results in the form of three theorems.

THEOREM 5.1. (Existence [11]) *If $\{\mu_n\}_{n=-\infty}^{\infty}$ is a regular and positive definite bisequence of real numbers, then there exists a strong moment distribution function (***SMDF***) $\psi(t)$ on $\mathbb{R}$ such that*

$$\mu_n = \int_{-\infty}^{\infty} t^n d\psi(t), \quad n \in \mathbb{Z} := [0, \pm 1, \pm 2, \ldots]; \quad (5.3)$$

*that is, the SHMP for $\{\mu_n\}_{-\infty}^{\infty}$ has a solution.*

THEOREM 5.2. (Uniqueness [22]) *If $\{\mu_n\}_{n=-\infty}^{\infty}$ is a regular and positive bisequence of real numbers, then the SHMP for $\{\mu_n\}_{-\infty}^{\infty}$ is determinate (i.e., has a unique solution) if and only if the APT-fraction corresponding to the pair of formal power series (5.2) is completely convergent (i.e., the limit point case holds). (See Theorems 4.5 and 4.6).*

Before stating Theorem 5.3, we define Nevanlinna functions. The class $\mathcal{N}$ of Nevanlinna functions is defined by

$$\mathcal{N} := [\phi : \phi \text{ is analytic on } U \text{ and } \phi(U) \subseteq \overline{U} - [\infty]]. \quad (5.4)$$

Here $U$ denotes the open upper half-plane in $\mathbb{C}$ and $\overline{U}-[\infty]$ denotes the closed finite upper half-plane. The extended class $\mathcal{N}^*$ of Nevanlinna functions is defined by

$$\mathcal{N}^* := \mathcal{N} \cup [\infty]. \tag{5.5}$$

One can see that

$$\mathcal{N} = \mathcal{N}_1 \cup \mathcal{N}_2, \quad \text{where} \tag{5.6a}$$

$$\mathcal{N}_1 := [\phi : \phi \text{ is analytic on } U \text{ and } \phi(U) \subseteq U], \tag{5.6b}$$

$$\mathcal{N}_2 := [\phi : \phi(z) \equiv r, \text{ for some } r \in \mathbb{R} \text{ and all } z \in U] \tag{5.6c}$$

and

$$\mathcal{N}^* = \mathcal{N}_1 \cup \mathcal{N}_2^*, \quad \text{where} \tag{5.7a}$$

$$\mathcal{N}_2^* := [\phi : \phi(z) \equiv r, \text{ for some } r \in \widehat{\mathbb{R}} := \mathbb{R} \cup [\infty] \text{ and all } z \in U]. \tag{5.7b}$$

For a given regular and positive definite bisequence $\{\mu_n\}_{-\infty}^{\infty}$ and corresponding normal APT-fraction (3.10), we define, for $z \in \mathbb{C} - [0]$,

$$\alpha(z) := \frac{(x_0 - z)}{x_0 z} \sum_{j=1}^{\infty} p_j(x_0) p_j(z), \tag{5.8a}$$

$$\beta(z) := \frac{1}{z} + \frac{(x_0 - z)}{x_0 z} \sum_{j=1}^{\infty} p_j(x_0) q_j(z), \tag{5.8b}$$

$$\gamma(z) := -z + (x_0 - z) \sum_{j=1}^{\infty} q_j(x_0) p_j(z), \tag{5.8c}$$

$$\delta(z) := (x_0 - z) \sum_{j=0}^{\infty} q_j(x_0) q_j(z), \tag{5.8d}$$

where $p_j(z)$ and $q_j(z)$ are defined by (3.7).

It is subsequently shown that the infinite series in (5.8) all converge locally uniformly on $\mathbb{C} - [0]$ to analytic functions, which satisfy

$$\alpha(z)\delta(z) - \beta(z)\gamma(z) \equiv 1, \quad \text{for } z \in \mathbb{C} - [0]. \tag{5.9}$$

THEOREM 5.3. (Nevanlinna–Njåstad [18]) *Let $\{\mu_n\}_{n=-\infty}^{\infty}$ be a regular and positive definite bisequence of real numbers for which the SHMP is indeterminate (i.e., there exist more solutions than one). Let $\alpha(z), \beta(z), \gamma(z), \delta(z)$*

*be defined by (5.8). Then there exists a one-to-one correspondence between the set* $\mathcal{N}^*$ *and the set*

$$M(\{\mu_n\}_{-\infty}^{\infty}) := [\mu(z) : \mu(z) \text{ is a solution to the SHMP for } \{\mu_n\}_{-\infty}^{\infty}]. \tag{5.10}$$

*This correspondence is given by*

$$G_\psi(z) := \int_{-\infty}^{\infty} \frac{z d\psi(t)}{z+t} = \frac{x_0 z\alpha(z) - \gamma(z)\phi(z)}{x_0 z\beta(z) - \delta(z)\phi(z)}. \tag{5.11}$$

**Proof Outlines**

In the following we make use of three theorems attributed to Helly; proofs can be found in [9, p. 56] and [17, pp. 232–241].

THEOREM 5.4 (Helly 1) *Let* $[a,b]$ *denote a closed interval in* $\mathbb{R}$ *(possibly* $[a,b] = \mathbb{R}$*). Let* $\{\phi_n(t)\}_{n=1}^{\infty}$ *be a sequence of real-valued, non-decreasing functions defined and uniformly bounded on* $[a,b]$*. Then there exists a subsequence* $\{\phi_{n_k}(t)\}_{k=1}^{\infty}$ *that converges to a bounded, non-decreasing function* $\phi(t)$ *on* $[a,b]$*.*

THEOREM 5.5 (Helly 2) *Let* $\{\phi_n(t)\}_{n=1}^{\infty}$ *be a sequence of real-valued, non-decreasing and uniformly bounded functions on a bounded closed interval* $[a,b]$ *in* $\mathbb{R}$ *and let* $\{\phi_n(t)\}_{n=1}^{\infty}$ *converge to a bounded function* $\phi(t)$ *at every* $t \in [a,b]$*. Let* $f(t)$ *be a real-valued function that is defined and continuous on* $[a,b]$*. Then*

$$\lim_{n\to\infty} \int_a^b f(t)d\phi_n(t) = \int_a^b f(t)d\phi(t). \tag{5.12}$$

THEOREM 5.6 (Helly 3) *Let* $\{\phi_n(t)\}_{n=1}^{\infty}$ *be a sequence of real-valued, non-decreasing and uniformly bounded functions on* $\mathbb{R}$ *and let* $\{\phi_n(t)\}_{n=1}^{\infty}$ *converge to a bounded function* $\phi(t)$ *at every* $t \in \mathbb{R}$*. Let* $f(t)$ *be a real-valued function that is defined and continuous on* $\mathbb{R}$ *and satisfies*

$$\lim_{t\to\infty} f(t) = \lim_{t\to-\infty} f(t) = 0. \tag{5.13}$$

*Then*

$$\lim_{n\to\infty} \int_{-\infty}^{\infty} f(t)d\phi_n(t) = \int_{-\infty}^{\infty} f(t)d\phi(t). \tag{5.14}$$

PROOF OF THEOREM 5.1. (Existence) Let (3.10) be the normal APT-fraction that corresponds to the regular and positive definite bisequence

$\{\mu_n\}_{-\infty}^{\infty}$. Let $\{\tau_n\}_{n=1}^{\infty}$ be a sequence in $\widehat{\mathbb{R}} := \mathbb{R} \cup [\infty]$ and let $\{R_n(z,\tau_n)\}_{n=1}^{\infty}$ denote the sequence of generalized approximants of the APT-fraction. Then by use of the partial fraction expansion (3.22) (Theorem 3.2), we can express $R_n(z,\tau_n)$ as a Stieltjes integral

$$R_n(z,\tau_n) = \sum_{j=1}^{\nu(n,\tau_n)} \frac{z\lambda_j^{(n)}(\tau_n)}{z+\zeta_j^{(n)}(\tau_n)} = \int_{-\infty}^{\infty} \frac{z\,d\psi_n(t)}{z+t}, \; n \in \mathbb{N}, \tag{5.15}$$

where $\{\psi_n(t)\}_{n=1}^{\infty}$ is the sequence of step-functions defined by

$$\psi_n(t) := \begin{cases} 0, \text{ if } -\infty < t \le -\zeta_1^{(n)}(\tau_n), \\ \sum_{j=1}^{m} \lambda_j^{(n)}(\tau_n), \text{ if } -\zeta_m^{(n)}(\tau_n) < t \le -\zeta_{m+1}^{(n)}(\tau_n), 1 < m < \nu(n,\tau_n), \\ \mu_0 = \sum_{j=1}^{\nu(n,\tau_n)} \lambda_j(\tau_n), \text{ if } \zeta_{\nu(n,\tau_n)}^{(n)}(\tau_n) < t < \infty. \end{cases} \tag{5.16}$$

Here $\nu(n,\tau_n)$, $\zeta_j^{(n)}(\tau_n)$ and $\lambda_j^{(n)}(\tau_n)$ are described by Theorem 3.2. It follows from (3.22b) and (5.16) that $\{\psi_n(t)\}$ is a sequence of real-valued, non-decreasing functions defined and uniformly bounded on $\mathbb{R}$. Hence by Theorem 5.4 (Helly 1) there exists a subsequence $\{\psi_{n_k}(t)\}_{k=1}^{\infty}$ that converges to a bounded, non-decreasing function $\psi(t)$ on $\mathbb{R}$. Let $z$ be an arbitrary (fixed) point in $\mathbb{C} - \mathbb{R}$, and let $f(t), f_1(t), f_2(t)$ be defined, for $t \in \mathbb{R}$, by

$$f(t) := \frac{z}{z+t}, \; f_1(t) := \operatorname{Re} f(z), \; f_2(t) := \operatorname{Im} f(z). \tag{5.17}$$

It is readily seen that $f_1(t)$ and $f_2(t)$ are real-valued and continuous on $\mathbb{R}$ and

$$\lim_{t\to\infty} f_k(t) = \lim_{t\to-\infty} f_k(t) = 0, \quad k = 1,2. \tag{5.18}$$

Thus by Theorem 5.6 (Helly 3),

$$\lim_{k\to\infty} R_{n_k}(z,\tau_{n_k}) = \int_{-\infty}^{\infty} \frac{z\,d\psi(t)}{z+t} =: G_\psi(z), \; z \in \mathbb{C} - \mathbb{R}. \tag{5.19}$$

By using (5.19) and the correspondence properties of $R_n(z,\tau)$ in (3.23) combined with an argument used in [16, Theorem 4.1], one can show that

$$\mu_n = \int_{-\infty}^{\infty} t^n\,d\psi(t), \quad \text{for} \quad n \in \mathbb{Z}. \tag{5.20}$$

From the theory of positive definite quadratic forms it can be shown that $\psi(t)$ has infinitely many points of increase (see, e.g., [12, Proposition 2.2]). $\square$

Before proceeding with a proof of Theorem 5.2, we state the following result [11, Theorem 6.1].

THEOREM 5.7. (Asymptotics) *Let* $\{\mu_n\}_{n=-\infty}^{\infty}$ *be a given regular and positive definite bisequence of real numbers. Let (3.10) be the normal APT-fraction corresponding to the pair* $(L_0, L_\infty)$ *of power series*

$$L_0(z) := -\sum_{k=1}^{\infty}(-1)^k \mu_k z^k, \quad L_\infty(z) := \sum_{k=0}^{\infty}(-1)^k \frac{\mu_k}{z^k}. \tag{5.21}$$

*Let* $\{R_{n_k}(z, \tau_{n_k})\}_{k=1}^{\infty}$ *be a sequence of generalized approximants of the normal APT-fraction that converges on* $\mathbb{C} - \mathbb{R}$ *to a Stieltjes integral as in (5.19). Let*

$$G_\psi(z) := \int_{-\infty}^{\infty} \frac{z d\psi(t)}{z+t} = \lim_{k\to\infty} R_{n_k}(z, \tau_{n_k}), \quad z \in \mathbb{C} - \mathbb{R}. \tag{5.22}$$

*Then:*

*(A)* $G_\psi(z)$ *is analytic in* $U$ *and in* $W$.

*(B) For* $\lambda > 0$, *let*

$$U_\lambda := \left[ z \in \mathbb{C} : Im(z) > 0 \text{ and } \left| \frac{Im\, z}{z} \right| > \lambda \right], \tag{5.23a}$$

$$W_\lambda := \left[ z \in \mathbb{C} : Im(z) < 0 \text{ and } \left| \frac{Im\, z}{z} \right| > \lambda \right]. \tag{5.23b}$$

*Then* $L_0(z)$ *is the asymptotic expansion of* $G_\psi(z)$ *as* $z \to 0$, *for* $z \in U_\lambda$ *and for* $z \in W_\lambda$ *and* $L_\infty(z)$ *is the asymptotic expansion of* $G_\psi(z)$ *as* $z \to \infty$, *for* $z \in U_\lambda$ *and* for $z \in W_\lambda$.

PROOF OF THEOREM 5.2. (Uniqueness) Let (3.10) be the normal APT-fraction corresponding to the bisequence $\{\mu_n\}_{-\infty}^{\infty}$. Let $R_n^*(z,\tau) = R_n(z,\tau)/z$ denote the $n$th modified generalized approximant of the APT-fraction (see (3.16)). For $z \in \mathbb{C} - \mathbb{R}$, let $\widehat{K}_n(z) := \partial K_n(z) \cup K_n(z)$, where $\partial K_n(z)$ and $K_n(z)$ are defined by (2.40) and let $\Delta_n(z) := \Gamma_n(z) \cup \operatorname{Int} \Gamma_n(z)$, where $\Gamma_n(z)$ is the circle defined by (4.7a), so that

$$\partial K_n(z) = z\Gamma_n(z) \text{ and } \widehat{K}_n(z) = z\Delta_n(z), \text{ for } n \in \mathbb{N}. \tag{5.24}$$

From (3.19) we obtain

$$R_n^*(z,\tau) \in \Delta_n(z), \quad \text{for all } z \in \mathbb{C}-\mathbb{R}, \tau \in \widehat{\mathbb{R}}, n \in \mathbb{N}. \tag{5.25}$$

(a) Suppose that *the limit point case holds*. Let $\{\tau_n\}$ be an arbitrary sequence in $\widehat{\mathbb{R}}$. Then by Theorem 4.6, $\{R_n^*(z,\tau_n)\}_1^\infty$ converges to a limit $L(z) \in \mathbb{C}$, for $z \in \mathbb{C}-\mathbb{R}$, and

$$L(z) := \lim_{n\to\infty} R_n^*(z,\tau_n) \in \bigcap_{n=1}^{\infty} \Delta_n(z), \quad z \in \mathbb{C}-\mathbb{R}; \tag{5.26}$$

Hence $L(z)$ is the limit point. By an argument used in the proof of Theorem 5.1, there exists a strong moment distribution function $\psi(t)$ satisfying (5.20) and

$$F_\psi(z) := \int_{-\infty}^{\infty} \frac{d\psi(t)}{z+t} = L(z), \quad z \in \mathbb{C}-\mathbb{R}. \tag{5.27}$$

An inversion of the Stieltjes transform in (5.27) shows that $\psi(t)$ is "essentially" uniquely determined by $L(z)$. Thus in the limit point case the SHMP for $\{\mu_n\}_{-\infty}^{\infty}$ is determinate (i.e., has an "essentially" unique solution). By essentially unique, it is meant that two solutions may differ in value at most at points of discontinuity.

(b) Suppose that *the limit circle case occurs*. Then (see (4.13)) $\Delta(z) := \bigcap_{n=1}^{\infty} \Delta_n(z)$ has radius $\rho(z) > 0$ for all $z \in \mathbb{C}-\mathbb{R}$. Let $p_1$ and $p_2$ be two distinct points on the boundary $\Gamma(z)$ of the disk $\Delta_n(z)$. Let $z_0 \in \mathbb{C}-\mathbb{R}$ be given. Then there exist sequences $\{\tau_n^{(1)}\}$ and $\{\tau_n^{(2)}\}$ in $\widehat{\mathbb{R}}$ such that

$$\lim_{n\to\infty} R_n(z_0,\tau_n^{(1)}) = p_1 \text{ and } \lim_{n\to\infty} R_n(z_0,\tau_n^{(2)}) = p_2. \tag{5.28}$$

By the argument used to prove Theorem 5.1, there exist subsequences $\{\tau_{n_k}^{(m)}\}$, $m = 1,2$, and there exist strong moment distribution functions $\psi_1(t)$ and $\psi_2(t)$ on $\mathbb{R}$ such that

$$F_{\psi_m}(z) := \int_{-\infty}^{\infty} \frac{d\psi_m(t)}{z+t} = \lim_{k\to\infty} R_n^*(z,\tau_{n_k}^{(m)}), \; m = 1,2, z \in \mathbb{C}-\mathbb{R}. \tag{5.29}$$

Both $F_{\psi_1}(z)$ and $F_{\psi_2}(z)$ are analytic in $U$ (and in $W$). Now we assume that $\psi_1(t)$ and $\psi_2(t)$ are "essentially" the same distribution functions. It follows from this and (5.29) that $F_{\psi_1}(z) = F_{\psi_2}(z)$ for all $z \in \mathbb{C}-\mathbb{R}$. Then from (5.28) and (5.29) we conclude that $p_1 = F_{\psi_1}(z_0) = F_{\psi_2}(z_0) = p_2$, since $z_0 \in \mathbb{C}-\mathbb{R}$. This contradicts the fact that $p_1 \neq p_2$. Therefore $\psi_1$ and $\psi_2$ are essentially equal. □

In the remaining part of this section we sketch a proof of Theorem 5.3 which emphasizes *continued fraction properties* of the corresponding APT-fraction. Nevertheless, our proof has many similarities with the one given by Njåstad in [18]. We begin with several lemmas that are subsequently employed and we make use of two forms (2.35) and (3.10)) of the APT-fraction which corresponds to the given bisequence $\{\mu_n\}_{-\infty}^{\infty}$. We also utilize the notations given in previous sections.

LEMMA 5.8. *For all* $z, \xi \in \mathbb{C}-[0]$ *and all* $m \in \mathbb{Z}_0^+$ *the following Christoffel–Darboux-type identities hold: (A)*

$$\xi D_{2m}(z)C_{2m-1}(\xi)-zD_{2m-1}(z)C_{2m}(\xi) = (\xi z)^m\left[(\xi - z)\sum_{k=0}^{2m-1} q_k(z)p_k(\xi) + \xi\right]; \tag{5.30a}$$

*(B)*

$$\xi C_{2m}(z)D_{2m-1}(\xi)-zC_{2m-1}(z)D_{2m}(\xi) = (\xi z)^m\left[(\xi - z)\sum_{k=0}^{2m-1} p_k(z)q_k(\xi) - z\right]; \tag{5.30b}$$

*(C)*

$$\xi D_{2m}(z)D_{2m-1}(\xi) - zD_{2m-1}(z)D_{2m}(\xi) = (\xi z)^m(\xi - z)\sum_{k=0}^{2m-1} q_k(z)q_k(\xi); \tag{5.30c}$$

*(D)*

$$\xi C_{2m}(z)C_{2m-1}(\xi) - zC_{2m-1}(z)C_{2m}(\xi) = (\xi z)^m(\xi - z)\sum_{k=0}^{2m-1} p_k(z)p_k(\xi); \tag{5.30d}$$

**Remarks.** For $C_n$ and $D_n$, see (2.20) and (2.35); for $p_k$ and $q_k$ see (3.7a and d); for connections see (3.8). For Christoffel–Darboux formulas see Theorems 4.1 and 4.2.

Let $S, S_1$ and $S_2$ be defined by

$$S_1 := \bigcup_{m=1}^{\infty}[z \in \mathbb{C} : e_{2m}D_{2m-1}(z) + zD_{2m-2}(z) = 0], \tag{5.31a}$$

$$S_2 := \bigcup_{m=1}^{\infty}[z \in \mathbb{C} : D_{2m-1}(z) = 0], \tag{5.31b}$$

$$S := \mathbb{R}^- - [S_1 \cup S_2 \cup [0]], \ \mathbb{R}^- := [x \in \mathbb{R} : x < 0]. \tag{5.31c}$$

It is clear that $S$ is not empty. Let $x_0 \in S$ be given. We define, for all $m \in \mathbb{N}$,

$$\alpha_m(z) := \frac{x_0 - z}{x_0 z} \sum_{k=0}^{2m-1} p_k(x_0) p_k(z), \tag{5.31d}$$

$$\beta_m(z) := \frac{1}{z} + \frac{x_0 - z}{x_0 z} \sum_{k=0}^{2m-1} p_k(x_0) q_k(z), \tag{5.31e}$$

$$\gamma_m(z) := -z + (x_0 - z) \sum_{k=0}^{2m-1} q_k(x_0) p_k(z), \tag{5.31f}$$

$$\delta_m(z) := (x_0 - z) \sum_{k=0}^{2m-1} q_k(x_0) q_k(z). \tag{5.31g}$$

From (2.7), (2.18) and (2.30) we obtain

$$\alpha_m(z)\delta_m(z) - \beta_m(z)\gamma_m(z) = 1, \text{ for } z \in \mathbb{C} - [0], m \in \mathbb{N}. \tag{5.32}$$

Solving for $C_{2m}(z), C_{2m-1}(z), D_{2m}(z), D_{2m-1}(z)$ from (5.30) and (5.31) yields, for $m \in \mathbb{N}$ and $z \in \mathbb{C} - [0]$,

$$C_{2m}(z) = \frac{z^m}{x_0^{m+1}} [x_0 z D_{2m}(x_0) \alpha_m(z) - C_{2m}(x_0) \gamma_m(z)], \tag{5.33a}$$

$$C_{2m-1}(z) = \frac{z^{m-1}}{x_0^m} [x_0 z D_{2m-1}(x_0) \alpha_m(z) - C_{2m-1}(x_0) \gamma_m(z)], \tag{5.33b}$$

$$D_{2m}(z) = \frac{z^m}{x_0^{m+1}} [x_0 z D_{2m}(x_0) \beta_m(z) - C_{2m}(x_0) \delta_m(z)], \tag{5.33c}$$

$$D_{2m-1}(z) = \frac{z^{m-1}}{x_0^m} [x_0 z D_{2m-1}(x_0) \beta_m(z) - C_{2m-1}(x_0) \delta_m(z)]. \tag{5.33d}$$

By substituting (5.33) into (2.42) and simplifying, we obtain, for $m \in \mathbb{N}$ and $z \in \mathbb{C} - [0]$,

$$\frac{C_{2m}(z, \tau)}{D_{2m}(z, \tau)} = \frac{x_0 z \alpha_m(z) - \gamma_m(z) \theta_m(\tau)}{x_0 z \beta_m(z) - \delta_m(z) \theta_m(\tau)}, \tag{5.34a}$$

where

$$\theta_m(\tau) = \frac{C_{2m}(x_0) + \tau x_0 C_{2m-1}(x_0)}{D_{2m}(x_0) + \tau x_0 D_{2m-1}(x_0, \tau)}. \tag{5.34b}$$

It follows from (3.18) and the argument used to prove Theorem 5.1 that there exists a sequence $\{\tau_{m_k}\}_{k=1}^{\infty}$ in $\mathbb{R}$ and there exists a solution $\psi(t)$ to the SHMP for $\{\mu_n\}_{-\infty}^{\infty}$ such that

$$G_\psi(z) := \int_{-\infty}^{\infty} \frac{z\,d\psi(t)}{z+t} = \lim_{k\to\infty} \frac{C_{2m_k}(z,\tau_{m_k})}{D_{2m_k}(z,\tau_{m_k})}, \quad z \in \mathbb{C} - \mathbb{R}, \tag{5.35}$$

and $G_\psi(z)$ is analytic in $U$ and in $W$.

LEMMA 5.9. *The sequences $\{\alpha_m(z)\}$, $\{\beta_m(z)\}$, $\{\gamma_n(z)\}$ and $\{\delta_m(z)\}$ converge locally uniformly on $\mathbb{C} - [0]$ to functions $\alpha(z), \beta(z), \gamma(z)$ and $\delta(z)$, respectively. These functions are analytic on $\mathbb{C} - [0]$ and they satisfy:*

$$\alpha(z) = \frac{x_0 - z}{x_0 z} \sum_{k=0}^{\infty} p_k(x_0) p_k(z), \quad z \in \mathbb{C} - [0], \tag{5.36a}$$

$$\beta(z) = \frac{1}{z} + \frac{x_0 - z}{x_0 z} \sum_{k=0}^{\infty} p_k(x_0) q_k(z), \quad z \in \mathbb{C} - [0], \tag{5.36b}$$

$$\gamma(z) = -z + (x_0 - z) \sum_{k=0}^{\infty} q_k(x_0) p_k(z), \quad z \in \mathbb{C} - [0], \tag{5.36c}$$

$$\delta(z) = (x_0 - z) \sum_{k=0}^{\infty} q_k(x_0) q_k(z), \quad z \in \mathbb{C} - [0], \tag{5.36d}$$

*and*

$$\alpha(z)\delta(z) - \beta(z)\gamma(z) = 1, \quad \textit{for} \quad z \in \mathbb{C} - [0]. \tag{5.36e}$$

*Moreover, there exists a $\theta \in \mathbb{R} \cup [\infty] =: \widehat{\mathbb{R}}$ and a solution $\psi$ to the SHMP for $\{\mu_n\}_{-\infty}^{\infty}$ such that*

$$G_\psi(z) := \int_{-\infty}^{\infty} \frac{z\,d\psi(t)}{z+t} = \frac{x_0 z \alpha(z) - \theta\gamma(z)}{x_0 z \beta(z) - \theta\delta(z)} \tag{5.37}$$

PROOF OF LEMMA 5.9. By hypothesis the SHMP for $\{\mu_n\}_{-\infty}^{\infty}$ is indeterminant. Hence by Theorems 4.5, 4.6 and 5.2 we have

$$\sum_{j=1}^{\infty} |p_j(z)|^2 < \infty \quad \text{and} \quad \sum_{j=1}^{\infty} |q_j(z)|^2 < \infty, \text{ for } z \in \mathbb{C} - [0]. \tag{5.38}$$

Let $K$ be an arbitrary compact subset of $\mathbb{C} - [0]$. Then by [11, pp. 374–375] there exist real numbers $M_1$ and $M_2$ (depending on $K$ but independent of $n$) such that, for all $z \in \mathbb{C} - \mathbb{R}$,

$$0 < M_1 < |u| < M_2 < \infty, \quad \text{for} \quad u \in K \tag{5.39a}$$

and

$$\log\left(1 + \sum_{j=1}^{n} |q_j(u)|^2\right) \leq T_n(M_1, M_2, z),$$

where

$$T_n(M_1, M_2, z) := \left(1 + \frac{M_2}{M_1}\right)\left[2\sum_{j=1}^{n-1} |q_j(z)|^2 \sum_{j=1}^{n-1} |p_j(z)|^2 + \sum_{j=1}^{n} |q_j(z)|^2\right] \tag{5.39b}$$

It follows that, for $u \in K$ and $z \in \mathbb{C} - \mathbb{R}$,

$$\sum_{j=1}^{\infty} |q_j(u)|^2 \leq A(z) := e^{T(M_1, M_2, z)} < \infty \tag{5.40a}$$

where

$$T(M_1, M_2, z) := \lim_{n\to\infty} T_n(M_1, M_2, z). \tag{5.40b}$$

Using (4.3) one can derive, for $u \in K$ and $z \in K$,

$$q_n(u) = \eta_n(u, z)\left[zq_n(z) + (z - u)\sum_{j=0}^{n-1} q_j(u)[q_n(z)p_j(z) - q_j(z)p_n(z)]\right], \tag{5.41a}$$

where

$$\eta_n(u, z) := \begin{cases} u^{-1}, & \text{if} \quad n = 2k+1 \\ z^{-1}, & \text{if} \quad n = 2k. \end{cases} \tag{5.41b}$$

An application of (5.41) yields, for $u \in K$ and $z \in K$,

$$|q_n(u)|^2 \leq \frac{8M_2^2}{M_1^2}\left[\left(\sum_{j=0}^{n-1} |q_j(u)||q_n(z)p_j(z) - p_n(z)q_j(z)|\right)^2 + |q_n(z)|^2\right]. \tag{5.42}$$

Then by the Schwarz inequality

$$|q_n(u)|^2 \leq N_n(z), \text{ for } u \in K \text{ and } z \in K \tag{5.43a}$$

where

$$N_n(z) := \frac{8M_2^2}{M_1^2}\left[A(z)\sum_{j=0}^{n-1}|q_n(z)p_j(z) - p_n(z)q_j(z)|^2 + |q_n(z)|^2\right]. \quad (5.43b)$$

It follows from (5.38) that

$$\sum_{n=1}^{\infty} N_n(z) < \infty, \quad \text{for} \quad z \in K.$$

Hence by the Weierstrass $M$-test, the series $\sum_{n=1}^{\infty}|q_n(u)|^2$ converges uniformly on $K$. A similar argument can be used to show that $\sum_{j=1}^{\infty}|p_j(u)|^2$ converges uniformly on $K$. From (5.31e) one can readily show that, for $\ell < k$ and $u \in K$,

$$|\beta_k(u) - \beta_\ell(u)| \leq \frac{M_2^2}{M_1^2}\sum_{j=2\ell}^{2k-1}(|q_j(u)|^2 + |q_j(x_0)|^2). \quad (5.44)$$

Since $\sum_{j=1}^{\infty}|q_j(u)|^2$ converges uniformly on $K$, it follows from (5.44) that $\{\beta_n(u)\}_{n=1}^{\infty}$ converges uniformly on $K$. Similar arguments hold for $\{\alpha_n(u)\}$, $\{\gamma_n(u)\}$ and $\{\delta_n(u)\}$.

Equation (5.36e) follows from (5.32). It follow from (5.34b), (3.18) and the argument used to prove Theorem 5.1 that a subsequence $\{\theta_{m_k}(\tau_{m_k})\}$ exists such that it is convergent to a value $\theta \in \mathbb{R} \cup [\infty] =: \widehat{\mathbb{R}}$ and such that there exists a solution $\psi$ of the SHMP for $\{\mu_n\}_{-\infty}^{\infty}$ such that (5.37) holds. □

LEMMA 5.10. *Let $\psi(t)$ be an arbitrary solution of the SHMP for $\{\mu_n\}_{-\infty}^{\infty}$. Then there exists a function $\phi \in \mathcal{N}^*$ (see (5.7)), such that for $z \in \mathbb{C} - \mathbb{R}$,*

$$G_\psi(z) := \int_{-\infty}^{\infty}\frac{z\,d\psi(t)}{z+t} = \frac{x_0 z\alpha(z) - \gamma(z)\phi(z)}{x_0 z\beta(z) - \delta(z)\phi(z)}. \quad (5.45)$$

PROOF. We define $\phi$ by

$$\phi(z) := \frac{x_0 z[\beta(z)G_\psi(z) - \alpha(z)]}{\delta(z)G_\psi(z) - \gamma(z)}, \quad \text{for } z \in \mathbb{C} - \mathbb{R}. \quad (5.46)$$

**Case a:** Suppose $G_\psi(z) = \gamma(z)/\delta(z)$, for $z \in \mathbb{C} - \mathbb{R}$. Then by (5.46), $\phi(z) \equiv \infty \in \mathcal{N}^*$ and hence (5.45) holds. **Case b:** Suppose that $G_\psi(z) \neq$

$\gamma(z)/\delta(z)$ for some $z \in \mathbb{C} - \mathbb{R}$. Then since both the numerator and the denominator in (5.46) are analytic for $z \in \mathbb{C} - \mathbb{R}$, $\phi(z)$ is meromorphic in $\mathbb{C} - \mathbb{R}$. For $m \in \mathbb{N}$, $\tau \in \widehat{\mathbb{R}} := \mathbb{R} \cup [\infty]$, $\theta \in \widehat{\mathbb{R}}$, and $z \in \mathbb{C} - \mathbb{R}$, we define (see (2.44))

$$w_m(z) := \frac{x_0 z \alpha_m(z) - \gamma_m(z)\theta}{x_0 z \beta_n(z) - \delta_m(z)\theta} = \frac{C_{2m}(z,\tau)}{D_{2m}(z,\tau)} = S_{2m}(z, \tau z), \tag{5.47a}$$

where

$$\theta = \frac{C_{2m}(x_0) + \tau x_0 C_{2m-1}(x_0)}{D_{2m}(x_0) + \tau x_0 D_{2m-1}(x_0)} = \frac{C_{2m}(x_0,\tau)}{D_{2m}(x_0,\tau)} \tag{5.47b}$$

and

$$\tau = -\frac{C_{2m}(x_0) - \theta D_{2m}(x_0)}{x_0[C_{2m-1}(x_0) - \theta D_{2m-1}(x_0)]}. \tag{5.47c}$$

It follows from (2.35), (5.47c) and the determinant formula for continued fractions (2.7) that

$$\frac{d\tau}{d\theta} = \frac{x_0^{2m-1}}{[C_{2m-1}(x_0) - \theta D_{2m-1}(x_0)]^2} < 0, \text{ for } m \in \mathbb{N}, \tag{5.48}$$

since $x_0 \in S \subseteq \mathbb{R}^-$ (see (5.31)). Therefore if $\theta$ moves from $-\infty$ to $+\infty$ along $\mathbb{R}$, then $\tau$ moves from $+\infty$ to $-\infty$ along $\mathbb{R}$ and $S_{2m}(z, \tau z)$ traverses the boundary $\partial K_{2m}(z)$ of the disk $K_{2m}(z)$ one time in the counter-clockwise direction. Hence $\theta \in \overline{U}$ if and only if $w_m(z) \in \widehat{K}_{2m}(z)$. Since the limit circle case occurs,

$$\widehat{K}_\infty(z) := \bigcap_{n=1}^{\infty} \widehat{K}_n(z) \tag{5.49}$$

is a (non-degenerate) closed circular disk. In (5.47) we let $m \to \infty$ to obtain

$$w = \frac{x_0 z \alpha(z) - \gamma(z)\theta}{x_0 z \beta(z) - \delta(z)\theta} \in \widehat{K}_\infty(z) \text{ if and only if } \theta \in \overline{U}. \tag{5.50}$$

By [11, Theorem 5.2] we have

$$G_\psi(z) \in K_n(z) \text{ for all } z \in \mathbb{C} - \mathbb{R} \text{ and } n \in \mathbb{N}. \tag{5.51}$$

Hence

$$G_\psi(z) \in \widehat{K}_\infty(z), \quad \text{for } z \in \mathbb{C} - \mathbb{R}. \tag{5.52}$$

For all $z \in U$ we let $\theta := \phi(z)$ in (5.50) to obtain (from (5.46) and (5.52))

$$\frac{x_0 z \alpha(z) - \gamma(z)\phi(z)}{x_0 z \beta(z) - \delta(z)\phi(z)} = G_\psi(z) \in \widehat{K}_\infty(z). \tag{5.53}$$

From this and (5.50) we conclude that

$$\phi(z) \in \overline{U} \quad \text{for all} \quad z \in U. \tag{5.54}$$

Since $\phi(z)$ is known to be meromorphic on $U$, we can apply the open mapping theorem for meromorphic functions [23, Theorem 4.7] to conclude that $\phi(U)$ is an open set. Hence $\infty \notin \phi(U)$ and so $\phi(z)$ is analytic on $U$. Therefore $\phi(z) \in \mathcal{N}$. □

We define

$$J_m(\phi; z) := -\frac{x_0 z \alpha_m(z) - \gamma_m(z)\phi(z)}{x_0 z \beta_m(z) - \delta_m(z)\phi(z)}, \text{ for } \phi \in \mathcal{N}^* \text{ and } z \in U. \tag{5.55}$$

For later use we state the Julia–Carathéodory lemma (see, e.g., [8]).

LEMMA 5.11. (Julia–Carathéodory) If $\phi(z) \in \mathcal{N}$, then there exist constants $C \geq 0$ and $D \leq 0$ and there exists a bounded function $\Phi(z)$ in $\mathcal{N}$ such that

$$\phi(z) = Cz + \frac{D}{z} + \Phi(z), \quad \text{for} \quad z \in U. \tag{5.56}$$

LEMMA 5.12. Let $\phi(z) \in \mathcal{N}^*$. Then

$$\frac{J_m(\phi; z)}{z} \in \mathcal{N} \tag{5.57}$$

PROOF. Since $\phi(z) \in \overline{U}$, by the argument in the proof of Lemma 5.10, $-J_m(\phi; z) \in \widehat{K}_{2m}(z) \subseteq \widehat{K}_1(z) \subseteq \widehat{V}_0(z)$. Thus, $-\frac{J_m(\phi;z)}{z} \in \frac{\widehat{V}_0(z)}{z} = \widehat{W}$, $\frac{J_m(\phi;z)}{z} \in \overline{U}$. Since $\frac{J_m(\phi;z)}{z}$ is meromorphic in $U$, $\frac{J_m(\phi;z)}{z} \neq \infty$ follows from the mapping property $\frac{J_m(\phi;U)}{U} \subseteq \overline{U}$. Hence $\frac{J_m(\phi;z)}{z}$ is analytic in $U$ and $\frac{J_m(\phi;z)}{z} \in \mathcal{N}$. □

LEMMA 5.13. *If $\phi(z) \in \mathcal{N}$ and $C > 0$, $D < 0$ in (5.56), then*
*(A)*

$$-J_m(\phi; z) - \frac{\gamma_m(z)}{\delta_m(z)} = O(z^{2m+2}) \text{ as } z \to 0, \tag{5.58a}$$

$$-J_m(\phi; z) - \frac{\gamma_m(z)}{\delta_m(z)} = O\left(\frac{1}{z^{2m}}\right) \text{ as } z \to \infty; \tag{5.58b}$$

*(B) There exists a distribution function $\psi_{2m}(t)$ such that*

$$-J_m(\phi; z) = z \int_{-\infty}^{\infty} \frac{d\psi_{2m}(t)}{z+t}, \tag{5.59a}$$

*and*

$$\mu_k = \int_{-\infty}^{\infty} t^k d\psi_{2m}(t) \quad k = -2m, \ldots, 2m-2, \tag{5.59b}$$

*(C) There exists a solution $\psi(t)$ of the SHMP such that*

$$\mu_k = \int_{-\infty}^{\infty} t^k d\psi(t), \quad k = 0, \pm 1, \pm 2, , \ldots.$$

*and*

$$z\int_{-\infty}^{\infty} \frac{d\psi(t)}{z+t} = \frac{x_0 z\alpha(z) - \gamma(z)\phi(z)}{x_0 z\beta(z) - \delta(z)\phi(z)}.$$

PROOF. (A) By (5.55), we get

$$-J_m(\phi; z) - \frac{\gamma_n(z)}{\delta_m(z)} = \frac{x_0 z}{(x_0 z\beta_m(z) - \delta_m(z)\phi(z))\delta_m(z)}. \tag{5.60}$$

The leading coefficient and the trailing coefficient of $\delta_m(z)$ are $-x_0^{-m}(e_{2m} \cdot \cdot D_{2m-1}(x_0) + x_0 D_{2m-2}(x)) \prod_{j=1}^{2m-1} d_j$, $x_0^{1-m} D_{2m-1}(x_0) \prod_{j=1}^{2m} e_j$ respectively. By the choice of $x_0$, those coefficients are not equal to zero. Then (5.58a) and (5.58b) follow from (5.60).

(B) By (5.31) and (3.23)

$$\begin{aligned}\frac{\gamma_m(z)}{\delta_m(z)} &= \frac{x_0 C_{2m}(z) D_{2m-1}(x_0) - zC_{2m-1}(z) D_{2m}(x_0)}{x_0 D_{2m}(z) D_{2m-1}(x_0) - zD_{2m-1}(z) D_{2m}(x_0)} \\ &= \frac{C_{2m}(z, \tau_{2m})}{D_{2m}(z, \tau_{2m})} \\ &= \begin{cases} -\sum_{k=1}^{2m} (-1)^k \mu_{-k} z^k + O(z^{2m+1}) \text{ at } 0, & (5.61a) \\ \sum_{k=0}^{2m-2} (-1)^k \frac{\mu_k}{z^k} + O\left(\frac{1}{z^{2m-1}}\right) \text{ at } \infty, & (5.61b) \end{cases}\end{aligned}$$

where

$$\tau_{2m} = \frac{-D_{2m}(x_0)}{x_0 D_{2m-1}(x_0)}.$$

By (5.58) and (5.61) we have

$$-J_m(\phi; z) = \begin{cases} -\sum_{k=1}^{2m} (-1)^k \mu_{-k} z^k + O(z^{2m+1}), & (5.62a) \\ \sum_{k=0}^{2m-2} (-1)^k \frac{\mu_k}{z^k} + O\left(\frac{1}{z^{2m+1}}\right), & (5.62b) \end{cases}$$

$$-\frac{J_m(\phi;z)}{z} = \begin{cases} -\sum_{k=1}^{2m}(-1)^k\mu_{-k}z^{k-1} + O(z^{2m}), & (5.63a) \\ \sum_{k=0}^{2m-2}(-1)^k\frac{\mu_k}{x^{k+1}} + O\left(\frac{1}{z^{2m}}\right). & (5.63b) \end{cases}$$

Then

$$\sup_{y\geq 1}\left|y\frac{J_m(\phi;iy)}{iy}\right| < \infty \tag{5.64}$$

Since $\frac{J_m(\phi;z)}{z} \in \mathcal{N}$, there exists a distribution function $\psi_{2m}(t)$ [2, p. 93] such that

$$\frac{J_m(z)}{z} = -\int_{-\infty}^{\infty}\frac{d\psi_{2m}(t)}{z+t}.$$

This is (5.59a). (5.59b) follows from (5.63) and the following asymptotic expansions

$$\int_{-\infty}^{\infty}\frac{d\psi_{2m}(t)}{z+t} \sim -\sum_{k=1}^{\infty}z^k\int_{-\infty}^{\infty}(-t)^{k+1}d\psi_m(t) \tag{5.65a}$$

$$\int_{-\infty}^{\infty}\frac{d\psi_{2m}(t)}{z+t} \sim \sum_{k=0}^{\infty}z^{-k-1}\int_{-\infty}^{\infty}(-t)^k d\psi_{2m}(t). \tag{6.65b}$$

(C) From (B), for each $m$, there exists $\psi_{2m}(t)$ such that

$$z\int_{-\infty}^{\infty}\frac{d\psi_{2m}(t)}{z+t} = -J_m(\phi;z) = \frac{x_0z\alpha_m(z) - \gamma_m(z)\phi(z)}{x_0z\beta_m(z) - \delta_m(z)\phi_m(z)} \tag{6.66}$$

By Helly's theorems (Theorem 5.4 and 5.6) and convergence of $\alpha_m(z), \beta_m(z), \gamma_m(z)$ and $\delta_m(z)$ we obtain the existence of $\psi(t)$. This completes the proof of Lemma 5.13. □

PROOF OF THEOREM 5.3. (Nevanlinna–Njåstad) For any $\psi \in M(\{\mu_n\}_{-\infty}^{\infty})$, by Lemma 5.10, there exists $\phi \in \mathcal{N}^*$ such that (5.11) holds. Conversely, for each $\phi \in \mathcal{N}^*$: First assume $\phi \not\equiv \infty$. If $C > 0$, $D < 0$ in (5.56), by Lemma 5.13 (C), there exists a solution $\psi \in M(\{\mu_n\}_{-\infty}^{\infty})$ such that (5.11) holds. If this is not the case, let

$$\phi_n(z) := \frac{1}{n}z - \frac{1}{nz} + \phi(z), \quad n \in \mathbb{N}.$$

By Lemma 5.13 (C), there exists a solution $\psi_n(t) \in M(\{\mu_n\}_{-\infty}^{\infty})$ such that

$$z\int_{-\infty}^{\infty}\frac{d\psi_n(t)}{z+t} = \frac{x_0z\alpha(z) - \gamma(z)\phi_n(z)}{x_0z\beta(z) - \delta(z)\phi_n(z)}.$$

Let $n \to \infty$, then $\phi_n(z) \to \phi(z)$ and by Theorems 5.4 and 5.6, we get a solution $\psi$ of the SHMP such that (5.11) holds

If $\phi(z) \equiv \infty$, then $J_m(\phi; z) = -\frac{\gamma_m(z)}{\beta_m(z)}$, (5.64) follows from (5.61). So there exists $\psi_{2m}(t)$ in Lemma 5.13 (B). Further there exists a solution $\psi(t)$ in Lemma 5.13 (C) such that (5.11) holds. Since $\psi(t)$ is essentially uniquely determined by the inversion of the integral transform $G_\psi(z) = \int_{-\infty}^{\infty} \frac{d\psi(t)}{z+t}$ taking into account formulas (5.46) and (5.45) then the correspondence is one-to-one. This completes the proof. □

## 6 NEVANLINNA PARAMETRIZATION FOR LOG-NORMAL MOMENTS

This section is used to give explicit expressions in the Nevanlinna parametrization for the regular indeterminate SHMP for the bisequence of log-normal moments. For each $q$, such that $0 < q < 1$, the *log-normal distribution function* $\psi(q;t)$ is defined by

$$\psi'(q;t) := \frac{\partial}{\partial t}\,\psi(q;t) := \begin{cases} \dfrac{\sqrt{q}}{2\kappa\sqrt{\pi}}\, e^{(\frac{\log t}{2\kappa})^2}, & 0 < t < \infty \\ 0, \quad -\infty < t \le 0, \end{cases} \tag{6.1a}$$

where

$$0 < q = e^{-2\kappa^2} < 1 \text{ and } \kappa := (\log q^{-1/2})^{1/2} \tag{6.1b}$$

Hence the log-normal moments $\mu_n(q)$ are given by

$$\mu_n(q) := \int_0^\infty t^n \psi'(q;t)dt = q^{\frac{-n^2}{2}-n}, \text{ for } n \in \mathbb{Z}. \tag{6.2}$$

Other distributions that generate the same moment bisequence $\{\mu_n(q)\}$ are found in [4] and [6]. For each integer $r \in \mathbb{Z}$ and $q$ satisfying (6.1b), there is a closely related distribution function $\psi_r(q;t)$ defined by

$$\psi_r'(q;t) := \frac{\partial}{\partial t}\,\psi_r(q;t) := t^r\psi'(q;t), \; -\infty < t < \infty, 0 < q < 1. \tag{6.3}$$

The moments $\mu_n^{(r)}(q)$ associated with $\psi_r(q;t)$ are given by

$$\mu_n^{(r)}(q) := \int_0^\infty t^n \psi_r'(q;t)dt = \int_0^\infty t^{n+r}\psi'(q;t)dt =: \mu_{n+r}(q), \tag{6.4}$$

for $n \in \mathbb{Z}$ and $r \in \mathbb{Z}$.

The bisequence $\{\mu_n^{(r)}(q)\}_{n=-\infty}^{\infty}$ is called the $r$-translate of $\{\mu_n(q)\}_{-\infty}^{\infty}$. Hereafter, statements involving $r$ will be understood to be valid for all $r \in \mathbb{Z}$ unless otherwise noted.

The Hankel determinants $H_k^{(m)}$ associated with the moment bisequence $\{\mu_n^{(m)}(q)\}_{n=-\infty}^{\infty}$ satisfy the conditions

$$H_{2n}^{(-2n)} > 0,\ H_{2n+1}^{(-2n)} > 0,\ H_{2n}^{(-2n+1)} > 0,\ H_{2n+1}^{(-2n-1)} > 0,\ \text{for } n \in \mathbb{N} \quad (6.5)$$

and hence the corresponding APT-fraction (2.34) has the form

$$\operatorname*{K}_{n=1}^{\infty}\left(\frac{F_n^{(r)} z}{1 + G_n^{(r)} z}\right), \text{ where } F_n^{(r)} > 0,\ G_n^{(r)} > 0, \text{ for } n \in \mathbb{N}. \quad (6.6)$$

A general T-fraction of the form (6.6) is called a *positive T-fraction*. Explicit formulas for the Hankel determinants $H_m^{(k)}$ can be found in [10, (3.2)]. The coefficients in the positive T-fraction (6.6) are given by

$$F_1^{(r)} = q^{\frac{1-r^2}{2}},\ F_n^{(r)} = q^{r+\frac{3}{2}-n}(1 - q^{n-1}),\ n \geq 2, \quad (6.7a)$$

$$G_n^{(r)} = q^{r+\frac{1}{2}}, \quad n \in \mathbb{N}. \quad (6.7b)$$

We let $A_n^{(r)}(z)$ and $B_n^{(r)}(z)$ denote the $n$th numerator and denominator of the positive T-fraction (6.6). Moreover, we let $U_{r,m}(z)$ and $V_{r,m}(z)$ be defined by

$$U_{r,2n}(z) := \gamma_{2n}^{(r)} \frac{A_{2n}^{(r)}(z)}{z^n},\ V_{r,2n}(z) := \gamma_{2n}^{(r)} \frac{B_{2n}^{(r)}(z)}{z^n},\ n \in \mathbb{N}, \quad (6.8a)$$

$$U_{r,2n+1}(z) := \gamma_{2n+1}^{(r)} \frac{A_{2n+1}^{(r)}(z)}{z^{n+1}},\ V_{r,2n+1}(z) := \gamma_{2n+1}^{(r)} \frac{B_{2n+1}^{(r)}(z)}{z^{n+1}}, n \in \mathbb{Z}_0^+, \quad (6.8b)$$

where

$$\gamma_{2n}^{(r)} := (-1)^n q^{-rn+n^2-\frac{n}{2}} \prod_{j=1}^{\infty}(1 - q^j), \quad n \in \mathbb{N}, \quad (6.8c)$$

$$\gamma_{2n+1}^{(r)} := (-1)^{n+1} q^{-r(n+1)+n^2+\frac{(n-1)}{2}} \prod_{j=1}^{\infty}(1 - q^j). \quad n \in \mathbb{Z}_0^+. \quad (6.8d)$$

It is readily seen that $U_{r,m}(z)/V_{r,m}(z)$ is the $m$th approximant of the positive T-fraction (6.6). It is known [10, (3.13)] that the sequences $\{U_{r,2n}\}_{n=0}^{\infty}$, $\{V_{r,2n}(z)\}_{n=0}^{\infty}$, $\{U_{r,2n+1}(z)\}_{n=0}^{\infty}$ and $\{V_{r,2n+1}(z)\}_{n=0}^{\infty}$ converge locally uniformly on $\mathbb{C} - [0]$ to analytic functions

$$U_r^{(0)}(z) := \lim_{n\to\infty} U_{r,2n}(z),\ V_r^{(0)}(z) := \lim_{n\to\infty} V_{r,2n}(z), \quad (6.9a)$$

$$U_r^{(1)}(z) := \lim_{n\to\infty} U_{r,2n+1}(z), \quad V_r^{(1)}(z) := \lim_{n\to\infty} V_{r,2n+1}(z), \tag{6.9b}$$

and that, for $z \in \mathbb{C} - [0]$,

$$U_r^{(0)}(z) = (-1)^r \mu_r \sum_{j=-\infty}^{\infty} q^{j^2} \left[\sum_{k=0}^{\infty}(-1)^k q^{\frac{k^2}{2}+(2j-\frac{1}{2})k}\right] (-q^{r+\frac{1}{2}}z)^j, \tag{6.10a}$$

$$U_r^{(1)}(z) = (-1)^r \mu_r \sum_{j=-\infty}^{\infty} q^{j^2} \left[\sum_{k=0}^{\infty}(-1)^k q^{\frac{k^2}{2}+(2j+\frac{1}{2})k}\right] (-q^{r+\frac{3}{2}}z)^j, \tag{6.10b}$$

$$V_r^{(0)}(z) = \sum_{j=-\infty}^{\infty} q^{j^2}(-q^{r+\frac{1}{2}}z)^j, \tag{6.10c}$$

$$V_r^{(1)}(z) = \sum_{j=-\infty}^{\infty} q^{j^2}(-q^{r+\frac{3}{2}}z)^j. \tag{6.10d}$$

We recall the theta function $\theta(w, q)$ is defined by

$$\theta(w, q) := \sum_{j=-\infty}^{\infty} q^{j^2}(-e^{2iw})^j \tag{6.11}$$

(see, e.g., [25, p. 462]). It follows from (6.10) and (6.11) that $V_r^{(0)}(z)$ and $V_r^{(1)}$ can be expressed in terms of theta functions

$$V_r^{(0)}(z) = \theta\left(\frac{1}{2i}\log(q^{r+\frac{1}{2}}z), q\right), \tag{6.12a}$$

and

$$V_r^{(1)}(z) = \theta\left(\frac{1}{2i}\log(q^{r+\frac{3}{2}}z), q\right). \tag{6.12b}$$

We also recall the Jacobi triple product formula [3, p. 348 (1.14)]

$$\begin{aligned} V(x) &:= \sum_{j=-\infty}^{\infty} (-1)^j q^{j^2} x^j = \theta\left(\frac{1}{2i}\log x, q\right) \\ &= \prod_{p=0}^{\infty}(1-q^{2p}) \prod_{p=0}^{\infty}(1-q^{2p+1}x) \prod_{p=0}^{\infty}(1-q^{2p+1}x^{-1}). \end{aligned} \tag{6.13}$$

Hence the zeros of $V(x)$ are $x = q^{2p-1}$, where $p \in \mathbb{Z}$; that is,

$$\ldots, q^5, q^3, q, \frac{1}{q}, \frac{1}{q^3}, \frac{1}{q^5}, \ldots. \tag{6.14}$$

It follows from (6.12) and (6.13) that

$$V_r^{(0)}(z) = V_r^{(0)}(q^{-r-\frac{1}{2}}w) = V(w), \;\; w := q^{r+\frac{1}{2}}z. \tag{6.15}$$

Hence the zeros of $V_r^{(0)}$ are all simple and they are given by

$$x_{r,p}^{(0)} := q^{2p-r-\frac{3}{2}}, \quad p \in \mathbb{Z}, \tag{6.16}$$

Similarly,

$$V_r^{(1)}(z) = V(w), \quad \text{where } w = q^{r+\frac{3}{2}}z. \tag{6.17}$$

Thus the zeros of $V_r^{(1)}(z)$ are all simple and they are given by

$$x_{r,p}^{(1)} := q^{2p-r-\frac{5}{2}}, \quad p \in \mathbb{Z}. \tag{6.18}$$

THEOREM 6.1. *Let $\{\mu_n^{(r)}(q)\}_{n=-\infty}^{\infty}$ be a regular bisequence defined as in (6.4) and (6.2). Let $\alpha^{(r)}(z), \beta^{(r)}(z), \gamma^{(r)}(z), \delta^{(r)}(z)$ be the associated functions, analytic in $\mathbb{C} - [0]$ and defined by (5.8), where $x_0 \in S$ (see (5.31)). Let*

$$Q(q) := \prod_{j=1}^{\infty}(1-q^j), \quad 0 < q < 1. \tag{6.19}$$

*Then*

$$\alpha^{(r)}(z) = \frac{q^{\frac{r^2}{2}+r}}{x_0 z Q^3(q)}[x_0 U_r^{(0)}(z)U_r^{(1)}(x_0) - zU_r^{(1)}(z)U_r^{(0)}(x_0)], \tag{6.20a}$$

$$\beta^{(r)}(z) = \frac{q^{\frac{r^2}{2}+r}}{x_0 z Q^3(q)}[x_0 V_r^{(0)}(z)U_r^{(1)}(x_0) - zV_r^{(1)}U_r^{(0)}(x_0)], \tag{6.20b}$$

$$\gamma^{(r)}(z) = \frac{q^{\frac{r^2}{2}+r}}{Q^3(q)}[x_0 U_r^{(0)}(z)V_r^{(1)}(x_0) - zU_r^{(1)}(z)V_r^{(0)}(x_0)], \tag{6.20c}$$

$$\delta^{(r)}(z) = \frac{q^{\frac{r^2}{2}+r}}{Q^3(q)}[x_0 V_r^{(0)}(z)V_r^{(1)}(x_0) - zV_r^{(1)}(z)V_r^{(0)}(x_0)]. \tag{6.20d}$$

PROOF. It is readily shown that

$$C_{r,2n}(z) := M_{r,2n}A_{2n}^{(r)}(z) = \frac{M_{r,2n}}{\gamma_{2n}^{(r)}}z^n U_{r,2n}(z), \; n \in \mathbb{N}, \tag{6.21a}$$

$$C_{r,2n-1}(z) := M_{r,2n-1}A_{2n-1}^{(r)}(z) = \frac{M_{r,2n-1}}{\gamma_{2n-1}^{(r)}} z^n U_{r,2n-1}(z), \; n \in \mathbb{N}, \quad (6.21b)$$

$$D_{r,2n}(z) := M_{r,2n}B_{2n}^{(r)}(z) = \frac{M_{r,2n}}{\gamma_{2n-1}^{(r)}} z^n V_{r,2n}(z), \; n \in \mathbb{N}, \quad (6.21c)$$

$$D_{r,2n-1}(z) := M_{r,2n-1}B_{2n-1}^{(r)}(z) = \frac{M_{r,2n-1}}{\gamma_{2n}^{(r)}} z^n V_{r,2n-1}(z), \; n \in \mathbb{N}, \quad (6.21d)$$

where

$$M_{r,2n} := \frac{q^{n^2}}{(q^{nr+\frac{n}{2}})(1-q)(1-q^3)\cdots(1-q^{2n-1})}, \quad (6.22a)$$

$$M_{r,2n-1} := \frac{q^{\frac{r^2}{2}+n^2}}{(q^{nr-r+\frac{3}{2}n})(1-q^2)(1-q^4)\cdots(1-q^{2n-2})}. \quad (6.22b)$$

It follows from (5.30) and (5.31) that

$$\begin{aligned} \alpha_m^{(r)}(z) &= \frac{1}{(x_0 z)^{m+1}}[x_0 C_{r,2m}(z)C_{r,2m-1}(x_0) - zC_{r,2m-1}(z)C_{r,2m(x_0)}] \\ &= \frac{1}{x_0 z}\,\frac{M_{r,2m}M_{r,2m-1}}{\gamma_{r,2m}\gamma_{r,2m-1}}[x_0 U_{r,2m}(z)U_{r,2m-1}(x_0) - \\ &\quad - zU_{r,2m-1}(z)U_{r,2m}(x_0)] \end{aligned}$$

Equation (6.20a) follows from (6.9) and (6.21). Similar arguments hold for (6.21 b,c,d). □

By (5.11) (Theorem 5.3) we can write

$$G_{\psi_r(q,t)}(z) := \int_0^\infty \frac{z d\psi_r(q;t)}{z+t} = \frac{x_0 z\alpha^{(r)}(z) - \gamma^{(r)}(z)\phi(z)}{x_0 z\beta^{(r)}(z) - \delta^{(r)}(z)\phi(z)}, \quad (6.23)$$

for some $\phi \in \mathcal{N}^*$. From (6.23) we obtain

$$\phi(z) = \frac{x_0 z[\beta^{(r)}(z)G_\psi(z) - \alpha^{(r)}(z)]}{\delta^{(r)}(z)G_\psi(z) - \gamma^{(r)}(z)}, \quad (6.24)$$

where

$$G_\psi(z) := \int_0^\infty \frac{z d\psi(t)}{z+t}, \quad (6.25)$$

for every solution $\psi$ of the SHMP (actually the strong Stieltjes moment problem (SSMP)) for $\{\mu_n^{(r)}(q)\}_{n=-\infty}^{\infty}$. In particular one obtains two such solutions $\psi_r^{(0)}(q;t)$ and $\psi_r^{(1)}(q,t)$ from the divergent positive T-fraction (6.6) (see [5, (3.9)]). These solutions are both step functions and they are given, for $\sigma = 0, 1$, by

$$\psi_r^{(\sigma)}(q;t) = \begin{cases} 0, & \text{for} \quad -\infty < t \leq 0, \\ \sum_{p=\ell+1}^{\infty} R_{r,p}^{(\sigma)}, & \text{for } x_{r,\ell+1}^{(\sigma)} \leq t < x_{r,\ell}^{(\sigma)},\ \ell \in \mathbb{Z}. \end{cases} \tag{6.26a}$$

Here

$$R_{r,p}^{(\sigma)} := \frac{U_r^{(\sigma)}(x_{r,p}^{(\sigma)})}{x_{r,p}^{(\sigma)} V_r'^{(\sigma)}(x_{r,p}^{(\sigma)})}, \quad \sigma = 0, 1, p \in \mathbb{Z}, \tag{6.26b}$$

and $x_{r,p}^{(\sigma)}$ are the zeros of $V_r^{(\sigma)}(z)$ (see (6.16) and (6.18)). Moreover, one can write

$$G_{\psi_r^{(\sigma)}(q;t)} = \int_0^\infty \frac{z d\psi_r^{(\sigma)}(q)}{z+t} = \frac{U_r^{(\sigma)}(z)}{V_r^{(\sigma)}(z)} = \sum_{p=-\infty}^{\infty} \frac{z R_{r,p}^{(\sigma)}}{z - x_{r,p}^{(\sigma)}}, \tag{6.27}$$

for $\sigma = 0, 1$. Substitution of (6.27) in (6.24) yields explicit expressions for the associated Nevanlinna functions

$$\phi_r^{(\sigma)}(q;z) = \frac{x_0 z[\beta^{(r)}(z) G_{\psi_r^{(\sigma)]}(q;t)}(z) - \gamma^{(r)}(z)]}{[\delta^{(r)}(z) G_{\psi_r^{(\sigma)}(q;t)}(z) - \gamma^{(r)}(z)]}, \quad \sigma = 0, 1. \tag{6.28}$$

We note in closing that

$$\sum_{p=-\infty}^{\infty} R_{r,p}^{(\sigma)} = q^{-\frac{r^2}{2} - r}, \quad \text{for} \quad r \in \mathbb{Z}. \tag{6.29}$$

## REFERENCES


1. Ahlfors, Lars V., *Complex Analysis*, Third Edition, McGraw-Hill Pub. Co., New York (1979).
2. Akhiezer, N.I., *The Classical Moment Problem and Some Related Questions in Analysis*, Hafner, New York (1965); translated by N. Kemmer).
3. Askey, R., Ramanujan's Extensions of the Gamma and Beta Functions, *Amer. Math. Monthly* 87 (May 1980), 335–426.
4. Askey, R., Beta Integrals and $q$-extensions, Proc. Ramanujan Centennial International Conference, Annamalainagar, 15–18 December 1987.
5. Bonan-Hamada, Catherine M., William B. Jones, Arne Magnus and W. J. Thron, Discrete distribution functions for Log-normal moments, in *Continued Fractions and Orthogonal Functions; Theory and Applications* (S.

Clement Cooper and W. J. Thron, eds.), Lecture Notes in Pure and Applied Math. 154, Marcel Dekker, Inc., New York (1994), 1–21.
6. Chihara, T. S., *Introduction to Orthogonal Polynomials*, (Mathematics and Its Applications Ser.), Gordon (1978).
7. Cochran, Lyle and S. Clement Cooper, Orthogonal Laurent Polynomials on the Real Line, *Continued Fractions and Orthogonal Polynomials: Theory and Applications* (S. C. Cooper and W. J. Thron, eds.) Lecture Notes in Pure and Applied Math. 154, Marcel Dekker Inc., New York (1994), 47–100.
8. Dinghas, A., *Vorlsungen über Funktionentheorie*, Springer–Verlag, Berlin (1961), p. 236.
9. Freud, Géza, *Orthogonal Polynomials*, Pergamon Press, New York (1971).
10. Jones, William B., Arne Magnus and W. J. Thron, PC-fractions and orthogonal Laurent polynomials for log-normal distributions, *J. Math. Anal. and Appl.* 170 (1) (October 1992), 225–244.
11. Jones, William B., Olav Njåstad and W. J. Thron, Continued Fractions and the Strong Hamburger Moment Problem, *Proc. London Math. Soc.* (3), 47 (1983), 363–384.
12. Jones, William B., Olav Njåstad and W.J. Thron, Orthogonal Laurent polynomials and the strong Hamburger moment problem, *J. Math. Anal. and Appl.* 98, No. 2 (February 1984), 528–554.
13. Jones, William B., Olav Njåstad, and W. J. Thron, Continued fractions associated with the trigonometric and other strong moment problems, *Constructive Approximation* 2 (1986), 197–211.
14. Jones, William B. and W. J. Thron, *Continued Fractions: Analytic Theory and Applications*, Encyclopedia of Mathematics and its Applications 11, Addison–Wesley Publishing Company, Reading, MA (1980), distributed now by Cambridge University Press, New York.
15. Jones, William B. and W. J. Thron, Survey of continued fraction methods of solving moment problems and related topics, *Analytic Theory of Continued Fractions* (W. B Jones, W. J. Thron and H. Waadeland, eds.), Lecture Notes in Mathematics 932, Springer–Verlag, New York (1982), 4–37.
16. Jones, William B., W. J. Thron, Haakon Waadeland, A strong Stieltjes moment problem, *Trans. Amer. Math. Soc.* 261, No. 2 (October 1980), 503–528.
17. Natanson, L. P., Theory of Functions of a Real Variable, Vol. 1 (translated by Leon F. Boron) Frederick Ungar Pub. Co., New York (1964).
18. Njåstad, Olav, Solutions of the strong Hamburger moment problem, *J. Math. Anal. and Applics.* 197 (1996), 227–248.
19. Njåstad, Olav, Remarks on Canonical Solutions of Strong Moment Problems, *Orthogonal Functions, Moment Theory and Continued Fractions:*

*Theory and Applications* (W. B Jones and A. Sri Ranga, eds.), Lecture Notes in Pure and Applied Mathematics, Marcel Dekker, Inc. (1998).

20. Njåstad, Olav and W. J. Thron, The theory of sequences of orthogonal L-polynomials, *Padé Approximants and Continued Fractions*, (Haakon Waadeland and Hans Wallin, eds.), Det Kongelige Norski Videnskabers Selskab Skrifter, Universitetsforlaget, Trondheim, Norway (1983), 54–91.
21. Njåstad, Olav and W. J. Thron, Completely convergent APT-fractions and strong Hamburger moment problems with unique solution, *Det Kongelige Norske Videnskabers Selskab Skrifter*, No. 2 (1984), 1–7, Universitetsforlaget, Trondheim, Norway.
22. Njåstad, Olav and W. J. Thron, Unique solvability of the strong Hamburger moment problem, *J. Austral. Math. Soc.*, Series A, 40 (1986), 5–19.
23. Palka, Bruce P., *An Introduction to Complex Function Theory*, Springer–Verlag, New York (1991).
24. Weyl, H. Über gewönliche Differentialgleichungen mit Singularitäten und die zugehörigen Entwickelungen willkürlicher Funktionen, *Math. Ann.* Vol 68 (1910), 220–269.
25. Whittaker, E. T and G. N. Watson, *A Course in Modern Analysis*, Cambridge University Press, Fourth Edition (1962).

# Asymptotic Behavior of the Continued Fraction Coefficients of a Class of Stieltjes Transforms Including the Binet Function

WILLIAM B. JONES* Department of Mathematics, University of Colorado, Boulder, CO 80309–0395, U.S.A.

WALTER VAN ASSCHE** Department of Mathematics, Katholieke Universiteit Leuven, Leuven, BELGIUM

## 1 INTRODUCTION

Let $G(z)$ be a Stieltjes transform

$$G(z) = \int_0^\infty \frac{v(t)}{z+t}\, dt, \quad z \in S_\pi := [z \in \mathbb{C} : |\arg z| < \pi], \tag{1.1}$$

where $v(t)$ is a non-negative weight function on $(0, \infty)$ such that the improper integrals

$$c_k := \int_0^\infty t^k v(t) dt, \qquad k = 0, 1, 2, \cdots \tag{1.2}$$

are convergent. It is well known (see, e.g., [4], [6] or [8]) that $G(z)$ is analytic in $S_\pi$ and admits an asymptotic expansion $L(z)$ defined by

$$G(z) \approx L(z) := \sum_{k=0}^{\infty} (-1)^k \frac{c_k}{z^{k+1}}, \quad \text{as } z \to \infty,\ z \in \hat{S}_\theta := [z \in \mathbb{C} : |\arg z| \le \theta], \tag{1.3}$$

*Research supported in part by the U.S. National Science Foundation under Grant Number DMS–9302584.

**Research Director of the Belgian National Fund for Scientific Research. Research supported in part by FWO research project G.0278.97: *Orthogonal Systems and their applications*.

for $0 \leq \theta < \pi$. Moreover, there exists a unique modified Stieltjes continued fraction (**MSF**)

$$\hat{G}(z) := \frac{b_1}{z} \, \frac{}{+} \, \frac{b_2}{1} \, \frac{}{+} \, \frac{b_3}{z} \, \frac{}{+} \, \frac{b_4}{1} \, \frac{}{+} \cdots, \quad b_n > 0 \text{ for } n = 1, 2, 3 \ldots \tag{1.4}$$

that corresponds to $L(z)$ at $z = \infty$ in the sense that, for $n = 1, 2, 3, \ldots$, the $n$th approximant $g_n(z)$ of (1.4) has (in a neighborhood of $z = \infty$) a convergent Laurent series of the form

$$g_n(z) = \sum_{k=0}^{n-1} (-1)^k \frac{c_k}{z^{k+1}} + \frac{\gamma_n^{(n)}}{z^{n+1}} + \frac{\gamma_{n+1}^{(n)}}{z^{n+2}} + \cdots . \tag{1.5}$$

In this paper we consider weight functions

$$v(t) = \frac{w_r(\sqrt{t})}{\sqrt{t}}, \tag{1.6}$$

where $w_r(x)$ is an even weight function on $(-\infty, \infty)$ of the form

$$w_r(x) = u(x)\psi(x)V(x)e^{-r|x|^\alpha}, \quad -\infty < x < \infty, \ r > 0, \ \alpha > 0 \tag{1.7a}$$

$$u(x) = \prod_{j=1}^{N} |x - x_j|^{\Delta_j}, \quad \Delta_j > -1, \ x_j \in \mathbb{R}, \ \text{(Jacobi factor)} \tag{1.7b}$$

$$\psi(x) \geq 0 \quad \text{for} \quad x \in \mathbb{R}, \quad \psi \in L_\infty(\mathbb{R}) \quad \text{and} \quad \lim_{x \to \pm\infty} \psi(x) = 1, \tag{1.7c}$$

$$V(x) > 0 \quad \text{for} \quad x \in \mathbb{R} \quad \text{and} \quad \lim_{x \to \pm\infty} \frac{\operatorname{Log} V(x)}{|x|^\alpha} = 0 \tag{1.7d}$$

together with some extra technical conditions on $V$ (see Theorem 2.4 in [9]): $V$ needs to be bounded from above in each finite interval, and given any sequence $\epsilon_n$, $n = 1, 2, 3, \ldots$, of real numbers satisfying $\lim_{n\to\infty} \epsilon_n = 0$, there exists for $n$ large enough a positive integer $\ell_n$ and a polynomial $S_n$ of degree $\ell_n$ such that $\lim_{n\to\infty}(\ell_n/n) = 0$,

$$\lim_{n\to\infty} |V(q_n^0 r^{-1/a}(1 + \epsilon_n)x)S_n(x)| = 1, \quad \text{a.e. in} \quad [-1, 1]. \tag{1.7e}$$

and for some $C > 0$ and all $n$ large enough

$$\sup_{-1 \leq x \leq 1} |V(q_n^0 r^{-1/\alpha}(1 + \epsilon_n)x)S_n(x)| \leq C, \tag{1.7f}$$

where $q_n^0 = n^{1/\alpha}(\sqrt{\pi}\Gamma(\alpha/2)/\Gamma((\alpha+1)/2))^{1/\alpha}$.

We show (Theorem 2.1(A)) that, if (1.6) and (1.7) hold, then the coefficients $b_n$ of the MSF (1.4) have the asymptotic behavior

$$\lim_{n\to\infty} \frac{b_n}{n^{2/\alpha}} = \frac{1}{4}\left[\frac{\sqrt{\pi}\Gamma(\frac{\alpha}{2})}{r\Gamma(\frac{\alpha+1}{2})}\right]^{2/\alpha}. \tag{1.8}$$

Our proof of (1.8) is based on Freud's conjecture on the asymptotic behavior of the recurrence coefficients of orthogonal polynomials with exponential weights on $(-\infty, \infty)$, which was proved in 1988 by Lubinsky, Mhaskar and Saff [9]. In addition, it is shown (Theorem 2.1(B)) by use of (1.8) and Thron's parabola theorem (see, e.g., [7] or [11]) that the continued fraction (1.4) converges to $G(z)$ if

$$1 \le \alpha < \infty. \tag{1.9}$$

Truncation error bounds for the approximants $g_n(z)$ and the asymptotic speed of convergence of $\{g_n(z)\}$ are also given (Theorem 2.1(D)). From the latter result we conclude that, for fixed $n$ and large $|z|$, $g_n(z)$ is a much better approximation of $G(z)$ than is the corresponding partial sum

$$L_{n-1}(z) := \sum_{k=0}^{n-1}(-1)^k \frac{c_k}{z^{k+1}} \tag{1.10}$$

of the asymptotic series $L(z)$ in (1.3).

Our principal application of Theorem 2.1 in the present paper is to the Binet function $J(z)$ which, for convenience, we define by

$$\frac{J(\sqrt{z})}{\sqrt{z}} := \int_0^\infty \frac{v(t)}{z+t}\,dt =: G^J(z), \quad \text{for} \quad z \in S_\pi, \tag{1.11a}$$

where

$$v(t) := \frac{1}{2\pi}\frac{1}{\sqrt{t}}\operatorname{Log}\frac{1}{1-e^{-2\pi\sqrt{t}}}, \quad 0 < t < \infty, \tag{1.11b}$$

and $\sqrt{z}$ denotes the principal branch of $z^{1/2}$. It is well known [4, §8.4] that $J(z)$ is related to the gamma function $\Gamma(z)$ by

$$\log\Gamma(z) = \operatorname{Log}\sqrt{2\pi} + \left(z - \frac{1}{2}\right)\operatorname{Log} z - z + J(z) \tag{1.12a}$$

or, equivalently,

$$\Gamma(z) = \sqrt{2\pi}\, z^{z-\frac{1}{2}}e^{-z}e^{J(z)}, \tag{1.12b}$$

where $\operatorname{Log} w$ denotes the principal branch and $\log \Gamma(z)$ is the analytic branch that is real if $z > 0$. For each $\theta$, with $0 \le \theta < \pi$, there exists a number $\kappa(\theta)$ such that

$$|J(z)| \le \frac{\kappa(\theta)}{|z|}, \quad \text{for} \quad z \in \hat{S}_\theta := [z \in \mathbb{C} : |\arg z| \le \theta]. \tag{1.13}$$

Using (1.13), one can obtain Stirling's approximation of $\Gamma(z)$, for large $|z|$ with $z \in \hat{S}_\theta$, by replacing $J(z)$ by 0 in (1.12b). The asymptotic power series $L^J(z)$ is given by

$$G^J(z) \approx L^J(z) := \sum_{k=0}^{\infty} (-1)^k \frac{c_k^J}{z^{k+1}}, \quad \text{as} \quad z \to \infty, \; z \in \hat{S}_\theta, \; 0 \le \theta < \pi, \tag{1.14a}$$

where

$$c_k^J := \frac{(-1)^k B_{2k+2}}{(2k+1)(2k+2)}, \quad (B_{2k+2} \text{ are Bernoulli numbers}). \tag{1.14b}$$

Since $|B_{2k}| \sim 2(2k)!/(2\pi)^{2k}$ as $k \to \infty$, one can show that Carleman's criterion

$$\sum_{k=1}^{\infty} \frac{1}{|c_k|^{\frac{1}{2k}}} = \infty \tag{1.15}$$

is satisfied and hence the MSF (1.4) corresponding to $L^J(z)$ at $z = \infty$ converges to $G^J(z)$ for $z \in S_\pi$. We express this by

$$G^J(z) := \frac{J(\sqrt{z})}{\sqrt{z}} = \frac{b_1^J}{z} \, \frac{}{+} \, \frac{b_2^J}{1} \, \frac{}{+} \, \frac{b_3^J}{z} \, \frac{}{+} \, \frac{b_4^J}{1} \, \frac{}{+} \cdots, \quad z \in S_\pi, \; b_n^J > 0. \tag{1.16}$$

Stieltjes [10, §6.4] computed a few of the coefficients $b_k^J$ of (1.16) (actually he computed $a_k$ such that $b_k^J = 1/(a_k a_{k+1})$ and $a_1 = 1/12$) and observed that "Le calcul des $a_k$ est tres pénible. La loi de ces nombres parait entrêment compliquée." In 1982, Cizek and Vrscay [2] investigated the behavior of the coefficients $b_k^J$ in this continued fraction and they formulated the following conjecture:

**Conjecture.** *The asymptotic behavior of the coefficients in the continued fraction (1.16) is given by*

$$\lim_{n \to \infty} \frac{b_n^J}{n^2} = \frac{1}{16}. \tag{1.17}$$

In fact, a stronger conjecture was formulated as

$$b_n^J = \frac{n^2}{16} + O(n), \quad \text{as} \quad n \to \infty. \tag{1.18}$$

More information about the conjecture (1.17) can be found in [7]. We prove (1.17) by showing that the weight function $v(t)$ in (1.11b) can be expressed by (1.6) in terms of an even weight function $w(x)$ on $(-\infty, \infty)$ which satisfies the hypotheses of Lemma 3.1 (which is a statement of the Lubinsky, Mhaskar, Saff theorem [9]) (see Theorem (2.2).

The main results of this paper are formulated in two theorems stated in Section 2. Proofs of these theorems are given in Section 3.

## 2 MAIN RESULTS

Our first result is applicable to a large class of Stieltjes transforms and their corresponding Stieltjes continued fractions.

THEOREM 2.1. *Let $\alpha$ and $r$ be given numbers satisfying $0 < \alpha < \infty$ and $0 < r < \infty$. Let $v(t)$ be a weight function on $(0, \infty)$ of the form (1.6), where $w_r(x)$ is an even weight function on $(-\infty, \infty)$ satisfying (1.7). Let $G$ and $L$ be defined by (1.1) and (1.3), respectively, and let $g_n(z)$ denote the nth approximant of the MSF (1.4) that corresponds to $L$ at $z = \infty$. Then:*

*(A)*

$$\lim_{n\to\infty} \frac{b_n}{n^{2/\alpha}} = \frac{1}{4}\left[\frac{\sqrt{\pi}\Gamma(\frac{\alpha}{2})}{r\Gamma(\frac{\alpha+1}{2})}\right]^{2/\alpha}. \tag{2.1}$$

*(B) If in addition $1 \le \alpha < \infty$, the MSF (1.4) converges locally uniformly on $S_\pi$ to the Stieltjes transform $G(z)$.*

*(C) For $n = 2, 3, 4, \ldots,$*

$$|G(z) - g_n(z)| \le K_n(z)|g_n(z) - g_{n-1}(z)|, \tag{2.2a}$$

*where*

$$K_n(\theta) := \begin{cases} 1, & \text{if} \quad 0 \le |\arg z| \le \frac{\pi}{2}, \\ \csc|\arg z|, & \text{if} \quad \frac{\pi}{2} < |\arg z| < \pi. \end{cases} \tag{2.2b}$$

*(D) For each $\theta$, with $0 \le \theta < \pi$, there exist positive constants $A, B$ and $C$, (independent of $|z|$ and $n$) with $C > 1$, such that, for $n = 1, 2, 3, \ldots,$ $0 \le |z| < \infty$ and $0 \le |\arg z| \le \theta$,*

$$|G(z) - g_n(z)| < \frac{A}{n^{B\sqrt{|z|}}}, \qquad \text{if} \quad \alpha = 1 \tag{2.3a}$$

*and*

$$|G(z) - g_n(z)| < \frac{A}{C^{(1-\frac{1}{\alpha})n\sqrt{|z|}}}, \quad \text{if} \quad 1 < \alpha < \infty. \tag{2.3b}$$

**Remark**

It is well known [4, p. 355] that, for each $\theta$, with $0 \le \theta < \pi$, and for each $n = 1, 2, 3, \ldots$, there exists a positive constant $\gamma_n$ such that, with $L_{n-1}(z)$ defined by (1.10),

$$|G(z) - L_{n-1}(z)| \le \frac{\gamma_n}{|z|^{n+1}}, \quad \text{for } 0 \le |z| < \infty,\ 0 \le |\arg z| \le \theta. \tag{2.4}$$

Comparison of the upper bounds in (2.3) with that in (2.4) yields

$$\lim_{|z|\to\infty} \frac{\frac{A}{n^{B\sqrt{|z|}}}}{\frac{\gamma_n}{|z|^{n+1}}} = \lim_{|z|\to\infty} \frac{A}{\gamma_n} \frac{e^{(n+1)\operatorname{Log}|z|}}{e^{B\sqrt{|z|}\operatorname{Log} n}} = 0 \tag{2.5a}$$

and

$$\lim_{|z|\to\infty} \frac{\frac{A}{C^{(1-\frac{1}{\alpha})n\sqrt{|z|}}}}{\frac{\gamma_n}{|z|^{n+1}}} = \lim_{|z|\to\infty} \frac{A}{\gamma_n} \frac{e^{(n+1)\operatorname{Log}|z|}}{e^{(1-\frac{1}{\alpha})n\sqrt{|z|}\operatorname{Log} C}} = 0. \tag{2.5b}$$

We conclude that, for fixed $n$ and $|\arg z|$, with $0 \le |\arg z| \le \theta < \pi$, and large $|z|$, the continued fraction approximant $g_n(z)$ yields a better approximation of $G(z)$ than does the partial sum $L_{n+1}(z)$ of the asymptotic series $L(z)$ in (1.3).

THEOREM 2.2. *Let $J(z)$ denote the Binet function defined by the Stieltjes transform (1.11a) with weight function $v(t)$ in (1.11b). Let*

$$G(z) = \frac{J(\sqrt{z})}{\sqrt{z}} = \frac{b_1^J}{z} + \frac{b_2^J}{1} + \frac{b_3^J}{z} + \frac{b_4^J}{1} + \cdots, \quad b_n^J > 0,\ z \in S_\pi \tag{2.6}$$

*denote the MSF representation. Then:*

*(A)*

$$\lim_{n\to\infty} \frac{b_n^J}{n^2} = \frac{1}{16}. \tag{2.7}$$

*(B) For each $\theta$, with $0 \le \theta < \pi$, there exist positive constants $A$ and $B$ (independent of $|z|$ and $n$) such that, for $n = 1, 2, 3, \ldots$, $0 \le |z| < \infty$ and $0 \le |\arg z| \le \theta$,*

$$\left| \frac{J(\sqrt{z})}{\sqrt{z}} - g_n(z) \right| < \frac{A}{n^{B\sqrt{|z|}}}. \tag{2.8}$$

## Remark

Replacing $z$ by $z^2$ in (2.8) and multiplying by $|z|$ yields

$$|J(z) - zg_n(z^2)| < \frac{A|z|}{n^{B|z|}}, \quad \begin{array}{l} n = 1,2,3,\ldots \\ 0 \le |z| < \infty \\ 0 \le |\arg z| \le \theta,\ 0 \le \theta < \frac{\pi}{2}. \end{array} \tag{2.9}$$

## 3 PROOFS OF MAIN RESULTS

Our proofs of results stated in Section 2 are based upon three lemmas that are stated for completeness with only references cited for proofs. The first is Freud's conjecture, proved in [9] by Lubinsky, Mhaskar and Saff.

Lemma 3.1. *Suppose $w$ is an even weight function on $(-\infty,\infty)$ of the form $w(x) = e^{-Q(x)}$ where $Q$ is continuous, $Q'$ exists for $x > 0$, and $xQ'(x)$ remains bounded for $x \to 0$. Furthermore, we assume that, for all $x$ sufficiently large, $Q'''(x)$ exists, $Q'(x) > 0$, $x^2|Q'''(x)|/Q'(x) \le C$ and*

$$\lim_{x\to\infty}\left(1 + \frac{xQ''(x)}{Q'(x)}\right) = \alpha.$$

*If $p_n(x)$ $(n = 0,1,2,\ldots)$ are the orthonormal polynomials for the weight function $w$, satisfying the recursion relations*

$$xp_n(x) = a_{n+1}p_{n+1}(x) + a_n p_{n-1}(x), \tag{3.1}$$

*then the recursion coefficients have asymptotic behavior given by*

$$\lim_{n\to\infty}\frac{a_n}{q_n} = \frac{1}{2},$$

*where $q_n$ is the positive root of the equation*

$$n = \frac{1}{\pi}\int_0^1 \frac{q_n x Q'(q_n x)}{\sqrt{1-x^2}}\,dx.$$

The Mhaskar–Rakhmanov–Saff number $q_n$, which determines the asymptotic behavior of the recursion coefficients $a_n$, can be found explicitly when $w(x) = e^{-|x|^\alpha}$, in which case

$$q_n^0 = n^{1/\alpha}\left[\frac{\sqrt{\pi}\,\Gamma(\frac{\alpha}{2})}{\Gamma(\frac{\alpha+1}{2})}\right]^{1/\alpha}.$$

Furthermore, if

$$w(x) = u(x)\psi(x)V(x)e^{-|x|^\alpha}, \tag{3.2}$$

where $u(x), \psi(x)$ and $V(x)$ satisfy conditions (1.7), then the Mhaskar–Rakhmanov–Saff number $q_n$ is asymptotically equivalent with this $q_n^0$, so that

$$\lim_{n\to\infty} \frac{a_n}{n^{1/\alpha}} = \frac{1}{2}\left[\frac{\sqrt{\pi}\Gamma(\frac{\alpha}{2})}{\Gamma(\frac{\alpha+1}{2})}\right]^{1/\alpha}. \tag{3.3}$$

For an even weight function $w_r(x)$ of the form (1.7a), with $u, \psi$ and $V$ as described earlier, the corresponding asymptotic behavior turns out to be

$$\lim_{n\to\infty} \frac{a_n}{n^{1/\alpha}} = \frac{1}{2}\left[\frac{\sqrt{\pi}\Gamma(\frac{\alpha}{2})}{r\Gamma(\frac{\alpha+1}{2})}\right]^{1/\alpha}. \tag{3.4}$$

This follows easily by considering $w_r(x) = w(xr^{1/\alpha})$, where $w$ is a weight function satisfying (3.2), for which the orthonormal polynomials are $\sqrt{r^{1/\alpha}}$ $p_n(xr^{1/\alpha})$, with recursion coefficients $a_n(r) = a_n/r^{1/\alpha}$.

We call $\varphi$ a moment distribution function (**MDF**) on $(-\infty, \infty)$ if $\varphi$ is a bounded, non-decreasing function with infinitely many points of increase on $(-\infty, \infty)$ such that all of the Riemann–Stieltjes improper integrals

$$\int_{-\infty}^{\infty} x^k d\varphi(x), \qquad k = 0, 1, 2 \ldots,$$

are convergent. If $\varphi$ is absolutely continuous, then $w(x) := \varphi'(x)$ is a weight function on $(-\infty, \infty)$. A MDF $\varphi$ on $(-\infty, \infty)$ gives rise to an inner product $\langle \cdot, \cdot \rangle_\varphi$ on the space $\Pi \times \Pi$ defined by

$$\langle P(x), Q(x)\rangle_\varphi := \int_{-\infty}^{\infty} P(x)Q(x)d\varphi(x) \quad \text{for} \quad P, Q \in \Pi,$$

where $\Pi$ denotes the space of polynomials in $x$ with real coefficients. In cases where $w(x) = \varphi'(x)$ we may write

$$(P(x), Q(x))_w := \langle P(x), Q(x)\rangle_\varphi := \int_{-\infty}^{\infty} P(x)Q(x)w(x)dx, \quad P, Q \in \Pi.$$

The second lemma summarizes properties of and connections between orthogonal polynomials and continued fractions called real $J$-fractions (Jacobi–continued fractions). Proofs of these results can be found, for example, in [1, Chapter III, Section 4] and [6, Section 7.2.2].

LEMMA 3.2. *Let*

$$\frac{\lambda_1}{z+\ell_1} - \frac{\lambda_2}{z+\ell_2} - \frac{\lambda_3}{z+\ell_3} - \cdots, \quad \begin{array}{l} \lambda_j > 0 \text{ for } j = 2,3,4,\ldots, \ 0 \neq \lambda_1 \in \mathbb{R}, \\ \ell_j \in \mathbb{R} \text{ for } j = 1,2,3,\ldots \end{array} \tag{3.5}$$

*be a given real J-fraction. Let $Q_n(z)$ and $t_n(z)$ denote the nth denominator and nth approximant, respectively, of (3.5). Then: (A) There exists a unique formal Laurent series* (**f.ℓ.s.**)

$$R(z) := \sum_{k=0}^{\infty} \frac{d_k}{z^{k+1}}, \quad d_k \in \mathbb{R} \text{ for } k = 0,1,2,\cdot, \tag{3.6}$$

*such that (3.5) corresponds to $R$ at $z = \infty$ in the sense that, for each $n = 1,2,3,\ldots$, the rational function $t_n(z)$ has (in a neighborhood of $z = \infty$) a convergent Laurent series of the form*

$$t_n(z) = \sum_{k=0}^{2n-1} \frac{d_k}{z^{k+1}} + \frac{d_{2n}^{(n)}}{z^{2n+1}} + \frac{d_{2n+1}^{(1)}}{z^{2n+2}} + \cdots. \tag{3.7}$$

*(B) There exists a moment distribution function* (**MDF**) *$\varphi(t)$ on $(-\infty,\infty)$ such that*

$$d_k = \int_{-\infty}^{\infty} t^k d\varphi(t), \qquad k = 0,1,2,\cdots. \tag{3.8}$$

*(C) If $\varphi$ is any MDF on $(-\infty,\infty)$ satisfying (3.8), then $\{Q_n(x)\}_{n=0}^{\infty}$ is the monic orthogonal polynomial sequence* (**OPS**) *with respect to $\varphi$.*

*(D)*

$$H_n^{(0)} > 0 \quad \text{for} \quad n = 1,2,3,\ldots, \tag{3.9a}$$

*where $H_k^{(m)}$ denotes the Hankel determinants (associated with $\{d_k\}_{k=0}^{\infty}$) defined by $H_0^{(m)} := 1$ and*

$$H_k^{(m)} := \begin{vmatrix} d_m & d_{m+1} & \cdots & d_{m+k-1} \\ d_{m+1} & d_{m+2} & \cdots & d_{m+k} \\ \vdots & \vdots & & \vdots \\ d_{m+k-1} & d_{m+k} & \cdots & d_{m+2k-2} \end{vmatrix}, \quad \begin{array}{l} m = 0,1,2,\cdots, \\ k = 1,2,3,\ldots. \end{array} \tag{3.9b}$$

*(E) For each $n = 1,2,3,\ldots$, $Q_n(z)$ can be expressed by the determinant*

$$Q_n(z) = \frac{1}{H_n^{(0)}} \begin{bmatrix} d_0 & d_1 & \cdot & d_{n-1} & d_n \\ d_1 & d_2 & \cdots & d_n & d_{n+1} \\ \vdots & \vdots & & \vdots & \vdots \\ d_{n-1} & d_n & \cdots & d_{2n-2} & d_{2n-1} \\ 1 & z & \cdots & z^{n-1} & z^n \end{bmatrix} \tag{3.10}$$

*and the* $Q_n(z)$ *satisfy the recurrence relation*

$$Q_n(z) = (z + \ell_n)Q_{n-1}(z) - \lambda_n Q_{n-2}(z), \quad n = 2, 3, 4, \ldots, \tag{3.11a}$$

*where*

$$Q_0(z) := 1 \quad \text{and} \quad Q_1(z) := z + \ell_1. \tag{3.11b}$$

**Remarks**

Correspondence of the real $J$-fraction to a f.$\ell$.s. (3.6) is a consequence of (3.11). (B) and (D) are equivalent statements and (D) is necessary and sufficient for the existence of a solution to the Hamburger moment problem for $\{d_k\}_{k=0}^{\infty}$. (E) can be derived from the correspondence (3.7). (C) can be proved quite easily from (3.10) or by induction using (3.11).

LEMMA 3.3. *Let*

$$\frac{b_1}{z} \, \frac{b_2}{+\,1} \, \frac{b_3}{+\,z} \, \frac{b_4}{+\,1} \, + \cdots, \quad b_n > 0 \textit{ for } n = 1, 2, 3, \ldots \tag{3.12}$$

*be a given MSF, with nth, approximant* $g_n(z)$. *Then: (A) There exists a unique f.$\ell$.s.*

$$L(z) := \sum_{k=0}^{\infty} (-1)^k \frac{c_k}{z^{k+1}}, \qquad c_k \in \mathbb{R} \textit{ for } k = 1, 2, 3, \ldots, \tag{3.13}$$

*such that (3.12) corresponds to (3.13) in the sense that, for* $n = 1, 2, 3, \ldots,$ *the rational function* $g_n(z)$ *has (in a neighborhood of* $z = \infty$*) a convergent Laurent series of the form*

$$g_n(z) = \sum_{k=0}^{n-1} (-1)^k \frac{c_k}{z^{k+1}} + \frac{\gamma_{n+1}^{(n)}}{z^{n+1}} + \frac{\gamma_{n+2}^{(n)}}{z^{n+2}} + \cdots . \tag{3.14}$$

*(B) There exists a MDF* $\varphi(t)$ *on* $(0, \infty)$ *such that*

$$c_k = \int_0^{\infty} t^k d\varphi(t), \qquad k = 0, 1, 2, \ldots . \tag{3.15}$$

*(C) The Hankel determinants* $H_k^{(m)}$ *associated with* $\{c_k\}_{k=0}^{\infty}$ *satisfy*

$$H_n^{(0)} > 0 \quad \text{and} \quad H_n^{(1)} > 0, \quad n = 1, 2, 3, \ldots . \tag{3.16}$$

See e.g., [4, §12.14], [6, Section 7.1.1 and Chapter 9], or [8, Chap. 4, Section 3] for proofs of Lemma 3.3.

PROOF OF THEOREM 2.1. (A): For the MSF (1.4) let $A_n(z), B_n(z)$ and $g_n(z)$ denote the $n$th numerator, denominator and approximant, respectively. By Lemma 3.3(A), there exists a unique f.$\ell$.s. $L(z)$ in (3.13) such that (1.4) corresponds to $L(z)$ in the sense of (3.14). By Lemma 3.3(B) and (C) there exists a MDF $\varphi$ on $(0, \infty)$ such that (3.15) and (3.16) hold. In (1.4), multiplying by $(-1)$ and replacing $z$ by $-z$ yields the continued fraction

$$\hat{H}(z) := -\hat{G}(-z) := \frac{-b_1}{(-z)} + \frac{b_2}{1} + \frac{b_3}{(-z)} + \frac{b_4}{1} + \cdots, \tag{3.17}$$

whose $n$th numerator $C_n(z)$, denominator $D_n(z)$ and approximant $h_n(z)$ satisfy

$$C_n(z) = -A_n(-z), \quad D_n(z) = B_n(-z), \quad h_n(z) = -g_n(-z). \tag{3.18}$$

The continued fraction (3.17) corresponds at $z = \infty$ to the f.$\ell$.s.

$$-L(-z) = -\sum_{k=0}^{\infty}(-1)^k c_k(-z)^{-k-1} = \sum_{k=0}^{\infty} \frac{c_k}{z^{k+1}}. \tag{3.19}$$

It is readily shown (see, e.g., [6, Section 2.4.2]) that the even part of (3.12) is

$$\frac{-b_1}{b_2 - z} + \frac{-b_2 b_3}{(b_3 + b_4) - z} + \frac{-b_4 b_5}{(b_5 + b_6) - z} + \cdots \tag{3.20}$$

The $n$th denominator $Q_n^*(z)$ and approximant $f_n^*(z)$ of (3.20) satisfy

$$Q_n^*(z) = D_{2n}(z) = B_{2n}(-z) \quad \text{and} \quad f_n^*(z) = h_{2n}(z) = -g_{2n}(-z). \tag{3.21}$$

The continued fraction (3.20) is equivalent to the real $J$-fraction

$$\frac{b_1}{z - b_2} + \frac{-b_2 b_3}{z - (b_3 + b_4)} + \frac{-b_4 b_5}{z - (b_5 + b_6)} + \cdots. \tag{3.22}$$

where the equivalence transformation is made using the multiplying factors $r_0 := 1$ and $r_n = -1$ for $n = 1, 2, 3, \cdots$ (see, e.g., [6, Section 2.3.1]). The $n$th denominator $Q_n(z)$ and approximant $f_n(z)$ then satisfy

$$Q_n(z) = \left(\sum_{j=0}^{n} r_j\right) Q_n^*(z) = (-1)^n Q_n^*(z) = (-1)^n D_{2n}(z) = (-1)^n B_{2n}(-z) \tag{3.23a}$$

and

$$f_n(z) = f_n^*(z) = h_{2n}(z) = -g_{2n}(-z). \tag{3.23b}$$

We now apply Lemma 3.2 to the real $J$-fraction (3.22), with,

$$\lambda_1 := b_1,\ \ell_1 := -b_2,\ \lambda_n := b_{2n-2}b_{2n-1},\ \ell_n := -(b_{2n-1} + b_{2n}), \tag{3.24}$$

for $n = 2, 3, 4, \ldots$ . It follows from (3.23b) and (3.19) that the real $J$-fraction (3.22) corresponds to the f.$\ell$.s.

$$R(z) = -L(-z) = \sum_{k=0}^{\infty} \frac{c_k}{z^{k+1}} \quad \text{at} \quad z = \infty. \tag{3.25}$$

Since the corresponding f.$\ell$.s. (3.25) is uniquely determined, it follows that the coefficients $c_k$ in (3.19) are given by (1.2). Therefore by Lemma 3.2(C), $\{Q_n(t)\}_{n=0}^{\infty}$ is the monic orthogonal polynomial sequence (**OPS**) with respect to the weight function $v(t)$ on $(0, \infty)$ given by (1.6) where $w(x)$ has the form (1.7), so that

$$w(x) := |x|v(x^2), \quad \text{for} \quad -\infty < x < \infty \tag{3.26}$$

and hence $w(x)$ is an even weight function on $(-\infty, \infty)$.

Now we let $\{p_n(x)\}_{n=0}^{\infty}$ denote the sequence of orthonormal polynomials with respect to the weight function $w$ and let $\{P_n(x)\}_{n=0}^{\infty}$ denote the monic OPS with respect to $w$. It follows that

$$p_n(x) = \kappa_n P_n(x), \quad \text{where} \quad \kappa_n := \frac{1}{\sqrt{(P_n, P_n)_w}}, \quad n = 0, 1, 2, \cdots . \tag{3.27}$$

The $p_n(x)$ satisfy three-term recurrence relations of the form

$$xp_n(x) = a_{n+1}p_{n+1}(x) + a_n p_{n-1}(x), \quad n = 1, 2, 3, \ldots, \tag{3.28a}$$

where

$$p_0 := \kappa_0 = \frac{1}{\sqrt{\gamma_0}}, \qquad p_1 := \kappa_1 x - \frac{\gamma_1}{\gamma_0}\kappa_1, \tag{3.28b}$$

$$a_n := \frac{\kappa_{n-1}}{\kappa_n} \quad \text{for} \quad n = 0, 1, 2, \ldots,\ \gamma_k := \int_{-\infty}^{\infty} x^k w(x) d\mu,\ k = 0, 1, 2, \cdots . \tag{3.28c}$$

Since $w(x)$ is an even function, one can readily show that

$$p_{2n}(-x) = p_{2n}(x) \quad \text{and} \quad p_{2n-1}(-x) = -p_{2n-1}(x) \text{ for } n = 1, 2, 3, \ldots . \tag{3.29}$$

We define $\{q_n(x)\}_{n=0}^{\infty}$ by

$$q_n(x) := p_{2n}(\sqrt{x}), \quad \text{for} \quad 0 < x < \infty,\ n = 0, 1, 2, \dots . \tag{3.30}$$

The recurrence coefficients for the polynomials $q_n(x)$ can easily be obtained in terms of the recurrence coefficients of the polynomials $p_n(x)$ (see, e.g., [1, Theorem 9.1]). A derivation is given here to make the paper more self-contained. Then, for $m, n = 0, 1, 2, \dots$, we obtain

$$\begin{aligned}(q_n(t), q_m(t))_v &:= \int_0^{\infty} q_n(t)q_m(t)v(t)dt \\ &= \int_0^{\infty} q_n(t)q_m(t)\frac{w(\sqrt{t})}{\sqrt{t}}\, dt, \quad \text{by (1.6)} \\ &= 2\int_0^{\infty} q_n(x^2)q_m(x^2)w(x)dx, \quad \text{with } t = x^2, \\ &= \int_{-\infty}^{\infty} p_{2n}(x)p_{2m}(x)w(x)dx \\ &= (p_{2n}(x), p_{2m}(x))_w = \delta_{m,n},\end{aligned} \tag{3.31a}$$

where

$$\delta_{m,n} := \begin{cases} 0, & \text{if} \quad m \neq n \\ 1, & \text{if} \quad m = n. \end{cases} \tag{3.31b}$$

Therefore $\{q_n(x)\}_{n=0}^{\infty}$ is the orthonormal polynomial sequence with respect to $v$ on $(0, \infty)$. It follows that

$$q_n(x) = \mu_n Q_n(x), \quad \text{where} \quad \mu_n = \kappa_{2n},\ n = 0, 1, 2, \cdots . \tag{3.32}$$

Since $\{Q_n(x)\}_{n=0}^{\infty}$ satisfies recursion relations (3.11), where the $\ell_n$ and $\lambda_n$ are given by (3.24), we obtain (using (3.32))

$$xq_n(x) = \frac{\mu_n}{\mu_{n+1}}\, q_{n+1}(x) - \ell_{n+1}q_n(x) + \lambda_{n+1}\frac{\mu_n}{\mu_{n-1}}\, q_{n-1}(x),\ n = 1, 2, 3, \dots . \tag{3.33}$$

Replacing $x$ by $x^2$ in (3.33) and using (3.30) and $\mu_n = \kappa_{2n}$ yields

$$x^2 p_{2n}(x) = \frac{\kappa_{2n}}{\kappa_{2n+2}}\, p_{2n+2}(x) - \ell_{n+1}p_{2n}(x) + \lambda_{n+1}\frac{\kappa_{2n}}{\kappa_{2n-2}}\, p_{2n-2}(x), \text{ for } n = 2, 3, 4, \cdots . \tag{3.34}$$

Next we multiply both sides of (3.28a) by $x$ to obtain

$$x^2 p_n(x) = a_{n+1}xp_{n+1}(x) + a_n xp_{n-1}(x), \quad n = 1, 2, 3, \dots . \tag{3.35}$$

Substituting (3.28a) in the right hand side of (3.35) (for $xp_{n+1}(x)$ and for $xp_{n-1}(x)$), then rearranging terms and replacing $n$ by $2n$ gives us

$$x^2p_{2n}(x) = a_{2n+1}a_{2n+2}p_{2n+2}(x) + (a_{2n+1}^2 + a_{2n}^2)p_{2n}(x) + a_{2n}a_{2n-1}p_{2n-2}(x),$$
$$n = 2, 3, 4 \cdots . \tag{3.36}$$

Equating coefficients of like terms in (3.34) and (3.36) results in the equations (for $n \geq 2$)

$$a_{2n+1}a_{2n+2} = \frac{\kappa_{2n}}{\kappa_{2n+2}}, \quad a_{2n+1}^2 + a_{2n}^2 = -\ell_{n+1}, \quad a_{2n}a_{2n-1} = \lambda_{n+1}\frac{\kappa_{2n}}{\kappa_{2n-2}}. \tag{3.37}$$

Elimination of the $\kappa$-terms from the first and third equations and using (3.24), we arrive at

$$a_{2n+1}^2 + a_{2n}^2 = b_{2n+1} + b_{2n+2}, \quad a_{2n-1}^2a_{2n}^2 = b_{2n}b_{2n+1}, \quad n = 2, 3, 4, \ldots,$$

from which we obtain the relation

$$b_n = a_{n-1}^2, \qquad n = 2, 3, 4, \ldots . \tag{3.38}$$

The final step in proving (A) is to apply (3.4) (derived from Lemma 3.1) to (3.38) to obtain

$$\lim_{n\to\infty} \frac{b_n}{n^{2/\alpha}} = \lim_{n\to\infty} \left(\frac{a_{n-1}}{n^{1/\alpha}}\right)^2 = \frac{1}{4}\left[\frac{\sqrt{\pi}\Gamma(\frac{\alpha}{2})}{r\Gamma(\frac{\alpha+1}{2})}\right]^{2/\alpha} . \tag{3.39}$$

(B): It is well known [6, Theorem 4.58] that a MSF (1.4) is convergent (for $|\arg z| < \pi$) if and only if at least one of the series

$$\sum_{n=1}^{\infty} \frac{b_2b_4\cdots b_{2n}}{b_1b_3\cdots b_{2n+1}}, \qquad \sum_{n=1}^{\infty} \frac{b_1b_3\cdots b_{2n-1}}{b_2b_4\cdots b_{2n}} \tag{3.40}$$

is divergent. One of the series (3.40) is divergent if $\sum_{n=1}^{\infty} h_n = \infty$, where $b_1 = 1/h_1$, and $b_n = 1/(h_nh_{n-1})$ for $n \geq 2$. Since $b_n^{-1/2} = (h_nh_{n-1})^{1/2} =$ geometric mean of $h_n$ and $h_{n-1}$, and since the arithmetic mean of two numbers is not less than their geometric mean, we have

$$b_n^{-1/2} \leq \frac{1}{2}(h_n + h_{n-1}). \tag{3.41}$$

Therefore $\sum b_n^{-1/2} = \infty$ implies $\sum h_n = \infty$; hence the continued fraction (1.4) is convergent if $\sum b_n^{-1/2} = \infty$. It follows from this and (3.39) that the continued fraction (1.4) is convergent if

$$1 \leq \alpha < \infty.$$

This proves (B).

(C) is due to Henrici and Pfluger [5] (see also [6, Theorem 8.9]). A somewhat sharper error bound can be found in [3].

(D) is a consequence of [7, Theorem 4.8 and (4.26)]. □

PROOF OF THEOREM 2.2. (A): Let

$$w_{2\pi}(x) := \frac{1}{2\pi}\operatorname{Log}\frac{1}{1-e^{-2\pi|x|}}, \tag{3.42}$$

then the inequalities

$$y \leq -\operatorname{Log}(1-y) \leq \frac{y}{1-y}, \qquad 0 < y < 1,$$

show that

$$e^{-2\pi|x|} \leq 2\pi w_{2\pi}(x) \leq \frac{e^{-2\pi|x|}}{1-e^{-2\pi|x|}},$$

so that $w_{2\pi}(x)$ behaves like $e^{-2\pi|x|}$ as $x \to \pm\infty$. Therefore we let $\alpha = 1$ and $r = 2\pi$. Unfortunately $w_{2\pi}(x)$ behaves like $-\operatorname{Log}|x|$ as $x \to 0$ so that this weight function is unbounded near 0. We can therefore try to write the weight as

$$w_{2\pi}(x) = u(x)\psi(x)V(x)e^{-2\pi|x|},$$

where the Jacobi factor $u(x) := |x|^{-1/2}$ takes care of the singularity near 0,

$$\psi(x) := -[\sqrt{|x|}\, e^{2\pi|x|}\operatorname{Log}(1-e^{-2\pi|x|})]/V(x),$$

and $V(x)$ is a continuous function satisfying (1.7d)–(1.7f) such that $V(0) = 1/2\pi$ and $V(x)$ behaves like $|x|^{1/2}/2\pi$ as $x \to \pm\infty$. To find such a function (and in particular to show that there exist polynomials $S_n$ of degree $\ell_n$ so that (1.7e) and (1.7f) hold) is not straightforward, and therefore we choose a more direct approach by using Lemma 3.1. We write $Q(x) = -\log w_{2\pi}(x)$, then $Q$ is continuous for $x > 0$. Clearly

$$Q'(x) = \frac{e^{-2\pi x}}{1-e^{-2\pi x}}\,\frac{1}{w_{2\pi}(x)}, \qquad x > 0, \tag{3.44}$$

so that $Q'(x)$ exists for $x > 0$ and $xQ'(x)$ remains bounded for $x \to 0$. Straightforward calculus gives

$$[xQ'(x)]' = \frac{e^{-2\pi x}}{(1-e^{-2\pi x})^2} \frac{1}{w_{2\pi}(x)} \left(1 - 2\pi x - e^{-2\pi x} + \frac{xe^{-2\pi x}}{w_{2\pi}(x)}\right), \quad (3.45)$$

hence by (3.43) we have

$$[xQ'(x)]' \geq \frac{e^{-2\pi x}}{(1-e^{-2\pi x})} \frac{1}{w_{2\pi}(x)} (1 - e^{-2\pi x} - 2\pi x e^{-2\pi x}) \geq 0, \qquad x > 0,$$

so that $xQ'(x)$ is increasing on $(0, \infty)$ and by (3.45)

$$\lim_{x\to\infty} \left(1 + \frac{xQ''(x)}{Q'(x)}\right) = \lim_{x\to\infty} \frac{[xQ'(x)]'}{Q'(x)} = 1. \quad (3.46)$$

Furthermore it is easily checked that $Q'''(x)$ exists for $x > 0$ and $x^2|Q'''(x)|/Q'(x) \leq C$ for $x$ large enough, so that $Q$ satisfies all the properties of Lemma 3.1 with $\alpha = 1$. Therefore the recurrence coefficients $a_n$ of the orthonormal polynomials with weight $w_{2\pi}(x)$ have the asymptotic behavior

$$\lim_{n\to\infty} \frac{a_n}{q_n} = \frac{1}{2},$$

where $q_n$ is the positive root of

$$n = \frac{1}{\pi} \int_0^1 \frac{q_n x Q'(q_n x)}{\sqrt{1-x^2}}\, dx, \quad (3.47)$$

and by (3.38) it follows that for the coefficients $b_n^J = a_{n-1}^2$ we have

$$\lim_{n\to\infty} \frac{b_n^J}{q_n^2} = \frac{1}{4}. \quad (3.48)$$

To get the asymptotic behavior of $q_n$ we observe that $xQ'(x)$ is increasing so that (3.47) gives the inequality

$$n \leq q_n Q'(q_n) \frac{1}{\pi} \int_0^1 \frac{dx}{\sqrt{1-x^2}} = q_n Q'(q_n)/2,$$

from which it follows that $q_n \to \infty$ as $n$ tends to infinity. Now write (3.47) as

$$\frac{n}{q_n} = \frac{1}{\pi} \int_0^1 \frac{xQ'(q_n x)}{\sqrt{1-x^2}}\, dx. \quad (3.49)$$

Since $xQ'(x)$ is increasing, we have $xq_nQ'(xq_n) \le q_nQ'(q_n)$ for $x \in [0,1]$, so that $xQ'(q_nx) \le Q'(q_n)$ on the range of integration. Since $\lim_{x\to\infty} Q'(x) = 2\pi$ and $q_n \to \infty$ we deduce that $Q'(q_n)$ remains bounded. Hence Lebesgue's dominated convergence theorem applied to (3.49) gives

$$\lim_{n\to\infty} \frac{n}{q_n} = \frac{1}{\pi}\int_0^1 \lim_{n\to\infty} \frac{xQ'(q_nx)}{\sqrt{1-x^2}}\,dx = 2\int_0^1 \frac{x}{\sqrt{1-x^2}}\,dx = 2.$$

Using this in (3.48) then gives the desired behavior

$$\lim_{n\to\infty} \frac{b_n^J}{n^2} = \frac{1}{16}.$$

(B) follows immediately from Theorem 2.1(D), (2.3a), since this only requires the asymptotic behavior given by (2.7). □

## REFERENCES

1. Chihara, T.S., *An Introduction to Orthogonal Polynomials*, (Mathematics and Its Applications Ser.), Gordon and Breach, New York 1978.
2. Cizek, J. and E.R. Vrscay, Asymptotic estimation of the coefficients of the continued fraction representing the Binet function, C.R. Math. Rep. Acad. Sci. Canada, Vol IV, No. 4, August 1982, 201–206.
3. Craviotto, Cathleen, William B. Jones, W.J. Thron, "A survey of truncation error analysis for Padé and continued fraction approximants," *Acta Applicandae Mathematicae* **33**, Nos. 2 and 3 (December 1993), 211-272.
4. Henrici, P., *Applied and Computational Complex Analysis*, Vol 2, Special Functions, Integral Transforms, Asymptotics and Continued Fractions, John Wiley and Sons, New York (1977).
5. Henrici, P. and Pia Pfluger, Truncation error estimates for Stieltjes fractions, *Numer. Math.* **9** (1966), 120–138.
6. Jones, William B. and W.J. Thron, *Continued Fractions: Analytic Theory and Applications*, Encyclopedia of Mathematics and its Applications, 11, Addison–Wesley Publishing Company, Reading, Mass. (1980), distributed now by Cambridge University Press, New York.
7. Jones, William B. and W.J. Thron, Continued fractions in numerical analysis, *Applied Numerical Math.* **4** (1988), 143–230.
8. Lorentzen, Lisa and Haakon Waadeland, *Continued Fractions with Applications*, Studies in Computational Math., Vol 3, North–Holland, New York (1992).
9. D.S. Lubinsky, H.N. Mhaskar, E.B. Saff, A proof of Freud's conjecture for exponential weights, *Constr. Approx.* **4** (1988), 65–83.
10. T.J. Stieltjes, Recherches sur les fractions continue, *Ann. Fac. Sci. Toulouse* **8** (1894), J, 1-122; **9** (1895), A, 1–47; (G. Van Dijk) (ed.),

Oeuvres Complètes–Collected Papers, Vol. II, Springer–Verlag, Berlin, 1993, pp. 406–570, English translation on pp. 609–745.
11. Thron, W.J., On parabolic convergence regions for continued fractions, *Math. Zeitschr.* **69** (1958), 173–182.
12. Valent, G. and W. Van Assche, The impact of Stieltjes' work on continued fractions and orthogonal polynomials: additional material *J. Comput. Appl. Math.* **65** (1995), 419–447.

# Uniformity and Speed of Convergence of Complex Continued Fractions $K(a_n/1)$

L. J. LANGE Department of Mathematics, University of Missouri, Columbia, Missouri 65211

## 1 INTRODUCTION

Using elementary tools and taking advantage of the fascinating underlying geometric behavior, our fundamental goal in this article is to simplify, yet enrich, the catalog of twin convergence region results for continued fractions (c.f.s) of the form

$$K(a_n/1) = \mathop{\mathrm{K}}_{n=1}^{\infty} \frac{a_n}{1} = \frac{a_1}{1} + \frac{a_2}{1} + \frac{a_3}{1} + \cdots, \qquad 0 \neq a_n \in \mathbb{C}. \tag{1.1}$$

Since nonempty intersections of twin convergence regions are simple convergence regions we also have in mind the aim that our results in this paper shed more light on convergence questions in general for continued fractions and their applications.

By a *region* we shall mean an open connected set plus a subset (possibly empty) of its boundary. A region $\Omega$ is said to be a *simple convergence region* for c.f.s (1.1) if $a_n \in \Omega$ implies the convergence of $K(a_n/1)$. A pair of nonempty regions $(\Omega_1, \Omega_2)$ are called *twin convergence regions* for c.f.s (1.1) if the conditions

$$a_{2n-1} \in \Omega_1, \qquad a_{2n} \in \Omega_2, \qquad a_n \neq 0, \qquad n = 1, 2, \ldots,$$

imply that $K(a_n/1)$ converges, that is, if $\lim f_n = f \in \mathbb{C} \cup \{\infty\}$, where $f_n$ is defined to be the $n$th approximant of $K(a_n/1)$. The first significant twin convergence region result was proved sixty years ago (1936) by Leighton and Wall [11]. They proved that the element conditions

$$|a_{2n-1}| \leq 1/4, \qquad |a_{2n}| \geq 25/4, \qquad n = 1, 2, \ldots$$

are sufficient for the convergence of (1.1). In the period from 1942 to 1959, Thron [14], [15], [16], among his other important works in this span, published three fundamental papers on this subject which, besides fostering our interest in convergence questions for continued fractions, exposed us to some of the techniques that might be attempted to establish convergence in twin region cases. The fact that the nested circle method was so successfully employed by Thron in [16] prompted Thron and this author in 1960 to attempt a proof of the Uniform Twin Limacon Theorem below using this method. The two of us ([6], [7]) were able to establish convergence for all permissible values of $\alpha$ with the added help of the Stieltjes-Vitali Convergence Extension Theorem, but we were only able to establish the uniform convergence for real values of $\alpha$ at that time. Six years later (1966), this author [8], by finally obtaining sharp enough estimates for certain ratios of denominators in the continued fractions involved, was able to push through a nested circle proof of Theorem 1.1

for complex values of $\alpha$. It is this theorem that is at the heart of the matter in this paper.

In 1970 in a very substantial paper on twin convergence regions for c.f.s Jones and Thron [3] developed and exposed a new convergence establishing method that yielded a large number of twin convergence regions, including most of the known results up to that time. The number was so large that these authors suggested that there was a need to come to grips with the problem of eliminating redundancies. In fact they devoted the closing section of their paper to this problem which they tabbed as difficult as they presented some preliminary results toward its solution. Though the convergence technique used in [3] has proven to be very fruitful in generating new convergence regions, it does have the drawback that it does not lend to the establishment of uniform convergence or to an estimation of the truncation error of the continued fractions over the regions. Having the latter information can be quite useful in the applications of this subject. In a recent paper, Lorentzen [13] expanded on the 1970 twin convergence region results of Jones and Thron and she presented a number of new convergence region results in the *general convergence* setting. The promising concept of general convergence was introduced by her in 1986 (see L. Jacobsen [1]).

Most of our results in this paper relate to the convergence region results given in [3] for the disk-disk, disk-half plane, and disk-complement of disk cases. We have dealt with the half plane-half plane case earlier in [9]. We show that a vast majority of the regions obtained by Jones and Thron in these cases are embeddable in regions generated by our Uniform Twin Limacon Theorem. Therefore, we are not only able to establish ordinary convergence by elementary methods, but we can also deduce that the convergence regions have the added property of being uniform convergence regions. Moreover, we are able to present speed of convergence estimates over the regions. We present our embedding theorems for the disk-half plane, disk-complement of disk, and disk-disk cases in Sections 2, 3, and 4, respectively. We wish to point out, also, that our results in Section 3 settle the conjecture that was given in §6. of [3]. We hope that our work here and in [9] establishes the fundamental, encompassing, and powerful nature of the Uniform Twin Limacon and the Uniform Twin Parabola Theorems in the convergence theory catalog for continued fractions.

We are now ready to state the theorem that we [8] proved in 1966 and that now, with the proper choices of $\alpha$ and $\sigma$ that we have recently learned to make, allows us to draw the principal conclusions throughout the remainder of this paper. We have placed the word 'limacon' in the title of the theorem to reflect that, in general, the bounded element region $L_1(\alpha, \sigma)$ is bounded by the inner loop of a limacon with two loops and the unbounded element region $L_2(\alpha, \sigma)$ has a limacon with a single loop as its boundary. A special property of these regions $L_i$ with fourth degree boundary curves, that aids in our handling of them, is that they are images under the mapping $w = z^2$ of regions in the $z$-plane bounded by arcs of a single circle and several of its translations. The uniform and bestness properties involved in the theorem will be explained below the statement of the theorem.

**THEOREM 1.1 (Uniform Twin Limacon Theorem)** *The c.f.* $K(c_n^2/1)$ *converges to a value* $f \in \{ z : |z - \alpha| \leq \sigma \}$ *if for* $n = 1, 2, 3, \ldots$,

$$c_{2n-1} \in C_1(\alpha, \sigma) := \{ z : |z \pm i\alpha| \leq \sigma \} \tag{1.2}$$

$$c_{2n} \in C_2(\alpha, \sigma) := \{ z : |z \pm i(1 + \alpha)| \geq \sigma \}, \tag{1.3}$$

*where $\alpha$ is a complex number and $\alpha$ and $\sigma$ satisfy the inequality*

$$|\alpha| < \sigma < |1+\alpha|. \tag{1.4}$$

*If $a_n = c_n^2$, then*

$$a_{2n-1} \in L_1(\alpha,\sigma), \qquad a_{2n} \in L_2(\alpha,\sigma), \qquad n = 1,2,3,\ldots, \tag{1.5}$$

*where*

$$L_1(\alpha,\sigma) \;:=\; \{\, w : |w\overline{\alpha} - \alpha(\sigma^2 - |\alpha|^2)| + \sigma|w| \le \sigma(\sigma^2 - |\alpha|^2)\,\} \tag{1.6}$$

$$L_2(\alpha,\sigma) \;:=\; \{\, w : |w(1+\overline{\alpha}) - (1+\alpha)(\sigma^2 - |1+\alpha|^2)| - \sigma|w| \ge \sigma(|1+\alpha|^2 - \sigma^2)\,\}. \tag{1.7}$$

*The convergence of $K(a_n/1)$ is uniform with respect to the regions $L_1(\alpha,\sigma)$ and $L_2(\alpha,\sigma)$ and these regions are best twin convergence regions. Furthermore, if $f_n$ denotes the nth approximant of $K(c_n^2/1)$ or $K(a_n/1)$, then for all $n \ge 1$*

$$|f_1 - f| \le 2\sigma, \;\; |f_{2n+1} - f| \le |f_{2n} - f| \le 2\sigma(1+n/d)^{-c}, \tag{1.8}$$

*where*

$$c = \frac{|1+\alpha|(\sigma - |\alpha|)(|1+\alpha| - \sigma)}{\sigma\left[1 + \Re(\alpha) + (\sigma^2 - (\Im(\alpha))^2)^{1/2}\right]}, \; d = c\left[1 + \frac{(\sigma+|\alpha|)(|1+\alpha|+\sigma)}{2|1+\alpha|(\sigma - |\alpha|)}\right]. \tag{1.9}$$

Twin convergence regions $(\Omega_1, \Omega_2)$ are said to be *best* if there do not exist twin convergence regions $(\Omega_1^*, \Omega_2^*)$ such that

$$\Omega_n \subseteq \Omega_n^*, \qquad n = 1,2,$$

where at least one of the sets is a proper subset of the other. Twin convergence regions $(\Omega_1, \Omega_2)$ are said to be *uniform twin convergence regions* if there exists a sequence $\{\epsilon_n\}$ of positive numbers depending only on $(\Omega_1, \Omega_2)$ with $\lim \epsilon_n = 0$ such that

$$a_{2n-1} \in \Omega_1, a_{2n} \in \Omega_2, n = 1, 2, 3, \ldots,$$

implies that

$$\lim f_n = f \in \mathbb{C} \text{ and } |f_n - f| \leq \epsilon_n, n = 1, 2, 3, \ldots,$$

where $f_n$ is the $n$th approximant of $K(a_n/1)$.

We find it convenient to introduce here some notation and formulas which will be useful throughout the remainder of this paper. If $f(z)$ is a function of a complex variable and if $\Omega$ is a set in the domain of definition of $f$, then we define $f(\Omega)$ by $f(\Omega) := \{ f(z) : z \in \Omega \}$. For any $c \in \mathbb{C}$ and any $r \geq 0$ we define the sets

$$K(c,r), N(c,r), M(c,r), C_1(c,r), C_2(c,r), L_1(c,r), L_2(c,r)$$

as follows:

$$\begin{aligned}
K(c,r) &:= \{ z : |z - c| = r \} \\
M(c,r) &:= \{ z : |z - c| \geq r \} \\
N(c,r) &:= \{ z : |z - c| \leq r \} \\
C_1(c,r) &:= N(ic,r) \cap N(-ic,r) \qquad (1.10) \\
C_2(c,r) &:= M(i(1+c),r) \cap M(-i(1+c),r) \\
L_1(c,r) &:= [C_1(c,r)]^2 \\
L_2(c,r) &:= [C_2(c,r)]^2
\end{aligned}$$

It is not difficult to see that the regions $C_n(\alpha, \sigma)$ and $L_n(\alpha, \sigma)$ , $n = 1, 2$, derived from these formulas after setting $c = \alpha$ and $r = \sigma$ are the same as those given in the statement of Theorem 1.1.

We now remind the reader of a basic inequality that we shall use often in this paper. If $a$ and $b$ are nonnegative real numbers, then the Arithmetic-Geometric Mean (AGM) inequality is

$$\sqrt{ab} \leq \frac{a+b}{2} \tag{1.11}$$

with equality occurring if and only if $a = b$. It is convenient for us, also, to present here a simple little lemma that will be employed at several spots in the following sections of this article.

**LEMMA 1.1** *Let $I \subseteq \mathbb{R}$ be an interval, let $f : I \to \mathbb{R}$ be a function, and let $g, F, G : I \to \mathbb{R}$ be nonnegative functions. Furthermore, let the conditions*

$$(i) \qquad f(y) - g(y) = F(y) - G(y),$$

$$(ii) \qquad [F(y)]^2 - [G(y)]^2 = (cy + d)^2,$$

*be satisfied for all $y \in I$, where $c, d \in \mathbb{R}$ are constants. Then $f(y) \geq g(y) \geq 0$ on $I$.*

**Proof:** If $cy + d = 0$, then either $F(y) = G(y) = 0$ or $F(y) - G(y) = 0$, but in both cases it follows from our hypotheses that $f(y) \geq g(y) \geq 0$ . If $cy + d \neq 0$, then $F(y) > G(y)$ and again it is easy to see that the conclusion in our lemma is true.

## 2 Disk-Half Plane Case

The following theorem is the relative part of [3, Theorem 5.2]. We have interchanged the subscripts of $E$ and the subscripts of $V$ that were used in [3], so that now $E_1$ is the bounded element region whose boundary is an ellipse and $E_2$ is the unbounded element region whose boundary is a hyperbola.

**THEOREM 2.1** **([3], Disk-Half Plane Case )** *A c.f. $K(a_n/1)$ converges if its elements satisfy*

$$a_{2n-1} \in E_1, \qquad a_{2n} \in E_2, \qquad a_n \neq 0, \qquad n = 1, 2, 3, \ldots, \tag{2.12}$$

*where*

$$E_1 = \{ w : |w - 2\Gamma e^{i\psi}(\cos\psi - p)| + |w| \leq 2\rho(\cos\psi - p) \}, \tag{2.13}$$

$$E_2 = \{ w : |w|[\rho - |1+\Gamma| \cos(\arg w - \arg(1+\Gamma) - \psi)] \leq p(|1+\Gamma|^2 - \rho^2) \}, \tag{2.14}$$

$$|\Gamma| < \rho < |1+\Gamma|, \qquad 0 \leq p < \cos\psi. \tag{2.15}$$

*The corresponding pre twin value regions are the closed half plane and the closed disk defined by*

$$V_1 = \{ z : \Re(ze^{-i\psi}) \geq -p \}, \qquad V_2 = \{ z : |z - \Gamma| \leq \rho \}. \tag{2.16}$$

We claim that the following theorem is true under the hypotheses of Theorem 2.1. Our theorem shows that not only does Theorem 2.1 follow from Theorem 1.1, but much more can be said about the nature of the convergence.

**THEOREM 2.2** **(Disk-Half Plane Embedding)** *If $E_1$ and $E_2$ are given by (2.13) and (2.14), respectively, and their defining parameters satisfy conditions (2.15), then*

$$E_1 \subseteq L_1(\alpha_n, \sigma_n), \;\; E_2 \subseteq L_2(\alpha_n, \sigma_n), \;\; |\alpha_n| < \sigma_n < |1+\alpha_n|, \;\; n = 1, 2,$$

*where*

$$\alpha_n = |n - 1 + \Gamma|(e^{i\arg(n-1+\Gamma)} + e^{i\psi}) + 1 - n, \tag{2.17}$$

$$\sigma_n = [(|n-1+\Gamma|+\rho)(|n-1+\Gamma|+|\Gamma|\cos(\psi-\arg\Gamma)+\cos\psi-p)]^{1/2}. \tag{2.18}$$

*The convergence of $K(a_n/1)$ is uniform with respect to each pair of regions*

$$L_1(\alpha_n,\sigma_n), L_2(\alpha_n,\sigma_n) \text{ and } E_1, E_2,$$

*and the first pair is a best pair of twin convergence regions. Furthermore, if $f_\nu$ denotes the $\nu$th approximant of $K(c_n^2/1)$ or $K(a_n/1)$, then for $\nu \geq 1$ and $n = 1, 2$*

$$|f_1 - f| \leq 2\sigma_n, \quad |f_{2\nu+1} - f| \leq |f_{2\nu} - f| \leq 2\sigma_n(1+\nu/d_n)^{-b_n},$$

*where*

$$b_n = \frac{|1+\alpha_n|(\sigma_n - |\alpha_n|)(|1+\alpha_n| - \sigma_n)}{\sigma_n[1+\Re(\alpha_n) + (\sigma_n^2 - (\Im(\alpha_n))^2)^{1/2}]}, \quad d_n = c_n\left[1 + \frac{(\sigma_n + |\alpha_n|)(|1+\alpha_n| + \sigma_n)}{2|1+\alpha_n|(\sigma_n - |\alpha_n|)}\right].$$

**Proof:** We have organized the proof of this theorem in the form of four lemmas followed by concluding remarks and explanations intended to tie together any loose ends. We find it convenient to deal with the embedding questions in the square root setting, and we make repeated use of formulas (1.10). Before we present our lemmas, we give a shaded plot illustrating the regions in a typical case whose images under the mapping $w = z^2$ are the regions involved in the theorem.

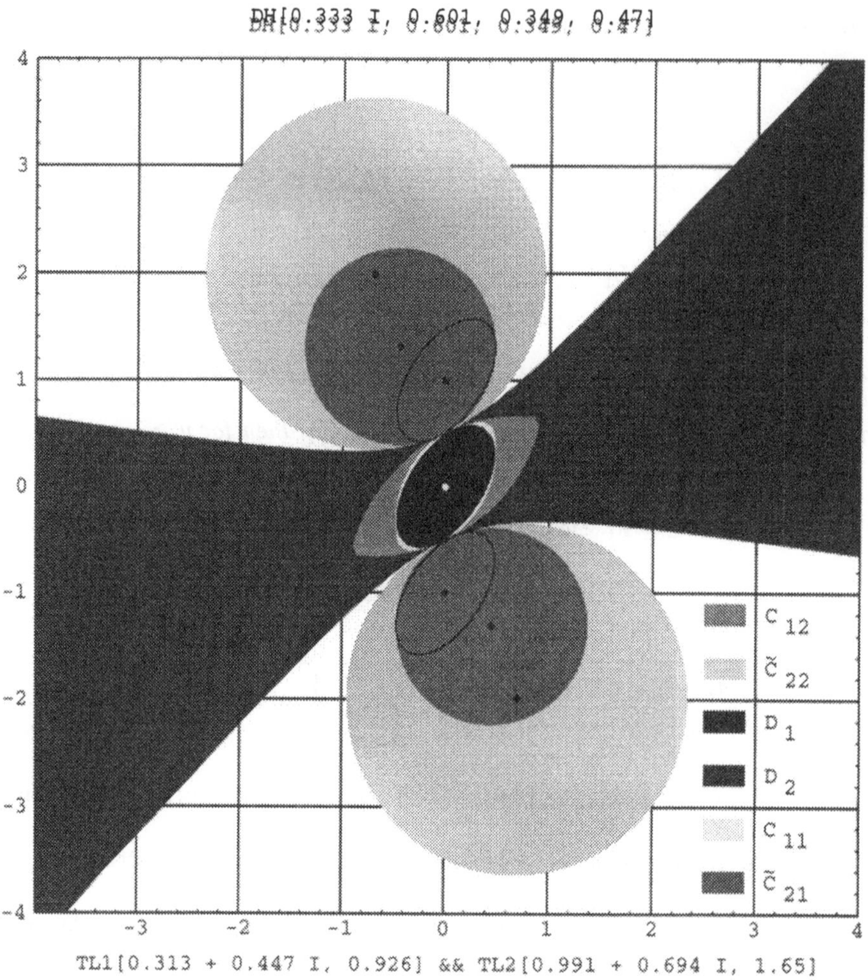


Figure 2.1

In the plot above, the arguments of $DH$ are $\Gamma$, $\rho$, $\psi$, $p$; and the arguments of the $TLn$ are $\alpha_n$, $\sigma_n$, $n = 1, 2$. Also,

$$C_{ij} := C_i(\alpha_j, \sigma_j), \quad D_i := \{\, z : z^2 \in E_i \,\}, \quad i, j = 1, 2.$$

A set with a tilde symbol as a cap is meant to denote the complement of the set with respect to the complex plane.

**LEMMA 2.1** *If $\alpha_n$, $\sigma_n$, $n = 1, 2$, are defined by (2.17) and (2.18), respectively, and (2.15) is satisfied, then $N(\alpha_1, \sigma_1) \subseteq N(\alpha_2, \sigma_2)$.*

**Proof:** It is sufficient to prove that $|\alpha_2 - \alpha_1| + \sigma_1 \le \sigma_2$, which in turn is equivalent to the inequality

$$2|\alpha_2 - \alpha_1|\sigma_1 \le \sigma_2^2 - \sigma_1^2 - |\alpha_2 - \alpha_1|^2. \tag{2.19}$$

After making the appropriate substitutions and dividing out the common factor $|1 + \Gamma| - |\Gamma|$, (2.19) simplifies to

$$2[(|\Gamma| + \rho)(|\Gamma| + |\Gamma| \cos(\psi - \arg \Gamma) + \cos \psi - p)]^{1/2}$$
$$\le 2|\Gamma| + \rho + |\Gamma| \cos(\psi - \arg \Gamma) + \cos \psi - p).$$

But this inequality reduces to

$$0 \le [(|\Gamma| + \rho)^{1/2} - (|\Gamma| + |\Gamma| \cos(\psi - \arg \Gamma) + \cos \psi - p)^{1/2}]^2,$$

so our lemma is true.

**LEMMA 2.2** *Let $D_1 = \{ z : z^2 \in E_1 \}$, where $E_1$ is given by (2.13) and conditions (2.15) are satisfied. Then $D_1 \subseteq C_1(\alpha_1, \sigma_1)$, where $\alpha_1$ and $\sigma_1$ are given by (2.17) and (2.18), respectively.*

**Proof:** The boundary of $D_1$ is also an ellipse whose polar equation is

$$r^2 = \frac{(\rho^2 - |\Gamma|^2)(\cos \psi - p)}{\rho - |\Gamma| \cos(2\theta - \arg(\Gamma) - \psi)}.$$

We perform a rotation so that the new ellipse is described by

$$r^2 = \frac{(\rho^2 - |\Gamma|^2)(\cos \psi - p)}{\rho - |\Gamma| \cos 2\theta}.$$

The latter equation is given in rectangular coordinates in a form convenient for our purposes by

$$x^2 = (\rho + |\Gamma|)(\cos\psi - p)\left[1 - \frac{y^2}{(\rho - |\Gamma|)(\cos\psi - p)}\right]. \qquad (2.20)$$

After performing the same rotation on the disks as was done on the ellipse, we choose the one with center $-|\alpha_1|$ and radius $\sigma_1$. A rectangular equation for its boundary is

$$x^2 = \sigma_1^2 - (y + |\alpha_1|)^2. \qquad (2.21)$$

Let $f(y)$ and $g(y)$ denote the polynomials on the right sides of (2.21) and (2.20), respectively. Then

$$f(y) - g(y) = \frac{2|\Gamma|}{\rho - |\Gamma|}y^2 - 2|\alpha_1|y + \sigma_1^2 - |\alpha_1|^2 - (\rho + |\Gamma|)(\cos\psi - p). \qquad (2.22)$$

In view of symmetry, it is sufficient to prove that $f(y) - g(y) \geq 0$ for all $y \in I$, where $I = [0, [(\rho - |\Gamma|)(\cos\psi - p)]^{1/2}]$. If $\Gamma = 0$, then $\alpha_1 = 0$ and $\sigma_1^2 = \rho(\cos\psi - p)$, so that $f(y) - g(y) \equiv 0$ on $I$ in this case. Since

$$\sigma_1^2 - |\alpha_1|^2 - (\rho + |\Gamma|)(\cos\psi - p) = \frac{|\alpha_1|^2(\rho - |\Gamma|)}{2|\Gamma|}$$

when $\Gamma \neq 0$, it follows from (2.22) that

$$f(y) - g(y) = \left[y\left(\frac{2|\Gamma|}{\rho - |\Gamma|}\right)^{1/2} - |\alpha_1|\left(\frac{\rho - |\Gamma|}{2|\Gamma|}\right)^{1/2}\right]^2,$$

which clearly shows that $f(y) \geq g(y) \geq 0$ on $I$ if $\Gamma$ is not zero. The equality above also tells us that the circle (2.21) and the ellipse (2.20) are tangent when

$$y = \frac{|\alpha_1|(\rho - |\Gamma|)}{2|\Gamma|} = (\rho - |\Gamma|)|\cos\frac{\psi - \arg(\Gamma)}{2}|$$

and $y \in I$. We have now finished the proof of this lemma.

**LEMMA 2.3** *Let $D_2 = \{ z : z^2 \in E_2 \}$, where $E_2$ is given by (2.14) and conditions (2.15) are satisfied. Then $D_2 \subseteq C_2(\alpha_2, \sigma_2)$, where $\alpha_2$ and $\sigma_2$ are given by (2.17) and (2.18) with $n = 2$.*

**Proof:** It is not difficult to verify that the polar equation for the boundary of $D_2$ is the hyperbola

$$r^2 = \frac{p(|1+\Gamma|^2 - \rho^2)}{\rho - |1+\Gamma|\cos(2\theta - \arg(1+\Gamma) - \psi)}. \tag{2.23}$$

We rotate this hyperbola so that the real and imaginary axes become its axes of symmetry and obtain

$$r^2 = \frac{p(|1+\Gamma|^2 - \rho^2)}{\rho - |1+\Gamma|\cos 2\theta}$$

as the polar equation of the new hyperbola. After changing to rectangular coordinates, this hyperbola can be defined by the equation

$$x^2 = (|1+\Gamma| + \rho)\left[\frac{y^2}{|1+\Gamma| - \rho} - p\right]. \tag{2.24}$$

The hyperbola reduces to two intersecting lines when $p = 0$. Let $\beta_2 = 1 + \alpha_2$ and recall that $C_2(\alpha_2, \sigma_2) = M(i\beta_2, \sigma_2) \cap M(-i\beta_2, \sigma_2)$. The boundary of the region $C_2(\alpha_2, \sigma_2)$, after it has undergone the same rotation as the hyperbola (2.23), is the union of the two circles $K(i|\beta_2|, \sigma_2)$ and $K(-i|\beta_2|, \sigma_2)$. Because of symmetry, we need to consider only one of these circles. We choose the first one and present the following equation for it in rectangular coordinates

$$x^2 = \sigma_2^2 - (y - |\beta_2|)^2. \tag{2.25}$$

Here we define the functions $f$ and $g$ by setting $f(y)$ and $g(y)$ equal to the polynomials on the right sides of (2.24) and (2.25), respectively. Our goal now is to show that $f(y) \geq g(y) \geq 0$ for all $y$ in the interval $I$, where $I = [|\beta| - \sigma_2, |\beta| + \sigma_2]$. By straightforward computations we arrive at

$$\begin{aligned} f(y) - g(y) &= \frac{2|1+\Gamma|}{|1+\Gamma|-\rho} y^2 - 2|\beta_2| y + |\beta_2|^2 - \sigma_2^2 - p(|1+\Gamma|+\rho) \\ &= \frac{2|1+\Gamma|}{|1+\Gamma|-\rho} y^2 - 2|\beta_2| y + \frac{|\beta_2|^2(|1+\Gamma|-\rho)}{2|1+\Gamma|} \\ &= \left[ \left( \frac{2|1+\Gamma|}{|1+\Gamma|-\rho} \right)^{1/2} y - |\beta_2| \left( \frac{|1+\Gamma|-\rho}{2|1+\Gamma|} \right)^{1/2} \right]^2 . \end{aligned} \quad (2.26)$$

Equation (2.26) clearly shows that we have achieved our goal, and it also leads us to the fact that the hyperbola (2.24) and the circle (2.25) are tangent when

$$y = \frac{|\alpha_2|(|1+\Gamma|-\rho)}{2|1+\Gamma|} = (|1+\Gamma|-\rho)|\cos\frac{\psi - \arg(1+\Gamma)}{2}|$$

and $y \in I$. Thus the proof of our lemma is complete.

**LEMMA 2.4** *If conditions (2.15) are satisfied and $\alpha_n$ and $\sigma_n$ are given by (2.17) and (2.18), respectively, then*

$$|\alpha_n| < \sigma_n < |1 + \alpha_n|, \quad n = 1, 2.$$

**Proof:** Straightforward calculations lead us to

$$\begin{aligned} \sigma_1^2 - |\alpha_1|^2 &= |\Gamma|(\rho - |\Gamma|)(1 + \cos(\psi - \arg(\Gamma))) + (|\Gamma| + \rho)(\cos\psi - p) \\ &\geq \rho(\cos\psi - p) > 0. \end{aligned}$$

Calculations of a similar nature lead us to the formula

$$|1+\alpha_2|^2 - \sigma_2^2 = |1+\Gamma|(|1+\Gamma|-\rho)(1+\cos(\psi - \arg(1+\Gamma))) + p(|1+\Gamma|+\rho)$$

which shows that $|1+\alpha_2|^2 - \sigma_2^2$ is clearly positive if $p > 0$. If $p = 0$, then $\cos\psi > 0$ so w.l.o.g. we can assume that $-\pi/2 < \psi < \pi/2$. Also, the hypotheses of our theorem guarantee that $\Re(1+\Gamma) > 1/2$ which implies that $-\pi/2 < \arg(1+\Gamma) < \pi/2$. Hence, $|\psi - \arg(1+\Gamma)| < \pi$ so that

$$|1+\alpha_2|^2 - \sigma_2^2 = |1+\Gamma|(|1+\Gamma| - \rho)(1+\cos(\psi - \arg(1+\Gamma))) > 0$$

when $p = 0$. The above results, along with the aid of Lemma 2.1, guarantee the validity of the inequality displayed in the statement of this lemma.

In our four lemmas above, we have established that

$$(i) \quad N(\alpha_1,\sigma_1) \subseteq N(\alpha_2,\sigma_2)$$

$$(ii) \quad D_1 \subseteq C_1(\alpha_1,\sigma_1)$$

$$(iii) \quad D_2 \subseteq C_2(\alpha_2,\sigma_2).$$

So to complete the proof of the embedding part of Theorem 2.2 we have only to prove that $D_1 \subseteq C_1(\alpha_2,\sigma_2)$ and $D_2 \subseteq C_2(\alpha_1,\sigma_1)$. But (i) $\Rightarrow N(i\alpha_1,\sigma_1) \subseteq N(i\alpha_2,\sigma_2)$ and $N(-i\alpha_1,\sigma_1) \subseteq N(-i\alpha_2,\sigma_2)$ so that $D_1 \subseteq C_1(\alpha_2,\sigma_2)$. Moreover, (i) $\Rightarrow N(i(1+\alpha_1),\sigma_1) \subseteq N(i(1+\alpha_2),\sigma_2)$ and $N(-i(1+\alpha_1),\sigma_1) \subseteq N(-i(1+\alpha_2),\sigma_2)$ so that $C_2(\alpha_1,\sigma_1) \supseteq C_2(\alpha_2,\sigma_2) \supseteq D_2$ and we are done with this part of our proof.

As for the uniform convergence and the truncation error estimates in Theorem 2.2, we can say the following. We have shown that $E_1 \subseteq L_1(\alpha_n,\sigma_n)$, $E_2 \subseteq L_2(\alpha_n,\sigma_n)$, $n = 1, 2$, and it follows from Theorem 1.1 that $L_1(\alpha_n,\sigma_n)$ and $L_2(\alpha_n,\sigma_n)$ are uniform convergence regions. Since subregions of uniform convergence regions also have the uniform property, $E_1$ and $E_2$ are uniform convergence regions. The formulas for $b_n$ and $d_n$ in the error estimates for the $f_\nu$ are also obtained from Theorem 1.1 by substituting $\alpha_n$ for $\alpha$ and $\sigma_n$ for $\sigma$ in formulas (1.8). This completes our proof of Theorem 2.2.

## 3 Disk-Complement of Disk Case

The following theorem is that part of Theorem 5.4 in [3] that is under consideration here. We have interchanged the subscripts of $E$ and $V$ from those used in [3].

**THEOREM 3.1** **([3], Disk-Complement of Disk Case)** *The c.f. $K(a_n/1)$ converges if its elements satisfy*

$$a_{2n-1} \in E_1, \qquad a_{2n} \in E_2, \qquad a_n \neq 0, \qquad n = 1, 2, 3, \ldots, \tag{3.1}$$

*where*

$$E_1 = \{ w : |w(1+\overline{\Gamma}_2) + \Gamma_1(\rho_2^2 - |1+\Gamma_2|^2)| + \rho_2|w| \leq \rho_1(\rho_2^2 - |1+\Gamma_2|^2) \}, \tag{3.2}$$

$$E_2 = \{ w : |w(1+\overline{\Gamma}_1) + \Gamma_2(\rho_1^2 - |1+\Gamma_1|^2)| - \rho_1|w| \geq \rho_2(|1+\Gamma_1|^2 - \rho_1^2) \}, \tag{3.3}$$

$$|\Gamma_1| < \rho_1 < |1+\Gamma_1|, \qquad |1+\Gamma_2| < \rho_2 \leq |\Gamma_2|. \tag{3.4}$$

*The corresponding pre twin value regions are given by*

$$V_1 = \{ z : |z - \Gamma_2| \geq \rho_2 \}, \qquad V_2 = \{ z : |z - \Gamma_1| \leq \rho_1 \}. \tag{3.5}$$

In general, $E_1$ is a bounded region with a Cartesian oval as its boundary, and $E_2$ is an unbounded region which also has a Cartesian oval as its boundary. Our embedding theorem below not only shows that the convergence in Theorem 3.1 follows from Theorem 1.1, but it also shows that the convergence is uniform. Moreover, our theorem settles a bestness conjecture that was made in the last section of [3] by Jones and Thron.

**THEOREM 3.2 (Disk-Complement of Disk Embedding)** *Let $E_1$ and $E_2$ be given by (3.2) and (3.3), respectively, and let conditions (3.4) be satisfied. Furthermore, let $s_0 = s_2$ for each symbol $s$ that follows, and let $K_n$ and $\tau_n$ be defined by*

$$K_n := \frac{|1+\Gamma_1+\Gamma_2|}{|\Gamma_n|+|1+\Gamma_{n-1}|}, \qquad \tau_n := \rho_n|1+\Gamma_{n-1}| - \rho_{n-1}|\Gamma_n|, \qquad n = 1,2. \quad (3.6)$$

*Then*

$$E_1 \subseteq L_1(\alpha_n, \sigma_n), \;\; E_2 \subseteq L_2(\alpha_n, \sigma_n), \;\; |\alpha_n| < \sigma_n < |1+\alpha_n|, \;\; n = 1,2,$$

*where*

$$\alpha_1 = 0 \text{ and } \sigma_1 = (\rho_1\rho_2)^{1/2} \text{ if } \Gamma_1 = -(1+\Gamma_2) = 0, \quad (3.7)$$

*and, otherwise,*

$$\alpha_n = \frac{|2-n+\Gamma_2|\Gamma_1 - |n-1+\Gamma_1|(1+\Gamma_2)}{|\Gamma_n|+|1+\Gamma_{n-1}|}, \quad (3.8)$$

$$\sigma_n = \begin{cases} [(\rho_1+|n-1+\Gamma_1|K_n)(\rho_2-|2-n+\Gamma_2|K_n)]^{1/2}, & \text{if } \tau_n \geq 0, \\ [(\rho_1-|n-1+\Gamma_1|K_n)(\rho_2+|2-n+\Gamma_2|K_n)]^{1/2}, & \text{if } \tau_n \leq 0 \end{cases} \quad (3.9)$$

*The convergence of $K(a_n/1)$ is uniform with respect to each pair of regions $L_1(\alpha_n, \sigma_n)$, $L_2(\alpha_n, \sigma_n)$ and $E_1$ , $E_2$, and the first pair is a best pair of twin convergence regions. Moreover, if $f_\nu$ denotes the $\nu$th approximant of $K(c_n^2/1)$ or $K(a_n/1)$, then for $\nu \geq 1$ and $n = 1,2$*

$$|f_1 - f| \leq 2\sigma_n, \;\; |f_{2\nu+1} - f| \leq |f_{2\nu} - f| \leq 2\sigma_n(1+\nu/d_n)^{-b_n},$$

*where*

$$b_n = \frac{|1+\alpha_n|(\sigma_n - |\alpha_n|)(|1+\alpha_n| - \sigma_n)}{\sigma_n[1+\Re(\alpha_n)+(\sigma_n^2-(\Im(\alpha_n))^2)^{1/2}]}, \; d_n = c_n\left[1+\frac{(\sigma_n+|\alpha_n|)(|1+\alpha_n|+\sigma_n)}{2|1+\alpha_n|(\sigma_n-|\alpha_n|)}\right].$$

**Proof:** We have broken up our proof into four lemmas followed by closing arguments. We start with a picture illustrating the regions in an important case of our theorem whose images under the $w = z^2$ mapping are the regions involved in the theorem.

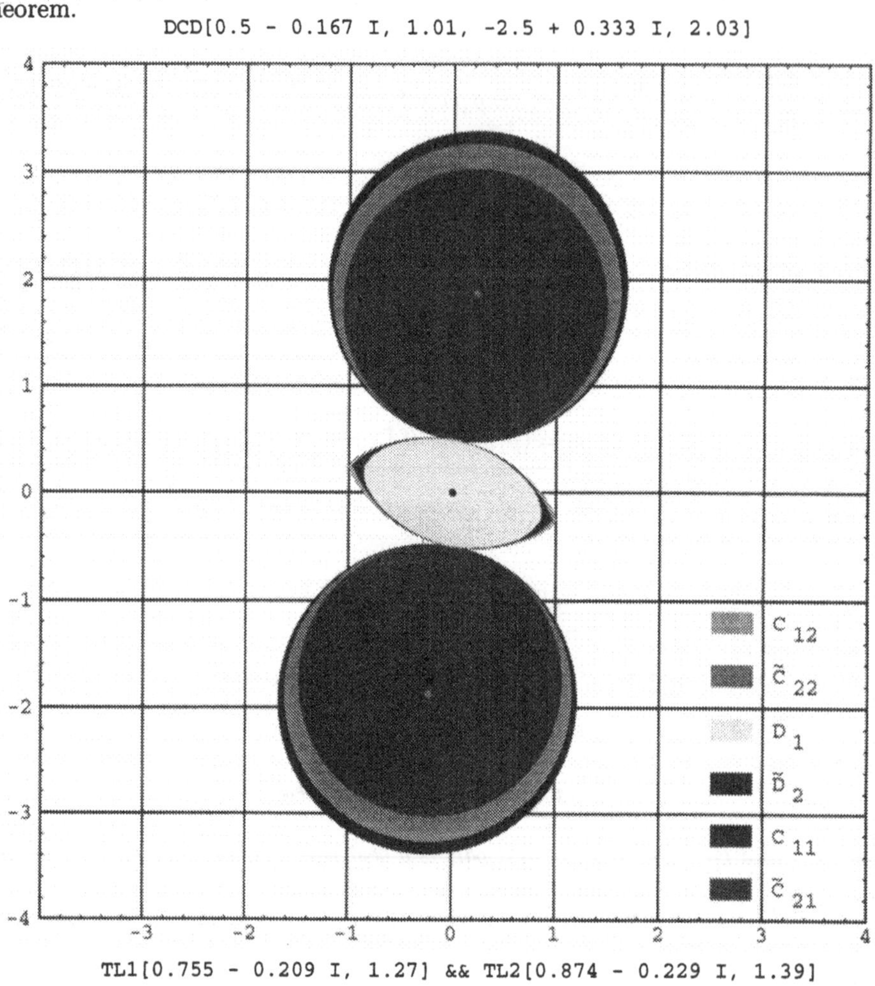


Figure 3.1

In Figure 3.1, the arguments of $DCD$ are $\Gamma_1$, $\rho_1$, $\Gamma_2$, $\rho_2$; and the arguments of the $TLn$ are $\alpha_n$, $\sigma_n$, $n = 1, 2$. Furthermore,

$$C_{ij} := C_i(\alpha_j, \sigma_j), \;\; D_i := \{\, z : z^2 \in E_2 \,\}, \;\; i, j = 1, 2.$$

The tilde cap on a symbol for a set denotes the complement of the set with respect to the set of complex numbers.

**LEMMA 3.1** *If $\alpha_1$ and $\sigma_1$ are determined by (3.7)- (3.9) and conditions (3.4) are satisfied, then*

$$\sigma_1 - |\alpha_1| \geq [(\rho_1 - |\Gamma_1|)(\rho_2 - |1 + \Gamma_2|)]^{1/2}. \tag{3.10}$$

**Proof:** It follows immediately from our hypotheses that $\sigma_1 > 0$. So inequality (3.10) will clearly be true if

$$\sigma_1^2 \geq |\alpha_1|^2 + (\rho_1 - |\Gamma_1|)(\rho_2 - |1 + \Gamma_2|) + 2|\alpha_1|[(\rho_1 - |\Gamma_1|)(\rho_2 - |1 + \Gamma_2|)]^{1/2} \tag{3.11}$$

is true. After substituting the values of $\alpha_1$ and $\sigma_1$, (3.11) becomes

$$\rho_2|\Gamma_1| + \rho_1|1+\Gamma_2| \geq 2|\Gamma_1||1+\Gamma_2| + 2|\alpha_1|[(\rho_1 - |\Gamma_1|)(\rho_2 - |1+\Gamma_2|)]^{1/2} + K_1T_1, \tag{3.12}$$

where

$$T_1 = |\rho_2|\Gamma_1| - \rho_1|1 + \Gamma_2||.$$

Suppose first that

$$\rho_1|1 + \Gamma_2| \geq \rho_2|\Gamma_1|.$$

Then $T_1$ can be written in the form

$$T_1 = |1 + \Gamma_2|(\rho_1 - |\Gamma_1|) - |\Gamma_1|(\rho_2 - |1 + \Gamma_2|).$$

We substitute this value for $T_1$ in (3.12) and obtain

$$|\Gamma_1|(\rho_2 - |1+\Gamma_2|)(1+K_1) + |1+\Gamma_2|(\rho_1 - |\Gamma_1|)(1-K_1) \geq 2|\alpha_1|[(\rho_1 - |\Gamma_1|)(\rho_2 - |1+\Gamma_2|)]^{1/2}.$$

Using the AGM inequality, the left side of this last inequality is $\geq$

$$2[|\Gamma_1||1 + \Gamma_2|(\rho_2 - |1 + \Gamma_2|)(\rho_1 - |\Gamma_1|)(1 - K_1^2)]^{1/2},$$

and this expression is equal to

$$2|\alpha_1|[(\rho_1 - |\Gamma_1|)(\rho_2 - |1+\Gamma_2|)]^{1/2}.$$

Thus inequality (3.11) is true. An argument paralleling the one above can be given showing the validity of (3.11) for the case

$$\rho_2|\Gamma_1| > \rho_1|1+\Gamma_2|.$$

This, along with the observation that (3.10) is easily seen to be true when $\Gamma_1 = -(1+\Gamma_2) = 0$, completes the proof of our lemma.

**LEMMA 3.2** *If $\alpha_2$ and $\sigma_2$ are given by (3.8) and (3.9) with $n = 2$, respectively, and if conditions (3.4) are satisfied, then*

$$|1+\alpha_2| + \sigma_2 \le [(|1+\Gamma_1| + \rho_1)(|\Gamma_2| + \rho_2)]^{1/2} \tag{3.13}$$

*and*

$$|1+\alpha_2| - \sigma_2 \ge [(|1+\Gamma_1| - \rho_1)(|\Gamma_2| - \rho_2)]^{1/2}. \tag{3.14}$$

*Furthermore,*

$$|1+\alpha_2| - \sigma_2 > 0$$

*even if $|\Gamma_2| = \rho_2$.*

**Proof:** First let us verify that $\sigma_2 > 0$ under our hypotheses. It is easily seen that this will be true if $\rho_2 - |\Gamma_2|K_2$ is positive when $\tau_2 \ge 0$ and $\rho_1 - |1+\Gamma_1|K_2$ is positive when $\tau_2 < 0$. Under the assumption $\tau_2 \ge 0$ we obtain

$$\begin{aligned}
\rho_2 - |\Gamma_2|K_2 &= \frac{\rho_2|1+\Gamma_1| + \rho_2|\Gamma_2| - |\Gamma_2||1+\Gamma_1+\Gamma_2|}{|1+\Gamma_1| + |\Gamma_2|} \\
&\ge \frac{\rho_1|\Gamma_2| + \rho_2|\Gamma_2| - |\Gamma_2|[|\Gamma_1| + |1+\Gamma_2|]}{|1+\Gamma_1| + |\Gamma_2|} \\
&= \frac{|\Gamma_2|[(\rho_1 - |\Gamma_1|) + (\rho_2 - |1+\Gamma_2|)]}{|1+\Gamma_1| + |\Gamma_2|} > 0.
\end{aligned}$$

Moreover, under the assumption $\tau_2 < 0$ we arrive at

$$\begin{aligned}\rho_1 - |1+\Gamma_1|K_2 &= \frac{\rho_1|1+\Gamma_1| + \rho_1|\Gamma_2| - |1+\Gamma_1||1+\Gamma_1+\Gamma_2|}{|1+\Gamma_1|+|\Gamma_2|} \\ &\geq \frac{\rho_1|1+\Gamma_1| + \rho_2|1+\Gamma_1| - |1+\Gamma_1|[|\Gamma_1|+|1+\Gamma_2|]}{|1+\Gamma_1|+|\Gamma_2|} \\ &= \frac{|1+\Gamma_1|[(\rho_1 - |\Gamma_1|) + (\rho_2 - |1+\Gamma_2|)]}{|1+\Gamma_1|+|\Gamma_2|} > 0.\end{aligned}$$

Hence, $\sigma_2 > 0$.

It is easy to verify that

$$|1+\alpha_2| \leq [(|1+\Gamma_1|+\rho_1)(|\Gamma_2|+\rho_2)]^{1/2},$$

so (3.13) is valid if

$$\sigma_2^2 \leq (|1+\Gamma_1|+\rho_1)(|\Gamma_2|+\rho_2)+|1+\alpha_2|^2 - 2|1+\alpha_2|[(|1+\Gamma_1|+\rho_1)(|\Gamma_2|+\rho_2)]^{1/2} \quad (3.15)$$

is valid. After substituting the values for $\alpha_2$ and $\sigma_2$, (3.15) reduces to

$$|\Gamma_2|(\rho_1+|1+\Gamma_1|)(1+K_2)+|1+\Gamma_1|(\rho_2+|\Gamma_2|)(1-K_2) \geq 2|1+\alpha_2|[(|1+\Gamma_1|+\rho_1)(|\Gamma_2|+\rho_2)]^{1/2}$$

if $\rho_2|1+\Gamma_1| \geq \rho_1|\Gamma_2|$ and to

$$|\Gamma_2|(\rho_1+|1+\Gamma_1|)(1-K_2)+|1+\Gamma_1|(\rho_2+|\Gamma_2|)(1+K_2) \geq 2|1+\alpha_2|[(|1+\Gamma_1|+\rho_1)(|\Gamma_2|+\rho_2)]^{1/2}$$

if $\rho_2|1+\Gamma_1| < \rho_1|\Gamma_2|$. But each of the last two displayed inequalities is true under the corresponding assumption by the AGM inequality, since

$$|1+\alpha_2|^2 = |\Gamma_2||1+\Gamma_1|(1-K_2^2). \quad (3.16)$$

Thus (3.13) is valid. Now we proceed to prove that (3.14) is valid. As a first step in our argument we prove that

$$|1+\alpha_2|^2 > (|1+\Gamma_1|-\rho_1)(|\Gamma_2|-\rho_2). \quad (3.17)$$

After substituting the value of $\alpha_2$, this inequality becomes equivalent to

$$\rho_1(|\Gamma_2| - \rho_2) + |1 + \Gamma_1|(\rho_2 - |\Gamma_2|K_2^2) > 0. \tag{3.18}$$

But

$$\begin{aligned}\rho_1(|\Gamma_2| - \rho_2) + |1 + \Gamma_1|(\rho_2 - |\Gamma_2|K_2^2) &\geq \rho_1(|\Gamma_2| - \rho_2) + |1 + \Gamma_1|(\rho_2 - |\Gamma_2|K_2) \\ &> 0,\end{aligned}$$

since $0 \leq K_2 \leq 1$, $|\Gamma_2| - \rho_2 \geq 0$, $|1 + \Gamma_1| > \frac{1}{2}$, and, as we have shown above, $\rho_2 - |\Gamma_2|K_2 > 0$. Thus (3.17) is true. Using this inequality it is sufficient to show that

$$(|1 + \alpha_2| - [(|1 + \Gamma_1| - \rho_1)(|\Gamma_2| - \rho_2)]^{1/2})^2 \geq \sigma_2^2 \tag{3.19}$$

holds in order to validate (3.14). After substituting for $\alpha_2$ and $\sigma_2$, (3.19) reduces to

$$|\Gamma_2|(|1+\Gamma_1|-\rho_1)(1+K_2)+|1+\Gamma_1|(|\Gamma_2|-\rho_2)(1-K_2) \geq 2|1+\alpha_2|[(|1+\Gamma_1|-\rho_1)(|\Gamma_2|-\rho_2)]^{1/2}$$

if $\rho_2|1 + \Gamma_1| \geq \rho_1|\Gamma_2|$ and to

$$|\Gamma_2|(|1+\Gamma_1|-\rho_1)(1-K_2)+|1+\Gamma_1|(|\Gamma_2|-\rho_2)(1+K_2) \geq 2|1+\alpha_2|[(|1+\Gamma_1|-\rho_1)(|\Gamma_2|-\rho_2)]^{1/2}$$

if $\rho_2|1 + \Gamma_1| < \rho_1|\Gamma_2|$ . Again it is easy to see that each of the last two displayed inequalities is true under the corresponding assumption using the simple AGM inequality and formula (3.16).

Finally, let us show that $|1 + \alpha_2| - \sigma_2 > 0$ when $|\Gamma_2| = \sigma_2$. It is not difficult to verify that

$$|1 + \alpha_2|^2 - \sigma_2^2 = \rho_2(1 - K_2)(|1 + \Gamma_1| - \rho_1)$$

under this added assumption. Since $\Re(\Gamma_2) < -\frac{1}{2}$ by our hypotheses, $\Gamma_2 \neq 0$, and therefore $\rho_2 > 0$. The factor $(1-K_2)$ is positive because the two inequalities $\Re(\Gamma_2) <$

$-\frac{1}{2}$ and $\Re(1+\Gamma_1) > \frac{1}{2}$ guarantee that $0 \le K_2 < 1$. The factor $(|1+\Gamma_1| - \rho_1)$ is positive also by our hypotheses. This completes our proof of this lemma.

**LEMMA 3.3** *If $\alpha_n$ and $\sigma_n$ , $n = 1, 2$, are determined by (3.7)-(3.9) and if conditions (3.4) are satisfied, then $N(\alpha_1, \sigma_1) \subseteq N(\alpha_2, \sigma_2)$.*

**Proof:** It is sufficient to prove that

$$|\alpha_2 - \alpha_1| + \sigma_1 \le \sigma_2. \tag{3.20}$$

This inequality is trivially true when $\Gamma_1 = -(1+\Gamma_2) = 0$, so from now on in our proof we shall assume that at least one of the two expressions $\Gamma_1$, $1+\Gamma_2$ is not zero.

For convenience, let

$$\Delta := |\alpha_2 - \alpha_1| \text{ and } \Sigma := \sigma_2^2 - \sigma_1^2 - \Delta^2.$$

Straightforward computation leads to the formula

$$\Delta = \frac{|1+\Gamma_1+\Gamma_2|\,||1+\Gamma_1||1+\Gamma_2| - |\Gamma_1||\Gamma_2||}{(|\Gamma_1| + |1+\Gamma_2|)(|1+\Gamma_1| + |\Gamma_2|)}. \tag{3.21}$$

Now verifying (3.20) is equivalent to verifying

$$2\sigma_1\Delta \le \Sigma. \tag{3.22}$$

Our method for showing that (3.22) is true under our hypotheses demands that we consider six different cases. To make it easier to express the conditions in each of these cases we set

$$\delta := |1+\Gamma_1||1+\Gamma_2| - |\Gamma_1||\Gamma_2|$$

and remind ourselves that

$$\begin{aligned} \tau_1 &:= \rho_1|1+\Gamma_2| - \rho_2|\Gamma_1| \\ \tau_2 &:= \rho_2|1+\Gamma_1| - \rho_1|\Gamma_2|. \end{aligned}$$

Case (1): $\tau_1 \geq 0$, $\tau_2 \geq 0$.

From these assumptions we obtain

$$\begin{aligned} |1+\Gamma_1|\rho_1|1+\Gamma_2| &\geq |1+\Gamma_1|\rho_2|\Gamma_1| \\ &\geq \rho_1|\Gamma_2||\Gamma_1|, \end{aligned}$$

so that

$$|1+\Gamma_1||1+\Gamma_2| - |\Gamma_1||\Gamma_2| = \delta \geq 0. \tag{3.23}$$

Using (3.23), it is tedious but not difficult to verify that

$$\begin{aligned} \Sigma &= [(\rho_1 + |\Gamma_1|K_1) + (\rho_2 - |1+\Gamma_2|K_1)]\Delta \\ &\geq 2[(\rho_1 + |\Gamma_1|K_1)(\rho_2 - |1+\Gamma_2|K_1)]^{1/2}]\Delta \\ &= 2\sigma_1\Delta. \end{aligned}$$

Thus our lemma is true in this case.

Case (2): $\tau_1 \leq 0$, $\tau_2 \geq 0$, $\delta \leq 0$.

In this case

$$\begin{aligned} \Sigma &= [(|1+\Gamma_2| - |\Gamma_1|)K_1 + \rho_2 + \rho_1]\Delta + 2(\rho_2|1+\Gamma_1| - \rho_1|\Gamma_2|)K_2 \\ &\geq [(|1+\Gamma_2| - |\Gamma_1|)K_1 + \rho_2 + \rho_1]\Delta \\ &\geq 2[(\rho_1 - |\Gamma_1|K_1)(\rho_2 + |1+\Gamma_2|K_1)]^{1/2}\Delta \\ &= 2\sigma_1\Delta. \end{aligned}$$

So our lemma is true in this case, also.

Case (3): $\tau_1 \leq 0$, $\tau_2 \geq 0$, $\delta \geq 0$.

In this case

$$\Sigma = [(\rho_1 + |\Gamma_1|K_1) + (\rho_2 - |1+\Gamma_2|K_1)]\Delta + 2K_1(\rho_2|\Gamma_1| - \rho_1|1+\Gamma_2|)$$

$$\geq \quad [(\rho_1 + |\Gamma_1|K_1) + (\rho_2 - |1+\Gamma_2|K_1)]\Delta \tag{3.24}$$

$$\geq \quad 2[(\rho_1 + |\Gamma_1|K_1)(\rho_2 - |1+\Gamma_2|K_1)]^{1/2}\Delta \tag{3.25}$$

$$\geq \quad 2[(\rho_1 - |\Gamma_1|K_1)(\rho_2 + |1+\Gamma_2|K_1)]^{1/2}\Delta \tag{3.26}$$

$$= \quad 2\sigma_1\Delta.$$

Inequality (3.25) follows from (3.24) by the AGM property, and (3.26) follows from (3.25) because $\rho_1|1+\Gamma_2| < \rho_2|\Gamma_1|$ .

Case (4): $\tau_1 \leq 0$, $\tau_2 \leq 0$.

These assumptions guarantee that

$$\rho_1|1+\Gamma_1||1+\Gamma_2| \leq \rho_2|1+\Gamma_1||\Gamma_1| \leq \rho_1|\Gamma_2||\Gamma_1|,$$

from which it follows that $\delta \leq 0$ since $\rho_1 > 0$. Computations similar to those in Case (1) will show that

$$\begin{aligned}\Sigma &= [(\rho_1 - |\Gamma_1|K_1) + (\rho_2 + |1+\Gamma_2|K_1)]\Delta \\ &\geq 2(\rho_1 - |\Gamma_1|K_1)(\rho_2 + |1+\Gamma_2|K_1)]^{1/2}\Delta \\ &= 2\sigma_1\Delta,\end{aligned}$$

so again we have achieved our desired result.

Case (5): $\tau_1 \geq 0$, $\tau_2 \leq 0$, $\delta \geq 0$.

Under the stated conditions

$$\begin{aligned}\Sigma &= [(|\Gamma_1| - |1+\Gamma_2|)K_1 + \rho_1 + \rho_2]\Delta + 2(\rho_1|\Gamma_2| - \rho_2|1+\Gamma_1|)K_2 \\ &\geq [(\rho_1 + |\Gamma_1|K_1) + (\rho_2 - |1+\Gamma_2|K_1)]\Delta \\ &\geq 2[(\rho_1 + |\Gamma_1|K_1)(\rho_2 - |1+\Gamma_2|K_1)]^{1/2}\Delta \\ &= 2\sigma_1\Delta,\end{aligned}$$

and we are done with this case.

Case (6): $\tau_1 \geq 0$, $\tau_2 \leq 0$, $\delta \leq 0$.

Our restrictions in this case lead, in a manner similar to that in Case (3), to

$$\begin{aligned}\Sigma &= [(|1+\Gamma_2| - |\Gamma_1|)K_1 + \rho_1 + \rho_2]\Delta + 2(\rho_1|1+\Gamma_2| - \rho_2|\Gamma_1|)K_1 \\ &\geq [(\rho_1 - |\Gamma_1|K_1) + (\rho_2 + |1+\Gamma_2|K_1)]\Delta \\ &\geq 2[(\rho_1 - |\Gamma_1|K_1)(\rho_2 + |1+\Gamma_2|K_1)]^{1/2}\Delta \\ &\geq 2[(\rho_1 + |\Gamma_1|K_1)(\rho_2 - |1+\Gamma_2|K_1)]^{1/2}\Delta \\ &= 2\sigma_1\Delta.\end{aligned}$$

With this the proof of our lemma is complete.

**LEMMA 3.4** *Let $D_n = \{z : z^2 \in E_n\}$, $n = 1, 2$, where $E_1$ and $E_2$ are given by (3.2) and (3.3), respectively, the $\alpha_n$ and $\sigma_n$, $n = 1, 2$, are defined by (3.7)-(3.9), and the constraints (3.4) are satisfied. Then*

$$D_1 \subseteq C_1(\alpha_1, \sigma_1) \text{ and } D_2 \subseteq C_2(\alpha_2, \sigma_2).$$

**Proof:** Let us prove the second conclusion first. After the appropriate rotation, an equation for the boundary of the element region $D_2$ is given by

$$x^2 = -y^2 + \rho_1\rho_2 - d_2^2 + (T_2^2 + 4d_2^2y^2)^{1/2}, \tag{3.27}$$

where

$$d_2 = (|1+\Gamma_1||\Gamma_2|)^{1/2} \text{ and } T_2 = |\rho_2|1+\Gamma_1| - \rho_1|\Gamma_2|| \geq 0.$$

This same rotation transforms the region $C_2(\alpha_2, \sigma_2)$ onto the region $C_2(|1+\alpha_2| - 1, \sigma_2)$ whose upper boundary is the circle $K(i|1+\alpha_2|, \sigma_2)$ defined in rectangular coordinates by

$$x^2 = \sigma_2^2 - (y - h_2)^2, \tag{3.28}$$

where $h_2 = |1+\alpha_2|$. Let $f$ and $g$ denote the functions of $y$ defined by setting $f(y)$ equal to the right side of (3.27) and $g(y)$ equal to the right side of (3.28). Making use of symmetry, it is sufficient to prove that $f(y) \geq g(y) \geq 0$ for all $y \in I$, where $I = [h_2 - \sigma_2, h_2 + \sigma_2]$, in order to reach our first goal. From (3.27) and (3.28) and the formulas for $\alpha_2$, $\sigma_2$, and $K_2$ we obtain

$$\begin{aligned} f(y) - g(y) &= (T_2^2 + 4d_2^2 y^2)^{1/2} + \rho_1\rho_2 - d_2^2 + h_2^2 - \sigma_2^2 - 2h_2 y \\ &= (T_2^2 + 4d_2^2 y^2)^{1/2} - (T_2 K_2 + 2h_2 y). \end{aligned} \tag{3.29}$$

Now let $F$ and $G$ be the real-valued functions defined by

$$F(y) = (T_2^2 + 4d_2^2 y^2)^{1/2} \text{ and } G(y) = T_2 K_2 + 2h_2 y.$$

Note that $F(y) \geq 0$ for all real $y$. Since

$$\begin{aligned} h_2^2 - \sigma_2^2 &= d_2^2 - \rho_1\rho_2 - T_2 K_2 \\ &\geq d_2^2 - \rho_1\rho_2 - T_2 \\ &= (|\Gamma_2| - \rho_2)(|1+\Gamma_1| + \rho_1) > 0, \end{aligned}$$

it follows that $y > 0$ if $y \in I$. It is certainly true, therefore, that $G(y) \geq 0$ on $I$. But

$$\begin{aligned} [F(y)]^2 - [G(y)]^2 &= T_2^2(1-K_2^2) - 4h_2 T_2 K_2 y + 4(d_2^2 - h_2^2)y^2 \\ &= T_2^2(1-K_2^2) - 4h_2 T_2 K_2 y + 4(d_2^2 K_2^2)y^2 \\ &= T_2^2(1-K_2^2) - 4h_2 T_2 K_2 d_2 (1-K_2^2)^{1/2} y + 4(d_2^2 K_2^2)y^2 \\ &= (T_2(1-K_2^2)^{1/2} - 2d_2 K_2 y)^2. \end{aligned} \tag{3.30}$$

Thus we have shown that the functions $f, g, F$, and $G$ satisfy the hypotheses of Lemma 1.1 on I so that we can conclude that $f(y) \geq g(y) \geq 0$ on $I$ as desired.

With the aid of equation (3.30) we can deduce that the curve (3.27) and the circle (3.28) will be tangent when

$$y = \frac{T_2}{2d_2}\left(\frac{1}{K_2^2} - 1\right)^{1/2}$$

and $y \in I$. It also follows from (3.30) and the above that (3.27) is a circle if both conditions $\Gamma_1 = -(1 + \Gamma_2)$ and $\rho_1 = \rho_2$ are satisfied.

Now we will proceed to prove the first assertion in our lemma under the added assumption that $\Gamma_1$ and $1 + \Gamma_2$ are not both zero.. We rotate $D_1$ onto the region whose boundary in rectangular coordinates is

$$x^2 = -y^2 + \rho_1\rho_2 - d_1^2 - (T_1^2 + 4d_1^2y^2)^{1/2}, \tag{3.31}$$

where

$$T_1 = |\rho_2|\Gamma_1| - \rho_1|1 + \Gamma_2|| \text{ and } d_1 = (|\Gamma_1||1 + \Gamma_2|)^{1/2}.$$

Under the same rotation the region $C_1(\alpha_1, \sigma_1)$ maps onto the region $C_1(|\alpha_1|, \sigma_1)$ whose upper boundary is an arc of the circle $K(-i|\alpha_1|, \sigma_1)$. A rectangular equation for this circle is

$$x^2 = \sigma_1^2 - (y + h_1)^2, \tag{3.32}$$

where $h_1 = |\alpha_1|$. As in the argument above, we again define functions and show that they meet the requirements of Lemma 1.1. We define $f$ by setting $f(y)$ equal to the polynomial on the right side of (3.32), and $g$ by setting $g(y)$ equal to the expression on the right side of (3.31). Again, in view of the symmetry of the regions involved, we have only to prove that $f(y) \geq g(y) \geq 0$ on the interval $I$, where $I$ in this case is given by

$$I = [0, [(\rho_1 - |\Gamma_1|)(\rho_2 - |1 + \Gamma_2|)]^{1/2}].$$

Here

$$
\begin{aligned}
f(y) - g(y) &= (T_1^2 + 4d_1^2y^2)^{1/2} + \sigma_1^2 - h_1^2 + d_1^2 - \rho_1\rho_2 - 2h_1y \\
&= (T_1^2 + 4d_1^2y^2)^{1/2} - (T_1K_1 + 2h_1y). \qquad (3.33)
\end{aligned}
$$

If we let $F$ denote the square root function in (3.33) and $G$ denote the function in parentheses that is subtracted from it, then

$$f(y) - g(y) = F(y) - G(y)$$

and both $F$ and $G$ are nonnegative on $I$. Moreover,

$$
\begin{aligned}
[F(y)]^2 - [G(y)]^2 &= T_1^2 - 4h_1T_1K_1y + 4(d_1^2 - h_1^2)y^2 \\
&= T_1^2 - 4h_1T_1K_1y + 4d_1^2K_1^2y^2 \\
&= T_1^2 - 4d_1(1 - K_1^2)^{1/2}T_1K_1y + 4d_1^2K_1^2y^2 \\
&= (T_1(1 - K_1^2)^{1/2} - 2d_1K_1y)^2. \qquad (3.34)
\end{aligned}
$$

Thus we have shown that the hypotheses of Lemma 1.1 are satisfied by the functions $f, g, F$, and $G$ we have defined for this case, so we can conclude that $f(y) \geq g(y) \geq 0$ for all $y \in I$. Futhermore, if $d_1K_1 \neq 0$, it follows from (3.34) that the circle and the curve are tangent for those values of $y$ for which

$$y = \frac{T_1}{2d_1}\left(\frac{1}{K_1^2} - 1\right)^{1/2}$$

and these values are in $I$. It remains to consider the case where $\Gamma_1 = -(1+\Gamma_2) = 0$. Under this assumption we have

$$f(y) = g(y) = \rho_1\rho_2 - y^2 \geq 0,$$

so we are done with the proof of this lemma.

Now let us return to completing the proof of Theorem 3.2. In our lemmas of this section we have proved that

$$\begin{array}{lrl} (i) & N(\alpha_1,\sigma_1) \subseteq & N(\alpha_2,\sigma_2), \\ (ii) & D_1 \subseteq & C_1(\alpha_1,\sigma_1), \\ (iii) & D_2 \subseteq & C_2(\alpha_2,\sigma_2). \end{array}$$

So to complete the proof of the embedding statements in Theorem 3.2 we have only to verify that $D_1 \subseteq C_1(\alpha_2,\sigma_2)$ and $D_2 \subseteq C_2(\alpha_1,\sigma_1)$ . But (i) $\Longrightarrow N(i\alpha_1,\sigma_1) \subseteq N(i\alpha_2,\sigma_2)$ and $N(-i\alpha_1,\sigma_1) \subseteq N(-i\alpha_2,\sigma_2)$, so that $D_1 \subseteq C_1(\alpha_2,\sigma_2)$. Moreover, (i) $\Longrightarrow N(i(1+\alpha_1),\sigma_1) \subseteq N(i(1+\alpha_2),\sigma_2)$ and $N(-i(1+\alpha_1),\sigma_1) \subseteq N(-i(1+\alpha_2),\sigma_2)$, so that $D_2 \subseteq C_2(\alpha_2,\sigma_2) \subseteq C_2(\alpha_1,\sigma_1)$.

It remains to verify the assertion $|\alpha_n| < \sigma_n < |1+\alpha_n|$, $n = 1,2$ and the uniform convergence statements in our Theorem 3.2. It follows from Lemmas 3.1 and 3.3 that $|\alpha_n| < \sigma_n$, and it follows from Lemmas 3.2 and 3.3 that $\sigma_n < |1+\alpha_n|$. The uniformity of the pairs of twin convergence regions $L_1(\alpha_n,\sigma_n)$, $L_2(\alpha_n,\sigma_n)$ and $E_1$, $E_2$ follows from Theorem 1.1 and the fact that a subregion of a uniform convergence region is a uniform convergence region. With this we have finished the proof of Theorem 3.2.

## 4 Disk-Disk Case

The following result is Theorem 5.1 in [3] with the subscripts of the element regions and the subscripts of the value regions interchanged.

**THEOREM 4.1 ([3], Disk-Disk Case)** *The c.f. $K(a_n/1)$ converges if its elements satisfy*

$$a_{2n-1} \in E_1, \qquad a_{2n} \in E_2, \qquad a_n \neq 0, \qquad n = 1,2,3,\ldots, \tag{4.1}$$

*where*

$$E_1 = \{ w : |w(1+\overline{\Gamma}_2) - \Gamma_1(|1+\Gamma_2|^2 - \rho_2^2)| + \rho_2|w| \leq \rho_1(|1+\Gamma_2|^2 - \rho_2^2) \}, \tag{4.2}$$

$$E_2 = \{w : |w(1+\overline{\Gamma}_1) - \Gamma_2(|1+\Gamma_1|^2 - \rho_1^2)| + \rho_1|w| \leq \rho_2(|1+\Gamma_1|^2 - \rho_1^2)\}, \quad (4.3)$$

$$|\Gamma_1| \neq \rho_1 < |1+\Gamma_1|, \qquad |\Gamma_2| \leq \rho_2 < |1+\Gamma_2|. \quad (4.4)$$

*The corresponding pre twin value regions are the closed disks*

$$V_1 = \{z : |z-\Gamma_2| \leq \rho_2\} \qquad V_2 = \{z : |z-\Gamma_1| \leq \rho_1\}. \quad (4.5)$$

The regions $E_1$ and $E_2$ in Theorem 4.1 are both bounded regions, each, in general, with a Cartesian oval as its boundary. We shall show in our first lemma below that the condition $\rho_1|1+\Gamma_2| \geq \rho_2|\Gamma_1|$ must be added to conditions (4.4) in order for $E_1$ to be nonempty. From now on we shall consider Theorem 4.1 with this condition added to the hypotheses.

**LEMMA 4.1** *Let $E_1$ and $E_2$ be given by (4.2) and (4.3), respectively, and let conditions (4.4) be satisfied. Then $E_1$ is nonempty if and only if*

$$\rho_1|1+\Gamma_2| \geq \rho_2|\Gamma_1|. \quad (4.6)$$

**Proof:** Let

$$c = \frac{1+\overline{\Gamma}_2}{|1+\Gamma_2|^2-\rho_2^2}, \quad d = \frac{1+\overline{\Gamma}_1}{|1+\Gamma_2|^2-\rho_1^2}, \quad r = \frac{\rho_2}{|1+\Gamma_2|}, \quad s = \frac{\rho_1}{|1+\Gamma_1|}.$$

Furthermore, let $V_1$ and $V_2$ be given by (4.5), and let

$$C = \{z : |z-c| \leq r|c|\}, \ D = \{z : |z-d| \leq s|d|\}.$$

Then it is easily verified that

$$\frac{E_1}{1+V_1} \subseteq V_2 \Longleftrightarrow \frac{1}{1+E_1C} \subseteq D.$$

From Theorems A and Theorem 2.1 of Lane [5] we have that $E_1$ is nonvacuous if and only if (a) $1 \in D$ or (b) $0 \in D$ and $0 \notin C$, or (c) $0 \notin C$, $0 \notin D$, and

$$|d-1|^2 \leq s^2|d|^2 + \frac{s^2(1-r^2)}{r^2(1-s^2)}.$$

Our next step is to show that, when conditions (4.4) are satisfied, $1 \in D$ and $\rho_1|1+\Gamma_2| < \rho_2|\Gamma_1|$ cannot hold simultaneously. For if so, $\rho_1 < |\Gamma_1|$ and

$$\left|\frac{1+\overline{\Gamma}_1}{|1+\Gamma_1|^2-\rho_1^2}-1\right| \leq \frac{\rho_1}{|1+\Gamma_1|^2-\rho_1^2}.$$

The last inequality leads to

$$(|\Gamma_1|^2-\rho_1^2)(|1+\Gamma_1|^2-\rho_1^2) \leq 0$$

which is false. Hence it is true that $1 \in D$ implies $\rho_1|1+\Gamma_2| \geq \rho_2|\Gamma_1|$. It is easy to see from our hypotheses that case (b) cannot occur, so it remains to examine case (c). Since $r < 1$ and $s < 1$ it is clear that $0 \notin C$ and $0 \notin D$. Straightforward computations will show, after the appropriate substitutions for $r$, $s$, and $d$ have been made in terms of $\rho_1$, $\rho_2$, $\Gamma_1$, and $\Gamma_2$, that the inequality in (c) simplifies to

$$\rho_2^2|\Gamma_1|^2 \leq \rho_1^2|1+\Gamma_2|^2.$$

Thus we have shown that $E_1$ is nonempty if and only if $\rho_1|1+\Gamma_2| \geq \rho_1|\Gamma_1|$.

We are now about to state our uniform convergence embedding theorem for the disk-disk case after we make a few remarks. This case, by far, has been the most difficult case for us to consider. The reason is that the embedding problems become more complicated when $\rho_1 < |\Gamma_1|$ in Theorem 4.1. In this case the origin is no longer contained in one of the two ovals, and, in fact, some of the twin convergence regions obtained under the condition $\rho_1 < |\Gamma_1|$ are not embeddable in the twin convergence regions of the Uniform Twin Limacon Theorem. For example, if we set $\Gamma_1 = 2$, $\rho_1 = 1.8$, $\Gamma_2 = 0$, and $\rho_2 = 0.9$, the conditions of Theorem 4.1 are satisfied and $E_1 = \{0.38\}$ and $E_2 = \{w : |w| \leq 1.08\}$. By looking at these regions under the mapping $w = \sqrt{z}$, it is fairly easy to see that the images are not embeddable in regions like $C_1(\alpha,\sigma)$ and $C_2(\alpha,\sigma)$ of Theorem 1.1 under the condition $\alpha < \sigma < |1+\alpha|$ that $\alpha$ and $\sigma$ must satisfy. As the reader can see, though, from the statement of our theorem below we have been successful in embedding the twin ovals in the twin limacons when $\rho_1 > |\Gamma_1|$ and in many cases when $\rho_1 < |\Gamma_1|$, along, of course, with the other restrictions placed on the parameters in Theorem 4.1.

**THEOREM 4.2 (Disk-Disk Embedding)** *Let $E_1$ and $E_2$ be given by (4.2) and (4.3), respectively, let conditions (4.4) and (4.6) be satisfied, and let $s_0 = s_2$ for any symbol $s$ that follows. Then at least one of the two inequalities*

$$|\Gamma_1| < |1+\Gamma_2|, \qquad |\Gamma_2| < |1+\Gamma_1| \tag{4.7}$$

*must hold. If $n = 1, 2$ and $|\Gamma_n| \neq |1+\Gamma_{n-1}|$, let $\alpha_n$, and $\sigma_n$ be defined by*

$$\alpha_n := \frac{|1+\Gamma_{n-1}|\Gamma_n + |\Gamma_n|(1+\Gamma_{n-1})}{|1+\Gamma_{n-1}| - |\Gamma_n|} \tag{4.8}$$

$$\sigma_n := [(\rho_n + |\Gamma_n|M_n)(|1+\Gamma_{n-1}|M_n - \rho_{n-1})]^{1/2}, \tag{4.9}$$

*where*

$$M_n := \frac{|1+\Gamma_1+\Gamma_2|}{||1+\Gamma_{n-1}| - |\Gamma_n||}. \tag{4.10}$$

*If $|\Gamma_1| < \rho_1$, then for $n = 1, 2$, $|\alpha_n| < \sigma_n < |1+\alpha_n|$, and*

$$|\Gamma_n| < |1+\Gamma_{n-1}| \implies E_n \subseteq L_1(\alpha_n,\sigma_n),\ E_{n-1} \subseteq L_2(\alpha_n,\sigma_n), \tag{4.11}$$

$$|\Gamma_{n-1}| > |1+\Gamma_n| \implies E_n \subseteq L_1(\alpha_n,\sigma_n),\ E_{n-1} \subseteq L_2(\alpha_n,\sigma_n). \tag{4.12}$$

*If $|\Gamma_1| > \rho_1$, then*

$$|\Gamma_1| < |1+\Gamma_2| \implies E_1 \subseteq L_1(\alpha_1,\sigma_1),\ E_2 \subseteq L_2(\alpha_1,\sigma_1), \tag{4.13}$$

$$|\Gamma_1| \leq |1+\Gamma_2| \text{ and } |\Gamma_2| < |1+\Gamma_1| \implies E_1 \subseteq L_2(\alpha_2,\sigma_2),\ E_2 \subseteq L_1(\alpha_2,\sigma_2), \tag{4.14}$$

$$|\Gamma_1| > |1+\Gamma_2| \text{ and } |\sigma_2| < |1+\alpha_2| \implies E_1 \subseteq L_2(\alpha_2,\sigma_2),\ E_2 \subseteq L_1(\alpha_2,\sigma_2), \tag{4.15}$$

$$|\alpha_n| < \sigma_n < |1+\alpha_n|,\ n = 1,2.$$

*The condition* $\sigma_2 < |1+\alpha_2|$ *in (4.15) is satisfied, in particular, if*

$$0 < |\Gamma_1| - \rho_1 \leq |1+\Gamma_2| - \rho_2.$$

*In each case above, the convergence of* $K(a_n/1)$ *is uniform with respect to each pair of regions* $E_1$, $E_2$ *and its corresponding embedding pair* $L_1(\alpha_n,\sigma_n)$, $L_2(\alpha_n,\sigma_n)$, *and the latter pair is a best pair of twin convergence regions. Furthermore, if* $f_\nu$ *denotes the* $\nu$*th approximant of* $K(c_n^2/1)$ *or* $K(a_n/1)$, *then for* $\nu \geq 1$ *and* $n = 1,2$

$$|f_1 - f| \leq 2\sigma_n,\ |f_{2\nu+1} - f| \leq |f_{2\nu} - f| \leq 2\sigma_n(1+\nu/d_n)^{-b_n},$$

*where*

$$b_n = \frac{|1+\alpha_n|(\sigma_n - |\alpha_n|)(|1+\alpha_n| - \sigma_n)}{\sigma_n[1+\Re(\alpha_n) + (\sigma_n^2 - (\Im(\alpha_n))^2)^{1/2}]},\ d_n = c_n\left[1 + \frac{(\sigma_n + |\alpha_n|)(|1+\alpha_n| + \sigma_n)}{2|1+\alpha_n|(\sigma_n - |\alpha_n|)}\right].$$

**Proof:** Our proof of this theorem (it seems by necessity) is fairly involved, so we have decided to prove a number of lemmas to help organize our arguments. After we have established these lemmas, we shall try to make clear how they can be pieced together to verify those parts of Theorem 4.2 that we have not pointed out as being established along the way. We begin by giving two example plots illustrating some of the region behavior that is possible in connection with Theorems 4.1 and 4.2. It would take a sizable number of graphs just to indicate the "pathological" cases that can occur. We follow the graphs with a proof of assertion (4.7) through Lemma 4.2.

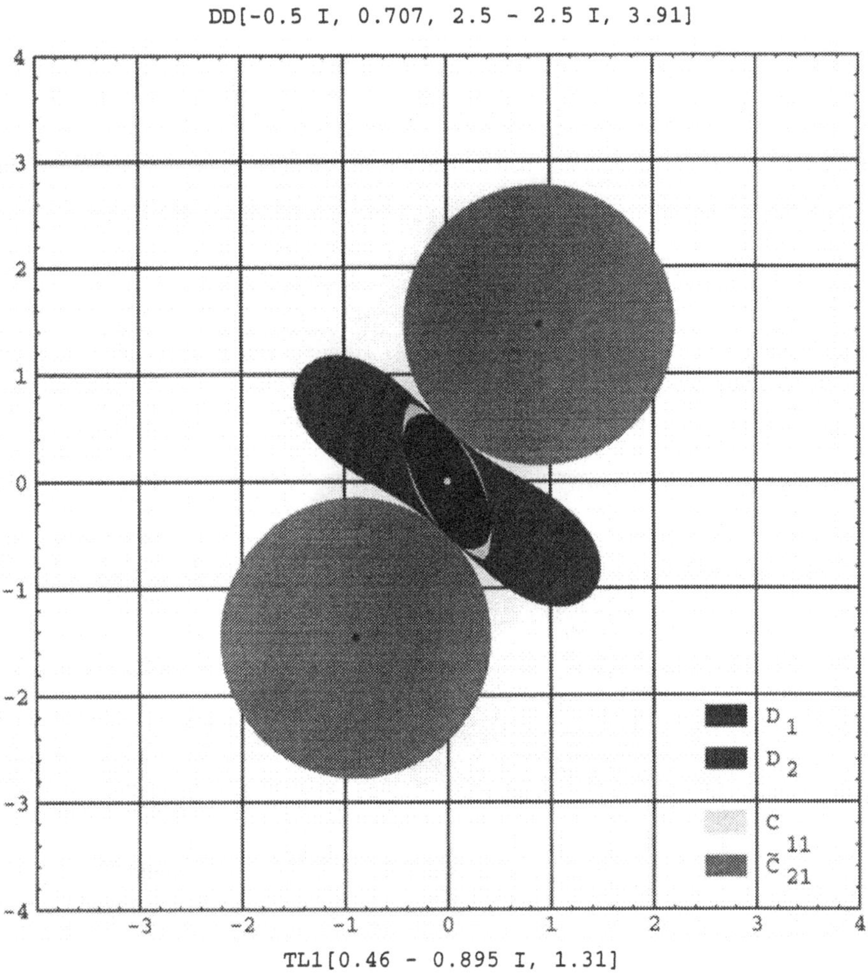


Figure 4.1

In Figure 4.1, the arguments of $DD$ are $\Gamma_1$, $\rho_1$, $\Gamma_2$, $\rho_2$; and the arguments of the $TLn$ are $\alpha_n$, $\sigma_n$, $n = 1, 2$. As in our earlier plots,

$$C_{ij} := C_i(\alpha_j, \sigma_j), \;\; D_i := \{\, z : z^2 \in E_i \,\}, \;\; i, j = 1, 2,$$

and the cap $\sim$ is used to denote the complement. Note the peanut shape of the region $D_2$ and the antisymmetric relationship of $D_1$ and $D_2$ with each other.

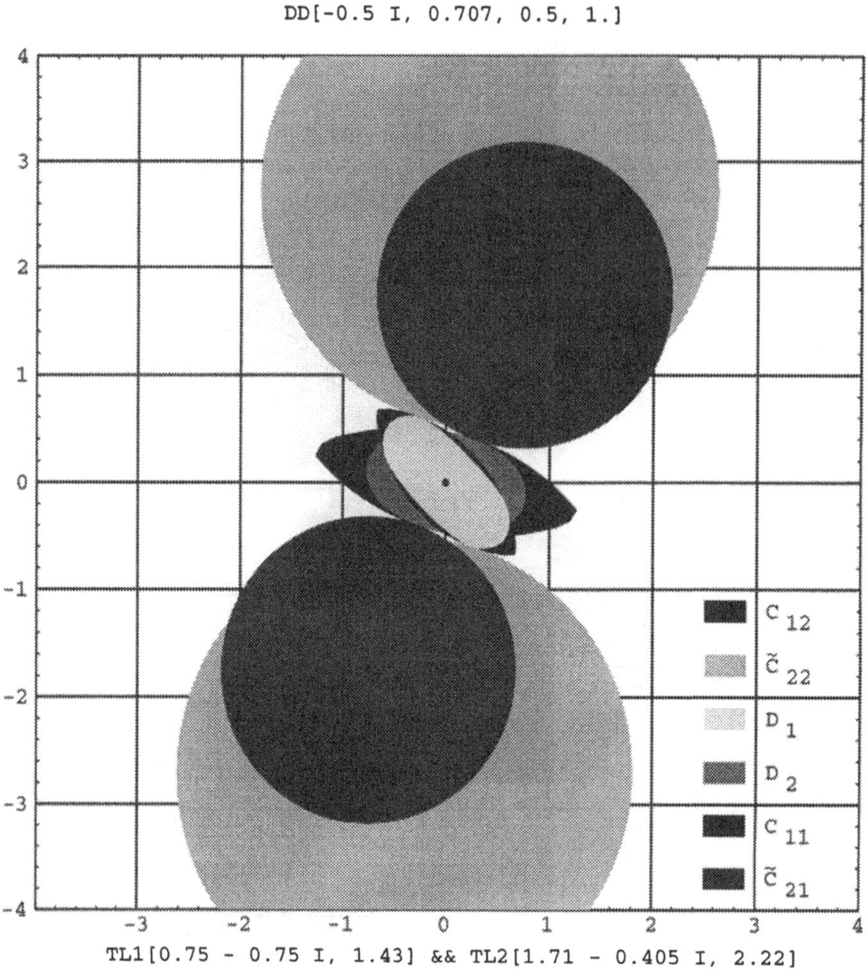


Figure 4.2

The symbols in Figure 4.2 have the same meaning as they had in Figure 4.1. Note how the axes of symmetry of $D_1$ and $D_2$ differ with respect to each other. Note, also, the two solutions of the embedding problem for the given values of the parameters in $DD$.

**LEMMA 4.2** *The inequalities* $|1+\Gamma_2| \leq |\Gamma_1|$ *and* $|1+\Gamma_1| \leq |\Gamma_2|$ *cannot be satisfied simultaneously under conditions (4.4).*

**Proof:** Assume that both inequalities in the statement of the lemma do hold simultaneously. Then the two inequalities

$$\begin{aligned} 1 + 2\Re(\Gamma_2) + |\Gamma_2|^2 &\leq |\Gamma_1|^2 \\ 1 + 2\Re(\Gamma_1) + |\Gamma_1|^2 &\leq |\Gamma_2|^2 \end{aligned}$$

are satisfied. By summing them, it follows easily that

$$1 + \Re(\Gamma_1) + \Re(\Gamma_2) \leq 0$$

must hold, which cannot be since $\Re(\Gamma_1) > -\frac{1}{2}$ and $\Re(\Gamma_2) > -\frac{1}{2}$ . Hence our lemma is true.

In our next lemma we verify that the inequality $|\alpha_n| < \sigma_n < |1+\alpha_n|$ for admissible $n$ holds true for each embedding case considered in Theorem 4.2.

**LEMMA 4.3** *Let conditions (4.4) and (4.6) be satisfied and let* $\alpha_n$ *and* $\sigma_n$ *be given by (4.8) and (4.9), respectively. Then for* $n = 1, 2$

$$|\Gamma_n| \neq |1+\Gamma_{n-1}| \implies |\alpha_n| < \sigma_n, \tag{4.16}$$

$$|\Gamma_1| < |1+\Gamma_2| \implies \sigma_1 < |1+\alpha_1|, \tag{4.17}$$

$$|\Gamma_2| < |1+\Gamma_1| \text{ and } \rho_1 > |\Gamma_1| \implies \sigma_2 < |1+\alpha_2|, \tag{4.18}$$

$$|\Gamma_2| < |1+\Gamma_1|, |\Gamma_1| \leq |1+\Gamma_2|, \text{ and } \rho_1 < |\Gamma_1| \implies \sigma_2 < |1+\alpha_2|, \tag{4.19}$$

$$|\Gamma_1| > |1+\Gamma_2| \text{ and } 0 < |\Gamma_1| - \rho_1 < |1+\Gamma_2| - \rho_2 \implies \sigma_2 < |1+\alpha_2|. \tag{4.20}$$

**Proof:** When $|\Gamma_n| \neq |1+\Gamma_{n-1}|$ it is not difficult to verify that

$$\sigma_n^2 - |\alpha_n|^2 = \tau_n M_n - \rho_{n-1}\rho_n + |\Gamma_n||1+\Gamma_{n-1}|. \tag{4.21}$$

It follows from our hypotheses that $M_n \geq 1$, $|1+\Gamma_{n-1}| > \rho_{n-1}$, $\tau_n \geq 0$, and $\rho_n$ and $|\Gamma_n|$ are not zero for the same value of $n$. Hence $\sigma_n > 0$, and from (4.21) we obtain

$$\begin{aligned}\sigma_n^2 - |\alpha_n|^2 &\geq \tau_n - \rho_{n-1}\rho_n + |\Gamma_n||1+\Gamma_{n-1}| \\ &= (\rho_n + |\Gamma_n|)(|1+\Gamma_{n-1}| - \rho_{n-1}) > 0.\end{aligned}$$

Next we shall verify that $|\Gamma_1| < |1+\Gamma_2| \Rightarrow |1+\alpha_1| > \sigma_1$. Simple calculations will show that $|1+\alpha_n|^2 - \sigma_n^2$

$$= \frac{\rho_{n-1}\rho_n(|1+\Gamma_{n-1}| - |\Gamma_n|) + |1+\Gamma_n|^2|1+\Gamma_{n-1}| - |\Gamma_{n-1}|^2|\Gamma_n| - \tau_n|1+\Gamma_1+\Gamma_2|}{|1+\Gamma_{n-1}| - |\Gamma_n|} \tag{4.22}$$

when $|\Gamma_n| < |1+\Gamma_{n-1}|$, where

$$\tau_n = \rho_n|1+\Gamma_{n-1}| - \rho_{n-1}|\Gamma_n|.$$

Taking $n = 1$ in (4.22) and using the fact that $\tau_1 \geq 0$ by (4.6) we arrive at

$|1+\alpha_1|^2 - \sigma_1^2$

$$\begin{aligned}&\geq \frac{\rho_1\rho_2(|1+\Gamma_2| - |\Gamma_1|) + |1+\Gamma_1|^2|1+\Gamma_2| - |\Gamma_2|^2|\Gamma_1| - \tau_1(|1+\Gamma_1| + |\Gamma_2|)}{|1+\Gamma_2| - |\Gamma_1|} \\ &= \frac{(\rho_2|\Gamma_1| + |1+\Gamma_1||1+\Gamma_2|)(|1+\Gamma_1| - \rho_1) + (\rho_1|1+\Gamma_2| + |\Gamma_2||\Gamma_1|)(\rho_2 - |\Gamma_2|)}{|1+\Gamma_2| - |\Gamma_1|} \\ &> 0.\end{aligned}$$

To prove (4.18) we put $n = 2$ in (4.22) and obtain $|1+\alpha_2|^2 - \sigma_2^2$

$$\begin{aligned}&= \frac{\rho_1\rho_2(|1+\Gamma_1| - |\Gamma_2|) + |1+\Gamma_2|^2|1+\Gamma_1| - |\Gamma_1|^2|\Gamma_2| - \tau_2|1+\Gamma_1+\Gamma_2|}{|1+\Gamma_1| - |\Gamma_2|} \\ &\geq \frac{\rho_1\rho_2(|1+\Gamma_1| - |\Gamma_2|) + |1+\Gamma_2|^2|1+\Gamma_1| - |\Gamma_1|^2|\Gamma_2| - \tau_2(|1+\Gamma_2| + |\Gamma_1|)}{|1+\Gamma_1| - |\Gamma_2|} \\ &= \frac{(|1+\Gamma_1||1+\Gamma_2| + \rho_1|\Gamma_2|)(|1+\Gamma_2| - \rho_2) + (\rho_2|1+\Gamma_1| + |\Gamma_1||\Gamma_2|)(\rho_1 - |\Gamma_1|)}{|1+\Gamma_1| - |\Gamma_2|} \\ &> 0.\end{aligned}$$

To prove (4.19) we use our assumption $|\Gamma_1| \leq |1+\Gamma_2|$ and condition (4.6) to establish that

$$|1+\Gamma_2| - \rho_2 \geq |1+\Gamma_2|(|\Gamma_1| - \rho_1)/|\Gamma_1| \geq |\Gamma_1| - \rho_1).$$

Then applying this inequality to the last line in the sequence of inequalities immediately above we obtain $|1+\alpha_2|^2 - \sigma_2^2$

$$\geq \frac{(|1+\Gamma_1||1+\Gamma_2| + \rho_1|\Gamma_2|)(|1+\Gamma_2| - \rho_2) + (\rho_2|1+\Gamma_1| - |\Gamma_1||\Gamma_2|)(|\Gamma_1 - \rho_1|)}{|1+\Gamma_1| - |\Gamma_2|}$$
$$\geq \frac{(|1+\Gamma_1||1+\Gamma_2| + \rho_1|\Gamma_2|)(|1+\Gamma_2| - \rho_2) + (\rho_2|1+\Gamma_1| - |\Gamma_1||\Gamma_2|)(|1+\Gamma_2| - \rho_2)}{|1+\Gamma_1| - |\Gamma_2|}$$
$$\geq (|1+\Gamma_2| - \rho_2)^2 > 0.$$

As for statement (4.20), our hypotheses guarantee that

$$|1+\Gamma_1| > |\Gamma_1| > |1+\Gamma_2| > \rho_2 \geq |\Gamma_2|.$$

With this and the hypothesis

$$0 < |\Gamma_1| - \rho_1| < |1+\Gamma_2| - \rho_2$$

it follows from the preceding argument that

$$|1+\alpha_2|^2 - \sigma_2^2 > (|1+\Gamma_2| - \rho_2)^2 > 0.$$

**LEMMA 4.4** *Let $\alpha_n$, $\sigma_n$, and $M_n$ be determined by their formulas in Theorem 3.2 and let*

$$\delta_n = [(\rho_{n-1} + |1+\Gamma_{n-1}|M_n)(|\Gamma_n|M_n - \rho_n)]^{1/2}, \quad n = 1, 2. \tag{4.23}$$

*Assume also that conditions (4.4), and (4.6) are satisfied and $s_0 = s_2$ for each symbol $s$. Then, for $n = 1, 2$, $N(1+\alpha_n, \sigma_n) \subseteq N(-\alpha_{n-1}, \delta_{n-1})$ if $|1+\Gamma_n| < |\Gamma_{n-1}|$. Moreover, $N(\alpha_n, \sigma_n) \subseteq M(-1-\alpha_{n-1}, \sigma_{n-1})$ if both inequalities (4.7) are satisfied.*

**Proof:** Let n=1,2. It is clear that $\delta_{n-1}$ is positive if the second factor in its formula is positive. Thus our first goal is to show that the second factor in the formula for $\delta_{n-1}$ is positive if $|1+\Gamma_n| < |\Gamma_{n-1}|$. With the aid of the triangle inequality we obtain

$$\begin{aligned} |\Gamma_{n-1}|M_{n-1} - \rho_{n-1} &= \frac{|\Gamma_{n-1}||1+\Gamma_{n-1}+\Gamma_n| - \rho_{n-1}(|\Gamma_{n-1}| - |1+\Gamma_n|)}{|\Gamma_{n-1}| - |1+\Gamma_n|} \\ &\geq \frac{|\Gamma_{n-1}|(|1+\Gamma_{n-1}| - |\Gamma_{n-1}|) - \rho_{n-1}(|\Gamma_{n-1}| - |1+\Gamma_n|)}{|\Gamma_{n-1}| - |1+\Gamma_n|} \\ &= \frac{|\Gamma_{n-1}|(|1+\Gamma_{n-1}| - \rho_{n-1}) + \rho_{n-1}|1+\Gamma_n| - |\Gamma_{n-1}||\Gamma_n|}{|\Gamma_{n-1}| - |1+\Gamma_n|}. \end{aligned} \tag{4.24}$$

It follows from (4.24) and (4.6) that

$$|\Gamma_1|M_1 - \rho_1 \geq \frac{|\Gamma_1|[(|1+\Gamma_1| - \rho_1) + (\rho_2 - |\Gamma_2|)]}{|\Gamma_1| - |1+\Gamma_2|} > 0.$$

To settle the $n=1$ case we use (4.24) and the fact that $\rho_2 \geq |\Gamma_2|$ to obtain

$$|\Gamma_2|M_2 - \rho_2 \geq \frac{|\Gamma_2|[(|1+\Gamma_2| - \rho_2) + (|1+\Gamma_1| - |\Gamma_1|)]}{|\Gamma_2| - |1+\Gamma_1|} > 0.$$

Hence $\delta_{n-1}$ is positive.

Now let

$$\Delta := |1+\alpha_1+\alpha_2|.$$

Then, if $|1+\Gamma_n| < |\Gamma_{n-1}|$ , using formula (4.8) for $\alpha_n$ we obtain

$$\Delta = \frac{|1+\Gamma_1+\Gamma_2|[|1+\Gamma_1||1+\Gamma_2| - |\Gamma_1||\Gamma_2|]}{(|\Gamma_2| - |1+\Gamma_1|)(|1+\Gamma_2| - |\Gamma_1|)}.$$

To prove the first part of our lemma it is sufficient to show that

$$\sigma_n + \Delta \leq \delta_{n-1}$$

or, equivalently, that

$$2\sigma_n\Delta \leq \delta_{n-1}^2 - \sigma_n^2 - \Delta^2.$$

However, somewhat tedious but straightforward computations lead to

$$\begin{aligned}\delta_{n-1}^2 - \sigma_n^2 - \Delta^2 &= [(\rho_n + |\Gamma_n|M_n) + (|1+\Gamma_{n-1}|M_n - \rho_{n-1})]\Delta \\ &\geq 2\sigma_n\Delta,\end{aligned}$$

where the last inequality follows from the AGM inequality .

To verify that $N(\alpha_n, \sigma_n) \subseteq M(-1-\alpha_{n-1}, \sigma_{n-1})$ when $|\Gamma_n| < |1+\Gamma_{n-1}|$ for both $n=1$ and $n=2$, it is sufficient to prove that

$$\sigma_1 + \sigma_2 \leq \Delta,$$

where in this case

$$\Delta = \frac{|1+\Gamma_1+\Gamma_2|[|1+\Gamma_1||1+\Gamma_2| - |\Gamma_1||\Gamma_2|]}{(|1+\Gamma_1| - |\Gamma_2|)(|1+\Gamma_2| - |\Gamma_1|)}.$$

Let

$$\sigma_1 = (A_1B_1)^{1/2} \text{ and } \sigma_2 = (A_2B_2)^{1/2},$$

where the $A_i$ and $B_i$ are the appropriate factors in parentheses in the indicated formulas for the $\sigma_i$. Then using the AGM inequality, and after making the obvious substitutions, we arrive at

$$\sigma_1 + \sigma_2 \leq (A_1 + B_1)/2 + (A_2 + B_2)/2 = \Delta.$$

With this the proof of our lemma is complete.

**LEMMA 4.5** *Let conditions (4.4) and (4.6) be satisfied and let the half plane $H_n$ be defined by*

$$H_n := \{ z : \Re(ze^{-i\phi_n}) \leq (\rho_n - \rho_{n-1})/2 \}, \quad n = 1, 2, \qquad \rho_0 = \rho_2, \tag{4.25}$$

*where*

$$\phi_n := (\arg(1+\Gamma_{n-1}) + \arg(\Gamma_n))/2, \quad n = 1, 2, \qquad \Gamma_0 = \Gamma_2. \tag{4.26}$$

*Then*

$$|1+\Gamma_{n-1}| = |\Gamma_n| \Longrightarrow H_n \subseteq M(1+\alpha_{n-1}, \sigma_{n-1}), \quad n = 1, 2, \tag{4.27}$$

*where $\alpha_n$ and $\sigma_n$ are given by (4.8) and (4.9), respectively.*

**Proof:** Let $z = x + iy$ and $n = 1, 2$. Also let $d_n$ denote the distance from $(1+\alpha_{n-1})$ to the line

$$x \cos\phi_n + y \sin\phi_n = (\rho_n - \rho_{n-1})/2.$$

To prove our lemma, it is sufficient to prove that $d_n \geq \sigma_{n-1}$ under the added assumption $|1+\Gamma_{n-1}| = |\Gamma_n|$. Note that by Lemma 4.2, $|1+\Gamma_n| \neq |\Gamma_{n-1}|$ when $|1+\Gamma_{n-1}| = |\Gamma_n|$ . After setting $\Delta_n = \rho_n - \rho_{n-1}$ and using the well known formula for the distance from a point to a line, it follows that

$$\begin{aligned}
d_n &= |\Re((1+\alpha_{n-1}))\cos\phi_n + \Im((1+\alpha_{n-1}))\sin\phi_n - \Delta_n/2| \\
&= \left| \frac{|1+\Gamma_n||1+\Gamma_{n-1}| + |\Gamma_n||\Gamma_{n-1}|}{|1+\Gamma_n| - |\Gamma_{n-1}|} \cos\left[(\arg(1+\Gamma_{n-1}) - \arg(\Gamma_n))/2\right] - \Delta_n/2 \right| \\
&= \frac{1}{2}[M_{n-1}(|1+\Gamma_n| + |\Gamma_{n-1}|) - \Delta_n] \\
&= \frac{1}{2}[(\rho_{n-1} + |\Gamma_{n-1}|M_{n-1})(|1+\Gamma_n|M_{n-1} - \rho_n)] \\
&\geq \sigma_{n-1},
\end{aligned}$$

where $M_n$ is given by (4.10). The last inequality follows from the AGM inequality.

**LEMMA 4.6** *Let $n = 1, 2$ and $D_n = \{z : z^2 \in E_n\}$, where $E_1$ and $E_2$ are given by (4.2) and (4.3) , respectively. Assume also that the constraints (4.4) and (4.6) are satisfied and $s_0 = s_2$ for each symbol $s$. Then*

$$\begin{aligned}
|\Gamma_n| < |1+\Gamma_{n-1}| &\Longrightarrow D_n \subseteq C_1(\alpha_n, \sigma_n), \\
|\Gamma_n| > |1+\Gamma_{n-1}| &\Longrightarrow D_n \subseteq C_2(-1-\alpha_n, \delta_n), \\
|\Gamma_n| = |1+\Gamma_{n-1}| &\Longrightarrow D_n \subseteq iH_n \cap -iH_n,
\end{aligned}$$

*where $H_n$ is given by (4.25) and $\alpha_n$, $\sigma_n$, and $\delta_n$ are determined by formulas (4.8), (4.9), and (4.23), respectively .*

**Proof:** It will be convenient throughout this proof to set

$$\beta_n = (|\Gamma_n||1+\Gamma_{n-1}|)^{1/2} \text{ and } \tau_n = \rho_n|1+\Gamma_{n-1}| - \rho_{n-1}|\Gamma_n|.$$

Under our hypotheses, $\tau_n$ is nonnegative. The polar equations for the boundaries of the regions $D_n$, $n = 1, 2$, are

$$r^4 + 2r^2(\rho_{n-1}\rho_n - \beta_n^2 \cos 2(\theta - \phi_n)) + \gamma_n^2 = 0, ;\ n = 1, 2,$$

respectively,, where

$$\phi_n = (\arg(1+\Gamma_{n-1}) + \arg(\Gamma_n))/2 \tag{4.28}$$

and

$$\gamma_n^2 = (|\Gamma_n|^2 - \rho_n^2)(|1+\Gamma_{n-1}|^2 - \rho_{n-1}^2),\ \ \Gamma_0 = \Gamma_2,\ \ \rho_0 = \rho_2. \tag{4.29}$$

After a rotation of $-\phi_n$ radians and a change to rectangular coordinates, these equations become

$$(x^2+y^2)^2 + 2(x^2+y^2)\rho_n\rho_{n-1} - 2(x^2-y^2)\beta_n^2 + \gamma_n^2 = 0,\ \ n = 1, 2. \tag{4.30}$$

We solve the latter equations for $x^2$ and arrive at

$$x^2 = -y^2 + \beta_n^2 - \rho_n\rho_{n-1} \pm (\tau_n^2 - 4\beta_n^2 y^2)^{1/2}. \tag{4.31}$$

It turns out that only the + sign needs to be chosen when

$$2\beta_n^2 \geq \rho_n|1+\Gamma_{n-1}| + \rho_{n-1}|\Gamma_n|$$

is satisfied. Otherwise, both branches are needed to determine the boundary of $D_n$. By implicitly differentiating (4.30) with respect to $x$ and solving for $yy'$, we obtain

$$yy' = \frac{x[\beta_n^2 - \rho_n\rho_{n-1} - (x^2+y^2)]}{(x^2+y^2) + \beta_n^2 + \rho_n\rho_{n-1}}. \tag{4.32}$$

Clearly, $y' = 0$ if $x = 0$ and in this case we are led to

$$\max(y^2) = (\rho_n - |\Gamma_n|)(|1 + \Gamma_{n-1}| - \rho_{n-1}) \tag{4.33}$$

if

$$\beta_n^2 - \rho_n\rho_{n-1} \leq 0. \tag{4.34}$$

But if

$$\beta_n^2 - \rho_n\rho_{n-1} > 0, \tag{4.35}$$

then $y' = 0$ if

$$x^2 + y^2 = \beta_n^2 - \rho_n\rho_{n-1}, \tag{4.36}$$

provided there is a point $(x, y)$ on the curve satisfying this equation. Making this substitution in (4.30) gives us

$$\max(y^2) = \frac{\tau_n^2}{4\beta_n^2}, \tag{4.37}$$

provided that there is a real value of $x \neq 0$ that satisfies equation (4.36) when $y^2$ is replaced by this value . This will be the case if and only if

$$\beta_n^2 - \rho_n\rho_{n-1} > \frac{\tau_n^2}{4\beta_n^2},$$

which is equivalent to the condition $V_n > 0$, where $V_n$ in this proof is defined by

$$V_n = 2\beta_n^2 - \rho_n|1 + \Gamma_{n-1}| - \rho_{n-1}|\Gamma_n|. \tag{4.38}$$

It is easily seen that inequality $V_n > 0$ holds, in particular, if $|\Gamma_n| \geq \rho_n$. Hence, $\max(y^2)$ is given by (4.37) in this case. But, if $|\Gamma_n| < \rho_n$, then

$$\begin{aligned}\frac{\tau_n^2}{4\beta_n^2} &= \frac{(\rho_n - |\Gamma_n||1 + \Gamma_{n-1}| + (|1 + \Gamma_{n-1}| - \rho_{n-1})|\Gamma_n|)^2}{4\beta_n^2} \\ &\geq (\rho_n - |\Gamma_n|)(|1 + \Gamma_{n-1}| - \rho_{n-1}),\end{aligned}$$

which shows that, in either case,

$$\max(y^2) \le \frac{\tau_n^2}{4\beta_n^2}. \tag{4.39}$$

Now we set out to prove $|\Gamma_n| < |1+\Gamma_{n-1}| \Longrightarrow D_n \subseteq C_1(\alpha_n, \sigma_n)$. We perform the same rotations on the circles $K(\pm i\alpha_n, \sigma_n)$ as we did on the $D_n$ and obtain

$$x^2 = \sigma_n^2 - (y \pm |\alpha_n)^2 \tag{4.40}$$

as rectangular equations for these circles. Let I be the closed interval defined by

$$I = [0, [(\rho_n - |\Gamma_n|)(|1+\Gamma_{n-1}| - \rho_{n-1})]^{1/2}] \text{ or } I = [0, \frac{\tau_n}{2\beta_n}]$$

according as $V_n \le 0$ or $V_n > 0$. For fixed $n$ we define the function $f$ by

$$f(y) = \sigma_n^2 - (y + |\alpha_n|)^2,$$

and we define the function g for all $y \in I$ by

$$g(y) = -y^2 + \beta_n^2 - \rho_n\rho_{n-1} + (\tau_n^2 - 4\beta_n^2 y^2)^{1/2}.$$

Note that $g(y)$ is equal to the right side of (3.19) with the + sign in front of the radical. In view of symmetry, it is sufficient to prove that $f(y) \ge g(y) \ge 0$ for all $y \in I$. To that effect

$$\begin{aligned} f(y) - g(y) &= \sigma_n^2 - |\alpha_n|^2 - \beta_n^2 + \rho_n\rho_{n-1} - 2|\alpha_n|y - (\tau_n^2 - 4\beta_n^2 y^2)^{1/2} \\ &= (\tau_n M_n - 2|\alpha_n|y) - (\tau_n^2 - 4\beta_n^2 y^2)^{1/2}. \end{aligned} \tag{4.41}$$

Now we define the functions the functions $F$ and $G$ on $I$ by setting $F(y)$ equal to the enclosed linear factor and $G(y)$ equal to the square root term in (4.41) . Clearly $G(y) \ge 0$ on $I$. Since we have shown above that $y \le \frac{\tau_n}{2\beta_n}$ in our domains of interest, and since

$$|\alpha_n|^2 = (M_n^2 - 1)\beta_n^2, \tag{4.42}$$

it is easily seen that $F(y) \geq 0$ on $I$, also. Moreover,

$$\begin{aligned}[F(y)]^2 - [G(y)]^2 &= 4(|\alpha_n|^2 + \beta_n^2)y^2 - 2|\alpha_n \tau_n M_n y + \tau_n^2(M_n^2 - 1) \\ &= [2\beta_n M_n y - \tau_n(M_n^2 - 1)^{1/2}]^2,\end{aligned}$$

where we have again employed the relation (4.42) . Thus the functions we have defined here meet the hypotheses of Lemma 1.1, so our desired inequalities $f(y) \geq g(y) \geq 0$ are true and we are done with this part.

Next we prove the assertion

$$|\Gamma_n| > |1 + \Gamma_{n-1}| \Longrightarrow D_n \subseteq C_2(-1 - \alpha_n, \delta_n). \tag{4.43}$$

After a rotation of $-\phi_n$ radians the boundary of $D_n$ is described by (4.31) and $C_2(-1 - \alpha_n, \delta_n)$ becomes a region whose circular boundaries can be defined in rectangular coordinates by

$$x^2 = \delta_n^2 - (y \pm |\alpha_n|)^2. \tag{4.44}$$

First we shall prove that our assertion is true if the additional inequality $V_n \leq 0$ is satisfied by the given parameters. In this case the curve (4.31) is oval shaped, symmetric about both the $x$ and $y$ axes and, as can be deduced from above, its maximum $y$-value of

$$(\rho_n - |\Gamma_n|)(|1 + \Gamma_{n-1}| - \rho_{n-1})]^{1/2}$$

is achieved on the $y$-axis. Thus it is sufficient to prove that

$$|\alpha_n| \geq \delta_n + [(\rho_n - |\Gamma_n|)(|1 + \Gamma_{n-1}| - \rho_{n-1})]^{1/2} \tag{4.45}$$

whenever $\rho_n \geq |\Gamma_n|$ in order to verify our assertion in this case. We square both sides of (4.45) and obtain the inequality

$$\tau_n M_n + U_n \geq 2|\sigma_n|[(\rho_n - |\Gamma_n|)(|1 + \Gamma_{n-1}| - \rho_{n-1})]^{1/2}, \tag{4.46}$$

where

$$U_n = 2\rho_n\rho_{n-1} - \rho_n|1+\Gamma_{n-1}| - \rho_{n-1}|\Gamma_n|.$$

We will prove that (4.45) is true by showing that (4.46) is true. Since

$$\begin{aligned}\tau_n M_n + U_n &\geq \tau_n + U_n \\ &= 2\rho_{n-1}(\rho_n - |\Gamma_n|) \geq 0\end{aligned}$$

when $\rho_n \geq |\Gamma_n|$, we can examine the validity of (4.46) by the examining the validity of the inequality we obtain by squaring both sides of (4.46) . Let $F$ and $G$ denote the left and right sides of (4.46) , respectively. After making use of the relations (4.42) and

$$\delta_n^2 = -\tau_n M_n - \rho_n\rho_{n-1} + \beta_n^2 M_n^2 \tag{4.47}$$

and a bit of algebraic maneuvering we obtain

$$F^2 - G^2 = (\tau_n + M_n V_n)^2 \geq 0,$$

where $V_n$ is given by (4.38) . Thus our assertion is true in this case in which we have made the assumption that $V_n \leq 0$. Recall that $\rho_n \geq |\Gamma_n|$ under this assumption.

Now let us set about the task of verifying (4.43) when $V_n \geq 0$ holds. In this case the curve (4.31) for a given $n$ is peanut shaped and symmetric about both the $x$ and $y$ axes. In the $n = 1$ case the peanut curve actually separates into two disjoint ovals not containing the origin when the condition $|\Gamma_1| > \rho_1$ holds. Under our assumptions, $\max(y) = \frac{\tau_n}{2\beta_n}$ for the curve (4.31) . So in view of (4.39) and what we have proved above, our desired result certainly holds if

$$|\alpha_n| \geq \delta_n + \frac{\tau_n}{2\beta_n}.$$

Otherwise, let

$$I := [|\alpha_n| - \delta_n, \frac{\tau_n}{2\beta_n}]$$

and let the functions $f$ and $g$ be defined for all $y \in I$ as follows:

$$\begin{aligned} f(y) &= -y^2 + \beta_n^2 - \rho_n\rho_{n-1} - (\tau_n^2 - 4\beta_n^2 y^2)^{1/2} \\ g(y) &= \delta_n^2 - (y - |\alpha_n|)^2. \end{aligned}$$

To complete our proof it is sufficient to show that $f(y) \geq g(y) \geq 0$ on I. Note the connections of $f$ and $g$ to the curves (4.31) and (4.40) , respectively. Simple calculations with the aid of the relations (4.42) and (4.47) will show that

$$f(y) - g(y) = (\tau_n M_n - 2|\alpha_n|y) - (\tau_n^2 - 4\beta_n^2 y^2)^{1/2}.$$

But the expression on the right side of this equation is the same as the expression on the right side of (4.41) which we have already shown is nonnegative on the interval $[0, \frac{\tau_n}{2\beta_n}]$. We have now finished proving this part of our lemma.

It remains to verify our assertion that $|\Gamma_n| = |1 + \Gamma_{n-1}| \Longrightarrow D_n \subseteq iH_n \cap -iH_n$. By rotating $D_n$ and $iH_n$ by $-\phi_n$ radians , then converting to rectangular coordinates and using (4.39) we have that

$$\max{(y^2)} \leq \frac{\tau_n^2}{4\beta_n^2} = (\frac{\rho_n - \rho_{n-1}}{2})^2.$$

Hence,

$$\max{|y|} \leq \frac{\rho_n - \rho_{n-1}}{2}$$

and it follows that what we set out to prove here is true. Note that if $|\Gamma_2| = |1 + \Gamma_1|$ then

$$\rho_2 \geq |\Gamma_2| = |1 + \Gamma_1| > \rho_1,$$

and, if $|\Gamma_1| = |1+\Gamma_2|$, then condition (4.6) implies that $\rho_1 \geq \rho_2$. Thus $\rho_n - \rho_{n-1} \geq 0$ in either case. This completes the proof of our lemma.

We are now ready to take up the task of proving the embedding statements in Theorem 4.2. For convenience we shall give the inclusions in terms of the D-regions rather than the E-regions. We shall divide up our argument into several different cases.

Case (1): $|\Gamma_1| < |1+\Gamma_2|$ and $|\Gamma_2| < |1+\Gamma_1|$.

These assumptions and Lemma 4.6 guarantee that

$$D_1 \subseteq C_1(\alpha_1, \sigma_1) \text{ and } D_2 \subseteq C_1(\alpha_2, \sigma_2). \tag{4.48}$$

By Lemma 4.4 and symmetry,

$$N(\pm\alpha_1, \sigma_1) \subseteq M(\mp(1+\alpha_2), \sigma_2) \text{ and } N(\pm\alpha_2, \sigma_2) \subseteq M(\mp(1+\alpha_1), \sigma_1).$$

Hence,

$$C_1(\alpha_1, \sigma_1) \subseteq C_2(\alpha_2, \sigma_2) \text{ and } C_1(\alpha_2, \sigma_2) \subseteq C_2(\alpha_1, \sigma_1). \tag{4.49}$$

It follows from (4.48) and (4.49) that

$$D_1 \subseteq C_1(\alpha_1, \sigma_1), D_2 \subseteq C_2(\alpha_1, \sigma_1) \text{ and } D_2 \subseteq C_1(\alpha_2, \sigma_2), D_1 \subseteq C_2(\alpha_2, \sigma_2).$$

Case (2): $|\Gamma_{n-1}| > |1+\Gamma_n|$ .

Our assumption and conditions (4.4) give us the relations

$$|1+\Gamma_{n-1}| > |\Gamma_{n-1}| > |1+\Gamma_n| > |\Gamma_n|.$$

It follows from Lemma 4.6 that

$$D_n \subseteq C_1(\alpha_n, \sigma_n) \text{ and } D_{n-1} \subseteq C_2(-1-\alpha_{n-1}, \sigma_{n+2}), \tag{4.50}$$

where

$$C_2(-1-\alpha_{n-1},\sigma_{n+2}) = M(i\alpha_{n-1},\sigma_{n+2}) \cap M(-i\alpha_{n-1},\sigma_{n+2}).$$

Using Lemma 4.4 and symmetry, we obtain

$$N(\pm i(1+\alpha_n),\sigma_n) \subseteq N(\mp i\alpha_{n-1},\sigma_{n+2}),$$

which implies that

$$M(\mp i\alpha_{n-1},\sigma_{n+2}) \subseteq M(\pm i(1+\alpha_n),\sigma_n).$$

After employing this result and (4.50), we have that

$$D_{n-1} \subseteq C_2(\alpha_n,\sigma_n).$$

Case (3): $|\Gamma_n| = |1+\Gamma_{n-1}|$ .

Our hypothesis and conditions (4.4) lead us to

$$|1+\Gamma_n| > |\Gamma_n| = |1+\Gamma_{n-1}| > |\Gamma_{n-1}|.$$

An application of Lemma 4.6 guarantees that

$$D_{n-1} \subseteq C_1(\alpha_{n-1},\sigma_{n-1}).$$

It also follows from Lemma 4.6 that

$$D_n \subseteq iH_n \cap -iH_n.$$

But from Lemma 4.5 we have that

$$iH_n \cap -iH_n \subseteq M(i(1+\alpha_{n-1}),\sigma_{n-1}) \cap M(-i(1+\alpha_{n-1}),\sigma_{n-1}),$$

so that

$$D_n \subseteq C_2(\alpha_{n-1},\sigma_{n-1}).$$

This completes our proof of Theorem 4.2.

## REFERENCES


1. L. Jacobsen, General convergence of continued fractions, *Trans. Amer. Math. Soc.* **294** (1986), 477-485.
2. W. B. Jones and W. J. Thron, Convergence of continued fractions, *Canad. J. Math.* **20** (1968), 1037-1055.
3. W. B. Jones and W. J. Thron, Twin-convergence regions for continued fractions $K(a_n/1)$, *Trans. Amer. Math. Soc.* **150** (1970), 93-119.
4. W. B. Jones and W. J. Thron, Continued Fractions: Analytic Theory and Applications, "Encyclopedia of Mathematics and Its Applications", Vol. 11, Addison-Wesley, Reading, MA, 1980.
5. R. E. Lane, The value region problem for continued fractions, *Duke Math. J.* **12** (1945), 207-216.
6. L. J. Lange, Divergence, Convergence, and Speed of Convergence of Continued Fractions $1 + K(a_n/1)$, "Doctoral Thesis", University of Colorado, Boulder, 1960.
7. L. J. Lange and W. J. Thron, A two parameter family of best twin convergence regions for continued fractions, *Math. Zeitschr.* **73** (1960), 295-311.
8. L. J. Lange, On a family of twin convergence regions for continued fractions, *Illinois J. Math.* **10** (1966), 97-108.
9. L. J. Lange, A uniform twin parabola convergence theorem for continued fractions, *J. Math. Anal. Appl.* **188** (1994), 985-998.
10. L. J. Lange, Convergence region inclusion theorems for continued fractions $K(a_n/1)$, *Constr. Approx.* **11** (1995), 321-329.

11. W. Leighton and H. S. Wall, On the transformation and convergence of continued fractions, *Amer. J. Math.* **58** (1936), 267-281.

12. L. Lorentzen and H. Waadeland, Continued Fractions With Applications, "Studies in Computational Mathematics", Vol. 3, North-Holland, Amsterdam, 1992.

13. L. Lorentzen, Circular twin value sets for continued fractions and how they imply convergence, "Continued Fractions and Orthogonal Functions: Theory and Applications", Lecture Notes in Pure and Applied Mathematics, **154**, Marcel-Dekker, New York, 1994, 305-344.

14. W. J. Thron, Two families of twin convergence regions for continued fractions, *Duke Math. J.* **10** (1943), 677-685.

15. W. J. Thron, Twin convergence regions for continued fractions $b_0 + K(1/b_n$, II, *Amer. J. Math.* **71** (1949), 112-120.

16. W. J. Thron, Zwillingskonvergenzgebeite für Kettenbrüche $1+K(a_n/1)$, deren eines die Kreisscheibe $|a_{2n-1}| \leq \rho^2$ ist, *Math. Zeitschr.* **70** (1959), 310-344.

# Separation Theorem of Chebyshev–Markov–Stieltjes Type for Laurent Polynomials Orthogonal on $(0, \infty)$

XIN LI Department of Mathematics, University of Central Florida, Orlando, FL 32816, USA

## 1 INTRODUCTION

Orthogonal Laurent polynomials (L-polynomials) are introduced and studied by Jones and Thron [1] in connection with the study of the strong moment problems, which is a generalization of the classical moment problems. See also [2, 3]. Since orthogonal L-polynomials and the ordinary orthogonal polynomials share a lot of similar properties, many results on orthogonal polynomials have been adapted to the orthogonal L-polynomial setting. For example, the recurrence relations, Favard's theorem, Christoffel-Darboux identities, the separation of zeros property, and Gaussian quadrature have all been established for orthogonal L-polynomials. For survey articles on orthogonal L-polynomials, see [4, 5, 6]. One of the items missing in the above list is a separation theorem of Chebyshev-Markov-Stieltjes type that relates the Christoffel numbers in the Gaussian quadrature, a solution to the strong moment problem, and the separation property of the zeros of orthogonal L-polynomials, although the validity of such separation theorem has been conjectured by many researchers. In this paper, we will establish such a separation theorem. The key of the proof is an existence theorem for one-sided interpolation L-polynomials (see Theorem 4 below), which is of interest in its own.

We will use $\mathcal{P}_n$ to denote the set of polynomials of degree at most $n$. The set of L-polynomials will be denoted by $\mathcal{R}$, or more precisely,

$$\mathcal{R} := \left\{ \sum_{j=m}^{n} r_j x^j \;:\; r_j \in \mathbb{C},\; -\infty < m \le n < +\infty \right\}.$$

Let, for $n = 0, 1, 2, \ldots,$

$$\mathcal{R}_{2n} := \left\{ \sum_{j=-n}^{n} r_j x^j \;:\; r_j \in \mathbb{C} \right\},\ \hat{\mathcal{R}}_{2n} := \left\{ \sum_{j=-n}^{n} r_j x^j \;:\; r_j \in \mathbb{C},\ r_n \neq 0 \right\},$$

$$\mathcal{R}_{2n+1} := \left\{ \sum_{j=-n-1}^{n} r_j x^j \;:\; r_j \in \mathbb{C} \right\},\ \hat{\mathcal{R}}_{2n+1} := \left\{ \sum_{j=-n-1}^{n} r_j x^j \;:\; r_j \in \mathbb{C},\ r_{-n-1} \neq 0 \right\}.$$

Then

$$\mathcal{R} = \bigcup_{n=0}^{\infty} \mathcal{R}_n = \bigcup_{n=0}^{\infty} \hat{\mathcal{R}}_n$$

and

$$\mathcal{R}_m \subseteq \mathcal{R}_n \text{ if } m \leq n, \text{ and } \hat{\mathcal{R}}_m \cap \hat{\mathcal{R}}_n = \emptyset \text{ if } m \neq n.$$

We will say that an L-polynomial $R$ has degree $n$ if $R \in \hat{\mathcal{R}}_n$.

Let $d\mu$ be a positive measure on $(0, \infty)$ with an infinite number of points in its support such that all moments

$$\int_0^\infty x^n d\mu(x),\ n = 0, \pm 1, \pm 2, \ldots$$

are finite (note that both positive and negative powers are assumed). Then $d\mu$ defines an inner product on $\mathcal{R}$:

$$\langle f, g \rangle := \int_0^\infty f(x)\overline{g(x)} d\mu(x),$$

for every $f, g \in \mathcal{R}$. The orthogonal L-polynomials $\{R_n\}_{n=0}^{\infty}$ with respect to $d\mu$ on $(0, \infty)$ are obtained by applying the Gram-Schmidt orthogonalization process to the sequence

$$1,\ x^{-1},\ x,\ x^{-2},\ x^2, \ldots,\ x^{-n},\ x^n, \ldots$$

with respect to the inner product. For each $n \geq 0$, we have that $R_n$ is of degree $n$, i.e., $R_n \in \hat{\mathcal{R}}_n$, and

$$\langle R_m, R_n \rangle = \delta_{mn}.$$

It is also known (cf., e.g., [2, 4]) that $R_n$ is real and has only real simple zeros $x_j^{(n)}$ $(j = 1, 2, \ldots, n)$ in $(0, \infty)$. Furthermore, the following Gaussian quadrature formula holds (cf., e.g., [2, 4]):

$$\int_0^\infty R(x)d\mu(x) = \sum_{j=1}^{n} \lambda_j R(x_j) \text{ for all } R \in \mathcal{R}_{2n-1}, \tag{1}$$

where the $\lambda_j$, $j = 1, 2, ..., n$, are positive and are called *Christoffel numbers* (associated with $R_n$). Based on Gaussian quadrature, using Christoffel numbers, we can define a sequence of increasing step functions (that induce discrete measures) as follows:

$$\psi_n(x) = \begin{cases} 0, & 0 < x \leq x_1 \\ \sum_{j=1}^{k} \lambda_j, & x_k < x \leq x_{k+1},\ k = 1, 2, ..., n-1 \\ \sum_{j=1}^{n} \lambda_j, & x_n < x < \infty \end{cases} \tag{2}$$

Since $\int_0^\infty d\psi_n(x) = 1$ for all $n = 1, 2, ...$, we see that $\{\psi_n\}$ has a subsequence, $\{\psi_{n_k}(x)\}$, such that

$$\psi_{n_k}(x) \overset{w*}{\to} \hat{\psi}(x)$$

for some increasing function $\hat{\psi}(x)$ (Here $\overset{w*}{\to}$ denotes convergence in the weak-star topology for unit measures on $(0, \infty)$). The measure $d\hat{\psi}$ induced by $\hat{\psi}$ has the same moments as the measure $d\mu$ does even though $d\hat{\psi}$ may be different from $d\mu$. When the strong moment problem is determinate, we must have $d\hat{\psi} = d\mu$ (and consequently, the whole sequence $\{\psi_n\}$ converges). A natural question is how $d\hat{\psi}$ is related to $d\mu$. In the classical orthogonal polynomials setting, a separation theorem of Chebyshev-Markov-Stieltjes (conjectured by Chebyshev in 1874 and proved by A. Markov and Stieltjes independently in 1884, see [7, §3.41]) relates Christoffel numbers, $d\hat{\psi}$, and the zeros of orthogonal (ordinary) polynomials by inequalities. Our goal is to establish a similar result for L-polynomials orthogonal on $(0, \infty)$. We state new results and their corollaries in next section, §2 and give their proofs in sections 3 and 4.

## 2 STATEMENTS OF NEW RESULTS

Let $d\mu$ be a positive measure on $(0, \infty)$ with an infinite number of points in its support such that its (strong) moments

$$\int_0^\infty x^n d\mu(x),\ n = 0, \pm 1, \pm 2, ...,$$

are all finite.

Let $\{R_n\}_{n=0}^{\infty}$ be orthogonal with respect to $d\mu$, i.e., $R_n \in \hat{\mathcal{R}}_n$, $n = 0, 1, 2, ...$, and

$$\int_0^\infty R_m(x)\overline{R_n(x)}d\mu(x) = \delta_{m,n}.$$

Let $x_1 < x_2 < ... < x_n$ be the $n$ simple zeros of $R_n$ in $(0, \infty)$, and let $\lambda_j > 0$ be the Christoffel numbers in Gaussian quadrature (1). Using the Christoffel numbers, we define $y_1, y_2, ..., y_{n-1}$ with $0 < y_1 < y_2 < ... < y_{n-1} < \infty$ such that

$$\lambda_k = \int_{y_{k-1}}^{y_k} d\mu(x), \ k = 1, 2, ..., n, \ y_0 := 0, \ y_n := \infty.$$

Finally, let $\psi$ be an increasing function, with at least $n$ points of increase on $(0, \infty)$, such that

$$\int_0^\infty x^j d\psi(x) = \int_0^\infty x^j d\mu(x) \text{ for } j = -n, -n+1, ..., n-1.$$

Now, we can state the main result of the paper.

THEOREM 1 (Separation Theorem) For $k = 1, 2, ..., n-1$, we have

$$x_k < y_k < x_{k+1} \tag{3}$$

and

$$\int_0^{x_k+0} d\psi(x) < \lambda_1 + \cdots + \lambda_k < \int_0^{x_{k+1}-0} d\psi(x) \tag{4}$$

As simple consequences, as in the classical setting, we have the following results from Theorem 1: First, by taking $d\psi = d\mu$ in (4), we obtain

COROLLARY 2 Each $R_n(x)$ has at most one zero in an open interval outside the support of $d\mu$.

In other words: In the open interval $(x_k, x_{k+1})$, between two neighboring zeros of $R_n$, there must be at least one point that is in the support of $d\mu$. Next, using $\psi_m$ (see (2)) with $m > n$ for $\psi$ in (4), we have

COROLLARY 3 Between any two neighboring zeros of the orthogonal L-polynomial $R_n$ there is at least one zero of each orthogonal L-polynomial $R_m$ with $m > n$.

A crucial step in proving the separation theorem is the following result on the existence of one-sided interpolation L-polynomials.

THEOREM 4 (One-sided Interpolation of L-polynomials) For $1 \le k \le n$, there exists a unique $R(x) \in \mathcal{R}_{2n-2}$ having the following properties:

(i) $R(x_j) = \begin{cases} 1, & 1 \le j \le k \\ 0, & k+1 \le j \le n \end{cases}$

(ii) $R'(x_j) = 0$ for $1 \le j \le n$ and $j \ne k$.

(iii) $R(x) \ge 0$ for $x \ge x_k$

(iv) $R(x) \ge 1$ for $0 < x \le x_k$.

Remarks: 1. The proof of Theorem 4 uses only the property that the $n$ points $\{x_j\}_{j=1}^n$ satisfy

$$0 < x_1 < x_2 < \dots < x_n.$$

2. Properties (i) and (ii) together uniquely determine $R(x)$.
3. The crux of Theorem 4 is that an L-polynomial $R$ having properties (i) and (ii) will satisfy (iii) and (iv) automatically!

Here is a plot of an L-polynomial $R(x) \in \mathcal{R}_{12}$ that corresponds to points

$$x_1 = 2, x_2 = 2.2, x_3 = 2.5, x_4 = 2.7, x_5 = 3, x_6 = 3.3, x_7 = 3.6$$

with $k = 4$ and $n = 7$:

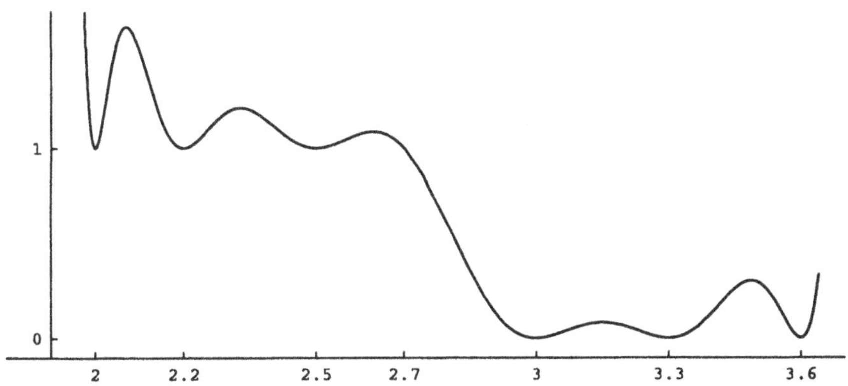


We also need a variation of Theorem 4 as follows.

THEOREM 5 There exists a unique $S(x) \in \mathcal{R}_{2n-2}$ satisfying the following conditions:

(i) $S(x_j) = \begin{cases} 1 & 1 \le j < k \\ 0 & k \le j \le n \end{cases}$

(ii) $S'(x_j) = 0,\ j \neq k$

(iii) $S(x) \leq 0,\ x \geq x_k$

(iv) $S(x) \leq 1,\ 0 < x \leq x_k$.

## 3 PROOF OF THEOREM 4

We will give two proofs of Theorem 4, our original proof and the shorter and more elegant proof suggested by the referee. We decided to provide both proofs in hopes that one of them could lead to the establishment of a separation theorem for Laurent polynomials orthogonal on the whole real line $(-\infty, \infty)$. We present our original proof first. This requires some preliminary auxiliary results.

LEMMA 6 If $w(x) := (x - x_{k+1})^2 \cdots (x - x_n)^2$, then, for $m = 0, 1, 2, \ldots$,

$$\frac{d^m}{dx^m}\left(\frac{x^{n-1}}{w(x)}\right) > 0 \text{ for } 0 < x \leq x_k.$$

Proof: By the Leibniz rule, we have

$$\frac{d^m}{dx^m}\left(\frac{x^{n-1}}{w(x)}\right) = \sum_{j=0}^{m} \binom{m}{j} \frac{d^{m-j}x^{n-1}}{dx^{m-j}} \frac{d^j}{dx^j}\left(\frac{1}{w(x)}\right).$$

So, to prove the lemma, it suffices to show that, for $j = 0, 1, 2, \ldots$,

$$\frac{d^j}{dx^j}\left(\frac{1}{w(x)}\right) > 0 \text{ for } 0 < x \leq x_k.$$

This can be verified by induction as follows.

When $j = 0$, the statement is obviously true. Now assume the statement is true for $0, 1, \ldots, j$. Then

$$\begin{aligned}
\frac{d^{j+1}}{dx^{j+1}}\left(\frac{1}{w(x)}\right) &= \frac{d^j}{dx^j}\left(\frac{d}{dx}\frac{1}{w(x)}\right) \\
&= \frac{d^j}{dx^j}\left(-\frac{w'(x)}{w(x)}\frac{1}{w(x)}\right) = \frac{d^j}{dx^j}\left(\sum_{i=k+1}^{n} \frac{2}{x_i - x}\frac{1}{w(x)}\right) \\
&= \sum_{h=0}^{j} \binom{j}{h} \frac{d^{j-h}}{dx^{j-h}}\left(\sum_{i=k+1}^{n} \frac{2}{x_i - x}\right) \frac{d^h}{dx^h}\left(\frac{1}{w(x)}\right) \\
&= \sum_{h=0}^{j} \left(\sum_{i=k+1}^{n} \frac{2(j-h)!}{(x_i - x)^{j-h+1}}\right) \frac{d^h}{dx^h}\left(\frac{1}{w(x)}\right) > 0,
\end{aligned}$$

for $0 < x \leq x_k$. ■

LEMMA 7 Let $n$ and $k$ be as in Theorem 4 and let $\nu \leq k$. Let

$$R_\nu(x) := \frac{T_\nu(x)}{x^{n-1}} \text{ for } \nu \geq 1$$

and

$$\tilde{R}_\nu(x) := \frac{\tilde{T}_\nu(x)}{x^{n-1}} \text{ for } \nu \geq 2$$

satisfy (i) $T_\nu \in \mathcal{P}_{2(n-\nu)}$ and $\tilde{T}_\nu \in \mathcal{P}_{2(n-\nu)+1}$, and (ii) the interpolation conditions

$$R_\nu(x_j) = \begin{cases} 1, & \nu \leq j \leq k \\ 0, & k+1 \leq j \leq n \end{cases}$$

$$R'_\nu(x_j) = 0 \text{ for } \nu \leq j \leq n \text{ and } j \neq k,$$

and

$$\tilde{R}_\nu(x_j) = \begin{cases} 1, & \nu - 1 \leq j \leq k \\ 0, & k+1 \leq j \leq n \end{cases}$$

$$\tilde{R}'_\nu(x_j) = 0 \text{ for } \nu \leq j \leq n \text{ and } j \neq k.$$

Then both $R_\nu(x)$ and $\tilde{R}_\nu(x)$ exist and are unique.

Proof: Note that $R_\nu(x) = T_\nu(x)/x^{n-1}$ and

$$R'_\nu(x) = \frac{xT'_\nu(x) - (n-1)T_\nu(x)}{x^n}. \tag{5}$$

So the the interpolation conditions for $R_\nu(x)$ can be written as the following interpolation conditions for $T_\nu(x)$:

$$T_\nu(x_j) = \begin{cases} x_j^{n-1}, & \nu \leq j \leq k \\ 0, & k+1 \leq j \leq n \end{cases} \tag{6}$$

$$T'_\nu(x_j) = \begin{cases} (n-1)x_j^{n-2}, & \nu \leq j \leq k-1 \\ 0, & k+1 \leq j \leq n. \end{cases} \tag{7}$$

Similarly, we can show that the interpolation conditions for $\tilde{R}_\nu(x)$ can be written like these for the numerator polynomial $\tilde{T}_\nu(x)$:

$$\tilde{T}_\nu(x_j) = \begin{cases} x_j^{n-1}, & \nu - 1 \le j \le k \\ 0, & k+1 \le j \le n \end{cases} \tag{8}$$

$$\tilde{T}'_\nu(x_j) = \begin{cases} (n-1)x_j^{n-2}, & \nu \le j \le k-1 \\ 0, & k+1 \le j \le n. \end{cases} \tag{9}$$

From the polynomial interpolation theory, we know that there exist uniquely determined polynomials $T_\nu \in \mathcal{P}_{2(n-\nu)}$ and $\tilde{T}_\nu \in \mathcal{P}_{2(n-\nu)+1}$ that satisfy the specified interpolation conditions. This, in turn, implies that both $R_\nu(x)$ and $\tilde{R}_\nu(x)$ exist and are uniquely determined. ■

LEMMA 8 Let $R_\nu(x)$ and $\tilde{R}_\nu(x)$ be defined as in Lemma 7. Then (a) $R'_\nu(x)$ has at most $2(n-\nu)$ zeros, and it has at least $2(n-\nu)-1$ zeros described as follows: one zero in each interval $(x_j, x_{j+1})$ for $\nu \le j \le n-1$ and $j \ne k$, and one zero at each $x_j$ for $\nu \le j \le n$ and $j \ne k$; (b) $\tilde{R}'_\nu(x)$ has at most $2(n-\nu)+1$ zeros, and it has at least $2(n-\nu)$ zeros described as follows: one zero in each interval $(x_j, x_{j+1})$ for $\nu - 1 \le j \le n-1$ and $j \ne k$, and one zero at each $x_j$ for $\nu \le j \le n$ and $j \ne k$.

Proof: Since $T_\nu \in \mathcal{P}_{2(n-\nu)}$, it follows from (5) that $R'_\nu$ has at most $2(n-\nu)$ zeros. By the interpolation conditions we have $R_\nu(x_j) = R_\nu(x_{j+1})$ for $\nu \le j \le n-1$ and $j \ne k$, so, by Rolle's theorem, $R'_\nu$ has a zero in $(x_j, x_{j+1})$ for $\nu \le j \le n-1$ and $j \ne k$. This completes the proof of (a).

The proof of (b) is similar. ■

LEMMA 9 With $R_\nu(x)$ and $\tilde{R}_\nu(x)$ defined as in Lemma 7, the following inequalities hold:

$$R_\nu(x) \ge 1 \text{ for } 0 < x \le x_k, \tag{10}$$

$$\tilde{R}_\nu(x) \ge 1 \text{ for } x_{\nu-1} \le x \le x_k, \tag{11}$$

and

$$\tilde{R}_\nu(x) \le 1 \text{ for } 0 < x \le x_{\nu-1}. \tag{12}$$

Furthermore,

$$\tilde{R}'_\nu(x_{\nu-1}) > 0. \tag{13}$$

Proof: From equations (6) and (7), there exist $P \in \mathcal{P}_{\max\{2(n-\nu),n-1\}-2(k-\nu)-1}$ and $Q \in \mathcal{P}_{2(k-\nu)}$ such that

$$T_\nu(x) = (x - x_{k+1})^2 \cdots (x - x_n)^2 Q(x) \tag{14}$$

and

$$T_\nu(x) = x^{n-1} + (x - x_\nu)^2 \cdots (x - x_{k-1})^2 (x - x_k) P(x). \tag{15}$$

Similarly, from equations (8) and (9), there exist $\tilde{P} \in \mathcal{P}_{\max\{2(n-\nu),n-2\}-2(k-\nu)-1}$ and $\tilde{Q} \in \mathcal{P}_{2(k-\nu)+1}$ such that

$$\tilde{T}_\nu(x) = (x - x_{k+1})^2 \cdots (x - x_n)^2 \tilde{Q}(x) \tag{16}$$

and

$$\tilde{T}_\nu(x) = x^{n-1} + (x - x_{\nu-1})(x - x_\nu)^2 \cdots (x - x_{k-1})^2 (x - x_k) \tilde{P}(x). \tag{17}$$

Write $w(x) = (x - x_{k+1})^2 \cdots (x - x_n)^2$. Then (14) - (17) imply

$$w(x)Q(x) = x^{n-1} + (x - x_\nu)^2 \cdots (x - x_{k-1})^2 (x - x_k) P(x).$$

and

$$w(x)\tilde{Q}(x) = x^{n-1} + (x - x_{\nu-1})(x - x_\nu)^2 \cdots (x - x_{k-1})^2 (x - x_k) \tilde{P}(x).$$

Solving for $Q$ and $\tilde{Q}$ we get

$$Q(x) = \frac{x^{n-1}}{w(x)} + \frac{P(x)}{w(x)} (x - x_\nu)^2 \cdots (x - x_{k-1})^2 (x - x_k) \tag{18}$$

and

$$\tilde{Q}(x) = \frac{x^{n-1}}{w(x)} + \frac{\tilde{P}(x)}{w(x)} (x - x_{\nu-1})(x - x_\nu)^2 \cdots (x - x_{k-1})^2 (x - x_k). \tag{19}$$

Equations (18) and (19) tell us that $Q$ and $\tilde{Q}$ (as polynomials of degrees $2(k-\nu)$ and $2(k-\nu)+1$ respectively) interpolate $x^{n-1}/w(x)$ at points

$$x_{\nu}, x_{\nu}, x_{\nu+1}, x_{\nu+1}, \ldots, x_{k-1}, x_{k-1}, x_k$$

and

$$x_{\nu-1}, x_{\nu}, x_{\nu}, x_{\nu+1}, x_{\nu+1}, \ldots, x_{k-1}, x_{k-1}, x_k$$

respectively. Using the remainder formula for the polynomial interpolation (see, e.g., [8, p. 35]), we get, with $\sigma(x) := (x - x_{\nu})^2 \cdots (x - x_{k-1})^2 (x - x_k)$,

$$\frac{x^{n-1}}{w(x)} = Q(x) + \frac{1}{(2(k-\nu)+1)!} \frac{d^{2(k-\nu)+1}}{dx^{2(k-\nu)+1}} \left( \frac{x^{n-1}}{w(x)} \right) \bigg|_{\xi} \sigma(x) \tag{20}$$

and

$$\frac{x^{n-1}}{w(x)} = \tilde{Q}(x) + \frac{1}{(2(k-\nu)+2)!} \frac{d^{2(k-\nu)+2}}{dx^{2(k-\nu)+2}} \left( \frac{x^{n-1}}{w(x)} \right) \bigg|_{\zeta} (x - x_{\nu-1}) \sigma(x) \tag{21}$$

for some $\xi \in (\min\{x, x_{\nu}\}, \max\{x, x_k\})$ and $\zeta \in (\min\{x, x_{\nu-1}\}, \max\{x, x_k\})$. Comparing equations (18) with (20), and (19) with (21), we find

$$\frac{P(x)}{w(x)} = -\frac{1}{(2(k-\nu)+1)!} \frac{d^{2(k-\nu)+1}}{dx^{2(k-\nu)+1}} \left( \frac{x^{n-1}}{w(x)} \right) \bigg|_{\xi}$$

and

$$\frac{\tilde{P}(x)}{w(x)} = -\frac{1}{(2(k-\nu)+2)!} \frac{d^{2(k-\nu)+2}}{dx^{2(k-\nu)+2}} \left( \frac{x^{n-1}}{w(x)} \right) \bigg|_{\zeta}$$

for some $\xi \in (\min\{x, x_{\nu}\}, \max\{x, x_k\})$ and $\zeta \in (\min\{x, x_{\nu-1}\}, \max\{x, x_k\})$. Now, by Lemma 6, we see that

$$P(x) < 0 \text{ for } 0 < x \leq x_k \tag{22}$$

and

$$\tilde{P}(x) < 0 \text{ for } 0 < x \leq x_k. \tag{23}$$

Using these in (15) and (17) respectively, we obtain

$$T_{\nu}(x) \geq x^{n-1} \text{ for } 0 < x \leq x_k,$$

$$\tilde{T}_\nu(x) \geq x^{n-1} \text{ for } x_{\nu-1} \leq x \leq x_k,$$

$$\tilde{T}_\nu(x) \leq x^{n-1} \text{ for } 0 < x \leq x_{\nu-1},$$

which imply (10), (11) and (12) respectively.

Now, (11) and (12) imply that

$$\frac{\tilde{R}_\nu(x) - \tilde{R}_\nu(x_{\nu-1})}{x - x_{\nu-1}} = \frac{\tilde{R}_\nu(x) - 1}{x - x_{\nu-1}} > 0 \ (x \neq x_{\nu-1}).$$

So $\tilde{R}'_\nu(x_{\nu-1}) \geq 0$. We now use an indirect argument to show that the equality can not hold in this relation. Assume that $\tilde{R}'_\nu(x_{\nu-1}) = 0$. Then (11) and (12) would force $\tilde{R}''_\nu(x_{\nu-1}) = 0$, and so $x_{\nu-1}$ is (at least) a double zero of $\tilde{R}'_\nu$. These two zeros, together with the $2(n-\nu)$ zeros described in Lemma 8, tell us that $\tilde{R}'_\nu$ has at least $2(n-\nu)+2$ zeros, contradicting Lemma 8. Hence, inequality (13) is verified. ∎

We are now ready to prove Theorem 4.

Proof of Theorem 4: It is easy to see that, when $k = 1$, the L-polynomial

$$R(x) := \frac{x_1^{n-1}(x-x_2)^2 \cdots (x-x_n)^2}{x^{n-1}(x_1-x_2)^2 \cdots (x_1-x_n)^2}$$

satisfies the conditions and conclusions of the theorem. On the other hand, when $k = n$, we see that $R(x) \equiv 1$ will do.

Let $2 \leq k \leq n-1$.

Taking $\nu = 1$ in Lemma 7 shows that $R(x) = R_1(x)$ will satisfy the interpolation conditions of the theorem, and that such an interpolation L-polynomial is unique. An application of Lemma 9 now yields that this L-polynomial satisfies $R(x) \geq 1$ for $0 < x \leq x_k$. It remains to show that

$$R(x) \geq 0 \text{ for } x \geq x_k. \tag{24}$$

Now, we use Lemma 9 (and the notations in the lemma and its proof) and apply a backward inductive argument to verify (24).

Obviously, with $w(x) = (x - x_{k+1})^2 \cdots (x - x_n)^2$,

$$R_k(x) = \frac{x_k^{n-1} w(x)}{x^{n-1} w(x_k)}.$$

From this expression and Lemma 9, we have

$$R_k(x) \geq 0 \text{ for } x \geq x_k$$

and

$$R_k(x) \geq 1 \text{ for } 0 < x \leq x_k.$$

We now show that if $2 \leq \nu \leq k$ and if

$$R_\nu(x) \geq 0 \text{ for } x \geq x_k$$

and

$$R_\nu(x) \geq 1 \text{ for } 0 < x \leq x_k,$$

then we also have

$$R_{\nu-1}(x) \geq 0 \text{ for } x \geq x_k \tag{25}$$

and

$$R_{\nu-1}(x) \geq 1 \text{ for } 0 < x \leq x_k. \tag{26}$$

In fact, from the interpolation properties, we have

$$\tilde{R}_\nu(x) = R_\nu(x)\left(1 + \frac{1 - R_\nu(x_{\nu-1})}{(x_{\nu-1} - x_\nu)R_\nu(x_{\nu-1})}(x - x_\nu)\right).$$

So we see that $\tilde{R}_\nu(x) \geq R_\nu(x) \geq 0$ for $x \geq x_k$. From the interpolation properties again, we have

$$R_{\nu-1}(x) = \tilde{R}_\nu(x) + \frac{A(x - x_{\nu-1})(x - x_\nu)(x - x_{\nu+1})^2 \cdots (x - x_n)^2}{x^{n-1}} \tag{27}$$

for some constant $A$. In the above equality, evaluating the derivative at $x_{\nu-1}$ on both sides yields

$$0 = \tilde{R}'_\nu(x_{\nu-1}) + A\left(\frac{(x_{\nu-1} - x_\nu)(x_{\nu-1} - x_{\nu+1})^2 \cdots (x_{\nu-1} - x_n)^2}{x_{\nu-1}^{n-1}}\right).$$

So

$$A = \frac{\tilde{R}'_\nu(x_{\nu-1})x_{\nu-1}^{n-1}}{(x_\nu - x_{\nu-1})(x_{\nu-1} - x_{\nu+1})^2 \cdots (x_{\nu-1} - x_{n2})^2},$$

which is positive by (13) in Lemma 9. It then follows from (27) that

$$R_{\nu-1}(x) \geq 0 \text{ for } x \geq x_k.$$

This, together with Lemma 9, verifies (25) and (26).

Using induction, we start with $\nu = k$ and eventually arrive at $\nu = 1$ and obtain, in particular,

$$R_1(x) \geq 0 \text{ for } x \geq x_k,$$

which is the same as (24). This completes our proof. ■

Now, we give the referee's proof.

Another Proof of Theorem 4: First, from Lemmas 7 and 8, we note that, there exists a unique $R \in \mathcal{R}_{2n-2}$ which satisfies conditions (i) and (ii), and that $R'$ has at least $2n-3$ zeros in $(0, \infty)$.

Write $R(x) = x^{-n+1}T(x)$ with $T(x) \in \mathcal{P}_{2n-2}$. Then the $2n-3$ zeros of $R'$ in $(0, \infty)$ are also zeros of $g(x) := xT'(x) + (-n+1)T(x)$. Thus, $g$ is a real algebraic polynomial of degree $2n-2$ and has at least $2n-3$ positive zeros. Then $g$ has one more real zero $\xi$. Observe that the coefficient of $x^{n-1}$ of $g$ is zero, which is equivalent to

$$g^{(n-1)}(0) = 0. \tag{28}$$

If $\xi$ is nonnegative, then a repeated application of Rolle's theorem implies that all the zeros of $g^{(n-1)}$ are strictly positive, which contradicts (28). Thus, $\xi < 0$ and, therefore, $R$ changes its monotonicity in $(0, \infty)$ only between the positive zeros of $g$. Moreover, $R$ is decreasing in $(x_k, x_{k+1})$ because $R(x_k) = 1$ and $R(x_{k+1}) = 0$. Therefore, $R$ satisfies conditions (ii) and (iv) as well. ■

## 4 PROOF OF THEOREM 1

Having established Theorem 4, we now can follow A. Markov's approach (see, e.g., [9, Chapter 2, §5] or [7, §3.411]) to prove Theorem 1 easily.

Let us take the $R(x)$ from Theorem 4. Then, by applying Gaussian quadrature to $R(x)$ and using the interpolation properties (i) and (ii) of $R(x)$ in Theorem 4, we have

$$\int_0^\infty R(x)d\psi(x) = \sum_{j=1}^{n} \lambda_j R(x_j) = \lambda_1 + \cdots + \lambda_k. \tag{29}$$

By the positivity property (iii) and $R \not\equiv 0$, we then have

$$\int_0^{+\infty} R(x)d\psi(x) > \int_0^{x_k+0} R(x)d\psi(x). \tag{30}$$

Finally, from property (iv) of $R(x)$ and $R \not\equiv 1$, we infer

$$\int_0^{x_k+0} R(x)d\psi(x) > \int_0^{x_k+0} d\psi(x) \tag{31}$$

So, by combining (29)-(31), we obtain

$$\lambda_1 + \cdots + \lambda_k > \int_0^{x_k+0} d\psi(x).$$

This is half of (4). Similarly, we can use Theorem 5 to obtain the remaining half of (4). Now (3) follows from (4), by applying (4) to $\psi = \psi_n$ defined in (2), as in the classical case (see, for example, [7]). ■

ACKNOWLEDGEMENT I am grateful to the referee for suggesting the shorter proof of Theorem 4. I also thank Sandy Cooper and Phil Gustafson for showing me numerical results that strongly suggested the formulation of Theorem 4. I am very grateful to Olav Njastad, Sandy Cooper, and Wolf Thron for discussions of my early attempts to prove the separation theorem. Finally, I want to thank Olav Njastad and Bill Jones for their encouragement and interest in this work.

# References

[1] WB Jones, WJ Thron. Orthogonal Laurent polynomials and Gaussian Quadrature. In: KE Gustafson and WP Reinhardt eds. Quantum Mechanics in Mathematics. Plenum. 1981, pp 449-445.

[2] WB Jones, O Njastad, WJ Thron. Orthogonal Laurent polynomials and the strong Hamburger moment problem. J Math Anal Appl 98:528-554, 1984.

[3] WB Jones, WJ Thron, H Waadeland. A strong Stieltjes moment problem. Trans Amer Math Soc 261:503-528, 1980.

[4] L Cochran, SC Cooper. Orthogonal Laurent polynomials on the real line. In: SC Cooper, WJ Thron eds. Continued Fractions and Orthogonal Functions. New York: Marcel Dekker, 1994, pp 47-100.

[5] WB Jones, WJ Thron. Survey of continued fraction methods of solving moment problems and related topics. In: WB Jones, WJ Thron, H Waadeland eds. Analytic Theory of Continued Fractions. Lecture Notes in Math 932:4-37, 1982.

[6] O Njastad, WJ Thron. The theory of sequences of orthogonal L-polynomials. In: H Waadeland, H Wallin eds. Pade Approximants and Contunued Fractions. Det Kongelige Norske Videnskabers Selskab. Skrifter. 1:54-91, 1983.

[7] G Szegő. Orthogonal Polynomials. 4th ed. Providence: Amer Math Soc Colloquium Publication. 23. 1975.

[8] MJD Powell. Approximation Theory and Methods. Cambridge: Cambridge University Press. 1981.

[9] NI Akhiezer. The Classical Moment Problem. New York: Hafner Publication, 1965.

# Orthogonal Polynomials Associated with a Nondiagonal Sobolev Inner Product with Polynomial Coefficients

MARÍA ÁLVAREZ DE MORALES[1], TERESA E. PÉREZ[1] & MIGUEL A. PIÑAR[1] Dto. de Matemática Aplicada, Universidad de Granada, Granada, Spain

ANDRÉ RONVEAUX Facultés Universitaires N. D. de la Paix, Namur, Belgium

## ABSTRACT

In this work, we will study the family of polynomials which are orthogonal with respect to the Sobolev inner product:

$$(f,g)_S^{(N)} = \sum_{m,k=0}^{N} \langle \lambda_{m,k} u, f^{(m)} g^{(k)} \rangle, \quad N \geq 1,$$

where $u$ is a semiclassical positive definite linear functional and $\lambda_{m,k} = \lambda_{m,k}(x)$ are polynomials for $m, k = 0, \ldots, N$, such that $\Lambda^N = (\lambda_{m,k}(x))_{m,k=0}^N$ is a symmetric and positive definite matrix. For this inner product, we deduce the expression of a symmetric linear operator which plays an essential role in the relation between the orthogonal polynomials associated to the linear functional $u$ and the orthogonal polynomials associated to $(.,.)_S^{(N)}$. Also, for these non–standard orthogonal polynomials, a difference–differential relation is deduced.

*AMS Subject Classification (1991)*: 33 C 45
*Key words*: Semiclassical linear functional, non diagonal Sobolev inner product, Sobolev orthogonal polynomials.

[1]This research was partially supported by Junta de Andalucía, Grupo de Investigación FQM 0229.

## 1 INTRODUCTION

In the last years (see [1]) several authors have devoted considerable attention to Sobolev inner products, that is, inner products involving derivatives. This kind of inner product can adopt the following general form

$$(f,g)_S^{(N)} = \sum_{m,k=0}^{N} \langle u_{m,k}, f^{(m)}g^{(k)}\rangle, \tag{1.1}$$

where $N \geq 1$ is a given integer, and $u_{m,k}$ are linear functionals, for $m, k = 0, \ldots, N$.

In this paper, we will consider in (1.1) a family of linear functionals $u_{m,k}$ defined in such a way that $u_{m,k} = \lambda_{m,k}(x)u$, where $u$ is a positive definite semiclassical linear functional and $\lambda_{m,k}(x)$ are polynomials in $x$, for $m, k = 0, \ldots N$, satisfying several restrictions which make the polynomial matrix $\Lambda^N(x) = (\lambda_{m,k}(x))_{m,k=0}^{N}$ a symmetric and positive definite matrix.

Orthogonal polynomials with respect to a Sobolev inner product appear for the first time in 1947, in a pioneering paper by Lewis ([8]). In this paper, Sobolev orthogonal polynomials were considered in relation with a problem of simultaneous approximation of a function and its derivatives.

Later, in the sixties, several German authors studied some particular cases of the inner product defined in (1.1). Althammer, in [2], considers the case when $N = 1$, the matrix $\Lambda^N(x)$ is diagonal and constant, and $u$ is the linear functional associated to the Lebesgue measure supported in $I = [-1, 1]$. This work was completed by Cohen ([5]), Gröbner ([6]) and Schäfke ([17]). Later, Brenner ([3]) makes an analogous study for the Laguerre case.

In a paper of Marcellán, Pérez and Piñar ([13]), the authors study the inner product

$$(f,g)_S^{(1)} = \sum_{i=0}^{1}\langle u_i, f^{(i)}g^{(i)}\rangle, \tag{1.2}$$

where $u_0$ and $u_1$ are positive definite semiclassical linear functional related by a rational expression. In particular, the authors make a more detailed study in the case when $u_0 = u_1$ is the linear functional associated to the Gegenbauer ([12]) or Laguerre–Sonine measure ([14]), extending in this way the above cited results.

The non–diagonal case for a Sobolev inner product, as defined in (1.1), has been considered in the case $N = 1$ by Schäfke and Wolf (see [18]). In this paper, the authors study a situation where $u$ is a classical linear functional and the polynomial matrix $\Lambda^N(x)$ has been defined in such a way that the Sobolev inner product is reduced to a standard inner product by means of an integration by parts, where the non–integrated terms vanish.

In ([16]), Pérez and Piñar give an example of an non–diagonal Sobolev inner product like (1.1), where $\Lambda^N(x)$ is a constant matrix. For this inner product the corresponding orthogonal polynomials are the generalized Laguerre polynomials $\{L_n^{(\alpha)}(x)\}_{n\geq 0}$ with $\alpha$ an arbitrary real number.

In a recent paper, Marcellán, Pérez, Piñar and Ronveaux ([15]) study a Sobolev inner product where $\Lambda^N(x)$ is a constant and diagonal matrix. For this inner product, the authors define a linear differential operator acting on the linear space of polynomials with real coefficients, which is symmetric with respect to the inner product (1.1). As a consequence, explicit relations between the Sobolev orthogonal polynomials and the sequence of the semiclassical orthogonal polynomials associated with the linear functional $u$ are obtained. Also, the Sobolev orthogonal polynomials satisfy a difference–differential relation and a $2N+2$ order differential equation. The results obtained in this work constitute a straightforward generalization to the non–diagonal case.

In this way, this paper provides a common framework for the study of the polynomials which are orthogonal with respect to the Sobolev inner product (1.1).

Next, we will describe the structure of the paper. In section 2, we will introduce the notations, definitions and previous results used throughout the paper. In section 3, we give the definition for the inner product (1.1). Next, we define a linear differential operator $\mathcal{F}^{(N)}$, acting on the linear space of the real polynomials, which is symmetric with respect to the inner product (1.1). From this property, we deduce an explicit relation between the Sobolev orthogonal polynomials and the semiclassical orthogonal polynomials associated with the linear functional $u$. Again, the Sobolev orthogonal polynomials satisfy a difference–differential relation and a $2N+2$ order differential equation. Finally, we present some interesting examples where $u$ is a positive definite classical linear functional.

## 2 DEFINITIONS AND PRELIMINARY RESULTS

Let $\mathbb{P}$ denote the linear space of the polynomials with real coefficients. For a given sequence of real numbers we can define a linear functional acting on $\mathbb{P}$ in the following way

DEFINITION 2.1. Let $\{\mu_n\}_n$ be a given sequence of real numbers. We can define a linear functional

$$\begin{array}{rccc} u: & \mathbb{P} & \longrightarrow & \mathbb{R} \\ & f & \longmapsto & \langle u, f\rangle \end{array}$$

such that $\langle u, x^n\rangle = \mu_n, \quad n = 0, 1, \ldots$ The linear functional $u$ is known as a moment functional associated to the sequence $\{\mu_n\}_n$ and the real number $\mu_n$ will be called the $n$-th order moment.

Now, we recall some well known definitions.

DEFINITION 2.2. An orthogonal polynomial sequence (in short OPS) associated with the linear functional $u$ is a family of polynomials $\{P_n\}_{n\geq 0}$ satisfying

i) $\deg P_n = n, \quad n \geq 0,$
ii) $\langle u, P_nP_m\rangle = k_n\delta_{nm}, \quad k_n \neq 0; \quad n, m = 0, 1, \ldots$

Obviously, an orthogonal polynomial sequence is unique up to a multiplicative factor, and in this way, to preserve the uniqueness we must standardize the polynomials in some way. In this paper, we will consider always a monic orthogonal polynomial sequence (MOPS).

DEFINITION 2.3. A moment functional $u$ will be said *regular* if there exists an OPS associated to $u$.

DEFINITION 2.4. A moment functional $u$ will be said *positive definite* if for every non null polynomial $p(x)$ we have

$$p(x) \geq 0, \quad \forall x \in \mathbb{R} \Longrightarrow \langle u, p(x) \rangle > 0.$$

We remember (see [4], p. 56), that if $u$ is positive definite then $u$ is regular and there exists a Borel measure $\mu$ supported on a real interval $I$ such that

$$\langle u, p \rangle = \int_I p(x) d\mu(x). \tag{2.1}$$

Next, we recall the definitions of some useful operations for moment functionals (see, for instance [11].)

- Let $u$ be a linear functional, and $\phi(x) \in \mathbb{P}$. We define the *left multiplication* of the polynomial $\phi(x)$ by the functional $u$ as the functional (denoted by $\phi(x)u$) such that:
$$\langle \phi(x)u, f(x) \rangle = \langle u, \phi(x)f(x) \rangle, \quad f(x) \in \mathbb{P}.$$
- The (distributional) *derivative* of the functional $u$, denoted by $\mathcal{D}u$, is the functional such that:
$$\langle \mathcal{D}u, f(x) \rangle = -\langle u, f'(x) \rangle, \quad f(x) \in \mathbb{P}.$$

From the two previous properties, we have

$$\mathcal{D}(\phi(x)u) = \phi(x)\mathcal{D}u + \phi'(x)u,$$

where $\phi(x)$ is a polynomial and $u$ is a functional. By taking successive derivatives we can deduce an analogous of the Leibnitz formula

$$\mathcal{D}^n(\phi(x)u) = \sum_{k=0}^{n} \binom{n}{k} \phi^{(k)}(x)\mathcal{D}^{n-k}u.$$

DEFINITION 2.5.[[7], [9], [10]] A linear functional $u$ is called semiclassical, if there exist two polynomials $\phi(x)$ and $\psi(x)$ such that $u$ satisfy the following distributional differential equation

$$\mathcal{D}(\phi(x)u) = \psi(x)u, \tag{2.2}$$

where $\phi(x) = a_p x^p + \dots, \quad a_p \neq 0$ and $\psi(x) = b_q x^q + \dots, \quad b_q \neq 0, \quad q \geq 1$.
If $u$ is regular, the MOPS associated with $u$ will be called a semiclassical MOPS.

DEFINITION 2.6. We will say that $s \geq 0$ is the class of $u$ when

$$s = \min\{\max\{p-2, q-1\}, \quad \forall(\phi(x), \psi(x)) \quad \text{satisfying} \quad (2.2)\}. \tag{2.3}$$

Obviously, equation (2.2) is equivalent to

$$\phi(x)\mathcal{D}u = (\psi(x) - \phi'(x))u. \tag{2.4}$$

As it is well known, this last equation can be generalized for higher order derivatives of $u$.

LEMMA 2.7.[15] Let $u$ be a semiclassical linear functional, then for every value of $n$ we have

$$\phi^n(x)\mathcal{D}^n u = \psi(x,n)u, \tag{2.5}$$

where the polynomials $\psi(x,n)$ are recursively defined by

$$\begin{aligned} \psi(x,0) &= 1, \\ \psi(x,n) &= \phi(x)\psi'(x,n-1) + \psi(x,n-1)[\psi(x) - n\phi'(x)], \quad n \geq 1. \end{aligned}$$

LEMMA 2.8. In the above conditions

$$\deg(\psi(x,n)) \leq n(s+1), \quad n \geq 0,$$

where $s$ denotes the class of $u$ which is defined in (2.3).

In the classical cases, the expressions for the polynomials $\phi(x)$ and $\psi(x,n)$, for $n \geq 0$, are given in the following table

| Name | $\phi(x)$ | $\psi(x,n)$ |
|---|---|---|
| **Hermite** | $1$ | $(-1)^n n! \sum_{m=0}^{[\frac{n}{2}]} \frac{(-1)^m}{m!} \frac{(2x)^{n-2m}}{(n-2m)!}$ |
| **Laguerre** | $x$ | $\sum_{m=0}^{n} \binom{n}{m} (\alpha+1-(n-m))_{n-m}(-x)^m$ |
| **Jacobi** | $1-x^2$ | $n! \sum_{m=0}^{n} \binom{\alpha+1}{m} \binom{\beta+1}{n-m} (1+x)^m (1-x)^{n-m}$ |

where $(a)_n$ denotes the Pochhammer symbol $(a)_n = a(a+1)\dots(a+n-1)$.
In the classical cases, a very easy computation shows that the degree of the polynomial $\psi(x,n)$ is always $n$, with the extra condition $\alpha+\beta \neq 0, 1, \dots, n-1$, in the Jacobi case.

In the next lemma, we recall a very useful formula relating the derivatives of two different functions.

LEMMA 2.9.[15] Let $f$ and $g$ be $m$–times and $(m+k)$–times differentiable functions, respectively. Then,

$$f^{(m)}(x)g^{(k)}(x) = \sum_{i=0}^{m} (-1)^i \binom{m}{i} \left(f(x)g^{(k+i)}(x)\right)^{(m-i)}.$$

## 3 THE NON–DIAGONAL SOBOLEV INNER PRODUCT

Let $u$ be a positive definite semiclassical linear functional, and $\phi(x)$ and $\psi(x)$ two polynomials satisfying

$$\mathcal{D}(\phi(x)u) = \psi(x)u,$$

with $p = \deg(\phi(x)) \geq 0$, $q = \deg(\psi(x)) \geq 1$, and $s = \max\{p-2, q-1\}$ the class of $u$.

We define the non–diagonal Sobolev inner product in the following way

$$(f,g)_S^{(N)} = \sum_{m,k=0}^{N} \langle \lambda_{m,k} u, f^{(m)} g^{(k)} \rangle, \quad N \geq 1, \quad \forall f,g \in \mathbb{P}, \tag{3.1}$$

where $\lambda_{m,k} = \lambda_{m,k}(x)$, for $m,k = 0,\dots N$, $x \in I$, are real polynomials satisfying several properties in order to obtain a positive definite and symmetric bilinear form. In fact, by using a matrix formalism for (3.1), we get

$$(f,g)_S^{(N)} = \langle u, \left(f, f', \dots, f^{(N)}\right) \begin{pmatrix} \lambda_{0,0}(x) & \lambda_{0,1}(x) & \dots & \lambda_{0,N}(x) \\ \lambda_{1,0}(x) & \lambda_{1,1}(x) & \dots & \lambda_{1,N}(x) \\ \vdots & \vdots & \ddots & \vdots \\ \lambda_{N,0}(x) & \lambda_{N,1}(x) & \dots & \lambda_{N,N}(x) \end{pmatrix} \begin{pmatrix} g \\ g' \\ \vdots \\ g^{(N)} \end{pmatrix} \rangle.$$

Thus, if we denote by

$$\Lambda^N(x) = \begin{pmatrix} \lambda_{0,0}(x) & \lambda_{0,1}(x) & \dots & \lambda_{0,N}(x) \\ \lambda_{1,0}(x) & \lambda_{1,1}(x) & \dots & \lambda_{1,N}(x) \\ \vdots & \vdots & \ddots & \vdots \\ \lambda_{N,0}(x) & \lambda_{N,1}(x) & \dots & \lambda_{N,N}(x) \end{pmatrix},$$

a sufficient condition in order to (3.1) being an inner product is that $\Lambda^N(x)$ be symmetric and positive definite for $x \in I$, the interval support of $u$ in (2.1).

In the case $N = 0$ and $\lambda_{0,0}(x) = 1$, the inner product (3.1) is the standard inner product defined from the semiclassical linear functional $u$, that is,

$$(f,g) = (f,g)_S^{(0)} = \langle u, fg \rangle.$$

Let $\{P_n\}_n$ denote the MOPS associated to the linear functional $u$, $\{Q_n\}_n$ the MOPS associated to the Sobolev inner product (3.1), and denote

$$k_n = \langle u, P_n^2 \rangle > 0, \quad \tilde{k}_n = (Q_n, Q_n)_S^{(N)} > 0, \quad \forall n \geq 0.$$

## 4 THE LINEAR OPERATOR $\mathcal{F}^{(N)}$

We define a linear differential operator $\mathcal{F}^{(N)}$ on the linear space $\mathbb{P}$ in the following way

$$\mathcal{F}^{(N)} = \sum_{m,k=0}^{N} (-1)^m \sum_{i=0}^{m} \binom{m}{i} \sum_{j=0}^{m-i} \binom{m-i}{j} \lambda_{m,k}^{(m-i-j)} \phi^{N-j} \psi(x,j) \mathcal{D}^{k+i}, \tag{4.1}$$

where $\mathcal{D}$ denotes the derivative operator.
REMARK 4.1. In the particular case of a semiclassical linear functional $u$ defined from a weight function, expression (4.1) can be written in a very compact form

$$\mathcal{F}^{(N)} = \frac{\phi^N(x)}{\rho(x)} \sum_{m,k=0}^{N} (-1)^m \mathcal{D}^m(\rho(x)\lambda_{m,k}(x)\mathcal{D}^k),$$

where $\rho(x)$ denotes the weight function associated to the semiclassical linear functional $u$.

REMARK 4.2. The linear operator $\mathcal{F}^{(N)}$ is a $2N$ order differential operator. In fact, if we apply the linear operator $\mathcal{F}^{(N)}$ to $x^n$, from (4.1) we get

$$\begin{aligned}\mathcal{F}^{(N)}x^n &= (-1)^N\lambda_{N,N}(x)\phi^N(x)\mathcal{D}^{2N}(x^n) + (-1)^N N[\lambda'_{N,N}(x)\phi^N(x) + \\ &+ \lambda_{N,N}(x)\phi^{N-1}(x)\psi(x,1)]\mathcal{D}^{2N-1}(x^n) + \dots,\end{aligned}$$

and thus the order is $2N$ if and only if the polynomial $\lambda_{N,N}(x)$ doesn't vanishes identically. However, from the positive definite character of the matrix $\Lambda^{(N)}(x)$, we conclude that the polynomial $\lambda_{N,N}(x)$ doesn't vanishes identically. Therefore, $\mathcal{F}^{(N)}$ is always a $2N$ order differential operator.

In the following proposition, we will show how the linear operator $\mathcal{F}^{(N)}$ allow us to obtain a representation for the Sobolev inner product (3.1), in terms of the consecutive derivatives of the linear functional $u$.

PROPOSITION 4.3. Let $f, g$ be arbitrary polynomials. Then, for $0 \leq h \leq N$, we have

$$(\phi^{N-h}(x)\psi(x,h)f, g)_S^{(N)} = \langle \mathcal{D}^h u, f\mathcal{F}^{(N)}g\rangle.$$

PROOF. From equation (2.5), lemma 2.9 and Leibnitz formula, we get

$$\begin{aligned}(\phi^{N-h}(x)\psi(x,h)f, g)_S^{(N)} &= \sum_{m,k=0}^{N} \langle \lambda_{m,k}u, \left(\phi^{N-h}\psi(x,h)f\right)^{(m)} g^{(k)}\rangle \\ &= \sum_{m,k=0}^{N}\sum_{i=0}^{m}(-1)^i\binom{m}{i}\langle\lambda_{m,k}u, \left(\phi^{N-h}\psi(x,h)fg^{(k+i)}\right)^{(m-i)}\rangle \\ &= \sum_{m,k=0}^{N}\sum_{i=0}^{m}(-1)^m\binom{m}{i}\langle\mathcal{D}^{m-i}\left(\lambda_{m,k}u\right), \phi^{N-h}\psi(x,h)fg^{(k+i)}\rangle \\ &= \sum_{m,k=0}^{N}\sum_{i=0}^{m}(-1)^m\binom{m}{i}\sum_{j=0}^{m-i}\binom{m-i}{j}\langle\lambda_{m,k}^{(m-i-j)}\mathcal{D}^j u, \phi^{N-h}\psi(x,h)fg^{(k+i)}\rangle\end{aligned}$$

$$= \sum_{m,k=0}^{N}\sum_{i=0}^{m}(-1)^m\binom{m}{i}\sum_{j=0}^{m-i}\binom{m-i}{j}\langle\psi(x,j)u,\lambda_{m,k}^{(m-i-j)}\phi^{N-h-j}\psi(x,h)fg^{(k+i)}\rangle$$

$$= \sum_{m,k=0}^{N}\sum_{i=0}^{m}(-1)^m\binom{m}{i}\sum_{j=0}^{m-i}\binom{m-i}{j}\langle\psi(x,h)u,\lambda_{m,k}^{(m-i-j)}\phi^{N-h-j}\psi(x,j)fg^{(k+i)}\rangle$$

$$= \sum_{m,k=0}^{N}\sum_{i=0}^{m}(-1)^m\binom{m}{i}\sum_{j=0}^{m-i}\binom{m-i}{j}\langle\mathcal{D}^h u,\lambda_{m,k}^{(m-i-j)}\phi^{N-j}\psi(x,j)fg^{(k+i)}\rangle$$

$$= \langle\mathcal{D}^h u, f\mathcal{F}^{(N)}g\rangle. \quad \square$$

THEOREM 4.4. The linear operator $\mathcal{F}^{(N)}$ is symmetric with respect to the Sobolev inner product (3.1),i. e.

$$(\mathcal{F}^{(N)}f,g)_S^{(N)} = (f,\mathcal{F}^{(N)}g)_S^{(N)}.$$

PROOF. From proposition 4.3 and lemma 2.9, we can deduce

$$(\mathcal{F}^{(N)}f,g)_S^{(N)} =$$

$$= \sum_{m,k=0}^{N}\sum_{i=0}^{m}(-1)^m\binom{m}{i}\sum_{j=0}^{m-i}\binom{m-i}{j}\left(\lambda_{m,k}^{(m-i-j)}\phi^{N-j}\psi(x,j)f^{(k+i)},g\right)_S^{(N)} =$$

$$= \sum_{m,k=0}^{N}\sum_{i=0}^{m}(-1)^m\binom{m}{i}\sum_{j=0}^{m-i}\binom{m-i}{j}\langle\mathcal{D}^j u,\lambda_{m,k}^{(m-i-j)}f^{(k+i)}\mathcal{F}^{(N)}g\rangle =$$

$$= \sum_{m,k=0}^{N}\sum_{i=0}^{m}(-1)^m\binom{m}{i}\langle\mathcal{D}^{m-i}\left(\lambda_{m,k}u\right), f^{(k+i)}\mathcal{F}^{(N)}g\rangle =$$

$$= \sum_{m,k=0}^{N}\langle\lambda_{m,k}u,\sum_{i=0}^{m}(-1)^i\binom{m}{i}\left(f^{(k+i)}\mathcal{F}^{(N)}g\right)^{(m-i)}\rangle =$$

$$= \sum_{m,k=0}^{N}\langle\lambda_{m,k}u, f^{(k)}\left(\mathcal{F}^{(N)}g\right)^{(m)}\rangle = (f,\mathcal{F}^{(N)}g)_S^{(N)}. \quad \square$$

Now, we will study the degree of the polynomial obtained by means of the application of $\mathcal{F}^{(N)}$ to a $n$–degree polynomial.

PROPOSITION 4.5. For every value of $N \geq 1$, we get

$$\deg\left(\mathcal{F}^{(N)}x^n\right) \leq n + \max_{0\leq m,k\leq N}\{\deg\left(\lambda_{m,k}(x)\right) + p(N-m) + m - k + ms\}.$$

PROOF. By using lemma 2.8, we obtain

$$
\begin{aligned}
&\deg\left(\mathcal{F}^{(N)}x^n\right) \leq \\
&\leq \max_{\substack{0\leq j\leq m-i\\ 0\leq i\leq m\\ 0\leq m,k\leq N}} \{\deg(\lambda_{m,k}(x)) - m + i + j + (N-j)p + j(1+s) + n - k - i\} \\
&\leq n + \max_{\substack{0\leq j\leq m-i\\ 0\leq i\leq m\\ 0\leq m,k\leq N}} \{\deg(\lambda_{m,k}(x)) - m + j + (N-j)p + j(1+s) - k\},
\end{aligned}
$$

and the result follows. □

In spite of the inequality which appears in the above proposition, the linear operator $\mathcal{F}^{(N)}$ never reduces the degree for all the polynomials.

PROPOSITION 4.6. For every $N \geq 1$, we have

$$\deg\left(\mathcal{F}^{(N)}x^n\right) \geq n.$$

PROOF. Let us suppose $\deg\left(\mathcal{F}^{(N)}x^n\right) < n$. Then, we can expand $\mathcal{F}^{(N)}Q_n(x)$ in terms of the polynomials $Q_n(x)$ in the following way

$$\mathcal{F}^{(N)}Q_n(x) = \sum_{i=0}^{n-1} a_{n,i}Q_i(x),$$

where the coefficients can be computed as follows

$$a_{n,i} = \frac{\left(\mathcal{F}^{(N)}Q_n, Q_i\right)_S^{(N)}}{(Q_i, Q_i)_S^{(N)}}, \quad i = 0, \ldots, n-1.$$

From the symmetry of $\mathcal{F}^{(N)}$ and the orthogonality of the polynomials $Q_n(x)$ we get

$$a_{n,i} = \frac{\left(Q_n, \mathcal{F}^{(N)}Q_i\right)_S^{(N)}}{(Q_i, Q_i)_S^{(N)}} = 0, \quad i = 0, \ldots, n-1.$$

and therefore $\mathcal{F}^{(N)}(Q_n(x)) = 0, \quad \forall n \geq 1$, which contradicts our assumptions. □

As we can see from propositions 4.5 and 4.6, the linear operator $\mathcal{F}^{(N)}$ usually increases the degree of the polynomials in a fixed quantity. In this way, there exists an non–negative integer number $t$ such that

$$\mathcal{F}^{(N)}x^n = F(N,n)x^{n+t} + \ldots,$$

where $F(N,n)$ denotes the leading coefficient of the polynomial $\mathcal{F}^{(N)}x^n$. We must notice that this coefficient can be zero for an specific value $n$.

## 5 ALGEBRAIC PROPERTIES FOR THE SOBOLEV ORTHOGONAL POLYNOMIALS

As a direct consequence of proposition 4.3, which relates the non–diagonal Sobolev inner product $(.,.)_S^{(N)}$ and the standard inner product defined from $u$, we can establish the relation between the Sobolev orthogonal polynomials $\{Q_n\}_n$ and the semiclassical orthogonal polynomials $\{P_n\}_n$, associated with the linear functional $u$.

PROPOSITION 5.1. We have

$$\text{(i)} \quad \phi^N(x)P_n(x) = \sum_{i=n-t}^{n+Np} \alpha_{n,i}Q_i(x), \quad n \geq t, \tag{5.1}$$

where $\alpha_{n,n+Np} = a_p^N, \quad \alpha_{n,n-t} = F(N,n)\frac{k_n}{\tilde{k}_{n-t}}$.

$$\text{(ii)} \quad \mathcal{F}^{(N)}Q_n(x) = \sum_{i=n-Np}^{n+t} \beta_{n,i}P_i(x), \quad n \geq Np, \tag{5.2}$$

where $\beta_{n,n+t} = F(N,n), \quad \beta_{n,n-Np} = a_p^N\frac{\tilde{k}_n}{k_{n-Np}}$.

PROOF.
(i) Expanding the polynomial $\phi^N(x)P_n(x)$ in terms of the Sobolev polynomials $\{Q_n\}_n$, we get

$$\phi^N(x)P_n(x) = \sum_{i=0}^{n+Np} \alpha_{n,i}Q_i(x).$$

Using proposition 4.3,

$$\alpha_{n,i} = \frac{(\phi^N P_n(x), Q_i)_S^{(N)}}{(Q_i,Q_i)_S^{(N)}} = \frac{\langle u, P_n\mathcal{F}^{(N)}Q_i\rangle}{\tilde{k}_i},$$

and, from the orthogonality of $\{P_n\}_n$, we deduce $\alpha_{n,i} = 0$, for $0 \leq i \leq n-t-1$.
(ii) If we expand $\mathcal{F}^{(N)}Q_n(x)$ in terms of the polynomials $\{P_n\}_n$, we have

$$\mathcal{F}^{(N)}Q_n(x) = \sum_{i=0}^{n+t} \beta_{n,i}P_i(x).$$

Again, the coefficients can be computed from proposition 4.3, and therefore

$$\beta_{n,i} = \frac{\langle u, P_i\mathcal{F}^{(N)}Q_n\rangle}{\langle u, P_iP_i\rangle} = \frac{(\phi^N P_i, Q_n)_S^{(N)}}{k_i}.$$

Finally, using the orthogonality of $\{Q_n\}_n$, we deduce $\beta_{n,i} = 0$ for $0 \leq i \leq n-Np-1$.
□

From the symmetrical character of the linear operator $\mathcal{F}^{(N)}$ with respect to the Sobolev inner product (3.1), we can obtain a difference–differential relation satisfied by the Sobolev polynomials.

PROPOSITION 5.2.[Difference–Differential Relation] For every $n \geq t$, the following relation holds

$$\mathcal{F}^{(N)}Q_n(x) = \sum_{i=n-t}^{n+t} \gamma_{n,i}Q_i(x), \tag{5.3}$$

where $\gamma_{n,n+t} = F(N,n)$, $\gamma_{n,n-t} = F(N,n)\dfrac{\tilde{k}_n}{\tilde{k}_{n-t}}$.

PROOF. Expand $\mathcal{F}^{(N)}Q_n(x)$ in terms of the polynomials $Q_n(x)$

$$\mathcal{F}^{(N)}Q_n(x) = \sum_{i=0}^{n+t} \gamma_{n,i}Q_i(x),$$

and, from theorem 4.4, we get

$$\gamma_{n,i} = \frac{\left(\mathcal{F}^{(N)}Q_n, Q_i\right)_S^{(N)}}{\left(Q_i, Q_i\right)_S^{(N)}} = \frac{\left(Q_n, \mathcal{F}^{(N)}Q_i\right)_S^{(N)}}{\tilde{k}_i}.$$

From the orthogonality of the polynomials $Q_n(x)$, we deduce $\gamma_{n,i} = 0$ for $0 \leq i \leq n-t-1$, and the result follows. □

In the following proposition, for every polynomial $Q_n(x)$, we can deduce a $2N+2$ order differential equation whose coefficients are polynomials depending on $n$, but with fixed degree.

PROPOSITION 5.3. Sobolev orthogonal polynomials $\{Q_n\}_n$ satisfy the following $2N+2$ order differential equation

$$A(x,n)\mathcal{D}^2\left(\mathcal{F}^{(N)}Q_n(x)\right) + B(x,n)\mathcal{D}\left(\mathcal{F}^{(N)}Q_n(x)\right) + C(x,n)\mathcal{F}^{(N)}Q_n(x) = 0,$$

where $A(x,n)$, $B(x,n)$ and $C(x,n)$ are polynomials with fixed degree.

PROOF. Using the three term recurrence relation for the sequence of orthogonal polynomials $\{P_n\}_n$ (see [4], p. 18), equation (5.2) can be written

$$\mathcal{F}^{(N)}Q_n(x) = M_0(x,n)P_n(x) + N_0(x,n)P_{n-1}(x), \tag{5.4}$$

where $M_0(x,n)$ and $N_0(x,n)$ are polynomials in $x$ depending on $n$.

Taking derivatives in (5.4) and multiplying by $\phi(x)$, we get

$$\begin{aligned}\phi(x)\mathcal{D}\left(\mathcal{F}^{(N)}Q_n(x)\right) &= \phi(x)\left[M_0'(x,n)P_n(x) + M_0(x,n)P_n'(x) + \right.\\ &+ \left. N_0'(x,n)P_{n-1}(x) + N_0(x,n)P_{n-1}'(x)\right].\end{aligned}$$

Using in this expression the structure relation for the semiclassical orthogonal polynomials $\{P_n\}_n$ ([9],[10])

$$\phi(x)P_n'(x) = \sum_{k=n-s-1}^{n+p-1} \delta_{n,k}P_k(x),$$

and the recurrence relation again, we obtain

$$\phi(x)\mathcal{D}\left(\mathcal{F}^{(N)}Q_n(x)\right) = M_1(x,n)P_n(x) + N_1(x,n)P_{n-1}(x), \tag{5.5}$$

where $M_1(x,n)$ and $N_1(x,n)$ are polynomials in $x$. Taking derivatives again in (5.5), multiplying by $\phi(x)$, and applying the structure relation, we obtain

$$\phi(x)\mathcal{D}\left(\phi(x)\mathcal{D}\left(\mathcal{F}^{(N)}Q_n(x)\right)\right) = M_2(x,n)P_n(x) + N_2(x,n)P_{n-1}(x), \tag{5.6}$$

where $M_2(x,n)$ and $N_2(x,n)$ are polynomials in $x$ depending on $n$. If we impose the compatibility between equations (5.4), (5.5) and (5.6), we deduce that

$$\begin{vmatrix} \mathcal{F}^{(N)}Q_n(x) & M_0(x,n) & N_0(x,n) \\ \phi(x)\mathcal{D}\left(\mathcal{F}^{(N)}Q_n(x)\right) & M_1(x,n) & N_1(x,n) \\ \phi(x)\mathcal{D}\left(\phi(x)\mathcal{D}\left(\mathcal{F}^{(N)}Q_n(x)\right)\right) & M_2(x,n) & N_2(x,n) \end{vmatrix} = 0,$$

and the differential equation follows from the expansion of this determinant. □

## 6 SOME EXAMPLES WITH $N = 1$

In this section we shall consider some interesting examples for the Sobolev inner product defined in (3.1) when $N = 1$. In this case, the inner product can be expressed as

$$(f,g)_S^{(1)} = \langle u, (f, \ f') \begin{pmatrix} \lambda_{0,0}(x) & \lambda_{0,1}(x) \\ \lambda_{1,0}(x) & \lambda_{1,1}(x) \end{pmatrix} \begin{pmatrix} g \\ g' \end{pmatrix} \rangle, \quad \forall f,g \in \mathbb{P}, \tag{6.1}$$

and the corresponding polynomial matrix

$$\Lambda^1(x) = \begin{pmatrix} \lambda_{0,0}(x) & \lambda_{0,1}(x) \\ \lambda_{1,0}(x) & \lambda_{1,1}(x) \end{pmatrix},$$

is a symmetric and positive definite matrix. That means $\lambda_{0,1}(x) = \lambda_{1,0}(x)$ and $\lambda_{0,0}(x)\lambda_{1,1}(x) - \lambda_{0,1}^2(x) > 0$ for $x \in I$. The linear operator $\mathcal{F}^{(1)}$ associated to the inner product (6.1) is given by

$$\begin{aligned} \mathcal{F}^{(1)} &= \left\{\phi(x)\left[\lambda_{0,0}(x) - \lambda_{0,1}'(x)\right] - \psi(x,1)\lambda_{0,1}(x)\right\}\mathcal{I} - \\ &\quad - \left\{\phi(x)\lambda_{1,1}'(x) + \psi(x,1)\lambda_{1,1}(x)\right\}\mathcal{D} - \phi(x)\lambda_{1,1}(x)\mathcal{D}^2. \end{aligned} \tag{6.2}$$

In this section, we are interested in Sobolev inner products such that the corresponding linear functional $\mathcal{F}^{(1)}$ preserves the degree of the polynomials. Such a situation is a very interesting one, since if $\mathcal{F}^{(1)}$ preserves the degree of the polynomials, as a consequence of the difference–differential relation, the corresponding Sobolev orthogonal polynomials are the eigenfunction of differential operator, that is, they satisfy a second order differential equation.

PROPOSITION 6.1. If the linear operator $\mathcal{F}^{(1)}$ preserves the degree of the polynomials then the linear functional $u$ is classical. In this case, $\lambda_{1,1}(x)$ is a constant.

PROOF. If we want a linear operator preserving the degree of the polynomials, from (6.2), we can deduce the three following inequalities

$$\begin{aligned} \deg\left\{\phi(x)\left[\lambda_{0,0}(x)-\lambda_{0,1}'(x)\right]-\psi(x,1)\lambda_{0,1}(x)\right\} &= 0, & (6.3)\\ \deg\left\{\phi(x)\lambda_{1,1}'(x)+\psi(x,1)\lambda_{1,1}(x)\right\} &\leq 1, & (6.4)\\ \deg\left\{\phi(x)\lambda_{1,1}(x)\right\} &\leq 2. & (6.5)\end{aligned}$$

If we apply lemma 2.8 to (6.4) we get $\deg\{\lambda_{1,1}(x)\}+1+s\leq 1$, and, since the degree of a polynomial is a non–negative integer, we deduce that $s=0$ and, therefore, $u$ is a classical functional. Observe that as a consequence of (6.4), $\lambda_{1,1}(x)=\lambda_{1,1}$ is a constant. □

Next, we are going to study conditions on the matrix $\Lambda^1(x)$ in order to obtain a preserving degree functional $\mathcal{F}^{(1)}$ in the classical cases.

**The Hermite case**

As it is well known, in this case the polynomials in the distributional differential equation (2.5) are $\phi(x)=1$ and $\psi(x,1)=-2x$, $x\in(-\infty,+\infty)$. In this way the expression of $\mathcal{F}^{(1)}$ is

$$\mathcal{F}^{(1)}=\left\{\lambda_{0,0}(x)-\lambda_{0,1}'(x)+2x\lambda_{0,1}(x)\right\}\mathcal{I}+2x\lambda_{1,1}\mathcal{D}-\lambda_{1,1}\mathcal{D}^2.$$

Therefore, the unique possible choice for $\Lambda^1(x)$ in order to obtain a linear operator $\mathcal{F}^{(1)}$ preserving the degree of the polynomials is

$$\Lambda^1(x)=\begin{pmatrix}\lambda_{0,0} & 0\\ 0 & \lambda_{1,1}\end{pmatrix},$$

with $\lambda_{0,0}$ and $\lambda_{1,1}$ positive constants. In this case, the Sobolev orthogonal polynomials are the Hermite polynomials themselves.

**The Laguerre case**

When $u$ is the Laguerre functional, then $\phi(x)=x$ and $\psi(x,1)=\alpha-x$, respectively (with $\alpha>-1$ and $x\in[0,+\infty)$).

Thus, the linear operator $\mathcal{F}^{(1)}$ can be expressed as

$$\begin{aligned}\mathcal{F}^{(1)} &= \left\{x\left[\lambda_{0,0}(x)-\lambda_{0,1}'(x)\right]-(\alpha-x)\lambda_{0,1}(x)\right\}\mathcal{I}-\\ &\quad -(\alpha-x)\lambda_{1,1}\mathcal{D}-x\lambda_{1,1}\mathcal{D}^2.\end{aligned}$$

If we want that $\mathcal{F}^{(1)}$ preserves the degree of the polynomials we need a constant matrix $\Lambda^1(x)$ like

$$\Lambda^1(x) = \begin{pmatrix} \lambda_{0,0} & -\lambda_{0,0} \\ -\lambda_{0,0} & \lambda_{1,1} \end{pmatrix},$$

with $\lambda_{1,1} > \lambda_{0,0} > 0$. In this case, the Sobolev orthogonal polynomials are the generalized Laguerre polynomials $L_n^{(\alpha-1)}(x)$ (see [16]).

**The Jacobi case**

Here $\phi(x) = 1 - x^2$, $\psi(x,1) = (\beta-\alpha) - (\alpha+\beta)x$, with $\alpha$, $\beta > -1$ and $x \in [-1,1]$. If we replace the above relations in the expression (6.2) of the linear operator $\mathcal{F}^{(1)}$, we get

$$\begin{aligned} \mathcal{F}^{(1)} &= \{(1-x^2)\left[\lambda_{0,0}(x) - \lambda'_{0,1}(x)\right] - [(\beta-\alpha)-(\alpha+\beta)x]\lambda_{0,1}(x)\}\mathcal{I} - \\ &- [(\beta-\alpha)-(\alpha+\beta)x]\lambda_{1,1}\mathcal{D} - (1-x^2)\lambda_{1,1}\mathcal{D}^2. \end{aligned}$$

In order to obtain a preserving degree functional $\mathcal{F}^{(1)}$, is necessary that $\deg(\lambda_{0,1}) = \deg(\lambda_{0,0}) + 1$. The most simple case appears when the polynomial $\lambda_{0,0}$ is a positive constant. In this case, $\lambda_{0,1}(x) = b_1 x + b_0$ and now, we can compute $b_0$ and $b_1$ in terms of $\lambda_{0,0}$. In this way, we have the following linear system:

$$\begin{aligned} (\alpha+\beta+1)b_1 &= \lambda_{0,0} \\ (\alpha+\beta)b_0 &= (\beta-\alpha)b_1 \\ \lambda_{0,0} - (\beta-\alpha)b_0 - b_1 &= \text{constant} \end{aligned}$$

If $\alpha+\beta+1 = 0$, then $\lambda_{0,0} = 0$ using the first equation. In this case, we never get a preserving degree functional.

We suppose $\alpha+\beta+1 \neq 0$, and there are two different situations:

(i) If $\alpha+\beta = 0$, using second equation $\alpha-\beta = 0$, and this is possible only when $\alpha = \beta = 0$. In this way, $b_1 = \lambda_{0,0}$, and $b_0$ is free. Now, we impose the positive definite character of the matrix

$$\Lambda^1(x) = \begin{pmatrix} \lambda_{0,0} & \lambda_{0,0}x + b_0 \\ \lambda_{0,0}x + b_0 & \lambda_{1,1} \end{pmatrix},$$

then, it is necessary that $-b_0 < |\sqrt{\lambda_{0,0}\lambda_{1,1}} - \lambda_{0,0}|$, and $\lambda_{1,1} > \lambda_{0,0}$.

Finally, if we take $\lambda_{0,0} = 1$ and $b_0 = 0$, we deduce that the Sobolev polynomials associated to the inner product:

$$(f,g)_S^{(1)} = \int_{-1}^{1} (f, \; f') \begin{pmatrix} 1 & x \\ x & \lambda_{1,1} \end{pmatrix} \begin{pmatrix} g \\ g' \end{pmatrix} dx,$$

are the *generalized Jacobi polynomials* (see [19], p. 63) $\{P_n^{(-1,-1)}\}_{n\geq 0}$, defined by means of the following expressions:

$$\begin{aligned} P_0^{(-1,-1)}(x) &= 1, \quad P_1^{(-1,-1)}(x) = x, \\ P_n^{(-1,-1)}(x) &= \binom{2n-2}{n}^{-1} \sum_{m=0}^{n} \binom{n-1}{m}\binom{n-1}{n-m}(x+1)^m(x-1)^{n-m}, \end{aligned}$$

for $n \geq 2$.

This assertion can be shown because these polynomials are the unique monic polynomial solution of degree $n$ for the Jacobi differential equation

$$(1-x^2)y'' + n(n-1)y = 0$$

for all $n \neq 1$, and this equation is the same as the equation generated from the difference–differential relation (5.3), since in this case,

$$\mathcal{F}^{(1)} = -(1-x)^2 \lambda_{1,1} \mathcal{D}^2.$$

(ii) If $\alpha + \beta \neq 0$, then

$$\lambda_{0,1}(x) = \frac{\lambda_{0,0}}{\alpha+\beta+1}\left(x + \frac{\beta-\alpha}{\alpha+\beta}\right),$$

and $\lambda_{1,1}$ is a positive constant such that $\Lambda^1(x)$ is positive definite. In this case, the lineal functional $\mathcal{F}^{(1)}$ is

$$\mathcal{F}^{(1)} = \frac{4\alpha\beta}{(\alpha+\beta+1)(\alpha+\beta)} \lambda_{0,0}\mathcal{I} - [(\beta-\alpha) - (\alpha+\beta)x]\,\lambda_{1,1}\mathcal{D} - (1-x^2)\lambda_{1,1}\mathcal{D}^2,$$

which, obiously, preserves the degree of the polynomials, and the Sobolev orthogonal polynomials satisfy a second order differential equation.

## REFERENCES


1. M. Alfaro, F. Marcellán and M.L. Rezola, Orthogonal Polynomials on Sobolev spaces: old and new directions, *J. Comp. Appl. Math. 48*:113-131 (1993).

2. P. Althammer, Eine Erweiterung des Orthogonalitätsbegriffes bei Polynomen und deren Anwendung auf die beste Approximation, *J. Reine Angew. Math. 211*: 192-204 (1962).

3. J. Brenner, Über eine Erweiterung des Orthogonalitätsbegriffes bei Polynomen, in *Proc. Conference on the Constructive Theory of Functions, Budapest 1969*,pp. 77-83, G. Alexits and S. B. Stechkin Eds., Akadémiai Kiadó, Budapest (1972).

4. T. S. Chihara, An Introduction to Orthogonal Polynomials, Gordon and Breach, New York (1978).

5. E. A. Cohen, Zero distribution and behavior of orthogonal polynomials in the Sobolev space $W^{1,2}[-1,1]$, *SIAM J. Math. Anal. 6*: 105–116 (1975).

6. W. Gröbner, Orthogonale Polynomsysteme, die gleichzeitig mit $f(x)$ auch deren Ableitung $f'(x)$ aproximieren, in "Funktionalanalysis, Approximations-theorie, Numerische Mathematik", pp. 24–32, *ISNM 7*, Birkhäuser, Basel (1967).

7. E. Hendriksen and H. Van Rossum, Semiclassical Orthogonal Polynomials, in "Polynômes Orthogonaux et Applications, Bar-le-Duc 1984",pp. 354–361, C. Brezinski et al. Eds. *Lecture Notes in Math. vol. 1171*, Springer-Verlag, Berlin (1985).

8. D.C. Lewis, Polynomial least square approximations, *Amer. J. Math. 69*:273–278 (1947).

9. P. Maroni, Prolégoménes à l'étude des polynômes orthogonaux semiclassiques, *Ann. Mat. Pur. Appl. 149 (4)*:165–184 (1987).

10. P. Maroni, Une Théorie Algébrique des Polynômes Orthogonaux. Applications aux Polynômes Orthogonaux Semiclassiques, in "Orthogonal Polynomials and their applications", C. Brezinski, L. Gori and A. Ronveaux Eds. *IMACS Annals on Comp. and Appl. Math.*, Vol 9, pp. 98–130, J. C. Baltzer. AG Publ. Basel. (1991).

11. F. Marcellán, A. Branquinho and J. Petronilho, Classical Orthogonal Polynomials: A Functional Approach, Acta Applicandae Mathematicae 34:283–303 (1994).

12. F. Marcellán, T.E. Pérez and M.A. Piñar, Gegenbauer–Sobolev Orthogonal Polynomials, in "Non Linear Numerical Methods and Rational Approximation II", pp. 71–82, A. Cuyt Ed., Kluwer Ac. Pub., Antwerpen (1994).

13. F. Marcellán, T.E. Pérez and M.A. Piñar, Orthogonal Polynomials on weighted Sobolev spaces: The semiclassical case, *Annals of Numerical Mathematics 2*:93–122 (1995).

14. F. Marcellán, T.E. Pérez and M.A. Piñar, Laguerre–Sobolev Orthogonal Polynomials, *J. Comp. Appl. Math. 71*:245–265 (1996).

15. F. Marcellán, T.E. Pérez, M.A. Piñar and A. Ronveaux, General Sobolev Orthogonal Polynomials, *J. Math. Ann. Appl. 200*:614–634 (1996).

16. T.E. Pérez, and M.A. Piñar, On Sobolev Orthogonality for the Generalized Laguerre Polynomials, J. App. Th. 86:278–285 (1996).

17. F.W. Schäfke, Zu den Orthogonalpolynomen von Althammer, *J. Reine Angew. Math. 252*:195–199 (1972).

18. F. W. Schäfke and G. Wolf, Einefache verallgemeinerte klassische Orthogonalpolynomen, *J. Reine Angew. Math. 262/263*:339-355 (1973).

19. G. Szegő, Orthogonal Polynomials, 4th ed., *Amer. Math. Soc. Colloq. Publ. 23*, Amer. Math. Soc., Providence, RI (1975).

# Remarks on Canonical Solutions of Strong Moment Problems

OLAV NJÅSTAD Department of Mathematical Sciences, Norwegian University of Science and Technology, Trondheim, Norway

Abstract: This paper gives a survey of some aspects of indeterminate strong moment problems, with a more detailed treatment of a few questions concerned with special types of canonical solutions. The main topics are Nevanlinna parameterization of solutions, the structure of canonical solutions, orthogonal, quasi-orthogonal and pseudo-orthogonal Laurent polynomials, natural, quasi-natural and pseudo-natural solutions, especially natural solutions in the Stieltjes case.

## 1 INDETERMINATE STRONG MOMENT PROBLEMS

A solution of the *strong Hamburger moment problem* (SHMP) for a given (doubly infinite) sequence $\{c_n : n = 0, \pm 1, \pm 2, \ldots\}$ of real numbers is a positive measure $\mu$ with support $\mathrm{Supp}(\mu)$ in $(-\infty, \infty)$ satisfying

$$\int_{-\infty}^{\infty} \lambda^n d\mu(\lambda) = c_n, \quad n = 0, \pm 1, \pm 2, \ldots \tag{1.1}$$

A solution of the *strong Stieltjes moment problem* (SSMP) for the same sequence is a positive measure $\mu$ with support $\mathrm{Supp}(\mu)$ in $[0, \infty)$ satisfying

$$\int_{0}^{\infty} \lambda^n d\mu(\lambda) = c_n, \quad n = 0, \pm 1, \pm 2, \ldots \tag{1.2}$$

We shall in this paper consider a given sequence $\{c_n\}$ for which the SHMP is solvable. The corresponding SSMP may or may not be solvable.

A moment problem (SHMP or SSMP) is said to be *determinate* if it has exactly one solution, *indeterminate* if it has more than one solution. We shall in this paper be concerned with the indeterminate case of the SHMP.

For general treatments of strong moment problems, see [5, 6, 10, 13, 14, 16, 17, 22, 23, 24, 25, 26, 28]. For related questions concerning the classical moment problems, see e.g. [1, 2, 3, 4, 7, 8, 9, 11, 18, 19, 20, 21, 22, 27, 28, 29, 30, 31, 32].

The classical Carleman conditions for determinancy of (classical) moment problems have analogs for strong moment problems. They may be formulated as follows (see e.g. [1,2,8]).

PROPOSITION 1.1. *The SHMP is determinate if*

$$\sum_{n=1}^{\infty}(c_{2n})^{-\frac{1}{2n}} = \infty \quad or \quad \sum_{n=1}^{\infty}(c_{-2n})^{-\frac{1}{2n}} = \infty. \tag{1.3}$$

*The SSMP (provided it is solvable) is determinate if*

$$\sum_{n=1}^{\infty}(c_{n})^{-\frac{1}{2n}} = \infty \quad or \quad \sum_{n=1}^{\infty}(c_{-n})^{-\frac{1}{2n}} = \infty. \tag{1.4}$$

PROOF: This result can essentially be found in [2]. See also [22]. □

We shall in this paper denote by $\mathbf{R}^*$ the two-point compactification of the real line $\mathbf{R}$. Thus $\mathbf{R}^* = \mathbf{R} \cup \{-\infty\} \cup \{\infty\}$. We shall denote by $\hat{\mathbf{R}}$ the one-point compactification of $\mathbf{R}$ and by $\pm\infty$ the point at infinity. Thus $\hat{\mathbf{R}} = \mathbf{R} \cup \{\pm\infty\}$.

THEOREM 1.2. *The support of an arbitrary solution of an indeterminate SHMP contains the point* 0 *and has the point* $\pm\infty$ *as a limit point in* $\hat{\mathbf{R}}$.

*If the corresponding SSMP is solvable, the support of an arbitrary solution of this problem contains the point* 0 *and has the point* $\infty$ *as a limit point in* $\mathbf{R}^*$.

PROOF: Assume that $\text{Supp}(\mu) \subset [-M, M]$, $M < \infty$. Then $|c_n| \leq M^n c_0$ and hence $(c_{2n})^{-\frac{1}{2n}} \geq \dfrac{1}{M c_0^{\frac{1}{2n}}}$ for $n = 1, 2, \ldots$ . It follows from Proposition 1.1 that the SHMP is determinate.

Next assume that $\text{Supp}(\mu) \subset (-\infty, -m] \cup [m, \infty)$. Then $|c_n| \leq m^n c_0$ and hence $(c_{2n})^{\frac{1}{2n}} \geq \dfrac{1}{m c_0^{\frac{1}{2n}}}$ for $n = -1, -2, \ldots$ . Again it follows from Proposition 1.1 that the SHMP is determinate.

We have thus shown that the SHMP is determinate if there is a solution with support which is bounded or bounded away from zero. This is equivalent to the first statemant of the theorem. The second statement follows immediately from the first. □

## 2 NEVANLINNA PARAMETRIZATION

The *Stieltjes transform* or *Markov transform* of a positive measure on $(-\infty, \infty)$ is the function $F_\mu$ defined by

$$F_\mu(z) = \int_{-\infty}^{\infty} \frac{d\mu(\lambda)}{\lambda - z} \tag{2.1}$$

The correspondence between measure and transform is one-to-one. The function $F_\mu$ is holomorphic in the complex plane outside $\mathrm{Supp}(\mu)$.

A *Nevanlinna function* or *Pick function* is a function $\varphi$ which is holomorphic in the open upper half-plane and maps this half-plane into the closed upper half-plane. Stieltjes transforms are Nevanlinna functions with special additional properties.

The Nevanlinna function $\varphi$ may be classified as degenerate if $\varphi(z) \equiv r$, where $r$ is a constant in $\hat{\mathbf{R}}$ (we include the constant function $\pm\infty$ in the class of Nevanlinna functions), as non-degenerate if $\varphi(z)$ belongs to the open upper half-plane for all $z$ in the open upper half-plane (these are the only two possibilities).

THEOREM 2.1. *Let the SHMP be indeterminate. Then there exists a one-to-one correspondence between all Nevanlinna functions $\varphi$ and all solutions $\mu$ of the moment problem. The correspondence is given by*

$$F_\mu(z) = -\frac{\alpha(z)\varphi(z) - \gamma(z)}{\beta(z)\varphi(z) - \delta(z)} \tag{2.2}$$

*where $\alpha, \beta, \gamma, \delta$ are certain functions holomorphic in $\mathbf{C} - \{0\}$, satisfying*

$$\alpha(z)\delta(z) - \beta(z)\gamma(z) = 1. \tag{2.3}$$

PROOF: See [25]. Cf. also [23, 24]. □

We point out that the set of functions $\{\alpha, \beta, \gamma, \delta\}$ is not uniquely defined. Each possible choice determines a correspondence as above, with different $\{\varphi, \mu\}$ corresponding to each other for different sets $\{\alpha, \beta, \gamma, \delta\}$.

The mapping

$$t \longrightarrow -\frac{\alpha(z)t - \gamma(z)}{\beta(z)t - \delta(z)} \tag{2.4}$$

maps for each fixed $z \in \mathbf{C} - \mathbf{R}$ the open upper half-plane onto an open disk $\Delta_\infty(z)$ and the extended real line $\hat{\mathbf{R}}$ onto the boundary circle $\partial\Delta_\infty(z)$. The closed disk $\bar{\Delta}_\infty(z) = \Delta_\infty(z) \cup \partial\Delta_\infty(z)$ consists of exactly all values of the Stieltjes transform at $z$ of solutions of the moment problem. I.e.,

$$\bar{\Delta}_\infty(z) = \{w = F_\mu(z) : \mu \quad \text{is a solution of the SHMP}\}. \tag{2.5}$$

The points on the boundary $\partial\Delta_\infty(z)$ correspond exactly to the degenerate Nevanlinna functions, and the correspondence is one-to-one. The corresponding measures are called *extremal* or *N-extremal* (Nevanlinna extremal) solutions of the moment problem. We note that this concept is independent of the set $\{\alpha, \beta, \gamma, \delta\}$, since the disk $\bar{\Delta}_\infty(z)$ and hence $\partial\Delta_\infty(z)$ is independent of this set.

## 3 CANONICAL SOLUTIONS

A solution $\mu$ of the SHMP is called a *canonical solution* if the corresponding Nevanlinna function $\varphi$ is a real rational function. (I.e., $\varphi$ is a rational function which is real on $\mathbb{R}$.) We shall in this situation write $\varphi(z) = \dfrac{P(z)}{Q(z)}$, where $P$ and $Q$ are polynomials without common factors. Note that the coefficients in $P$ and $Q$ are real.

We say that the canonical solution is of order $\rho$ if $\rho = \max(\deg P, \deg Q)$. The canonical solutions of order 0 are then exactly the extremal solutions, corresponding to the constant functions $r$, $r \in \hat{\mathbf{R}}$.

PROPOSITION 3.1. *The concept "canonical solution of order* $\rho$*" is independent of the set* $\{\alpha, \beta, \gamma, \delta\}$.

PROOF: Let $\{\alpha, \beta, \gamma, \delta\}$ and $\{a, b, c, d\}$ be two function sets giving rise to Nevanlinna parametrizations. Let

$$F_\mu = -\frac{\alpha(z)\varphi(z) - \gamma(z)}{\beta(z)\varphi(z) - \delta(z)} = -\frac{a(z)f(z) - c(z)}{b(z)f(z) - d(z)}. \tag{3.1}$$

We then find that

$$f(z) = \frac{A(z)\varphi(z) - C(z)}{B(z)\varphi(z) - D(z)} \tag{3.2}$$

$$\varphi(z) = \frac{-D(z)f(z) + C(z)}{-B(z)f(z) + A(z)} \tag{3.3}$$

where

$$A(z) = \alpha(z)d(z) - \beta(z)c(z) \tag{3.4}$$

$$B(z) = \alpha(z)b(z) - \beta(z)a(z) \tag{3.5}$$

$$C(z) = \gamma(z)d(z) - \delta(z)c(z) \tag{3.6}$$

$$D(z) = \gamma(z)b(z) - \delta(z)a(z). \tag{3.7}$$

We know that if $\varphi$ is a constant in $\hat{\mathbf{R}}$, then $f$ is also a constant, and vice versa. Setting $\varphi(z) \equiv \pm\infty$ and $\varphi(z) \equiv 0$ in (3.2) we find that

$$\frac{A(z)}{B(z)} = k_1, \quad \frac{C(z)}{D(z)} = k_2, \tag{3.8}$$

$k_1$ and $k_2$ constants. Similarly by setting $f(z) \equiv \pm\infty$ and $f(z) \equiv 0$ in (3.3) we find that

$$\frac{D(z)}{B(z)} = c_1, \quad \frac{C(z)}{A(z)} = c_2, \tag{3.9}$$

$c_1$ and $c_2$ constants.

We may write

$$f(z) = \frac{A(z)}{B(z)} \left[ 1 + \left( \frac{D(z)}{B(z)} - \frac{C(z)}{A(z)} \right) \frac{1}{\varphi(z) - \frac{D(z)}{B(z)}} \right]. \tag{3.10}$$

It follows that

$$f(z) = K_1 + \frac{K_2}{\varphi(z) - K_3}, \tag{3.11}$$

$K_1, K_2, K_3$ constants, which shows that $f$ is a rational function if and only if $\varphi$ is a rational function, and the maximum degree of the numerator and denominator is the same. □

It follows from (2.3) that $\alpha(z)c - \gamma(z)$ and $\beta(z)c - \delta(z)$ have no common zeros for $c \in \mathbf{C}$, hence $\alpha(z)\varphi(z) - \gamma(z)$ and $\beta(z)\varphi(z) - \delta(z)$ have no common zeros for any Nevanlinna function $\varphi$.

Let $\mu$ be a canonical solution with corresponding Nevanlinna function $\varphi = \frac{P}{Q}$, $P$ and $Q$ polynomials without common factors. The Stieltjes transform of $\mu$ may be written as

$$F_\mu(z) = -\frac{P(z)\alpha(z) - Q(z)\gamma(z)}{P(z)\beta(z) - Q(z)\delta(z)}. \tag{3.12}$$

Since the numerator and the denominator have no common factors, the singularities of $F_\mu$ are the zeros of $P\beta - Q\delta$, and in addition 0 and $\pm\infty$. The zeros of $P\beta - Q\delta$ are simple poles, 0 and $\pm\infty$ are accumulation points of these poles. The support Supp($\mu$) contains all the finite singularities of $F_\mu$.

THEOREM 3.2. *Let $\mu$ be a canonical solution of the SHMP. Then* Supp($\mu$) *consists of all the poles of $F_\mu$ and* 0. *The poles $\{z_k\}$ are simple and have* 0 *and*

$\pm\infty$ *as limit points. The measure* $\mu$ *has a mass of magnitude* $-\operatorname*{Res}_{z=z_k} F_\mu(z)$ *at* $z_k$*, and a point of continuity at* 0.

PROOF: A proof can be based on the Perron-Stieltjes inversion formula, see [1]. For details, see [25]. □

It also follows from (2.3) that $\beta(z)c_1 - \delta(z)$ and $\beta(z)c_2 - \delta(z)$ have no common zeros when $c_1 \neq c_2$. From this and Theorem 3.2 we immediately obtain the following result:

PROPOSITION 3.3. *If* $\mu_1$ *and* $\mu_2$ *are two distinct extremal solutions of the SHMP, then* $\operatorname{Supp}(\mu_1) \cap \operatorname{Supp}(\mu_2) = \{0\}$.

# 4 ORTHOGONAL LAURENT POLYNOMIALS

The moment sequence $\{c_n\}$ determines an inner product $\langle\ ,\ \rangle$ on the space of all Laurent polynomials, given by

$$\langle f, g\rangle = \int_{-\infty}^{\infty} f(\lambda)\overline{g(\lambda)}d\mu(\lambda). \tag{4.1}$$

(This definition is independent of the solution $\mu$ of the moment problem.) The sequence $\{1, z^{-1}, z, z^{-2}, z^2, \dots\}$ gives by orthogonalization rise to a sequence of *orthonormal Laurent polynomials* $\varphi_n$. These may be written as

$$\varphi_{2m}(z) = \frac{q_{2m,-m}}{z^m} + \cdots + q_{2m,m}z^m, \quad q_{2m,m} > 0, \quad m = 0, 1, 2, \dots \tag{4.2}$$

$$\varphi_{2m+1}(z) = \frac{q_{2m+1,-(m+1)}}{z^{m+1}} + \cdots + q_{2m+1,m}z^m, \quad q_{2m+1,-(m+1)} > 0, \quad m = 0, 1, 2, \dots \tag{4.3}$$

For general introductions to orthogonal Laurent polynomials, see [10, 12, 13, 15, 22, 23, 24, 25].

We shall assume regularity, i.e.,

$$q_{2m,-m} \neq 0, \quad q_{2m+1,m} \neq 0 \quad \text{for all } m. \tag{4.4}$$

This is always the case when the SSMP is solvable.

The *associated orthogonal Laurent polynomials* $\psi_n$ are defined by

$$\psi_n(z) = \int_{-\infty}^{\infty} \frac{\varphi_n(\lambda) - \varphi_n(z)}{\lambda - z} d\mu(\lambda). \tag{4.5}$$

(Also this definition is independent of the solution $\mu$ of the moment problem.) *Quasi-orthogonal Laurent polynomials* $\varphi(z,\tau)$ are defined for $\tau \in \hat{\mathbf{R}}$

$$\varphi_{2m}(z,\tau) = \varphi_{2m}(z) - \tau z \varphi_{2m-1}(z) \tag{4.6}$$

$$\varphi_{2m+1}(z,\tau) = \varphi_{2m+1}(z) - \frac{\tau}{z}\varphi_{2m}(z), \tag{4.7}$$

and *associated quasi-orthogonal Laurent polynomials* by

$$\psi_{2m}(z,\tau) = \psi_{2m}(z) - \tau z \psi_{2m-1}(z) \tag{4.8}$$

$$\psi_{2m+1}(z,\tau) = \psi_{2m+1}(z) - \frac{\tau}{z}\psi_{2m}(z). \tag{4.9}$$

(Here $\varphi_{2m}(z,\tau)$ means $-\varphi_{2m-1}(z)$ for $\tau = \pm\infty$, etc.)

Let $x_0$ be a fixed point in $(-\infty, 0)$. We define functions $\alpha_n, \beta_n, \gamma_n, \delta_n$ by

$$\alpha_n(z) = (z - x_0) \sum_{j=1}^{n-1} \psi_j(x_0)\psi_j(z) \tag{4.10}$$

$$\beta_n(z) = -1 + (z - x_0) \sum_{j=1}^{n-1} \psi_j(x_0)\varphi_j(z) \tag{4.11}$$

$$\gamma_n(z) = 1 + (z - x_0) \sum_{j=1}^{n-1} \varphi_j(x_0)\psi_j(z) \tag{4.12}$$

$$\delta_n(z) = (z - x_0) \sum_{j=0}^{n-1} \varphi_j(x_0)\varphi_j(z). \tag{4.13}$$

By using Christoffel-Darboux type formulas for orthogonal Laurent polynomials it can be shown that $\beta_{2m}(z), \delta_{2m}(z), z^{-1}\beta_{2m+1}(z),\ z^{-1}\delta_{2m+1}(z)$ are quasi-orthogonal Laurent polynomials and $\alpha_{2m}(z),\ \gamma_{2m}(z),\ z^{-1}\alpha_{2m+1}(z),\ z^{-1}\gamma_{2m+1}(z)$ are associated quasi-orthogonal Laurent polynomials.

When the SHMP is indeterminate, the functions $\alpha_n, \beta_n, \gamma_n, \delta_n$ converge locally uniformly in $\mathbf{C} - \{0\}$ to functions $\alpha, \beta, \gamma, \delta$, holomorphic in $\mathbf{C} - \{0\}$ and satisfying (2.3). This set of functions $\{\alpha, \beta, \gamma, \delta\}$ determines a Nevanlinna parametrization as in Theorem 2.1. Thus to each $x_0 \in (-\infty, 0)$ there exists a set of functions $\{\alpha, \beta, \gamma, \delta\}$ determining a Nevanlinna parametrization.

## 5 QUASI-NATURAL SOLUTIONS

We define *quasi-approximants* $R_n(z,\tau)$ and $T_n(z,t)$ by

$$R_n(z,\tau) = \frac{\psi_n(z,\tau)}{\varphi_n(z,\tau)}, \quad \tau \in \hat{\mathbf{R}} \tag{5.1}$$

$$T_n(z,t) = \frac{\alpha_n(z)t - \gamma_n(z)}{\beta_n(z)t - \delta_n(z)}, \quad t \in \hat{\mathbf{R}}. \tag{5.2}$$

Then

$$T_n(z,t) = R_n(z,\tau) \tag{5.3}$$

where $t$ is obtained from $\tau$ by the transformation $t \to t_n(\tau)$, with

$$t_{2m}(\tau) = \frac{\varphi_{2m}(x_0) - \tau x_0 \varphi_{2m-1}(x_0)}{\psi_{2m}(x_0) - \tau x_0 \psi_{2m-1}(x_0)}, \tag{5.4}$$

$$t_{2m+1}(\tau) = \frac{x_0 \varphi_{2m+1}(x_0) - \tau \varphi_{2m}(x_0)}{x_0 \psi_{2m+1}(x_0) - \tau \psi_{2m}(x_0)}. \tag{5.5}$$

A quasi-orthogonal Laurent polynomial $\varphi_n(z,\tau)$ has $n$ simple zeros $\zeta_1^{(n)}(\tau)$, $\ldots, \zeta_n^{(n)}(\tau)$, and there exist positive weights $\lambda_1^{(n)}(\tau), \ldots, \lambda_n^{(n)}(\tau)$ such that the following representation of the quasi-approximant is valid:

$$R_n(z,\tau) = -\sum_{k=1}^{n} \frac{\lambda_k^{(n)}(\tau)}{\zeta_k^{(n)}(\tau) - z}. \tag{5.6}$$

Let $\mu_n(\lambda,\tau)$ denote the discrete measure having a mass $\lambda_k^{(n)}(\tau)$ at the point $\zeta_k^{(n)}(\tau), \quad k = 1, \ldots, n$. Then we may write

$$R_n(z,\tau) = -\int_{-\infty}^{\infty} \frac{d\mu_n(\lambda,\tau)}{\lambda - z}. \tag{5.7}$$

It follows by Helly's theorems that every subsequence of $\{\mu_n(\lambda,\tau_n)\}$ has a subsequence converging to a solution $\mu$ of the moment problem, and the corresponding subsequence of $R_n(z,\tau_n)$ converges locally uniformly outside $\mathbf{R}$ to $-F_\mu(z)$.

Solutions obtained in this way we call *quasi-natural solutions*. Taking into account (5.2)-(5.3) and the fact that $\alpha_n, \beta_n, \gamma_n, \delta_n$ converge to $\alpha, \beta, \gamma, \delta$, we conclude:

PROPOSITION 5.1 *The quasi-natural solutions are exactly the extremal solutions. They can be obtained as limits of convergent subsequences of the sequences of discrete measures* $\mu_n(\lambda, \tau_n)$, *where* $\tau_n = t_n^{-1}(t)$ *for a fixed* $t$.

## 6 PSEUDO-NATURAL SOLUTIONS

In this section we make the general assumption that the SSMP is solvable. We define *pseudo-orthogonal Laurent polynomials* $\Phi_n(z, \tau)$ and *associated pseudo-orthogonal Laurent polynomials* $\Psi_n(z, \tau)$ for all $\tau \in \hat{\mathbf{R}}$ by

$$\Phi_n(z, \tau) = \varphi_n(z) - \tau\varphi_{n-1}(z) \tag{6.1}$$

$$\Psi_n(z, \tau) = \psi_n(z) - \tau\psi_{n-1}(z). \tag{6.2}$$

We define *pseudo-approximants* $S_n(z, \tau)$ by

$$S_n(z, \tau) = \frac{\Psi_n(z, \tau)}{\Phi_n(z, \tau)}. \tag{6.3}$$

Then we may write

$$S_{2m}(z, \tau) = \frac{\alpha_{2m}(z)t_{2m}(\tau, z) - \gamma_{2m}(z)}{\beta_{2m}(z)t_{2m}(\tau, z) - \delta_{2m}(z)} \tag{6.4}$$

$$S_{2m+1}(z, \tau) = \frac{\alpha_{2m+1}(z)t_{2m+1}(\tau, z) - \gamma_{2m+1}(z)}{\beta_{2m+1}(z)t_{2m+1}(\tau, z) - \delta_{2m+1}(z)}, \tag{6.5}$$

where

$$t_{2m}(\tau, z) = \frac{z\varphi_{2m}(x_0) - \tau x_0\varphi_{2m-1}(x_0)}{z\psi_{2m}(x_0) - \tau x_0\psi_{2m-1}(x_0)} \tag{6.6}$$

$$t_{2m+1}(\tau, z) = \frac{x_0\varphi_{2m+1}(x_0) - \tau z\varphi_{2m}(x_0)}{x_0\psi_{2m+1}(x_0) - \tau z\psi_{2m}(x_0)}. \tag{6.7}$$

A pseudo-orthogonal Laurent polynomial $\Phi_n(z, \tau)$ has $n$ simple zeros $z_1^{(n)}(\tau), \ldots, z_n^{(n)}(\tau)$, at least $n-1$ of them positive. (Recall that we have assumed the SSMP solvable.) There exist weights $\kappa_1^{(n)}(\tau), \ldots, \kappa_n^{(n)}(\tau)$ such that the following representation of the pseudo-approximant is valid:

$$S_n(z, \tau) = -\sum_{k=1}^{n} \frac{\kappa_k^{(n)}(\tau)}{z_k^{(n)}(\tau) - z}. \tag{6.8}$$

Furthermore there exist constants $\{g_n\}$ such that

$$z_k^{(n)}(\tau) > 0, \quad \kappa_k^{(n)}(\tau) > 0, \quad k = 1, \ldots n, \tag{6.9}$$

when $\tau \in (-\infty, g_n)$ for $n$ even, $\tau \in (g_n, \infty)$ for $n$ odd. For these values of $\tau$ we let $\nu_n(\lambda, \tau)$ denote the discrete measure having a mass $\kappa_k^{(n)}(\tau)$ at the point $z_k^{(n)}(\tau)$, $k = 1, 2, \ldots n$. Then we may write

$$S_n(z, \tau) = -\int_0^\infty \frac{d\nu_n(\lambda, \tau)}{\lambda - z}. \tag{6.10}$$

Again it follows by Helly's theorems that every subsequence of $\nu_n(\lambda, \tau_n)$ has a subsequence converging to a solution $\nu$ of the SSMP, and the corresponding subsequence of $S_n(z, \tau_n)$ converges locally uniformly outside $\mathbf{R}$ to $-F_\nu(z)$. Solutions obtained in this way we call *pseudo-natural solutions*.

Let $t(z)$ be a function of the form

$$t(z) = \frac{az + b}{cz + d}, \quad ad - bc \neq 0, \tag{6.11}$$

and let $\tau_n$ be defined by

$$t_n(\tau_n, z) = t(z) \tag{6.12}$$

where $t_n$ is given by (6.6)-(6.7). Since $\alpha_n, \beta_n, \gamma_n, \delta_n$ converge to $\alpha, \beta, \gamma, \delta$, we conclude from (6.4)-(6.5) that $S_n(z, \tau_n)$ converge, hence that $\nu_n(\lambda, \tau_n)$ converge to a solution of the SSMP if the function (6.11) is such that $\tau_n$ satisfies $\tau_{2n} \in (-\infty, g_{2n})$, $\tau_{2n+1} \in (g_{2n+1}, \infty)$. Conversely, for every subsequence $\{\tau_{n_k}\}$ such that $\nu_{n_k}(\lambda, \tau_{n_k})$ converge to a solution it follows from (6.4)-(6.5) that $t_{n_k}(\tau_{n_k}, z)$ converge to a function of the form (6.11). (It is known that for every $x_0 \in (-\infty, 0)$ the sequences $\left\{\frac{\varphi_{2m}(x_0)}{\psi_{2m}(x_0)}\right\}$ and $\left\{\frac{\varphi_{2m+1}(x_0)}{\psi_{2m+1}(x_0)}\right\}$ converge to values in $\mathbf{R} - \{0\}$, cf. Section 8.)

We may thus conclude:

PROPOSITION 6.1. *The pseudo-natural solutions are those canonical solutions of order 1 that are solutions of the SSMP. They can be obtained as limits of convergent subsequences of the sequences of discrete measures $\nu_n(\lambda, \tau_n)$, where $\tau_n$ is determined by (6.12) for a fixed Nevanlinna function (6.11).*

We mention that the moment sequence $\{c_n\}$ determines for each $z \in \mathbf{C} - R$ an open disk $D_\infty(z)$ with boundary $\partial D_\infty(z)$ with the property

$$\{\omega = F_\nu(z) : \quad \nu \text{ is a pseudo-natural solution of the SSMP}\} = \partial D_\infty(z) \cap \Delta_\infty(z). \tag{6.13}$$

# 7 NATURAL SOLUTIONS

Solutions of the SHMP that are obtained as limits of convergent subsequences of the sequence $\{\mu_n(\lambda, 0)\}$ are called *natural solutions*. When the SSMP is solvable, the natural solutions can also be obtained as limits of convergent subsequences of the sequence $\{\nu_n(\lambda, 0)\}$, (since $\nu_n(\lambda, 0) = \mu_n(\lambda, 0)$) and they are then solutions of the SSMP. The natural solutions are thus special cases of the quasi-natural solutions, and of the pseudo-natural solutions when the SSMP is solvable.

When a measure $\mu$ is obtained as the limit of a sequence $\{\mu_i\}$ of discrete measures, then $\mathrm{Supp}(\mu)$ consists of all limit points of the sequence $\{\mathrm{Supp}(\mu_i) : i = 1, 2, \ldots\}$. Let $\zeta_k^{(n)}$ denote the zeros of $\varphi_n$ ordered by size,

$$\zeta_1^{(n)} < \zeta_2^{(n)} < \cdots < \zeta_{n-1}^{(n)} < \zeta_n^{(n)}. \tag{7.1}$$

The support of a natural solution of the SHMP obtained from a convergent sequence $\{\mu_{n_p}(\lambda, 0)\}$ then equals the set of limit points for the sequence $\{\zeta_1^{(n_p)}, \ldots, \zeta_{n_p}^{(n_p)} : p = 1, 2, \ldots\}$. Similar relationships hold for quasi-natural and pseudo-natural solutions.

THEOREM 7.1 *Assume that the limit* $\zeta = \lim_n \zeta_{k_n}^{(n)}$ *exists in* $\mathbf{R}^*$ *for some sequence* $\{k_n : n = 1, 2, \ldots\}$, $k_n \in \{1, 2, \ldots n\}$. *Further assume that the sequence* $\left\{\frac{\psi_n(z)}{\varphi_n(z)}\right\}$ *does not converge. Then* $\zeta \in \{-\infty, 0, \infty\}$

PROOF: Since $\left\{\frac{\psi_n(z)}{\varphi_n(z)}\right\}$ does not converge, there is more than one natural solution. If $\zeta$ is finite, it belongs to the support of all these natural solutions by the considerations above. The result then follows from Proposition 3.3. □

In the case of the classical moment problems, the zeros $\zeta_k^{(n)}$ of the orthogonal polynomial of degree $n$ separate completely the zeros $\zeta_k^{(n+1)}$ of the orthogonal polynomial of degree $n + 1$. Consequently the sequences $\{\zeta_k^{(n)} : n = 1, 2, \ldots\}$ and $\{\zeta_{n-j}^{(n)} : n = 1, 2, \ldots\}$ for arbitrary fixed $k$ and arbitrary fixed $j$ are monotonic and hence converge in $\mathbf{R}^*$. This is in general not the case for the SHMP. The situation can be described as follows. (Recall that we have assumed regularity of the orthogonal Laurent polynomials.)

The orthogonal Laurent polynomials satisfy Christoffel-Darboux type formulas which in differential form can be written as

$$\varphi_{2m}(z)\frac{d}{dz}(z\varphi_{2m-1}(z)) - z\varphi_{2m-1}(z)\frac{d}{dz}(\varphi_{2m}(z)) = \frac{q_{2m,-m}}{q_{2m-1,-m}} \sum_{j=0}^{2m-1} [\varphi_j(z)]^2 \tag{7.2}$$

$$\varphi_{2m}(z)\frac{d}{dz}(z\varphi_{2m+1}(z)) - z\varphi_{2m+1}(z)\frac{d}{dz}(\varphi_{2m}(z)) = \frac{q_{2m+1,m}}{q_{2m,m}}\sum_{j=0}^{2m}[\varphi_j(z)]^2. \tag{7.3}$$

By standard arguments we find from these formulas:

Let $a, b$ be two consecutive zeros of $\varphi_n$, both to the left or both to the right of 0. Then $(a, b)$ contains exactly one zero of $\varphi_{n-1}$.

Let $A$ be the largest zero of $\varphi_{n-1}$ to the left of 0, $B$ the smallest zero of $\varphi_{n-1}$ to the right of 0. Then $\varphi_n$ has two zeros in $(A, B)$ if $n = 2m$, $\varphi_n$ has one zero in $(A, B)$ if $n = 2m + 1$.

When the SSMP is solvable, all the zeros of all $\varphi_n$ are positive. It follows that we then have complete separation:

$$\zeta_1^{(n+1)} < \zeta_1^{(n)} < \zeta_2^{(n+1)} < \cdots < \zeta_n^{(n+1)} < \zeta_n^{(n)} < \zeta_{n+1}^{(n+1)}. \tag{7.4}$$

In the case of the SHMP in general this conclusion does not hold.

## 8 NATURAL SOLUTIONS IN THE STIELTJES CASE

We assume in this section that the SSMP is solvable. We state as a theorem some basic results on natural solutions in this situation. We write $\mu_n(\lambda)$ for $\mu_n(\lambda, 0)$.

THEOREM 8.1. *Assume that the SSMP is solvable. Then the following hold:*

*(i)* *The sequences* $\{\mu_{2m} : m = 1, 2, \ldots\}$ *and* $\{\mu_{2m+1} : m = 0, 1, 2, \ldots\}$ *converge to measures* $\mu^{(0)}$ *and* $\mu^{(\infty)}$ *both with support in* $[0, \infty)$. *Thus there are two (possibly coinciding) natural solutions of the SHMP, both of which are solutions of the SSMP.*

*(ii)* *The sequences* $\left\{\frac{\psi_{2m}(z)}{\varphi_{2m}(z)} : m = 1, 2, \ldots\right\}$ *and* $\left\{\frac{\psi_{2m+1}(z)}{\varphi_{2m+1}(z)} : m = 0, 1, 2, \ldots\right\}$ *converge locally uniformly outside* $[0, \infty)$ *to the holomorphic functions* $-F_{\mu^{(0)}}(z)$ *and* $-F_{\mu^{(\infty)}}(z)$.

*(iii)* *For a suitable sequence* $\{\kappa_n\}$ *of constants, the sequences* $\{\kappa_{2m}\varphi_{2m}(z)\}$, $\{\kappa_{2m+1}\varphi_{2m+1}(z)\}$, $\{\kappa_{2m}\psi_{2m}(z)\}$, $\{\kappa_{2m+1}\psi_{2m+1}(z)\}$ *converge locally uniformly in* $\mathbf{C} - \{0\}$ *to holomorvarphic functions* $\varphi^{(0)}(z)$, $\varphi^{(\infty)}(z)$, $\psi^{(0)}(z)$, $\psi^{(\infty)}(z)$.

PROOF: For proof of these results, see [6, 17, 23]. □

We saw in Section 7 that the sequence $\{\zeta_k^{(n)} : n = 1, 2, \ldots\}$ is non-increasing for arbitrary fixed $k$ and the sequence $\{\zeta_{n-j}^{(n)} : n = 1, 2, \ldots\}$ is non-decreasing for arbitrary fixed $j$. It follows that the limits

$$\xi_k = \lim_n \zeta_k^{(n)}, \quad k = 1, 2, \ldots \tag{8.1}$$

$$\eta_j = \lim_n \zeta_{n-j}^{(n)}, \quad j = 1, 2, \ldots \tag{8.2}$$

exist in $\mathbf{R}^*$. Note that $\xi_k < \eta_j$ for all $j, k$.

THEOREM 8.2 *Assume that the SSMP is solvable, and assume that the two natural solutions $\mu^{(0)}$ and $\mu^{(\infty)}$ are distinct. Let $\xi_k$ and $\eta_j$ be defined by (8.1)-(8.2). Then either*

*(i) $\xi_k = -\infty$ for all $k$ and $\eta_j \in \{0, \infty\}$ for all $j$,*

*or*

*(ii) $\xi_k \in \{-\infty, 0\}$ for all $k$ and $\eta_j = \infty$ for all $j$.*

PROOF: This follows from Theorem 7.1, the fact that $\left\{ \frac{\psi_n(z)}{\varphi_n(z)} \right\}$ does not converge, and the observation that $\xi_k < \eta_j$ for all $j$ and $k$. □

## REFERENCES

1. N.I. Akhiezer, The Classical Moment Problem and Some Related Qustions in Analysis, Hafner, New York, 1965.

2. E. Aldén, A survey of weak and strong moment problems with generalizations, Department of Mathematics, University of Umeå, *Preprint 2* (1985) 1–38.

3. C. Berg and H.L. Pedersen, On the order and type of the entire functions associated with an indeterminate Hamburger moment problem, *Ark. Mat. 32* (1994) 1–11.

4. C. Berg, Markov's theorem revisited, *J. Approximation Theory 78* (1994) 260–275.

5. C.M. Bonan-Hamada, W.B. Jones and O. Njåstad, Natural solutions of indeterminate strong Stieltjes moment problems derived from PC-fractions, In: *Orthogonal Functions, Moment Theory and Contiuned Fractions: Theory and Applications* (W.B. Jones and A. Sri Ranga, editors) Marcel Dekker, 1997.

6. C.M. Bonan-Hamada, W.B. Jones, O. Njåstad and W.J. Thron, A class of indeterminate strong Stieltjes moment problems with discrete distributions, In: *Orthogonal Functions, Moment Theory and Contiuned Fractions: Theory and Applications* (W.B. Jones and A. Sri Ranga, editors) Marcel Dekker, 1997.

7. T.S. Chihara, On indeterminate Hamburger moment problems, *Pacific J. Math.* *27* (1968), 475–484.

8. T.S. Chihara, An Introduction to Orthogonal Polynomials, Gordon and Breach, New York, 1978.

9. T.S. Chihara, Indeterminate symmetric moment problems, *J. Math. Anal. Appl.* *85* (1982), 331–346.

10. L. Cochran and S. Clement Cooper, Orthogonal Laurent polynomials on the real line, In: *Continued Fractions and Orthogonal Functions*, (S. Clement Cooper and W.J. Thron, eds.), Marcel Dekker, New York, 1994, pp. 47–100.

11. H. Hamburger, Über eine Erweiterung des Stieltjesschen Momentproblems, Parts I, II, III, *Math. Annalen 81* (1920), 235–319; 82 (1921), 120–164; 82 (1921), 168–187.

12. E. Hendriksen and H. van Rossum, Orthogonal Laurent polynomials, *Indag. Math.* *89* (1986), 17–36.

13. W.B. Jones, O. Njåstad, and W.J. Thron, Orthogonal Laurent polynomials and the strong Hamburger moment problem, *J. Math. Anal. Appl.* *98* (1984), 528–554.

14. W.B. Jones, O. Njåstad and W.J. Thron, Continued fractions and strong Hamburger moment problems, *Proc. Lond. Math. Soc.*, *47* (1983), 363–384.

15. W.B. Jones and W.J. Thron, Orthogonal Laurent polynomials and Gaussian quadrature, In: *Quantum Mechanics in Mathematics, Chemistry and Physics*, (K. Gustafson and W.P. Reinhardt, eds.), Plenum Publishing Corp., New York, 1981, pp. 449–455.

16. W.B. Jones and W.J. Thron, Survey of continued fractions methods of solving moment problems and related topics, In: *Analytic Theory of Continued Fractions*, (W.B. Jones, W.J. Thron, and H. Waadeland, eds.), Lecture Notes in Mathematics No. 932, Springer Verlag, Berlin, 1982, pp. 4–37.

17. W.B. Jones, W.J. Thron and H. Waadeland, A strong Stieltjes moment problem, *Trans. Amer. Math. Soc. 261* (1980), 503–528.

18. H.J. Landau, The classical moment problem: Hilbertian proofs, *J. Functional Analysis 38* (1980), 255–272.

19. H.J. Landau (ed.), Moments in Mathematics, *Proc. Symposium Appl. Math. 37, Amer. Math. Soc.*, Providence, RI, 1987.

20. R. Nevanlinna. Über beschränkte analytische Funktionen, *Ann. Acad. Sci. Fenn. A 32* (1929), 1–75.

21. O. Njåstad, Solutions of the Stieltjes moment problem, *Communications in the Analytic Theory of Continued Fractions 3* (1994) 32–51.

22. O. Njåstad, Classical and strong moment problems, *Communications in the Analytic Theory of Continued Fractions 4* (1995) 4–38.

23. O. Njåstad, Solutions of the strong Stieltjes moment problem, *Methods and Applications of Analysis, 2* (1995) 320–347.

24. O. Njåstad, Extremal solutions of the strong Stieltjes moment problem, *J. Comp. Appl. Math. 65* (1995) 309–318.

25. O. Njåstad, Solutions of the strong Hamburger moment problem, *J. Math. Anal. Appl. 197* (1996) 227–248.

26. O. Njåstad and W.J. Thron, Unique solvability of the strong Hamburger moment problem, *J. Austral. Math. Soc. (Series A) 40* (1986), 5–19.

27. H.L. Pedersen, The Nevenlinna matrix of entire functions associated with a shifted indeterminate Hamburger moment problem, *Math. Scand. 74* (1994) 152–160.

28. A. Sri Ranga, $\hat{J}$-fractions and strong moment problems, In: *Analytic Theory of Continued Fractions*, (W.J. Thron, Ed.), Lecture Notes in Mathematics No. 1199, Springer Verlag, Berlin, (1986), pp. 269–284.

29. M. Riesz, Sur le problème des moments, I, II, III, *Arkiv för Mathematik, Astronomi och Fysik 16* (12) (1922); 16 (19) (1922); 17 (16) (1923).

30. J.A. Shohat and J.D. Tamarkin, The Problem of Moments, *Mathematical Surveys No. 1, Amer. Math. Soc.*, Providence, R.I., 1943.

31. T.J. Stieltjes, Recherches sur les fractions continues, *Ann. Fac. Sci. Toulouse 8* (1894), J. 1–122, 9 (1894), A 1–47; Oeuveres 2, 402–566.

32. M.H. Stone, Linear transformations in Hilbert space and their applications to analysis, *Amer. Math. Soc. Coll. Publ. 15*, New York, 1932.

# Sobolev Orthogonality and Properties of the Generalized Laguerre Polynomials

TERESA E. PÉREZ[1] & MIGUEL A. PIÑAR[1] Departamento de Matemática Aplicada, Universidad de Granada, Granada, Spain

## ABSTRACT

The generalized Laguerre polynomials, $\{L_n^{(\alpha)}(x)\}_{n\geq 0}$, for any value of the parameter $\alpha$, are orthogonal with respect to some inner product involving derivatives, that is, a Sobolev inner product. This property was proved in a previous paper. In this paper, we use the Sobolev orthogonality to recover properties about the generalized Laguerre polynomials. For instance, we can obtain relations with the classical Laguerre polynomials and localization properties for the zeros of the generalized Laguerre polynomials.

*AMS Subject Classification (1991)*: 33 C 45
*Key words*: Laguerre polynomials, non diagonal Sobolev inner product, Sobolev orthogonal polynomials.

## 1 INTRODUCTION

Classical Laguerre polynomials constitute the main subject of a very extensive literature. If we denote by $\{L_n^{(\alpha)}(x)\}_{n\geq 0}$ the sequence of monic Laguerre polynomials, their crucial property is the orthogonality with respect to the weight function $\rho(x) = x^\alpha e^{-x}$, where the parameter $\alpha$ satisfies the condition $-1 < \alpha$ in order to assure the convergence of the integrals.

For the monic classical Laguerre polynomials, an explicit representation is well known (see G. Szegő [7], page 102):

[1]This research was partially supported by Junta de Andalucía, Grupo de Investigación FQM 0229.

$$L_n^{(\alpha)}(x) = (-1)^n n! \sum_{j=0}^{n} \frac{(-1)^j}{j!} \binom{n+\alpha}{n-j} x^j, \qquad n \geq 0, \tag{1.1}$$

where $\binom{a}{b}$ denotes the generalized binomial coefficient

$$\binom{a}{b} = \frac{\Gamma(a+1)}{\Gamma(b+1)\Gamma(a-b+1)}.$$

Expression (1.1) can be used to define a family of monic polynomials, for an arbitrary value of the parameter $\alpha \in \mathbb{R}$. These polynomials will be denoted by $\{L_n^{(\alpha)}(x)\}_{n\geq 0}$, and will be called *generalized Laguerre polynomials.* Obviously, they constitute a basis for the linear space of real polynomials, since $\deg(L_n^{(\alpha)}) = n, \quad \forall n \geq 0$.

These polynomials satisfy a three term recurrence relation, which can be deduced directly from their explicit representation

$$\begin{aligned} L_{-1}^{(\alpha)}(x) &= 0, \qquad L_0^{(\alpha)} = 1, \\ xL_n^{(\alpha)}(x) &= L_{n+1}^{(\alpha)}(x) + \beta_n^{(\alpha)} L_n^{(\alpha)}(x) + \gamma_n^{(\alpha)} L_{n-1}^{(\alpha)}(x), \end{aligned} \tag{1.2}$$

where

$$\beta_n^{(\alpha)} = 2n + \alpha + 1, \qquad \gamma_n^{(\alpha)} = n(n+\alpha),$$

From Favard's theorem (see T. S. Chihara [1], page 21), we conclude that the family of polynomials $\{L_n^{(\alpha)}(x)\}_{n\geq 0}$ for $\alpha \notin \{-1, -2, \ldots\}$ constitutes a monic orthogonal polynomial sequence with respect to a quasi-definite linear functional. For $-1 < \alpha$, the linear functional is positive definite. However, for $\alpha \in \{-1, -2, \ldots\}$, no orthogonality results can be deduced from Favard's theorem since $\gamma_n^{(\alpha)}$ vanishes for some value of $n$.

K. H. Kwon and L. L. Littlejohn in [2] have shown the orthogonality of the generalized Laguerre polynomials $\{L_n^{(-k)}(x)\}_{n\geq 0}, k = 1, 2, \ldots$ with respect to an inner product involving some Dirac masses and derivatives. But, in fact, the generalized Laguerre polynomials, for any value of the parameter $\alpha$, are orthogonal with respect to some inner product involving derivatives, that is a Sobolev inner product. This property was proved in a previous paper ([5]). Our objective in this paper is to show how the Sobolev orthogonality can be used to recover properties about the generalized Laguerre polynomials.

In section 2, we will define the monic generalized Laguerre polynomials, for $\alpha \in \mathbb{R}$, by using their explicit expression, also, we give some of their well known properties, which will be used in this paper.

In [5], by using a recursive method, we have shown that for an arbitrary real number $\alpha$, the sequence $\{L_n^{(\alpha)}(x)\}_{n\geq 0}$ is the monic orthogonal polynomial sequence, (in short, the MOPS) with respect to the non–diagonal Sobolev inner product

$$(f(x), g(x))_S^{(k,\alpha+k)} = \int_0^{+\infty} F(x) M(k) G(x)^T x^{\alpha+k} e^{-x} dx,$$

where $k = \max\{0, [-\alpha]\}$, $[\alpha]$ denotes the greatest integer less than or equal to $\alpha$, $F(x), G(x)$ are two vectors defined by

$$\begin{aligned} F(x) &= (f(x), f'(x), \ldots, f^{(k)}(x)), \\ G(x) &= (g(x), g'(x), \ldots, g^{(k)}(x)), \end{aligned}$$

and $M(k)$ is a matrix whose $(i, j)$ entry is defined by

$$m_{i,j}(k) = \sum_{p=0}^{\min\{i,j\}} (-1)^{i+j} \binom{k-p}{i-p} \binom{k-p}{j-p}, \qquad 0 \le i, j \le k.$$

In section 3, we extend the previous result to a more general situation where $k + \alpha > -1$, obtaining a very easy proof for the orthogonality property. As a consequence of this result, we can get a global sight of the generalized Laguerre polynomials as a MOPS with respect to a family of non–diagonal Sobolev inner products.

In section 4, we recall some of the properties obtained for the orthogonal polynomials associated to a non–diagonal Sobolev inner product, that is, an inner product involving derivatives defined by means of an expression

$$(f(x), g(x))_S^{(k)} = \int_{\mathbb{R}} F(x) A G(x)^T,$$

where $A$ is a symmetric positive definite matrix whose elements are signed Borel measures (see in this volume the chapter by Álvarez de Morales, Pérez, Piñar and Ronveaux). The existence of a linear differential operator which is symmetric with respect to the Sobolev inner product will allow us to relate the Sobolev orthogonal polynomials with the standard orthogonal polynomials.

Finally, in the last section, by using the Sobolev orthogonality as starting point, we recover some of the properties of the generalized Laguerre polynomials. For instance, we obtain the explicit representation for $L_n^{(-k)}(x)$ for any integer $k \le n$ or we deduce some localization property for the zeros of the generalized Laguerre polynomials.

## 2 GENERALIZED LAGUERRE POLYNOMIALS

Let $\alpha \in \mathbb{R}$, the *n-th degree monic generalized Laguerre polynomial* is defined in [7], p. 102, by means of the following expression

$$L_n^{(\alpha)}(x) = (-1)^n n! \sum_{j=0}^{n} \frac{(-1)^j}{j!} \binom{n+\alpha}{n-j} x^j, \qquad n \ge 0, \tag{2.1}$$

where $\binom{a}{b}$ denotes the generalized binomial coefficient

$$\binom{a}{b} = \frac{\Gamma(a+1)}{\Gamma(b+1)\Gamma(a-b+1)}.$$

Since $\deg(L_n^{(\alpha)}) = n, n \geq 0$, for a given $\alpha \in \mathbb{R}$, the family of the generalized Laguerre polynomials constitutes a basis for the linear space of the real polynomials.

In the following lemma, we give some of the usual properties of the generalized Laguerre polynomials. The three first properties are well known and can be deduced from the explicit expression for these polynomials. Property (iv) can be shown by using induction on $k$, and applying properties (ii) and (iii).

LEMMA 2.1. Given $\alpha \in \mathbb{R}$, let $k \geq 0$, $n \geq 1$ be integer numbers. Then
(i) (Recurrence relation)

$$\begin{aligned} L_{-1}^{(\alpha)}(x) &= 0, \qquad L_0^{(\alpha)}(x) = 1, \\ xL_n^{(\alpha)}(x) &= L_{n+1}^{(\alpha)} + \beta_n^{(\alpha)} L_n^{(\alpha)}(x) + \gamma_n^{(\alpha)} L_{n-1}^{(\alpha)}(x), \end{aligned} \tag{2.2}$$

where $\beta_n^{(\alpha)} = 2n + \alpha + 1, \qquad \gamma_n^{(\alpha)} = n(n + \alpha).$

(ii) $(L_n^{(\alpha)})'(x) = nL_{n-1}^{(\alpha+1)}(x),$ (2.3)

(iii) $L_n^{(\alpha)}(x) = L_n^{(\alpha+1)}(x) + nL_{n-1}^{(\alpha+1)}(x),$ (2.4)

(iv) $L_n^{(\alpha)}(x) = \sum_{i=0}^{k} (-1)^i \binom{k}{i} (L_n^{(\alpha-k)})^{(i)}(x).$ (2.5)

From Favard's theorem (T. S. Chihara [1], p. 21) and property (i), one can deduce that the generalized Laguerre polynomials constitute a MOPS with respect to a regular linear functional if and only if

$$\gamma_n^{(\alpha)} \neq 0, \qquad n \geq 1,$$

that is, if and only if $\alpha \notin \{-1, -2, \ldots\}$. Moreover, for $\alpha > -1$, the generalized Laguerre polynomials are the classical Laguerre polynomials which are orthogonal with respect to a positive definite linear functional.

## 3 SOBOLEV ORTHOGONALITY

In this section, we are going to show that Generalized Laguerre polynomials are orthogonal with respect to a inner product involving derivatives, that is, a Sobolev inner product.

Let $k \geq 0$ be an integer number. Let us define a lower triangular matrix $L(k)$, with dimension $k+1$

$$L(k) = \begin{pmatrix} 1 & 0 & 0 & \cdots & 0 \\ -\binom{k}{1} & 1 & 0 & \cdots & 0 \\ \binom{k}{2} & -\binom{k-1}{1} & 1 & \cdots & 0 \\ \vdots & \vdots & \vdots & \ddots & \vdots \\ (-1)^k\binom{k}{k} & (-1)^{k-1}\binom{k-1}{k-1} & (-1)^{k-2}\binom{k-2}{k-2} & \cdots & 1 \end{pmatrix}$$

From this, we can define a symmetric matrix by means of the expression

$$M(k) = L(k)L(k)^T. \tag{3.1}$$

Each element in $M(k)$ is given by

$$m_{i,j}(k) = \sum_{p=0}^{\min\{i,j\}} (-1)^{i+j} \binom{k-p}{i-p}\binom{k-p}{j-p}, \qquad 0 \le i,j \le k.$$

Obviously, $M(k)$ is positive definite since expression (3.1) constitutes the Cholesky factorization for $M(k)$ ([6], p. 174).

From this matrix, we can define the non–diagonal Sobolev inner product. Let $k \geq 0$ be an integer number and $\alpha > -k-1$ a real number, we define the symmetric bilinear form $(\cdot,\cdot)_S^{(k,\alpha+k)}$ by means of the expression

$$(f,g)_S^{(k,\alpha+k)} = \int_0^{+\infty} F(x)M(k)G(x)^T x^{\alpha+k}e^{-x}dx, \tag{3.2}$$

where $F(x)$ and $G(x)$ are two vectors defined by

$$\begin{aligned} F(x) &= (f(x), f'(x), \ldots, f^{(k)}(x)), \\ G(x) &= (g(x), g'(x), \ldots, g^{(k)}(x)). \end{aligned}$$

Since $\alpha + k > -1$, all the integrals in this expression are finite, and as a consequence of the positive definite character of the symmetric matrix $M(k)$, we conclude that (3.2) is an inner product, in fact, a non–diagonal Sobolev inner product.

The Sobolev orthogonality of the generalized Laguerre polynomials was stated in [5], by using an inductive reasoning and in the particular case $k = \max\{0, [-\alpha]\}$. However, an straightforward proof can show that the result remains true for every value of $k$ satisfying $\alpha + k > -1$, as we can see in the following

THEOREM 3.1. Let $\alpha \in \mathbb{R}$. Then the monic generalized Laguerre polynomials $\{L_n^{(\alpha)}(x)\}_{n\geq 0}$ constitute a MOPS with respect to the non–diagonal Sobolev inner product

$$(\cdot,\cdot)_S^{(k,\alpha+k)},$$

where $k \geq \max\{0, [-\alpha]\}$.

PROOF. We will multiply two generalized Laguerre polynomials with different degrees,say $L_n^{(\alpha)}(x)$ and $L_m^{(\alpha)}(x)$, where $n \neq m$. By using the expression (3.1) for the matrix $M(k)$, we get

$$(L_n^{(\alpha)}, L_m^{(\alpha)})_S^{(k,\alpha+k)} =$$

$$\int_0^{+\infty} (L_n^{(\alpha)}, (L_n^{(\alpha)})', \ldots, (L_n^{(\alpha)})^{(k)})L(k).L(k)^T \begin{pmatrix} L_m^{(\alpha)} \\ (L_m^{(\alpha)})' \\ \cdots \\ (L_m^{(\alpha)})^{(k)} \end{pmatrix} x^{\alpha+k}e^{-x}dx$$

and, from properties (2.3) and (2.5),we deduce

$$(L_n^{(\alpha)}, (L_n^{(\alpha)})', \ldots, (L_n^{(\alpha)})^{(k)})L(k) =$$

$$\left(\sum_{i=0}^{k}(-1)^i\binom{k}{i}(L_n^{(\alpha)})^{(i)},\sum_{i=0}^{k-1}(-1)^i\binom{k-1}{i}((L_n^{(\alpha)})')^{(i)},\dots,(L_n^{(\alpha)})^{(k)}\right)=$$

$$=\left(L_n^{(\alpha+k)},nL_{n-1}^{(\alpha+k)},\dots,n(n-1)\dots(n-k+1)L_{n-k}^{(\alpha+k)}\right)$$

where we assume $L_i^{(\alpha+k)}(x)=0$ for $i<0$. Finally, the result follows from the orthogonality of $L_i^{(\alpha+k)}$ with respect to the weight function $x^{\alpha+k}e^{-x}$. □

REMARK 3.2. In the case when $\alpha\in\{-1,-2,\dots\}$ and $k=[-\alpha]=-\alpha$, and the non–diagonal Sobolev inner product can be written as

$$(f,g)_S^{(k,0)}=\int_0^{+\infty}F(x)M(k)G(x)^Te^{-x}dx.$$

and, integrating by parts, we get

$$\begin{aligned}(f,g)_S^{(k,0)} &= \frac{1}{2}\sum_{i=0}^{k-1}\sum_{j=0}^{i}m_{i,j}(k)\left[f^{(i)}(0)g^{(j)}(0)+f^{(j)}(0)g^{(i)}(0)\right]+\\ &+\int_0^{+\infty}f^{(k)}(x)g^{(k)}(x)e^{-x}dx.\end{aligned}$$

which is the inner product introduced by K. H. Kwon and L. L. Littlejohn ([2]) in order to study the Sobolev orthogonality of the generalized Laguerre polynomials $\{L_n^{(-k)}(x)\}_{n\geq 0}$.

## 4 THE LINEAR OPERATOR $\mathcal{F}^{(k)}$

In a recent work a general theory for non-diagonal Sobolev inner product was constructed (see in this volume the chapter by Álvarez de Morales, Pérez, Piñar and Ronveaux). In fact, in this paper, we consider a positive definite semiclassical linear functional $u$ satisfying a distributional differential equation

$$\mathcal{D}(\phi(x)u)=\psi(x)u,$$

where $\phi(x)$ and $\psi(x)$ are polynomials, with $p=\deg(\phi(x))\geq 0$, $q=\deg(\psi(x))\geq 1$, and $s=\max\{p-2,q-1\}$ the class of $u$, (for definitions and properties of semiclassical linear functionals, see [4]).

Associated with the linear functional $u$, we can define a non–diagonal Sobolev inner product in the following way

$$(f,g)_S^{(k)}=\langle u,\left(f,f',\dots,f^{(k)}\right)\begin{pmatrix}\lambda_{0,0}(x) & \lambda_{0,1}(x) & \dots & \lambda_{0,k}(x)\\ \lambda_{1,0}(x) & \lambda_{1,1}(x) & \dots & \lambda_{1,k}(x)\\ \vdots & \vdots & \ddots & \vdots\\ \lambda_{k,0}(x) & \lambda_{k,1}(x) & \dots & \lambda_{k,k}(x)\end{pmatrix}\begin{pmatrix}g\\ g'\\ \vdots\\ g^{(k)}\end{pmatrix}\rangle. \tag{4.1}$$

where $\lambda_{m,h}(x)$, for $m, h = 0, \ldots k$, are real polynomials satisfying several properties in order to obtain a positive definite and symmetric bilinear form, for $x \in I$, the support of the linear functional $u$.

The main characteristic of such a kind of inner product is the existence of a linear differential operator $\mathcal{F}^{(k)}$, defined on the linear space of the real polynomials $\mathbb{P}$, which is symmetric with respect to the Sobolev inner product (4.1). The expression of the linear differential operator $\mathcal{F}^{(k)}$ is given by

$$\mathcal{F}^{(k)} = \sum_{m,h=0}^{k} (-1)^m \sum_{i=0}^{m} \binom{m}{i} \sum_{j=0}^{m-i} \binom{m-i}{j} \lambda_{m,h}^{(m-i-j)}(x)\phi^{k-j}(x)\psi(x,j)\mathcal{D}^{h+i}, \tag{4.2}$$

where $\mathcal{D}$ denotes the derivative operator, and the polynomials $\psi(x,j)$ are defined by means of the following recursion

$$\begin{aligned} \psi(x,0) &= 1, \\ \psi(x,n) &= \phi(x)\psi'(x,n-1) + \psi(x,n-1)[\psi(x) - n\phi'(x)], \quad n \geq 1. \end{aligned}$$

In the particular case of a semiclassical linear functional $u$ defined from a weight function $\rho(x)$, expression (4.2) can be written in a very compact form

$$\mathcal{F}^{(k)} = \frac{\phi^k(x)}{\rho(x)} \sum_{i,j=0}^{k} (-1)^i \mathcal{D}^i(\rho(x)\lambda_{i,j}(x)\mathcal{D}^j). \tag{4.3}$$

The linear operator $\mathcal{F}^{(k)}$ allows us to obtain a representation for the Sobolev inner product (4.1), in terms of the consecutive derivatives of the linear functional $u$.

PROPOSITION 4.1. Let $f, g$ be arbitrary polynomials. Then, for $0 \leq h \leq k$, we have

$$(\phi^{k-h}(x)\psi(x,h)f, g)_S^{(k)} = \langle \mathcal{D}^h u, f\mathcal{F}^{(k)} g\rangle.$$

And, as a consequence, we get the symmetry of the linear operator

THEOREM 4.2. The linear operator $\mathcal{F}^{(k)}$ is symmetric with respect to the Sobolev inner product (4.1), i. e.

$$(\mathcal{F}^{(k)} f, g)_S^{(k)} = (f, \mathcal{F}^{(k)} g)_S^{(k)}.$$

As we can see, the linear operator $\mathcal{F}^{(k)}$ usually increases the degree of the polynomials in a fixed quantity. In this way, there exists an non–negative integer number $t$ such that

$$\mathcal{F}^{(k)} x^n = F(k,n)x^{n+t} + \ldots,$$

where $F(k,n)$ denotes the leading coefficient of the polynomial $\mathcal{F}^{(k)}x^n$.

As a direct consequence of proposition 4.1, we can establish the relation between the Sobolev orthogonal polynomials $\{Q_n\}_n$ and the semiclassical orthogonal polynomials $\{P_n\}_n$, associated with the linear functional $u$. For this, denote by

$$k_n = \langle u, P_n^2\rangle > 0, \quad \tilde{k}_n = (Q_n, Q_n)_S^{(k)} > 0, \quad \forall n \geq 0.$$

PROPOSITION 4.3. We have

$$\text{(i)} \quad \phi^k(x)P_n(x) = \sum_{i=n-t}^{n+kp} \alpha_{n,i}Q_i(x), \quad n \geq t, \tag{4.4}$$

where $\alpha_{n,n+kp} = a_p^k, \quad \alpha_{n,n-t} = F(k,n)\dfrac{k_n}{\tilde{k}_{n-t}}$ and $a_p$ is the leading coefficient of the polynomial $\phi(x)$.

$$\text{(ii)} \quad \mathcal{F}^{(k)}Q_n(x) = \sum_{i=n-kp}^{n+t} \beta_{n,i}P_i(x), \quad n \geq kp, \tag{4.5}$$

where $\beta_{n,n+t} = F(k,n), \quad \beta_{n,n-kp} = a_p^k \dfrac{\tilde{k}_n}{k_{n-kp}}$.

From theorem 4.2, we can deduce a difference–differential relation satisfied by the Sobolev orthogonal polynomials.

PROPOSITION 4.4.(Difference–Differential Relation) For every $n \geq t$, the following relation holds

$$\mathcal{F}^{(k)}Q_n(x) = \sum_{i=n-t}^{n+t} \gamma_{n,i}Q_i(x), \tag{4.6}$$

where $\gamma_{n,n+t} = F(k,n), \quad \gamma_{n,n-t} = F(k,n)\dfrac{\tilde{k}_n}{\tilde{k}_{n-t}}$.

## 5 PROPERTIES OF THE GENERALIZED LAGUERRE POLYNOMIALS

Now, we are going to show that some of the properties of the generalized Laguerre polynomials can be deduced from the Sobolev orthogonality.

In our case, the Sobolev inner product is given by the expression (3.2), the Sobolev orthogonal polynomials are the generalized Laguerre polynomials $Q_n(x) = L_n^{(\alpha)}(x)$ and the standard orthogonal polynomials are the Laguerre polynomials $P_n(x) = L_n^{(\alpha+k)}(x)$, with $\alpha + k > -1$.

According to the previous section, there exists a linear differential operator, which is symmetric with respect to the Sobolev inner product. From (4.3) we can deduce the explicit expression for this linear differential operator

$$\mathcal{F}^{(k)} = \frac{x^k}{x^{\alpha+k}e^{-x}} \sum_{i,j=0}^{k} (-1)^i m_{ij}(k)\mathcal{D}^i(x^{\alpha+k}e^{-x}\mathcal{D}^j). \tag{5.1}$$

A direct calculation shows that the linear operator $\mathcal{F}^{(k)}$ preserves the degree of the polynomials, that is

PROPOSITION 5.1. For every polynomial $p(x)$, $\deg(p) = n$ implies $\deg(\mathcal{F}^{(k)}(p)) = n$.

In consequence, the difference–differential relation contains only one term in its right member, and we can conclude

PROPOSITION 5.2. The generalized Laguerre polynomials $L_n^{(\alpha)}$ are the eigenfunctions of the differential operator $\mathcal{F}^{(k)}$, for $\alpha + k > -1$, that is

$$\mathcal{F}^{(k)} L_n^{(\alpha)}(x) = c_n L_n^{(\alpha)}(x).$$

We must notice that the equality in proposition 5.2 can be considered as a differential equation of order $2k$ for the generalized Laguerre polynomials.

Now, using proposition 5.2 and relations (4.4) and (4.5), we can deduce some interesting relations between Laguerre polynomials

PROPOSITION 5.3. We have

(i) $x^k L_n^{(\alpha+k)}(x) = \sum_{i=n}^{n+k} a_{n,i} L_i^{(\alpha)}(x), \quad n \geq 0,$

(ii) $L_n^{(\alpha)}(x) = \sum_{i=n-k}^{n} b_{n,i} L_i^{(\alpha+k)}(x), \quad n \geq k,$

An interesting case appears when $\alpha \in \{-1, -2, ...\}$. In this situation, the generalized Laguerre polynomials $L_n^{(-k)}(x)$ are orthogonal with respect to the Sobolev inner product $(\cdot,\cdot)_S^{(k,0)}$, and from expression (3.1) for the matrix $M(k)$, we can deduce a very compact form for the linear operator $\mathcal{F}^{(k)}$

$$\mathcal{F}^{(k)} = x^k(k+1)(\mathcal{I} - \mathcal{D})^k \mathcal{D}^k. \tag{5.2}$$

Therefore, proposition 5.2 gives

$$x^k(k+1)(\mathcal{I} - \mathcal{D})^k \mathcal{D}^k L_n^{(-k)}(x) = (k+1)\frac{n!}{(n-k)!} L_n^{(-k)}(x),$$

and, since, from properties (2.3) and (2.5), we have

$$(\mathcal{I}-\mathcal{D})^k\mathcal{D}^k L_n^{(-k)}(x) = \frac{n!}{(n-k)!}L_{n-k}^{(k)}(x),$$

we conclude

$$L_n^{(-k)}(x) = x^k L_{n-k}^{(k)}(x),$$

for $n \geq k$, which, in fact, is the explicit representation given in [7], p. 102.

Finally, we can obtain some properties of the zeros of the generalized Laguerre polynomials as a consequence of the Sobolev orthogonality.

PROPOSITION 5.4. For every $n \geq k = [-\alpha]$, $L_n^{(\alpha)}(x)$ has at least $n-k$ real and simple zeros contained in the interval $[0, +\infty)$.

PROOF. Let $x_1, x_2, \ldots, x_s$ be the positive and simple roots of $L_n^{(\alpha)}(x)$, and define

$$h(x) = \prod_{i=1}^{s}(x - x_i)$$

then, using the representation formula 4.1, we get

$$\begin{aligned}(x^k h(x), L_n^{(\alpha)}(x))_S^{(k,k+\alpha)} &= \int_0^{+\infty} h(x)\mathcal{F}^{(k)}L_n^{(\alpha)}(x)x^{\alpha+k}e^{-x}dx = \\ &= c_n \int_0^{+\infty} h(x)L_n^{(\alpha)}(x)x^{\alpha+k}e^{-x}dx \neq 0,\end{aligned}$$

and therefore $s \geq n-k$. □

This property is shown in Szegő's book ([7], p. 151) by using a different technique based on the calculation of the discriminant for the Laguerre polynomials.

## REFERENCES

1. T. S. Chihara, An Introduction to Orthogonal Polynomials, Gordon and Breach, New York, (1978).

2. K. H. Kwon and L. L. Littlejohn, The Orthogonality of the Laguerre Polynomials $\{L_n^{(-k)}(x)\}$ for positive integers $k$, *Annals of Numerical Mathematics 2*:289–304 (1995).

3. F. Marcellán, T.E. Pérez, M.A. Piñar and A. Ronveaux, General Sobolev Orthogonal Polynomials, *J. Math. Anal. Appl. 200*:614–634 (1996).

4. P. Maroni, Une Théorie Algébrique des Polynômes Orthogonaux. Applications aux Polynômes Orthogonaux Semiclassiques, in "Orthogonal Polynomials and their applications", C. Brezinski, L. Gori and A. Ronveaux Eds. *IMACS Annals on Comp. and Appl. Math.*, Vol 9, pp. 98-130, J. C. Baltzer. AG Publ. Basel. (1991).

5. T.E. Pérez and M.A. Piñar, On Sobolev Orthogonality for the Generalized Laguerre Polynomials, *J. Approx. Theory 86*:278–285 (1996).

6. J. Stoer and R. Burlirsch, Introduction to Numerical Analysis, Springer-Verlag, New York, (1980).

7. G. Szegő, Orthogonal Polynomials, 4th ed., *Amer. Math. Soc. Colloq. Publ. 23*, Amer. Math. Soc., Providence, RI, (1975).

# A Combination of Two Methods in Frequency Analysis: The $R(N)$-Process

VIGDIS PETERSEN Sør-Trøndelag College, School of Teacher Education, Rotvoll Allé, N-7005, Trondheim, Norway

## 1 THE PROBLEM

The frequency analysis problem is to determine the unknown normalized frequencies $\omega_1, \omega_2, \ldots, \omega_I$ of a trigonometric signal $x_N(m)$ with a sample of size $N$.

Recently there has been established a method for determining those frequencies by using families of sequences of Szegő polynomials,$\rho_n(\psi_N; z)$. The polynomials are orthogonal on the unit circle with respect to a certain absolutely continuous distribution function constructed from the given signal. The Szegő polynomials in question may be expressed as a ratio of determinants, where the denominator is a Toeplitz determinant. The unknown frequencies are determined from asymptotic properties of the zeros of the Szegő polynomials.

There are two versions of the method. The one that came up first, is called the $N$-process. The latest version contains one additional parameter $R$, and it is called the $R$-process. Both versions of the method has its disadvantages.

The $N$-process has the disadvantage that it is necessary to go to subsequences to obtain convergence of the zeros of the Szegő polynomials. For the $R$-process this is not so. On the other hand the $R$-process has the disadvantage that an additional parameter is involved. One possible way of combining the advantage of both versions, is to let $R$ be a function $R(N)$ of $N$ such that

$$N \rightarrow \infty \Longleftrightarrow R \rightarrow 1. \tag{1.1}$$

This idea is mentioned in the final remarks in a paper from 1994 by Jones, Njåstad, Waadeland [7].

The purpose of the present paper is to prove a theorem showing that a properly chosen function $R(N)$ may be used to create convergence of the zeros of the

This work was supported by The Nansen Fund and Affiliated Funds, The Norwegian Academy of Science and Letters.

Szegő polynomials without going to subsequences and without having the additional parameter $R$.

# 2 THE METHOD

Both versions of the method start with observations of a continous trigonometric signal at equally spaced times $m\Delta t$. The observations lead to a discrete version of the signal given by

$$x_N(m) = \begin{cases} \sum_{j=-I}^{I} \alpha_j e^{i\omega_j m} & \text{if } m = 0, 1, 2, \ldots, N-1, \\ 0 & \text{otherwise,} \end{cases} \tag{2.1}$$

where $|\alpha_j|$ are the amplitudes, $\omega_j$ are the normalized frequencies, $N$ is the number of observations, $1 \leq N < \infty$, $I < \infty$, $\alpha_0 \geq 0$,
$0 \neq \alpha_{-j} = \overline{\alpha}_j \in \mathbf{C}$, $\omega_{-j} = -\omega_j \in \mathbf{R}, j = 1, 2, \ldots, I$.

The frequencies are ordered such that

$$0 = \omega_0 < \omega_1 < \cdots < \omega_I < \pi. \tag{2.2}$$

We have assumed that there is no noise.

The method to be outlined here has its root in the Wiener-Levinson method [8], [14], and has recently been dealt with in several papers ([2],[3], [4], [5],[6],[11]). For a detailed description see [7].

# 3 THE $N$-PROCESS

The start of the method is to construct, from the observations, a positive Borel measure with infinite support:

$$d\psi_N(\theta) = \frac{1}{2\pi} \left|X_N(e^{i\theta})\right|^2 d\theta, \; -\pi \leq \theta \leq \pi \tag{3.1}$$

where

$$X_N(z) = \sum_{m=0}^{N-1} x_N(m) z^{-m}. \tag{3.2}$$

For any fixed $N$ this measure $\psi_N$ gives rise to a double sequence of moments

$$\mu_m^{(N)} = \int_{-\pi}^{\pi} e^{-mi\theta} d\psi_N(\theta), \;\; m = 0, \pm 1, \pm 2, \ldots . \tag{3.3}$$

The moments can be computed as autocorrelation coefficients of the signal:

$$\mu_m^{(N)} = \sum_{k=0}^{N-m-1} x_N(k) x_N(k+m), \;\; 1 < N < \infty, \; m = 0, 1, 2, \ldots , \tag{3.4}$$

$$\mu_{-m}^{(N)} = \mu_m^{(N)} . \tag{3.5}$$

This sequence is hermitian positive definite, meaning that $\mu_{-m}^{(N)} = \bar{\mu}_m^{(N)}$, and that certain Toeplitz determinants are positive. In our case the moments are, in fact, real. Essential in the $N$-process is the use of the moments

$$\tilde{\mu}_m^{(N)} := \frac{\mu_m^{(N)}}{N}.$$

The Szegő polynomials may be determined in different ways. From a practical point of view the Levinson algorithm is by far the best method. Another method is connected with continued fractions where the Szegő polynomials are the denominators of the odd order approximants of the corresponding positive Perron-Carathéodory continued fraction (PPC-fraction). A third method, the one of greatest interest in the present paper, is by explicit determinant formulas. (See [1], [4],[13]).

In the N-process the Szegő polynomials are given by

$$\rho_n(\psi_N; z) = \frac{1}{\Delta_{n-1}^{(N)}} \begin{vmatrix} \tilde{\mu}_0^{(N)} & \tilde{\mu}_{-1}^{(N)} & \cdots & \tilde{\mu}_{-n}^{(N)} \\ \tilde{\mu}_1^{(N)} & \tilde{\mu}_0^{(N)} & \cdots & \tilde{\mu}_{-n+1}^{(N)} \\ \vdots & \vdots & & \vdots \\ \tilde{\mu}_{n-1}^{(N)} & \tilde{\mu}_{n-2}^{(N)} & \cdots & \tilde{\mu}_{-1}^{(N)} \\ 1 & z & \cdots & z^n \end{vmatrix} \tag{3.6}$$

where $\Delta_{n-1}^{(N)}$ is the Toeplitz determinant of order $n$. The expression for the Toeplitz determinant is

$$\Delta_n^{(N)} = \begin{vmatrix} \tilde{\mu}_0^{(N)} & \tilde{\mu}_{-1}^{(N)} & \tilde{\mu}_{-2}^{(N)} & \cdots & \tilde{\mu}_{-n}^{(N)} \\ \tilde{\mu}_1^{(N)} & \tilde{\mu}_0^{(N)} & \tilde{\mu}_{-1}^{(N)} & \cdots & \tilde{\mu}_{-n+1}^{(N)} \\ \tilde{\mu}_2^{(N)} & \tilde{\mu}_1^{(N)} & \tilde{\mu}_0(N) & \cdots & \tilde{\mu}_{-n+2}^{(N)} \\ \vdots & \vdots & \vdots & & \vdots \\ \tilde{\mu}_n^{(N)} & \tilde{\mu}_{n-1}^{(N)} & \tilde{\mu}_{n-2}^{(N)} & \cdots & \tilde{\mu}_0^{(N)} \end{vmatrix}. \tag{3.7}$$

Let

$$n_0 = 2I + L \text{ where } L = \begin{cases} 0 & \text{if } \alpha_0 = 0\,, \\ 1 & \text{otherwise.} \end{cases} \tag{3.8}$$

If $n_0$ is known, the frequencies $\omega_j$ are determined by using the following property:

$$\lim_{N\to\infty} \rho_{n_0}(\psi_N; z) = (z-1)^L \prod_{j=1}^{I} (z - e^{i\omega_j})(z - e^{-i\omega_j}). \tag{3.9}$$

Normally $n_0$ is not known, but for any fixed $n$ the sequence $\{\rho_n(\psi_N; z)\}_{N=1}^{\infty}$ is normal. For $n > n_0$ *any* convergent subsequence has a limit polynomial with the right-hand side of (3.9) as a factor. By Hurwitz' theorem, the zeros of the limit polynomial will be such that $n_0$ of them will converge to the frequency points $e^{\pm i\omega_j}$. The additional $(n - n_0)$ zeros will be inside the unit disc.

The results mentioned above are shown independently by Jones, Njåstad, Thron, Waadeland [4] and Pan, Saff [11].

# 4 THE $R$-PROCESS

In the $R$-process we replace $\left\{\tilde{\mu}_m^{(N)}\right\}$ from the $N$-process by

$$\left\{\tilde{\mu}_m^{(N)} R^{|m|}\right\} \quad \text{where} \quad 0 < R < 1. \tag{4.1}$$

We will again have a positive definite hermitian sequence [7]. This leads to a measure $\psi_N^{(R)}$ and the Szegő polynomials $\rho_n(\psi_N^{(R)}; z)$. By taking the $N$-limit ($N \to \infty$) *first* and *then* the $R$-limit ($R \uparrow 1$), we get a limit polynomial without going to subsequences. This limit polynomial contains as a factor the right-hand side of (3.9). In the $R$-process the moments are

$$\mu_m^{(R)} := (\lim_{N\to\infty} \frac{\mu_m^{(N)}}{N})(1-d)^m = (\sum_{j=-I}^{I} |\alpha_j|^2 e^{i\omega_j m})(1-d)^m \tag{4.2}$$

where $R = 1-d$. (See for instance [7], equation (2.12).) The corresponding measure shall be denoted $\psi^{(R)}$. The $R$-process is described by Jones, Njåstad, Waadeland in a paper from 1994 [7].

# 5 THE $R(N)$-PROCESS

We will now illustrate a method of combining the $N$-process and the $R$-process. We know that for a fixed $n$ the corresponding Szegő polynomials $\rho_n(\psi_N; z)$ and $\rho_n(\psi^{(R)}; z)$ may be expressed as a ratio of two determinants with a certain Toeplitz determinant as the denominator. From the uniform boundedness it follows in each case that the numerator and the denominator will be polynomials of the *same lowest power* of $(1/N)$ or $d$. Then this $(1/N)$ or $d$ may be cancelled in the fraction. In the $N$-process

$$\lim_{N\to\infty} \rho_n(\psi_N; z) \tag{5.1}$$

does not exist for a fixed $n > n_0$. We have to go to subsequences to get convergence. In the $R$-process this is not so. For a fixed $n > n_0$ the limit polynomial (described in 4) exists. In the combination of the two processes we shall be aiming at an $R(N)$ such that

$$\lim_{N\to\infty} \rho_n(\psi_N^{R(N)}; z) \tag{5.2}$$

exists.

To give an illustration of the problem of combining the two methods, we will first consider the moments in the $N$-process. From (2.1) and (3.4) we find

$$\tilde{\mu}_m^{(N)} = (\alpha_0^2 + 2\sum_{j=1}^{I} |\alpha_j|^2 \cos m\omega_j) + \frac{1}{N} f_m(N) \tag{5.3}$$

where $f_m(N)$ is a bounded expression of $N$.

We can see the cause of the non-convergence clearly if we consider a simple signal with two frequencies $\pm\omega$ and $\alpha_{-1} = \alpha_1 = 1$, $\alpha_m = 0$ otherwise. The expression $f_m(N)$ is easily computed and we find

$$f_m(N) = \frac{\sin(2N-m-1)\omega}{\sin\omega} - \frac{\sin(m-1)\omega}{\sin\omega} - 2m\cos m\omega. \tag{5.4}$$

In this expression the term

$$\sin(2N - m - 1)\omega \tag{5.5}$$

is the "trouble-maker", being in a way the cause for the non-convergence of the families of Szegő polynomials. The expression is bounded, but if we let $N \to \infty$, we will have no limit. We need to go to subsequences to get convergence of the Szegő polynomials. A similar problem occurs in (5.3).

For the sake of simplicity we let $\alpha_0 = 0$ in the rest of the paper. The moments in the $N$-process are

$$\tilde{\mu}_m^{(N)} = (2\sum_{j=1}^{I} |\alpha_j|^2 \cos m\omega_j) + \frac{1}{N} f_m(N). \tag{5.6}$$

In the $R$-process we find the moments (4.2)

$$\begin{aligned}
\mu_m^{(R)} &= (\lim_{N\to\infty} \tilde{\mu}_m^{(N)})(1-d)^m \\
&= (1-d)^m \lim_{N\to\infty} [(2\sum_{j=1}^{I} |\alpha_j|^2 \cos m\omega_j) + \frac{1}{N} f_m(N)] \\
&= (1-d)^m 2\sum_{j=1}^{I} |\alpha_j|^2 \cos m\omega_j. 
\end{aligned} \tag{5.7}$$

The "difficult" term $f_m(N)$ is taken care of in such a way that we get convergence of the Szegő polynomials since $\lim_{N\to\infty} \frac{f_m(N)}{N} = 0$.

The moments in the $R(N)$-process will be:

$$\begin{aligned}
\mu_m^{(R(N))} &:= \tilde{\mu}_m^{(N)}(1-d)^m \\
&= (1-d)^m [(2\sum_{j=1}^{I} |\alpha_j|^2 \cos m\omega_j) + \frac{1}{N} f_m(N)],
\end{aligned} \tag{5.8}$$

where $d$ is some function of $N$ which will be explained later. Here the difficult term $f_m(N)$ is intact, and the moments consist of the parameters $R$ and $N$. We do not expect the Szegő polynomials

$$\rho_n(\psi_N^{(R(N))}; z) \tag{5.9}$$

to converge.

For an arbitrary signal it is proved that the numerator and the denominator of the family of Szegő polynomials $\rho_n(\psi^{(R)}; z)$ are of *lowest power at least*

$$d^{n-n_0}$$

in the $R$-process [12]. The proof of this theorem was based upon certain row operations and lead to the possibility of pulling out factors $d$ from $(n - n_0)$ of the rows in the determinants of the Szegő polynomials. Thus we get a Toeplitz determinant

$$\Delta_{n-1}^{(R)} = d^{n-n_0} \cdot \text{polynomial}. \tag{5.10}$$

From [12] it can be extracted that a similar result holds for the $N$- process. Following the same steps we can pull out factors $(1/N)$ from $(n-n_0)$ rows in the determinants of $\rho_n(\psi_N;z)$.

Recently it is actually proved [10] that the numerator and the denominator in $\rho_n(\psi^{(R)};z)$ are of *exact lowest* power

$$d^{n-n_0}$$

which means that when $d^{n-n_0}$ is extracted, then the constant term in the polynomial (5.10) is different from zero. The main idea of this result is from Lillevold [9], but the proof has been worked out by Njåstad and Waadeland [10].

In view of [12] it is important to choose the function $R(N)$ in such a way that the "difficult" term $f_m(N)$ can be taken care of such that we get convergence of $\rho_n(\psi_N^{(R(N))};z)$ when $N\to\infty$.

To achieve convergence of the combined process it seems to be a good idea to let the $R$-process be dominating in the sense that $R\to 1\ \ (d\to 0)$ more slowly than $(1/N)\to 0$.

Let us get back to the moments in the $R(N)$-process. They may be written

$$\begin{aligned}\mu_m^{(R(N))} &= [(2\sum_{j=1}^{I}|\alpha_j|^2\cos m\omega_j)+\frac{1}{N}f_m(N)](1-d)^m \\ &= 2\sum_{j=1}^{I}|\alpha_j|^2\cos m\omega_j-(d)\cdot 2m\sum_{j=1}^{I}|\alpha_j|^2\cos m\omega_j \\ &\quad +\frac{1}{N}f_m(N)+\text{higher ord.terms in } d \text{ and } \frac{1}{N}. \end{aligned}\tag{5.11}$$

A crucial step is to choose $R(N)$ in such a way that the term

$$(d)\cdot 2m\sum_{j=1}^{I}|\alpha_j|^2\cos m\omega_j \tag{5.12}$$

will be the term of *lowest* power in $d$ and $(1/N)$. In this case it will be dominant compared to the term $\frac{1}{N}f_m(N)$. We are thus in the situation discussed in [12]. Hence the numerator and the denominator of

$$\rho_n(\psi_N^{(R(N))};z) \tag{5.13}$$

will be

$$d^{n-n_0}\cdot\text{polynomial}+O(d^{n-n_0+1}) \tag{5.14}$$

where the polynomial in (5.13) will not contain the term $f_m(N)$. These expressions have to be in the $O$-terms. The reason why will become clearer in the proof of the later Theorem 1.

# 6 THE MAIN RESULT

We will formulate a theorem for the $R(N)$-process:

THEOREM 1. *For any* $\alpha \in (0,1)$ *and* $R(N) = 1 - \frac{1}{N^\alpha}$ ,

$$\lim_{N\to\infty} \rho_n(\psi_N^{(R(N))}; z) \text{ exists.}$$

Proof: Let us start considering the moments in the different processes.
N-process (5.6):

$$\tilde{\mu}_m^{(N)} = (2\sum_{j=1}^{I} |\alpha_j|^2 \cos m\omega_j) + \frac{1}{N} f_m(N). \tag{6.1}$$

R-process (5.7):

$$\begin{aligned} \mu_m^{(R)} &= (1-d)^m 2\sum_{j=1}^{I} |\alpha_j|^2 \cos m\omega_j. \\ &= 2\sum_{j=1}^{I} |\alpha_j|^2 \cos m\omega_j - d(2m\sum_{j=1}^{I} |\alpha_j|^2 \cos m\omega_j) + O(d^2). \end{aligned} \tag{6.2}$$

R(N)-process (5.8):

$$\begin{aligned} \mu_m^{(R(N))} &:= \tilde{\mu}_m^{(N)}(1-d)^m \\ &= (1-d)^m[(2\sum_{j=1}^{I} |\alpha_j|^2 \cos m\omega_j + \frac{1}{N} f_m(N)] \\ &= (2\sum_{j=1}^{I} |\alpha_j|^2 \cos m\omega_j) + \frac{1}{N} f_m(N) - d(2m\sum_{j=1}^{I} |\alpha_j|^2 \cos m\omega_j) \\ &\quad -d\frac{1}{N}(mf_m(N)) + O(d^2). \end{aligned} \tag{6.3}$$

In the proof of Theorem 1 in [12] an essential step was to rewrite the Toeplitz determinant in such a way that the *first* terms

$$2\sum_{j=1}^{I} |\alpha_j|^2 \cos m\omega_j \tag{6.4}$$

of the moments $\mu_m^{(R)}$ vanish from the $(n - n_0)$ first rows. This was done by row operations, and the result was possible as a consequence of the recursion formula for Tchebycheff polynomials. Then the *first* and *dominating* term in each element in the $(n - n_0)$ first rows of the rewritten Toeplitz determinant will be

$$d \cdot \text{polynomial.} \tag{6.5}$$

The polynomial in question will consist of combinations of the *second* term (without $d$) in (6.2), and we get

$$d \cdot (\text{comb. of } 2m\sum_{j=1}^{I} |\alpha_j|^2 \cos m\omega_j) \tag{6.6}$$

Hence we can pull out $d$ from $(n-n_0)$ rows in the rewritten determinant.

In the *last* $n_0$ rows the *first* and *dominating* term is the same as in (6.4). Since we know that the Toeplitz determinant is of *exact lowest* power $d^{n-n_0}$ [10] we obtain the result

$$\Delta_{n-1}^{(R)} = d^{n-n_0} \text{ (comb. of } 2m\sum_{j=1}^{I} |\alpha_j|^2 \cos m\omega_j) + O(d^{n-n_0+1}). \tag{6.7}$$

Since we know that the factor $d^{n-n_0}$ can be cancelled in the fraction

$$\lim_{R\uparrow 1}[\lim_{N\to\infty} (\psi_N^{(R)}; z)] \tag{6.8}$$

exists.

We will now discuss what happens when we use the same type of argument in the $N$-process. Since the first term of $\tilde{\mu}_m^{(N)}$ (6.1) is the same as in the $R$-process, we can use the same recursion formula and thus the *first* and *dominating* terms

$$2\sum_{j=1}^{I} |\alpha_j|^2 \cos m\omega_j \tag{6.9}$$

will vanish in the first $(n-n_0)$ rows of the Toeplitz determinant. The elements in these rows will in the rewritten form be

$$\frac{1}{N} \cdot \text{(polynomial)}. \tag{6.10}$$

The polynomial in question is combinations of the *second* term (without $1/N$) in (6.2).

We thus get

$$\frac{1}{N} \text{ [comb. of } f_m(N)]. \tag{6.11}$$

Thus we can extract $(1/N)$ in the $(n-n_0)$ first rows. The last $n_0$ rows will still have combinations of

$$2\sum_{j=1}^{I} |\alpha_j|^2 \cos m\omega_j \tag{6.12}$$

as the *first* and *dominating* term. Thus we get

$$\Delta_{n-1}^{(N)} = (\frac{1}{N^{n-n_0}})[\text{comb. of } (f_m(N) \text{ and } 2\sum_{j=1}^{I} |\alpha_j|^2 \cos m\omega_j) + O(\frac{1}{N})]. \tag{6.13}$$

That is the reason why we do not expect

$$\lim_{N\to\infty} \rho_n(\psi_N; z) \tag{6.14}$$

to converge in the $N$-process. The term $f_m(N)$ is the "trouble-maker".

In the $R(N)$-process we can use the same argument in the first part, but then it will differ from the $N$-process and the $R$-process. Hence the $(n-n_0)$ first rows in the rewritten Toeplitz determinant will have elements of the form(see (6.3))

$$\frac{1}{N}[\text{ comb. of } f_m(N)] + d(\text{ comb. of } m\sum_{j=1}^{I}|\alpha_j|^2\cos m\omega_j)$$

$$+ \quad d\frac{1}{N}(\text{ comb. of } 2mf_m(N)) + O(d^2). \tag{6.15}$$

The crucial step is to let $d$ be the *dominating* term in the expression above and therefore we choose

$$d = \frac{1}{N^\alpha}, \quad \alpha \in (0,1). \tag{6.16}$$

Thus the rewritten elements in the $(n-n_0)$ first rows of $\Delta_{n-1}^{(R(N))}$ will be

$$\frac{1}{N}[\text{ comb. of } f_m(N)] + \frac{1}{N^\alpha}(\text{ comb. of } 2m\sum_{j=1}^{I}|\alpha_j|^2\cos m\omega_j)$$

$$+ \quad O(\frac{1}{N^{2\alpha}}). \tag{6.17}$$

We extract $1/N^\alpha$ and rewrite. Then we get

$$\frac{1}{N^\alpha}[(\text{ comb. of } 2m\sum_{j=1}^{I}|\alpha_j|^2\cos m\omega_j$$

$$+ \quad (\frac{1}{N^{1-\alpha}})(\text{ comb. of } f_m(N)) + O(\frac{1}{N^\alpha})]. \tag{6.18}$$

The important thing is that the *first* and *dominating* term in each element is

$$\frac{1}{N^\alpha}(\text{ comb. of } 2m\sum_{j=1}^{I}|\alpha_j|^2\cos m\omega_j). \tag{6.19}$$

since it does not contain the "trouble-making" term $f_m(N)$.

The $n_0$ last rows will have combinations of (see (6.3))

$$2\sum_{j=1}^{I}|\alpha_j|^2\cos m\omega_j \tag{6.20}$$

as the *first* and *dominating* term.

The reason why we get $\Delta_{n-1}^{(R(N))}$ of the exact lowest power $(\frac{1}{N^{\alpha(n-n_0)}})$ is shown in the appendix (see 6.25 - 6.29). We get

$$\Delta_{n-1}^{(R(N))} = (\frac{1}{N^{\alpha(n-n_0)}})[(\text{comb. of } 2m\sum_{j=1}^{I}|\alpha_j|^2\cos m\omega_j)$$

$$+O(\frac{1}{N^\alpha})]. \tag{6.21}$$

The same arguments can be used for the other determinants in $\rho_n(\psi_N^{(R(N))}; z)$ and the we can cancel $(1/N^{\alpha(n-n_0)})$ in the fraction. Because the dominating terms do not contain $f_m(N)$, then

$$\lim_{N\to\infty} [\rho_n(\psi_N^{(R(N))}; z)] \tag{6.22}$$

exists. □

COROLLARY 1 *For any* $\alpha \in (0,1)$ *and* $R(N) = 1 - \frac{1}{N^\alpha}$,

$$\lim_{N\to\infty} \rho_n(\psi_N^{(R(N))}; z) = \lim_{R\uparrow 1}[\lim_{N\to\infty} \rho_n(\psi_N^{(R)}; z)].$$

Proof: Consider the moments (5.8) in the $R(N)$-process. We see that with $\alpha \in (0,1)$ the term

$$(\frac{1}{N^\alpha})2m\sum_{j=1}^{I} |\alpha_j|^2 \cos m\omega_j \tag{6.23}$$

will be the *lowest* power of $(1/N)$, and hence the dominating term. Then we are back to the $R$-process since the lowest power of $(1/N^\alpha)$ will be $(1/N^\alpha)^{n-n_0} = d^{n-n_0}$. □

COROLLARY 2 *For* $\alpha = 1$ *and* $R(N) = 1 - \frac{1}{N}$

$$\lim_{N\to\infty} [\rho_n(\psi_N^{(R(N))}; z)]$$

*does not normally exist.*

Proof: Consider the moments in the $R(N)$-process (5.8). We see that $\alpha = 1$ leads to the moments

$$\mu_m^{(R(N))} = 2[\sum_{j=1}^{I} |\alpha_j|^2 \cos m\omega_j] + \frac{1}{N}[f_m(N) - 2m\sum_{j=1}^{I} |\alpha_j|^2 \cos m\omega_j] + \text{higher order terms} \tag{6.24}$$

Then the Szegő polynomial will contain combinations of the $f_m(N)$-terms and the sum $\sum_{j=-I}^{I} |\alpha_j|^2 \cos m\omega_j$. Hence the limit will not exist without going to subsequences. □

REMARK Observe that because of the additional $(1/N)$-terms from the sum we are *not* back to the $N$-process.

COROLLARY 3 *For any* $\alpha > 1$ *and* $R(N) = 1 - \frac{1}{N^\alpha}$

$$\lim_{N\to\infty} [\rho_n(\psi_N^{(R(N))})]$$

*does not normally exist. If we go to subsequences of* $\{N\}$ *then the limit exists and furthermore*

$$\lim_{N\to\infty}[\rho_n(\psi_N^{(R(N))})] = \lim_{N\to\infty}[\rho_n(\psi_N; z)].$$

Proof: Consider the moments in the $R(N)$-process (5.8). We notice that with $\alpha > 1$ the term $(1/N)f_m(N)$ will be the term of the *lowest* power of $(1/N)$, and hence the dominating term. Then we are back to the $N$-process. □

REMARK In [9] there are numerical examples with special signals illustrating the results in Theorem 1, and the Corollaries 1-3.

APPENDIX We include here the proof [10] that numerator and denominator of $\rho_n(\psi^{(R)}; z), n \geq n_0$ are of exact lowest power $(n - n_0)$ in $d$.
Here $d = 1 - R$, $0 < d < 1$.

It suffices to prove it for the denominator, i.e. for the Toeplitz determinant $\Delta_{n-1}(\psi^{(R)})$ with elements $\mu_k(1-d)^{|k|}$. Keep in mind that the sequence $\{\mu_n\}$ is $n_0$-definite, and hence that $\{\mu_n(1-d)^{|n|}\}$ is positive definite.

We will need the following well known connection between inner product and Toeplitz determinant

$$\langle \rho_n^*(\psi^{(R)}; z), 1\rangle = \frac{\Delta_n(\psi^{(R)})}{\Delta_{n-1}(\psi^{(R)})}. \tag{6.25}$$

Moreover we have

$$\begin{aligned}\langle \rho_n^*(\psi^{(R)}; z), 1\rangle &= \langle \rho_n(\psi^{(R)}; z), \rho_n\psi^{(R)}; z)\rangle \\ &= \| \rho_n(\psi^{(R)}; z) \|^2 .\end{aligned} \tag{6.26}$$

The key to the proof is the following [9]

$$\| \rho_n(\psi^{(R)}; z) \|^2 \geq \frac{\mu_0}{2} d. \tag{6.27}$$

Application to our Toeplitz determinants:
We know that for $n \leq n_0$

$$\lim_{d\to 0} \Delta_{n-1}(\psi^{(R)}) > 0. \tag{6.28}$$

For $n \geq n_0$ we have

$$\begin{aligned}\Delta_{n-1}(\psi^{(R)}) &= \Delta_{n_0-1}(\psi^{(R)}) \cdot \langle \rho_{n_0}^*(\psi^{(R)}; z), 1\rangle \cdots \langle \rho_{n-1}^*(\psi^{(R)}; z), 1\rangle \\ &> K \cdot d^{n-n_0},\end{aligned} \tag{6.29}$$

where $K$ is a positive constant. □

# REFERENCES

[1] N.I. Akhiezer, "The classical moment problem and some related questions in analysis", Hafner, New York, 1965.

[2] W.B. Jones and O. Njåstad, Applications of Szegő Polynomials to Digital Signal Processing, *Rocky Mountain Journal of Mathematics*, 21, (1992), 387-436.

[3] W.B. Jones, O. Njåstad and E.B. Saff, Szegő Polynomials Associated with Wiener-Levinsons Filters, *Journal of Computational and Applied Mathematics*, 32, (1990), 387-406.

[4] W.B. Jones, O. Njåstad, W.J. Thron and H. Waadeland, Szegő Polynomials Applied to Frequency Analysis, *Journal of Computational and Applied Mathematics*, 46, (1993), 217-228.

[5] W.B. Jones, O. Njåstad and H. Waadeland, Asymptotics for Szegő Polynomial Zeros, *Numerical Algorithms*, 3, (1992), 255-264.

[6] W.B. Jones, O. Njåstad and H. Waadeland, Applications of Szegő Polynomials to Frequency Analysis, *SIAM Journal of Mathematical Analysis*, Vol. 25, No.2, (1994), 491-512.

[7] W.B. Jones, O. Njåstad and H. Waadeland, An Alternative Way of Using Szegő Polynomials in Frequency Analysis, in: *Continued Fractions and Orthogonal Functions, theory and applications*, Marcel Dekker Inc., 141-151, (1994).

[8] N. Levinson, The Wiener RMS (root mean square) Error Criterion in Filter Design and Prediction, *Journal of Mathematics and Physics*, 25, (1947), 261-278.

[9] F.J. Lillevold, "Grenseprosesser med Szegő polynomer av fiksert grad", Master thesis, Norwegian University of Science and Technology, (1996).

[10] F.J. Lillevold, O. Njåstad, H. Waadeland, Personal communication, (1996).

[11] K. Pan and E.B. Saff, Asymptotics for Zeros of Szegő Polynomials Associated with Trigonometric Polynomial Signals, *Journal of Approximation Theory*, 71, (1992), 239-251.

[12] V. Petersen, A theorem on Toeplitz determinants with elements containing Tchebycheff polynomials of the first kind, *The Royal Norwegian Society of Sciences and Letters, Transactions*, 4, (1996)

[13] O. Toeplitz, *Über die Fourierische Entwicklung positiver Funktionen*, Rend. Circ. Mat. Palermo, 32, (1911), 191-192.

[14] N. Wiener, *Extrapolation, Interpolation and Smoothing of Stationary Time Series*, MIT Press, Cambridge, MA/Wiley, New York, 1949.

# Zeros of Szegö Polynomials Used in Frequency Analysis

VIGDIS PETERSEN Sør-Trøndelag College, School of Teacher Education, Rotvoll Allé, N-7005, Trondheim, Norway

Recently there has been established a method for determining unknown frequencies in a trigonometric signal by using families of Szegő polynomials. The Szegő polynomials are associated with a signal for which we have reason to believe that a proper mathematical model is of the form

$$G(t) = \sum_{j=-I}^{I} \alpha_j e^{i2\pi f_j t}, \tag{1.1}$$

where $t$ denotes time.

$$f_j = -f_{-j} > 0, \quad f_0 = 0, \quad f_j \quad \text{are the frequencies,} \tag{1.2}$$

$$0 \neq \alpha_j = \overline{\alpha}_{-j} \in \mathbf{C}, \quad \alpha_0 \geq 0, \ |\alpha_j| \ \text{are the amplitudes,} \tag{1.3}$$

for $j = 0, 1, 2, \ldots, I < \infty$.

The set of $N$ observations are made at times

$$t_m = m \cdot \Delta t, \quad \text{where } m = 0, 1, 2, \ldots, N-1, \tag{1.4}$$

where $\Delta t$ is chosen such that we have reason to believe that

$$2\pi f_j \cdot \Delta t < \pi \quad \text{for} \quad j = 1, 2, \ldots, I. \tag{1.5}$$

For the same frequencies we define

$$\omega_j = 2\pi f_j \cdot \Delta t. \tag{1.6}$$

The numbers $\omega_j$ are sometimes called *normalized* frequencies. For the sake of simplicity we shall call them frequencies.

[1]This work was supported by The Nansen Fund and Affiliated Funds, The Norwegian Academy of Science and Letters.

Furthermore we let

$$x_N(m),\ m = 0,1,2\ldots,N-1 \tag{1.7}$$

denote the value of the signal at the time $m \cdot \Delta t$. For all other values of $m$ we define $x_N(m) = 0$.

Our observations are thus given by

$$x_N(m) = \begin{cases} \sum_{j=-I}^{I} \alpha_j e^{i\omega_j m} & \text{if } m = 0,1,2,\ldots,N-1, \\ 0 & \text{otherwise,} \end{cases} \tag{1.8}$$

where

$\omega_{-j} = -\omega_j \in \mathbf{R},\ j = 1,2,\ldots,I$ and $1 \leq N < \infty$.

The frequencies are ordered such that

$$0 = \omega_0 < \omega_1 < \cdots < \omega_I < I. \tag{1.9}$$

We have assumed that there is no noise and that $x_N(0) \neq 0$. In this discrete form the frequency analysis problem will be to determine $\omega_j$.

The start of the method is to construct, from the observations, a positive Borel measure with infinite support. For any fixed $N$ this measure gives rise to a double sequence of moments which can be computed as autocorrelation coefficients of the signal:

$$\mu_m^{(N)} = \sum_{k=0}^{N-m-1} x_N(k)x_N(k+m),\ \ 1 < N < \infty,\ m = 0,1,2,\ldots\ , \tag{1.10}$$

$$\mu_{-m}^{(N)} = \mu_m^{(N)}\ . \tag{1.11}$$

This sequence is hermitian positive definite, meaning that $\mu_{-m}^{(N)} = \bar{\mu}_m^{(N)}$, and that certain Toeplitz-determinants are positive. In our case the moments are, in fact, real.

Furthermore, the measure gives rise to an inner product and hence to a sequence of monic orthogonal polynomials $\rho_n(\psi_N; z)$, which are the Szegő polynomials in question. The zeros of these polynomials are crucial in the method since the unknown frequencies are determined from asymptotic properties of these zeros. It is therefore important that the zeros can be computed easily, which is in fact the case.

From a practical point of view the Levinson algorithm [3] is by far the best method. Another method is connected with continued fractions where the Szegő polynomials are the denominators of the odd order approximants of the corresponding positive Perron-Carathéodory continued fraction (PPC-fraction) which was introduced by Jones, Njåstad, Thron in 1986. A continued fraction of the form

$$\delta_0 - \frac{2\delta_0}{1} \, + \frac{1}{\overline{\delta_1} z} \, + \frac{(1-|\delta_1|^2)z}{\delta_1} \, + \frac{1}{\overline{\delta_2} z} \, + \frac{(1-|\delta_2|^2)z}{\delta_2} \, + \cdots, \tag{1.12}$$

is called a PPC-fraction if

$$\delta_0 > 0 \text{ and } |\delta_k| < 1 \text{ for } k = 1,2,3,\ldots\ . \tag{1.13}$$

The elements $\delta_k$ are called the reflection coefficients. They may be computed in different ways, for example by Levinson's algorithm or determinant formulas [2].

A third method is by explicit determinant formulas. From a *numerical* point of view the determinants are of little use, but in the present paper they will play an important role, especially the formula for the reflection coefficients. The formulas are:

For the Toeplitz determinant:

$$\Delta_n = \begin{vmatrix} \mu_0 & \mu_{-1} & \mu_{-2} & \cdots & \mu_{-n} \\ \mu_1 & \mu_0 & \mu_{-1} & \cdots & \mu_{-n+1} \\ \mu_2 & \mu_1 & \mu_0 & \cdots & \mu_{-n+2} \\ \vdots & \vdots & \vdots & & \vdots \\ \mu_n & \mu_{n-1} & \mu_{n-2} & \cdots & \mu_0 \end{vmatrix}. \tag{1.14}$$

For the reflection coefficients:

$$\delta_n = \frac{(-1)^n}{\Delta_{n-1}} \begin{vmatrix} \mu_{-1} & \mu_{-2} & \cdots & \mu_{-n} \\ \mu_0 & \mu_{-1} & \cdots & \mu_{-n+1} \\ \vdots & \vdots & & \vdots \\ \mu_{n-2} & \mu_{n-3} & \cdots & \mu_{-1} \end{vmatrix}. \tag{1.15}$$

For the Szegő polynomials:

$$\rho_n(\psi; z) = \frac{1}{\Delta_{n-1}} \begin{vmatrix} \mu_0 & \mu_{-1} & \cdots & \mu_{-n} \\ \mu_1 & \mu_0 & \cdots & \mu_{-n+1} \\ \vdots & \vdots & & \vdots \\ \mu_{n-1} & \mu_{n-2} & \cdots & \mu_{-1} \\ 1 & z & \cdots & z^n \end{vmatrix}. \tag{1.16}$$

The unknown frequencies can be determined as follows:

Let $n_0$ be the number of unknown frequencies such that

$$n_0 = 2I + L \text{ where } L = \begin{cases} 0 & \text{if } \alpha_0 = 0, \\ 1 & \text{otherwise.} \end{cases} \tag{1.17}$$

If the value of $n_0$ is known, then the following holds uniformly on compact subsets of $\mathbf{C}$ (see [1], Lemma 3.2.C).

$$\lim_{N\to\infty} \rho_{n_0}(\psi_N; z) = (z-1)^L \prod_{j=1}^{I} \left(z - e^{i\omega_j}\right)\left(z - e^{-i\omega_j}\right). \tag{1.18}$$

This solves the frequency analysis problem when $n_0$ is known. Usually this is not the case. If $n_0$ is unknown, then we take an $n$ for which we have reason to believe that $n > n_0$. Then the sequence $\{\rho_n(\psi_N; z)\}_{N=1}^{\infty}$ cannot be expected to converge. For any fixed $n$ this sequence is uniformly bounded on $|z| \le 1$, and hence normal on $|z| < 1$. Then by going to subsequences of $\{N\}$ we can make it converge.

Important is that regardless of which convergent sequence of Szegő polynomials we have, the limit polynomial will always have the polynomial (1.18) as a factor. This implies, for instance by using Hurwitz' theorem that $n_0$ of the zeros will tend to the points $e^{\pm i\omega_j}$ and possibly 1 (for $L = 1$).

If we have a subsequence of $\{N\}$ such that the Szegő polynomials converge, then we have an additional $(n - n_0)$ zeros of the limit polynomial. These zeros will normally depend upon the subsequences chosen. These results have been shown independently by Jones, Njåstad, Thron, Waadeland [1] and Pan, Saff [6].

The method outlined above is called the N-process after the parameter $N$ involved. A later version which contains one additional parameter $R$, is called the R-process.

A disadvantage of the N-process is that we have to go to subsequences to obtain convergence of the Szegő polynomials. To avoid this problem we replace $\tilde{\mu}_m^{(N)} := \mu_m^{(N)}/N$ from the N-process with $\left\{\tilde{\mu}_m^{(N)} R^{|m|}\right\}$ where $R \in (0,1)$ and $d = 1 - R$.

We will again have a positive definite hermitian sequence. This leads to a measure $\psi_N^{(R)}$ and the Szegő polynomials $\rho_n\left(\psi_N^{(R)}; z\right)$. It can be proved that this implies existence of the limit

$$\lim_{R\uparrow 1}\left[\lim_{N\to\infty} \rho_n(\psi_N^{(R)}; z)\right] = \tilde{\rho}_n(z). \tag{1.19}$$

For $n > n_0$ the polynomial (1.19) is of the form

$$\tilde{\rho}_n(z) = (z-1)^L \prod_{j=1}^{I}\left(z - e^{i\omega_j}\right)\left(z - e^{-i\omega_j}\right) \prod_{p=1}^{n-n_0}\left(z - z_p^{(n)}\right). \tag{1.20}$$

From [1] it is known that the moments in the R-process are

$$\begin{aligned} \mu_m^{(R)} &:= \left(\lim_{N\to\infty} \tilde{\mu}_m^{(N)}\right)(1-d)^m \\ &= \left(\sum_{j=-I}^{I} |\alpha_j|^2 e^{im\omega_j}\right)(1-d)^m. \end{aligned} \tag{1.21}$$

The $(n - n_0)$ last zeros are sometimes called the "uninteresting" zeros. The R-process is described by Jones, Njåstad, Waadeland [2].

It is important to have information about the additional $(n - n_0)$ zeros of the Szegő polynomial. In the N-process these zeros will normally depend upon the subsequences chosen, but since the zeros of the Szegő polynomials are always located in $|z| < 1$ , those zeros have to be

$$\left|z_p^{(n)}\right| \leq 1. \tag{1.22}$$

From [6] it can be extracted that to any $n > n_0$ there exists a number

$$K_n \in (0,1) \tag{1.23}$$

depending only upon $n$ and *the given signal* such that

$$\left|z_p^{(n)}\right| \leq K_n \tag{1.24}$$

for all "uninteresting" zeros $z_p^{(n)}$. This knowledge, that the "uninteresting" zeros stay away from the unit circle, is very useful in the process of determining the unknown frequencies. Furthermore we know from [4] that the same knowledge about the $(n-n_0)$ additional zeros of the Szegő polynomials holds in the R-process.

# 2 A SIMPLE SIGNAL

In this section we want to see what may happen to the values of the "uninteresting" zeros if we let $n \to \infty$. Then we will have an infinite number of "uninteresting" zeros for a *fixed* $n_0$. Will there exist a $K_n \in (0,1)$ such that we can *really* separate the additional $(n - n_0)$ zeros from the zeros tending to the frequency points?

To find out what may happen we want to consider a simple signal with two frequencies $\pm\omega$. Without loss of generality we let

$$|\alpha_{-1}| = |\alpha_1| = \frac{1}{\sqrt{2}}, \quad \alpha_j = 0 \text{ otherwise.} \tag{2.1}$$

We will use the R-process. Then the moments may be computed from (1.21)

$$\begin{aligned} \mu_m^{(R)} &= \left(\sum_{j=-I}^{I} |\alpha_j|^2 e^{i\omega_j m}\right)(1-d)^m \\ &= \frac{1}{2}(e^{i\omega m} + e^{-i\omega m})(1-d)^m \\ &= \cos m\omega(1-d)^m. \end{aligned} \tag{2.2}$$

In the later proof we will use the Tchebycheff polynomials of the first kind, $T_m(x)$. The moments may then be written

$$\mu_m^{(R)} = T_m(x)(1-d)^m, \quad \mu_m^{(R)} = \mu_{-m}^{(R)} \tag{2.3}$$

where $T_m(x) = \cos m\omega$, $x = \cos\omega$. In the rest of the paper we use a short version $T_m := T_m(x)$.

A crucial idea in the following part is to use the determinant formula for the reflection coefficients (1.15). We will choose the parameter $d$ as superscript instead of $R$ since we are dealing with $d$. We know that the reflection coefficient $\delta_n^{(d)}$ is the product of the $n$ zeros in the Szegő polynomial of degree $n$ except for the sign. In the following part we will neglect the sign.

The denominator is the Toeplitz determinant $\Delta_{n-1}^{(d)}$ (1.14). In both cases $\mu_m$ is replaced by $\mu_m^{(R)}$. We are aiming at an expression for the numerator and denominator for the reflection coefficients.

The numerator determinant which we will denote by $K_{n-1}^{(d)}$ will be:

$$K_{n-1}^{(d)} = \begin{vmatrix} T_1(1-d) & T_2(1-d)^2 & \dots & T_n(1-d)^n \\ T_0 & T_1(1-d) & \dots & T_{n-1}(1-d)^{n-1} \\ T_1(1-d) & T_0 & \dots & T_{n-2}(1-d)^{n-2} \\ \vdots & \vdots & & \vdots \\ T_{n-2}(1-d)^{n-2} & T_{n-3}(1-d)^{n-3} & \dots & T_1(1-d) \end{vmatrix}. \tag{2.4}$$

In the following part the *recursion* formula for the Tchebycheff polynomials of the first kind will play an important role:

$$T_m = 2xT_{m-1} - T_{m-2}, \quad m = 2, 3, 4, \dots . \tag{2.5}$$

LEMMA 1 *For* $n \geq 3$

$$K_{n-1}^{(d)} = d(1-d)(2-d)(-x)K_{n-2}^{(d)}$$

*where*

$$K_1^{(d)} = (1-d)^2(1-x^2).$$

Proof: Replace Row(1) in (2.4) with

$$\text{Row}(1) - 2x(1-d)\text{Row}(2) + (1-d)^2\text{Row}(3).$$

Then we have the possibilities
(i) $T_m(1-d)^m - 2xT_{m-1}(1-d)^m + T_{m-2}(1-d)^m = 0$ for $m = 2, 3, \ldots, n$,
(ii) $T_1(1-d) - 2xT_0(1-d) + T_1(1-d)^3 = d(1-d)(2-d)(-x)$.
We get a new determinant where the first element in the first row is

$$d(1-d)(2-d)(-x).$$

All the other elements in the first row are zero. Expanding the determinant by the first row, we get Lemma 1, where

$$K_1^{(d)} = [T_1(1-d)]^2 - T_0T_2(1-d)^2 = (1-d)^2(1-x^2).$$

□

PROPOSITION 1 *For* $n \geq 3$

$$K_{n-1}^{(d)} = d^{n-2}(1-d)^n(2-d)^{n-2}(-x)^{n-2}(1-x^2).$$

Proof: By Lemma 1 we get for $n \geq 3$

$$\begin{aligned} K_{n-1}^{(d)} &= [d(1-d)(2-d)(-x)]^{n-2}(1-d)^2(1-x^2) \\ &= d^{n-2}(1-d)^n(2-d)^{n-2}(-x)^{n-2}(1-x^2). \end{aligned}$$

□

COROLLARY 1 *For* $n \geq 2$ *we have*

$$k_{n-1} := \lim_{d \to 0} \frac{K_{n-1}^{(d)}}{d^{n-2}(1-x^2)} = (-2x)^{n-2}. \tag{2.6}$$

Proof: Use proposition 1 and the limit is obtained. □

Now we will consider the denominator determinant, $\Delta_{n-1}^{(d)}$ (1.14). It will be

$$\Delta_{n-1}^{(d)} = \begin{vmatrix} T_0 & T_1(1-d) & \ldots & T_{n-1}(1-d)^{n-1} \\ T_1(1-d) & T_0 & \ldots & T_{n-2}(1-d)^{n-2} \\ T_2(1-d)^2 & T_1(1-d) & \ldots & T_{n-3}(1-d)^{n-3} \\ \vdots & \vdots & & \vdots \\ T_{n-1}(1-d)^{n-1} & T_{n-2}(1-d)^{n-2} & \ldots & T_0 \end{vmatrix}. \tag{2.7}$$

LEMMA 2 *For* $n \geq 3$

$$\Delta_{n-1}^{(d)} = d(2-d)(2-2d+d^2)\Delta_{n-2}^{(d)} - d^2(1-d)^2(2-d)^2x^2\Delta_{n-3}^{(d)} \tag{2.8}$$

*where*

$$\Delta_0^{(d)} = 1, \ \ \Delta_1^{(d)} = 1 - x^2(1-d)^2. \tag{2.9}$$

Proof: Follow the same step as in the proof of Lemma 1. Replace Row(1) with

$$\text{Row}(1) - 2x(1-d)\text{Row}(2) + (1-d)^2\text{Row}(3).$$

Then we get (i) and (ii) as in Lemma 1. In addition we get

$$\begin{aligned} \text{(iii)} \qquad & T_0 - 2xT_1(1-d)^2 + T_2(1-d)^4 \\ = \ & (4-4x^2)d + (10x^2-6)d^2 + (4-8x^2)d^3 + (2x^2-1)d^4 \\ := \ & f(d). \end{aligned}$$

We get a new determinant where the first element in the first row is (iii), the second is (ii) and all the other elements are zero.

$$\begin{vmatrix} f(d) & d(1-d)(2-d)(-x) & \cdots & 0 \\ T_1(1-d) & T_0 & \cdots & T_{n-2}(1-d)^{n-2} \\ T_2(1-d)^2 & T_1(1-d) & \cdots & T_{n-3}(1-d)^{n-3} \\ \vdots & \vdots & & \vdots \\ T_{n-1}(1-d)^{n-1} & T_{n-2}(1-d)^{n-2} & \cdots & T_0 \end{vmatrix}. \tag{2.10}$$

Now we replace Column(1) with

$$\text{Column}(1) - 2x(1-d)\text{Column}(2) + (1-d)^2\text{Column}(3)$$

Then we get (i) and (ii) as in Lemma 1. From the first row we get

$$\begin{aligned} \text{(iv)} \qquad & [(4-4x^2)d + (10x^2-6)d^2 + (4-8x^2)d^3 + (2x^2-1)d^4] \\ & -2x(1-d)[d(1-d)(2-d)(-x)] \\ = \ & d(2-d)(2-2d+d^2). \end{aligned}$$

Then the determinant (2.10) may be written

$$\begin{vmatrix} d(2-d)(2-2d+d^2) & d(1-d)(2-d)(-x) & \cdots & 0 \\ d(1-d)(2-d)(-x) & T_0 & \cdots & T_{n-2}(1-d)^{n-2} \\ 0 & T_1(1-d) & \cdots & T_{n-3}(1-d)^{n-3} \\ \vdots & \vdots & & \vdots \\ 0 & T_{n-2}(1-d)^{n-2} & \cdots & T_0 \end{vmatrix}. \tag{2.11}$$

Expanding the determinant by the first row (or the first column) we get Lemma 2, where

$$\Delta_1^{(d)} = T_0^2 - [T_1(1-d)]^2 = 1 - x^2(1-d)^2.$$

□

COROLLARY 2 *For $n \geq 2$ the limit*

$$t_{n-1} := \lim_{d \to 0} \frac{\Delta_{n-1}^{(d)}}{d^{n-2}(1-x^2)} \tag{2.12}$$

*exists.*

*For $n \geq 4$ the recurrence relation*

$$t_{n-1} = 4t_{n-2} - 4x^2 t_{n-3} \tag{2.13}$$

*holds with the initial conditions $t_1 = 1$, $t_2 = 4$.*

Proof: From (2.8) and (2.9) we get

$$\Delta_1^{(d)} = (1-x^2) + O(d),$$

$$\Delta_2^{(d)} = 4d(1-x^2) + O(d^2).$$

Then it follows immediatly that the limit (2.12) exixts for $n = 2$ and $n = 3$ and $t_1 = 1$, $t_2 = 4$. Let $m \geq 4$ be such that the limit (2.12) exixts for all $n \leq m-1$.

In the recurrence relation (2.8) we let $n = m$ and divide by $d^{m-2}(1-x^2)$. Then the right hand side tends to

$$4t_{m-2} - 4x^2 d t_{m-3}$$

when $d \to 0$.

Hence the left hand side of (2.8) which is

$$\frac{\Delta_{m-1}^{(d)}}{d^{m-2}(1-x^2)}$$

will tend to a limit $t_{m-1}$. This proves (2.12) and (2.13) by induction. □

REMARK From (2.8) it follows directly by induction that $\Delta_{n-1}^{(d)}$ has $d^{n-2}$ as a factor. This illustrates a result on Toeplitz determinants [8], [5].

For $n \geq 2$ we get by solving the recurrence relation (2.13) by standard methods

$$t_{n-1} = \frac{1}{4\sqrt{1-x^2}}[(2+2\sqrt{1-x^2})^{n-1} - (2-2\sqrt{1-x^2})^{n-1}]. \tag{2.14}$$

COROLLARY 3 *For $n \geq 3$*

$$\begin{aligned} \lim_{d \to 0} |\delta_n^{(d)}| &= \lim_{d \to 0} \left|\frac{K_{n-1}^{(d)}}{\Delta_{n-1}^{(d)}}\right| \\ &= \frac{|2x^{n-2}\sqrt{1-x^2}|}{|(1+\sqrt{1-x^2})^{n-1} - (1-\sqrt{1-x^2})^{n-1}|}. \end{aligned} \tag{2.15}$$

Proof: We know from (2.6) that

$$\lim_{d \to 0} \frac{K_{n-1}^{(d)}}{d^{n-2}(1-x^2)} = (-2x)^{n-2}.$$

From (2.12) and (2.14) we know that

$$\begin{aligned}\lim_{d\to 0}\frac{\Delta_{n-1}^{(d)}}{d^{n-2}(1-x^2)} &= t_{n-1}\\ &= \frac{1}{4\sqrt{1-x^2}}[(2+2\sqrt{1-x^2})^{n-1}-(2-2\sqrt{1-x^2})^{n-1}].\end{aligned}$$

From these two equations we get Corollary 3. □

To obtain more information about the $(n-2)$ "uninteresting" zeros we want to find the absolute value of the geometric mean of the limit of the product of those zeros. We know that the product of the two zeros tending to the frequency points is

$$\lim_{d\to 0} e^{i\omega}e^{-i\omega} = 1.$$

We further know that the geometric mean has the property that *at least one* of the factors has an absolute value as large as the absolute value of the geometric mean. We thus want to take a look at

$$\lim_{n\to\infty}(\lim_{d\to 0}|\delta_n^{(d)}|)^{\frac{1}{n-2}}.$$

Then we have neglected the two zeros tending to the frequency points since we know this limit to be 1.

PROPOSITION 2 *For* $|x|<1$ *we have*

$$\lim_{n\to\infty}(\lim_{d\to 0}|\delta_n^{(d)}|)^{\frac{1}{n-2}} = \frac{|x|}{|1+\sqrt{1-x^2}|} = \frac{|\cos\omega|}{|1+\sin\omega|}. \tag{2.16}$$

Proof: By (2.15) we have

$$\begin{aligned}\lim_{n\to\infty}|\delta_n^{(d)}|^{\frac{1}{n-2}} &= \lim_{n\to\infty}\frac{|2x^{n-2}\sqrt{1-x^2}|^{\frac{1}{n-2}}}{(1+\sqrt{1-x^2})^{\frac{n-1}{n-2}}(1-(\frac{1-\sqrt{1-x^2}}{1+\sqrt{1-x^2}})^{n-1})^{\frac{1}{n-2}}}\\ &= \frac{|x|}{1+\sqrt{1-x^2}}\\ &= \frac{|\cos\omega|}{1+\sin\omega}.\end{aligned}$$

□

REMARK From (2.16) follows

$$\lim_{\omega\to 0}\left[\lim_{n\to\infty}\left(\lim_{d\to 0}\left|\delta_n^{(d)}\right|\right)^{\frac{1}{n-2}}\right] = \lim_{\omega\to 0}\frac{|\cos\omega|}{|1+\sin\omega|} = 1.$$

This implies that the limit of the geometric mean of the absolute value of the zeros in the $R$-process can come arbitrarily close to 1. This has as a consequence that there are "uninteresting" zeros of absolute value arbitrarily close to 1. This is shown for signals with two frequencies and hence it is true in the set of all trigonometric signals. From this we get the not surprising observation:

OBSERVATION In the class of all trigonometric signals there are "uninteresting" zeros of Szegő polynomials arbitrarily close to 1 in absolute value. This means that there is no $K \in (0,1)$ such that all possible "uninteresting" zeros are located in

$$|z| \leq K.$$

With $K_n$ defined as in (1.23) and (1.24) this means that

$$\sup K_n = 1.$$

For specific signals or families of signals we may of course have bounds properly less than 1. One extreme case is when $\omega = \frac{\pi}{2}$ in the signal with two frequencies. In this case all "uninteresting" zeros are in the origin, as proved in [7].

# REFERENCES


[1] W.B. Jones, O. Njåstad, W.J. Thron and H. Waadeland, Szegő Polynomials Applied to Frequency Analysis, *Journal of Computational and Applied Mathematics*, 46, (1993), 217-228.

[2] W.B. Jones, O. Njåstad and H. Waadeland, An Alternative Way of Using Szegő Polynomials in Frequency Analysis, in: *Continued Fractions and Orthogonal Functions, theory and applications*, Marcel Dekker Inc., 141-151, (1994).

[3] N. Levinson, The Wiener RMS (root mean square) Error Criterion in Filter Design and Prediction, *Journal of Mathematics and Physics*, 25, (1947), 261-278.

[4] X. Li, Asymptotics of columns in the table of orthogonal polynomials with varying measures, in: M. Alfaro, A. Garcia, C. Jagels and F. Marcellan, eds., *Orthogonal Polynomials on the Unit Circle: Theory and Applications*, Universidad Carlos III de Madrid, (1994), 79-95.

[5] F.J. Lillevold, O. Njåstad, H. Waadeland, Personal communication, (1996).

[6] K. Pan and E.B. Saff, Asymptotics for Zeros of Szegő Polynomials Associated with Trigonometric Polynomial Signals, *Journal of Approximation Theory*, 71, (1992), 239-251.

[7] V. Petersen, *Bruk av Szegő - polynomer i frekvensanalyse*, Thesis for the degree cand. scient, University of Trondheim, Norway, 1995.

[8] V. Petersen, A theorem on Toeplitz-determinants with elements containing Tchebycheff-polynomials of the first kind, *The Royal Norwegian Society of Sciences and Letters, Transactions*, 4, (1996)

## Some Probabilistic Remarks on the Boundary Version of Worpitzky's Theorem

HAAKON WAADELAND, Department of Mathematical Sciences, Norwegian University of Science and Technology, N- 7034 Trondheim, Norway

The old and celebrated theorem of Worpitzky states that all continued fractions

$$\underset{n=1}{\overset{\infty}{\mathrm{K}}}(a_n/1), \tag{1}$$

where

$$|a_n| \leq \frac{1}{4}, \quad n = 1, 2, 3, \ldots \tag{2}$$

converge [4]. Moreover, it can easily be proved that all approximants are in the closed disk

$$|w| \leq \frac{1}{2}. \tag{3}$$

It is well known and easy to prove, that for any positive $\rho \leq \frac{1}{2}$, the above statement holds if (2) and (3) are replaced by, respectively

$$|a_n| \leq \rho(1-\rho) \tag{2'}$$

and

$$|w| \leq \rho. \tag{3'}$$

It is also well known, that any point in the disk (3) (or (3')), except $w = 0$, is the value of some continued fraction satisfying (2) (or (2')). Keep in mind that the definition itself of a continued fraction excludes $a_n = 0$.

In the paper [3] was raised the question about the set of values if the condition (2') is replaced by

$$|a_n| = \rho(1-\rho), \tag{4'}$$

and the answer was that the set of values was the annulus

$$\rho\frac{1-\rho}{1+\rho} \leq |w| \leq \rho. \qquad (5')$$

For $\rho = \frac{1}{2}$ the condition (4') reduces to

$$|a_n| = \frac{1}{4}, \qquad (4)$$

and the annulus (5') to

$$\frac{1}{6} \leq |w| \leq \frac{1}{2}. \qquad (5)$$

In the rest of the paper we shall restrict ourselves to the case when $\rho = \frac{1}{2}$. It is, however, not difficult to carry out the discussion for general $\rho$.

The discussion to be presented in this note was initiated by a special computer graphics observation: An attempt to illustrate the above annulus result produced on the screen an annulus

$$r_o \leq |w| \leq R_o, \qquad (6)$$

where $R_o$ was substantially smaller than .5 (actually closer to .45) even for a number of continued fraction values as large as $5 \cdot 10^7$. A similar effect was not visible on the screen for the smaller radius $\frac{1}{6}$. In the computer experiment the continued fraction values were replaced by approximants of high order, high enough to make the truncation error negligible in view of the resolution of the screen.

It does not take much heuristic reasoning to expect decreased probability for values near the boundary of the annulus, but the observation made here at the larger circle was still a bit surprising. The purpose of the present note is to throw some light on this observation.

We shall in the following assume that for every fixed $n$ the parameter $a_n$ is arbitrarily picked on the circle $|w| = \frac{1}{4}$ , such that

$$a_n = \frac{1}{4} \cdot e^{i\phi},$$

where $\phi$ is uniformly distributed. For the continued fraction values this gives rise to a probability measure $\mu$ , which is invariant with respect to rotations around the origin. In our discussion we shall take for granted that the measure $\mu$ is absolutely continuous with respect to Lebesgue measure in $R^2$ . We shall even assume that the Radon-Nikodym derivative, i. e. the probability density $H(r,\phi)$, is continuous. The present note does not include any *proof* of these assumptions. The results of the discussion are only claimed to be true under the condition that the assumptions mentioned hold.

The probability density $H(r,\phi)$ has the annulus (5) as its support, but can be described by the probability density $h(r)$ on the radius, with $[1/6, 1/2]$ as its support:

$$H(r,\phi) = \frac{h(r)}{2r\pi} \quad , \iint_{(5)} H(r,\phi) r dr d\phi = \int_{\frac{1}{6}}^{\frac{1}{2}} h(r) dr = 1.$$

Probabilistic discussions in the analytic theory of continued fractions are rare. One example is [**2**]. See also the book [**1**].

Since truncation error estimates in many cases are based on knowledge of value regions, the observation in the present paper may lead to probabilistic truncation error bounds.

Let $f$ be the value of an arbitrary continued fraction in the family we are studying, and let the $a_1$–value be

$$a_1 = \frac{1}{4} \cdot e^{i\phi}. \tag{7}$$

Then, with

$$f = \frac{\frac{1}{4} \cdot e^{i\phi}}{1+g}, \tag{8}$$

$g$ is also the value of a continued fraction in the same family. Actually, with $\phi$ distributed uniformly on $[0, 2\pi)$ , $f$ and $g$ will have the same distribution on the same set (5) , hereafter called $L$. We have in particular

$$L = \left\{ \frac{\frac{1}{4} \cdot e^{i\phi}}{1+L}; \quad 0 \leq \phi < 2\pi \right\}. \tag{9}$$

The set

$$\frac{\frac{1}{4}}{1+L}$$

consists of all points on and between the circles

$$\left| w - \frac{9}{35} \right| = \frac{3}{70} \quad , \quad \left| w - \frac{1}{3} \right| = \frac{1}{6}, \tag{10}$$

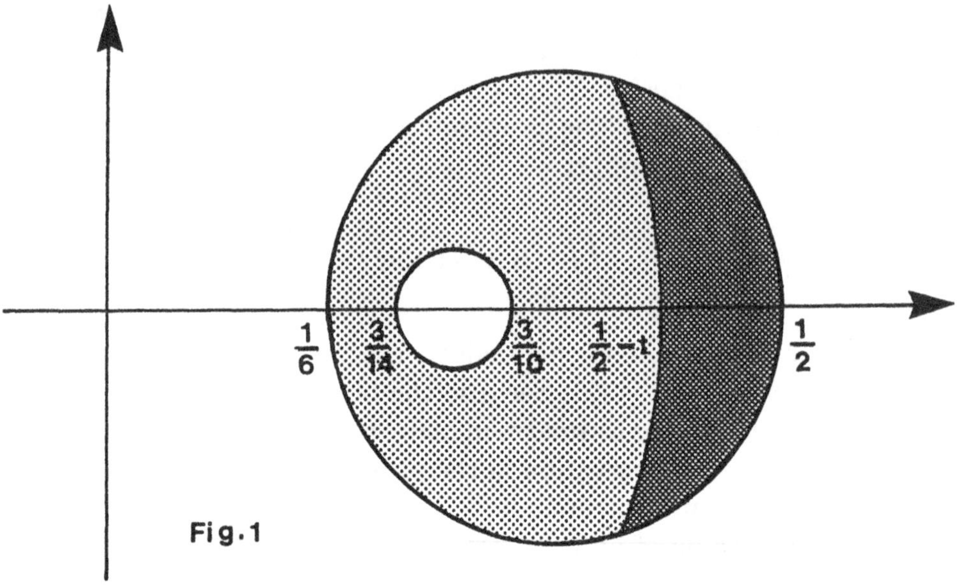


Fig. 1

as illustrated in Fig. 1.

Let $t$ be a positive number $\leq \frac{1}{2} - \frac{3}{10} = \frac{1}{5}$. Then

$$\frac{1}{2} - t \leq |f| \leq \frac{1}{2}$$

if and only if

$$\frac{\frac{1}{4}}{1+g}$$

is in the closed moon-shaped set

$$\left\{ w; |w| \geq \frac{1}{2} - t \quad \text{and} \quad |w - \frac{1}{3}| \leq \frac{1}{6} \right\} \quad , \tag{11}$$

illustrated in Fig. 1 , the darker part. This holds if and only if $g$ is in the set obtained from this set by the mapping

$$w \to \frac{1}{4w} - 1 \quad .$$

We shall call this set $A$. We have

$$A = \left\{ w; |w| \leq \frac{1}{2} \quad \text{and} \quad |w+1| \leq \frac{1}{2-4t} \right\} \quad . \tag{12}$$

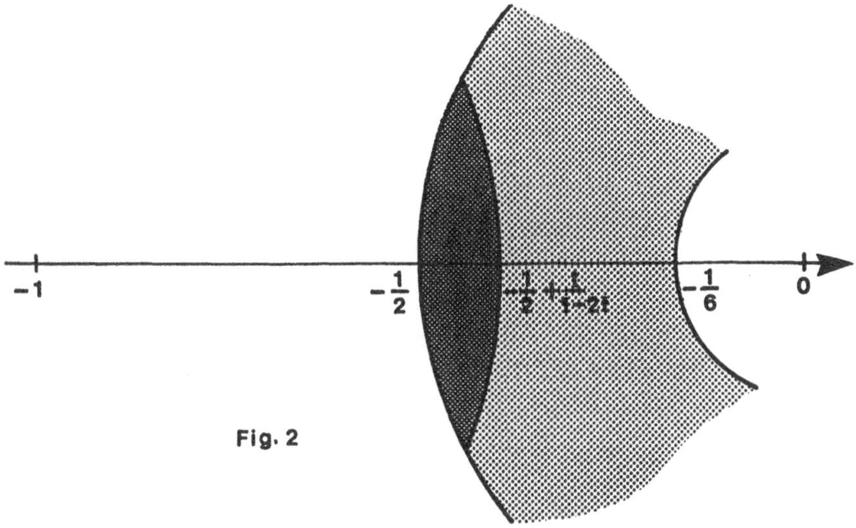


Fig. 2

This set is illustrated in Fig. 2, the darker part. Since $f$ and $g$ both have the same probability distribution we have for $0 \le t \le \frac{1}{5}$

$$\iint_{\frac{1}{2}-t\le|w|\le\frac{1}{2}} H(r,\phi)rdrd\phi \quad = \iint_A H(r,\phi)rdrd\phi \quad . \qquad (13)$$

The left-hand side reduces to

$$\int_{\frac{1}{2}-t}^{\frac{1}{2}} h(r)dr \quad .$$

On the right-hand side $r$ goes from $\frac{1}{2} - \frac{t}{1-2t}$ to $\frac{1}{2}$. For an $r$ in this interval let $2\alpha(t,r)$ be the angular measure of the part of the arc of the circle $|w| = r$ which intersects the set $A$. The right-hand side may therefore be written as

$$\int_{\frac{1}{2}-\frac{t}{1-2t}}^{\frac{1}{2}} \frac{\alpha(t,r)}{\pi} h(r)dr.$$

An elementary argument on the triangle with corners at $w = 0$, $w = -1$ and $w = r \cdot e^{i(\pi-\alpha)}$ shows that

$$\alpha(t,r) = \quad Arccos\frac{1+r^2-\frac{1}{(2-4t)^2}}{2r}. \qquad (14)$$

Since

$$\alpha(t,r) \le \alpha(t,\frac{1}{2}) \le Arccos(\frac{5}{4} - \frac{1}{(2-4t)^2}),$$

the equation (13) gives rise to the inequality

$$\int_{\frac{1}{2}-t}^{\frac{1}{2}} h(r)dr \le \frac{1}{\pi} Arccos\left(\frac{5}{4} - \frac{1}{(2-4t)^2}\right) \int_{\frac{1}{2}-\frac{t}{1-2t}}^{\frac{1}{2}} h(r)dr. \qquad (15)$$

for $0 \le t \le \frac{1}{5}$ . Choose in particular $t = \frac{1}{5}$ . With

$$T = \frac{t}{1-2t} \qquad (16)$$

we get $T = \frac{1}{3}$ , and the integral on the right-hand side is

$$\int_{\frac{1}{6}}^{\frac{1}{2}} h(r)dr = 1,$$

and (15) takes the form

$$\int_{\frac{1}{2}-\frac{1}{5}}^{\frac{1}{2}} h(r)dr \le \frac{1}{\pi} Arccos(\frac{5}{4} - \frac{1}{(2-\frac{4}{5})^2}) = \frac{1}{\pi} Arccos\frac{5}{9} < \frac{1}{3}. \qquad (17)$$

The inverse transformation of (16) is

$$t = \frac{T}{1+2T}. \qquad (16')$$

This is a parabolic linear fractional transformation with fixed point at $T = 0$ . It can be represented in the form

$$\frac{1}{t} = \frac{1}{T} + 2. \qquad (16")$$

Starting from $T_o = \frac{1}{3}$ we define a sequence $\{T_n\}$ by

$$T_{n+1} = \frac{T_n}{1+2T_n},$$

or

$$\frac{1}{T_{n+1}} = \frac{1}{T_n} + 2,$$

which gives

$$T_n = \frac{1}{2n+3}.$$

The inequality (15), with these choices of $t$−values, gives rise to a sequence of inequalities

$$\int_{\frac{1}{2}-T_n}^{\frac{1}{2}} h(r)dr \leq \frac{1}{\pi} Arccos(\frac{5}{4} - \frac{1}{(2-4T_n)^2}) \int_{\frac{1}{2}-T_{n-1}}^{\frac{1}{2}} h(r)dr.$$

By combining them we find, for every $N \geq 1$, an inequality

$$\int_{\frac{1}{2}-\frac{1}{2N+3}}^{\frac{1}{2}} h(r)dr \leq \prod_{n=1}^{N} \frac{Arccos(\frac{5}{4} - \frac{1}{4}\left(\frac{2n+3}{2n+1}\right)^2)}{\pi}, \qquad (18)$$

providing an upper estimate for the probability that the continued fraction value $f$ be in the annulus

$$\frac{1}{2} - \frac{1}{2N+3} \leq |f| \leq \frac{1}{2}.$$

For $n = 1$ the factor on the right-hand side is

$$\frac{1}{\pi} Arccos \frac{5}{9} = .3125... < \frac{1}{3},$$

and the factors of (18) decrease to 0 by increasing $n$ . A simple, but not very strong corollary of (18) is thus

$$\int_{\frac{1}{2}-\frac{1}{2N+3}}^{\frac{1}{2}} h(r)dr \leq \frac{1}{3^N}.$$

Better estimates are easily available, for instance as follows: By computing the first five factors in (18) and using the monotonicity we find, for $N \geq 4$

$$\int_{\frac{1}{2}-\frac{1}{2N+3}}^{\frac{1}{2}} h(r)dr \leq .0021 \cdot (.1430)^{N-4}. \qquad (19)$$

We have in particular for $N = 9$ :

$$\int_{\frac{1}{2}-\frac{1}{21}}^{\frac{1}{2}} h(r)dr \leq 1.25 \cdot 10^{-7}. \qquad (20)$$

This is likely to be a rough estimate, but it is strong enough to make the observation on the screen less surprising.

## REFERENCES.

1. K.O.Bowman and L.R.Shenton, "Continued fractions in statistical applications," Series Statistics: Textbooks and Monographs, Vol 103, Marcel Dekker, New York, 1989.
2. L.Jacobsen, W.J.Thron and H.Waadeland, *Some observations on the distribution of values of continued fractions*, Numerische Mathematik 55, 711-733 (1989).
3. H.Waadeland, *Boundary versions of Worpitzky's theorem and of parabola theorems*, Lecture Notes in Mathematics 1406 (ed. L. Jacobsen), Springer-Verlag, Berlin, 135-142 (1989).
4. J.D.Worpitzky, *Untersuchung über die Entwickelung der monodromen und monogenen Funktionen durch Kettenbrüche*, Friedrichs- Gymnasium und Realschule, Jahresbericht, Berlin, 3-39 (1865).

See also Progress Report, Communications in the Analytic Theory of Continued Fractions, Vol. 1, 81-82, Spring 1992.

Printed and bound by CPI Group (UK) Ltd, Croydon, CR0 4YY
21/10/2024
01777097-0015